Richardson's Guide to the Fossil Fauna of Mazon Creek

Richardson's Guide to

The Fossil Fauna of Mazon Creek

Charles W. Shabica and Andrew A. Hay
Editors

Christopher T. Ledvina, *Graphics Editor*
Donald Auler, *Principal Artist*

Northeastern Illinois University

Editorial production service: Lyceum Books, Inc.
Design: Cameron Poulter Design
Composition: Graphic Composition, Inc.
Printing: Thomson-Shore, Inc.

Published by Northeastern Illinois University, 5500 North St. Louis Avenue,
Chicago, Illinois 60625

Library of Congress Cataloging-in-Publication Data

Richardson's guide to the fossil fauna of Mazon Creek / Charles W. Shabica and Andrew A. Hay, editors.
p. cm.
Includes bibliographical references and indexes.
ISBN 0-925065-21-8
1. Paleontology—Pennsylvanian. 2. Animals, Fossil—Illinois—Mazon River Watershed. I. Shabica, Charles W.
II. Hay, Andrew A.
QE729.3.R53 1997
560′.9773′265—dc21 97-1162
CIP

This book is printed on acid-free recycled paper.

Gene Richardson was a man of many hats—paleontologist, poet, and printer. Those who knew him professionally may never have guessed how he spent his spare time, while bookmen who so admired the beautiful presswork of The Vanishing Press and the writings of E. Scumas Rory would have been surprised to see him in the field with hammer, knapsack, and muddy boots. But all would have attested to the great generosity and far-reaching vision of their friend and colleague.

Contents

Foreword

The Mazon Creek biota of northeastern Illinois provides the most complete record known of late Paleozoic life. It ranks among the world's outstanding fossil assemblages. Occurring in the 280-million-year-old Francis Creek Shale, the Mazon Creek fossil assemblage consists of more than 200 plant and 300 animal species, including representatives of no less than 11 phyla and 23 classes of animals from terrestrial through marginal marine environments.

Richardson's Guide is the first comprehensive, systematic account of the animal fossils from the Mazon Creek area. The plant fossils have been described several times, most recently by Jennings (1990). So, for the first time we can get a proper appreciation of the great variety of Mazon Creek animals, particularly those from the marine (Essex) fauna, and to a lesser extent from the freshwater (Braidwood) fauna. Many of the fossils are beautifully preserved. The Mazon Creek fauna thus can be ranked alongside the other extraordinary fossil assemblages in the geological record: the Cambrian Burgess Shale fauna of Canada, the Devonian Bundenbach and Jurassic Solnhofen faunas of Germany, and the Cretaceous Santana fauna of Brazil.

Paleoecologically, the Mazon Creek animal assemblage is important because it is similar to communities observed in modern euryhaline (variable salinity) coastal settings—in particular, river-influenced near-shore habitats that support large numbers of small crustaceans, fish hatchlings, and benthic forms, including polychaetes, infaunal bivalves, and gastropods. The *Guide* presents new concepts on the evolution of several groups, including fossil spiders, and the development of insect flight.

Richardson's Guide also represents one of the greatest collaborative efforts ever between professional paleontologists and amateur collectors. It grew out of the response of Gene Richardson from the Field Museum of Natural History in Chicago to requests from amateur collectors to help identify the many new fossils they were finding in concretions from the Pit 11 spoil heaps of the Peabody Coal Companies, opened up in the early 1950s. Richardson began compiling a photographic record of the species he saw for an atlas of Mazon Creek fossils. This project was cut short by his untimely death in 1983.

Charles W. Shabica of Northeastern Illinois University, a friend and colleague of Richardson's, completed and expanded the compilation. It has proved to be a monumental task, requiring the expertise of the 25 academics and amateurs listed as authors and the help and donation of specimens from dozens of amateur collectors from northeastern Illinois. Many of these amateurs had been collecting from Pit 11 since soon after it opened and took their finds to Gene Richardson. These contributions and those of Charlie Shabica, Andrew A. Hay, and other unnamed colleagues and friends of Gene Richardson have been a labor of love in remembrance of him.

Richardson's Guide to the Fossil Fauna of Mazon Creek is thus, simultaneously, a comprehensive account and reference for students and collectors of this extraordinary fossil fauna and a permanent tribute in memory of a remarkable and much-loved paleontologist.

Desmond Collins
Royal Ontario Museum
Toronto, Canada

Preface

This book documents through photographs, drawings, and written descriptions many members of the remarkably diverse Mazon Creek fossil fauna of northeastern Illinois.

The Coal Age Mazon Creek fossils have been studied for over 100 years, and more than 250 papers have been published describing them in detail. The first specimens collected along the banks of the Mazon River were plants, a few freshwater clams and crustacea, and, rarely, insects.

A large collection was established at the Field Museum of Natural History primarily through the efforts of George Langford, Curator of Fossil Plants. However, it was not until the 1950s and 1960s that more marine animals began to appear on mine dumps at Peabody Coal Companies Pit 11, southwest of the Mazon River.

Intensive collecting by amateur fossil collectors followed, many of whom had previously unknown fossils named in their honor. Probably the most famous is *Tullimonstrum gregarium,* named after Francis Tully, a soft-spoken farmer turned enthusiastic fossil collector.

In 1967 a National Science Foundation study was organized by Dr. Eugene Stanley Richardson, Jr., Field Museum Curator of Invertebrate Paleontology, and Dr. Ralph Gordon Johnson, University of Chicago professor. The study's objectives were to collect systematically from the seven square miles of Pit 11 spoil heaps, describe formally the many newly discovered fossil species, and explain the environments in which the Mazon Creek biota once lived.

The first "professional" expedition lasted through the summer of 1969 and yielded several hundred new specimens for the Field Museum's growing collection of Mazon Creek animals. The core of the collecting team included Peter Kranz, Paul Lund, Ida Thompson, Art Zangerl, and myself. Ralph Johnson, Gene Richardson, Riccardo Levi-Setti, Bert Woodland, and Rainer Zangerl made up the "irregulars." Peabody Coal Companies graciously gave us complete access to the mine and Melbourne McKee, Peabody's mine chemist, alternately attended to our needs and arranged for our "field vehicle" to be pulled out of the mud (see photo). The fossils we collected and censused at Pit 11 were "marine" as compared to the more terrestrial forms found to the northeast. The division of the Mazon Creek fossils into a marine association (Essex fauna) and a freshwater association (Braidwood fauna) suggested by Johnson and Richardson in 1966 was confirmed. Many species of worms, crustaceans, and fish were described by Johnson's students Ida Thompson and Fred Schram and others in the decade that followed.

As a graduate student of Johnson's and Richardson's, and an aspiring coastal geologist, I was given the task of attempting to unravel the question of how the Mazon Creek fauna lived and died. With additional help from Fred Ziegler at the University of Chicago, Bill

Mired University of Chicago field vehicle, 1969. Pictured from left to right are Paul Lund, Charles Shabica, Riccardo Levi-Setti, and Ed Nash. Photograph Ida Thompson.

Smith, M. E. Hopkins, and Roger Nance at the Illinois State Geological Survey, I examined hundreds of drill cores, driller's logs, natural outcrops, and mine highwalls and spoil heaps and found the Mazon Creek fossils preserved in a highly variable complex of rapidly deposited deltaic sediments. A Ph.D. thesis and several publications documented the work.

While Ralph Johnson's untimely death in the summer of 1976 came as a great shock to us, the cooperative research continued unabated. Detailed stratigraphic sampling combined with fossil census sampling by Gordon Baird allowed us to divide the Mazon Creek biota into several deltaic associations including 1) terrestrial plants and animals from coastal swamp, levee, and floodplain settings; 2) freshwater animals; and 3) animals inhabiting variable salinity waters near delta distributaries similar to the present Ganges-Brahmaputra Delta. The fossil census data, computerized by John Anderson at Argonne National Laboratory, showed a distinct regional boundary between estuarine (Essex) and freshwater and terrestrial (Braidwood) faunal assemblages that paralleled the ancient shoreline.

It was common knowledge among graduate students and coworkers at the Field Museum that Gene Richardson was amassing an extensive file of Mazon Creek fossil photos that we all presumed would someday coalesce into an atlas. When Gene died in 1983 I volunteered to "finish The Book." Because Gene had exemplified the cooperative spirit at the Field Museum it seemed reasonable that the book should be a shared project by those who worked with him and loved him. Thus professional paleontologists as well as amateur collectors were asked (or volunteered) to work on the book as contributors of fossils, photos, words, or labor.

Found among the thousands of photographs was a two-page typed and handwritten (in Gene's distinctive hand) statement about Mazon Creek:

Mazon Creek fossils hold a special place in the paleontological record. Because conditions of life, of death, and of burial were unusually propitious, we find a more complete record of plant and animal populations than in most other fossil occurrences. The fossils occur in ironstone concretions in the Francis Creek Shale member of the Middle Pennsylvania Carbondale formation.

I suppose that these elegant fossils have been collected—or casually picked up—for centuries. They are plentiful objects in the bed of the stream, and with every spate of high water the supply is renewed. But we have no inkling of any collections made by the Indians. The tradition that continues to the present was begun in the mid-nineteenth century, when there were already well-established towns and farms in the Illinois Valley.

The stream is identified on most maps and in the water-supply records of the United States as Mazon River, but it is stubbornly known as a creek in the paleontological literature. River or Creek, it flows north through 500 square miles of Illinois farmland, and empties into the Illinois River opposite Morris. It is a small stream, contributing less than 1% of the flow of the Illinois River, and often in the late summer no water at all moves past the gauging station near Coal City. But in a rainy season the water rises as much as 20 feet, roaring along at a rate of more than 7,000 cubic feet per second. It is then that the current cuts into the banks and liberates the fossils.

By great good fortune, those high banks along the creek west of Coal City happen to be exposures of the fossil-bearing Francis Creek Shale. And by further good fortune, the fossils are individually packaged in ironstone concretions embedded in a soft gray shale. In the rushing water, the shale quickly breaks down into mud and is carried away but the heavy concretions are left behind as a lag concentrate in the stream bed.

No sooner was this part of the country well settled, with peaceful farms and established villages, than the local people found that they had a mighty interesting way to spend a summer afternoon: they could collect fossils. There wasn't any point trying to dig them out of the vertical bluffs. Not with that great supply lying in the bed of Mazon Creek. A standard technique of collecting soon developed. With a wary eye on the deep pools, and a cautious look for snapping turtles, they would wade along, feeling for concretions with their toes. Soon, many of the collectors learned to take along potato forks, ideal for scooping up concretions from a few feet under water. This was presumably before the days when broken beer bottles were an occupational hazard. Nowadays one is well advised to wear sneakers but the fossils are still there!

Eugene S. Richardson
Gurnee, Illinois, 1983

The story of the Mazon Creek fossils would not be complete without a discussion of how we think they lived, died, were preserved in the geological record, and finally were exhumed by coal miners and amateur as well as professional paleontologists. Thus in the beginning chapters we tell the Mazon Creek story as it unfolded through collectors' love of the "find," industrial America's need for coal, and scientists' desire to understand how tropical fossil plants and animals came to be buried under the rolling farmland of Illinois.

Chapters 1 and 2 include a discussion of coal mining and fossil collecting. These chapters are followed by a review of the current state of our knowledge of Mazon Creek's geologic setting, fossil distribution, and environmental setting. Following this, individual chapters deal with the fossil groups represented in the Mazon Creek biota. There is no discussion of the fossil plants. For an excellent review of the Mazon Creek fossil flora, the reader is referred to the Illinois State Geological Survey *Guide to Pennsylvanian Fossil Plants of Illinois* (1990), by James R. Jennings.

The chapters are separately authored. Emphasis is placed on photographs and representative drawings of fossil species. Although we recommended a specific format for the writing of the chapters, the authors' styles vary from easily readable prose to detailed scientific description. Our goal was to make the book useful to laymen as well as professionals; we hope we have succeeded.

In general, specimens for the book were chosen by the chapter authors, and nearly all are housed in museum or university collections. The specimen numbers include a prefix indicating the source collection (a key to the abbreviations follows Appendix C). In cases where the specimens were in private collections, we encouraged the collectors to donate the specimens to a museum or institution of their choice. A number of fossils made their way into institutional collections this way. For example, the spider on the book cover was kindly donated to Northeastern Illinois University by Rita and Walter Lietz.

Photos are by the authors or by Chris Ledvina, who spent long hours with Andrew A. Hay, Eric Slusser, and Ralph Minnis documenting specimens for Steve Sroka, Fred Schram, and others. Drawings of specimens are by the authors unless otherwise stated. Many were drawn by Don Auler, a retired computer technician and a gentleman with a lot of patience. In general, the drawings are meant to facilitate scientific identification; all reflect each author's approach to his or her chapter. Thus a drawing may reveal internal structures or have some parts eliminated or may be a detailed reproduction of an actual fossil specimen as viewed by the author. For example, while wing venation is necessary for identifying many insects, body form may not be. Spider legs are generally not shown, as they are not considered necessary for identification. In contrast, sharks are drawn as viewed by the author.

We decided to show what the animals might have looked like in their original habitats; Gordon Baird and Don Auler kindly provided environmental reconstructions for Braidwood and Essex conditions.

It is our hope that this book will make the fossils of Mazon Creek accessible to more people and will encourage the cooperation between amateur collectors and professional paleontologists nurtured so well by Gene Richardson.

Charles Wright Shabica
Winnetka, Illinois, 1997

Acknowledgments

This book is the result of the cooperative effort of many people, including amateur collectors, amateur and professional paleontologists, artists, typists, and photographers.

Many professionals have assisted in the preparation of this book. We should like to mention a few by name and offer our gratitude: Fred Collier, Rob Garber, Michael Gottfried, Eric Gyllenhaall, Gerald Haddock, James R. Jennings, Peter Laraba, Thomas Miller, and Janet Waddington.

We appreciate the talents of the word processing team of Gloria Gonzalez, Vickie Kalinowski, and Syed Kazmi, working under the direction of Fran Skrabacz in the preparation of the final manuscript. We thank Mary R. Carman, Nina Cummings, Ellin Beltz, Bleue Benton, Ron Stoch, Penny Blubaugh, Lee Leibik, and David Ziegler for the retrieval and upgrading of data files.

At Northeastern Illinois University, David Jonaitis oversaw the *Guide*'s transformation from manuscript to printed book, and John Pennisi tied up many loose ends. At Lyceum Books, David Follmer coordinated the book's production, and Diana Gillooly edited the text that flowed into Cameron Poulter's thoughtful design.

Much of the artwork was supplied by the individual authors. The cover concept was provided by Angele Wexler. New artwork for this book was drawn by Don Auler, utilizing sketches and reconstructions from various sources. The accuracy of several reconstructions appearing in this volume owes much to the critical comments of Desmond Collins (*Tullimonstrum gregarium*), N. Gary Lane (the crinoid holotype), and Simon Conway Morris (*Priapulites konecniorum*).

We acknowledge the kindness of Carleton University, the Cleveland Museum, Field Museum, Illinois State Museum, Museum of Comparative Zoology, National Museum, Redpath Museum, Royal Ontario Museum, the Commonwealth Edison Co., McLanac Properties, Illinois State Geological Survey, McGill University, and David Bergman for the generous loan of materials and specimens.

We extend our thanks for continued support to Des Plaines Valley Geological Society, Earth Science Club of Northern Illinois, Mid America Paleontology Society, Chicago Area Paleontology Society, Chicago Rocks and Minerals Society, and Frank Pranschke.

Special thanks go to the collectors who donated fossil concretions to be photographed and added to the Mazon Creek Collection. They include the late George Agazzi, Don Auler, Roger Bohn, Craig Brown, Chris Cozart, Kathy and Judy Dedina, Ann Ederheimer, Betty and Gordon van Erden, Albert Forslev, Calvin George, the late Josephine M. Hay, Mary Helmus Smith, Keith Holm, Jim and Sylvia Konecny, Rita and Walter Lietz, Steven LeMay, Kathleen Morner, the late Mary Phillips, Helen and Ted Piecko, Joe Pohl, Bill Popp, Sharon Powell, John Prinos, Karlene and Steve Ramsdell, Daniel Sroka, Steve Sroka, Tom Testa, and many others.

Ralph Minnis, Janet Howe, Eric Slusser, and Steve Sroka should be commended for caring for many of the Northeastern specimens photographed by Chris Ledvina.

To our wives, Susan Ewing Shabica and Josephine M. Hay, we are especially grateful for their support and patience during the long process of assembling this volume.

C. W. S.
A. A. H.

Geologic calendar, biological events, and key lagerstätten

Millions of Years before Present	Eon	Era	Period	Biological Events	Key Preservation "Bonanzas" or Konservat Lagerstätten
0	Phanerozoic	Cenozoic	Neogene	**Appearance of modern mammals**	• La Brea tar pits
65			Paleogene	**Diversification of mammals** • Mass extinction of dinosaurs, flying reptiles, ammonites disappeared.	• Eocene Green River localities; Grube Messel locality
		Mesozoic	Cretaceous	• Appearance of flowering plants.	
			Jurassic	**Dinosaurs, small mammals**	• Solnhofen Limestone • Holzmaden
225			Triassic	• First dinosaurs, mammals, mammallike animals.	
		Paleozoic	Permian	• Mass extinction event. Diversification of reptiles.	
			Carbon-iferous: Pennsylvanian	**Coal swamps, first reptiles**	• Montceau-les-Mines ★ Mazon Creek
350			Carbon-iferous: Mississippian	**First coal swamps, amphibians**	• Bear Gulch
			Devonian	• First amphibians. • Mass extinction event. **Diversification of fish, numerous reefs**	• Hunsrück Slate
390 470			Silurian	**Primitive fish, first coral reefs**	• Waukesha locality
500			Ordovician	• Mass extinction event. **Abundant bryozoans, brachiopods, graptolites, first corals**	• Beecher's trilobite bed
570			Cambrian	• First boney agnathan fish. **Abundant trilobites, archaic invertebrates in seas**	• Alum Shale "Orsten" fauna • Burgess Shale (numerous problematic marine invertebrates)
	Proterozoic	Precambrian	Vendian	• Appearance and diversification of shelled organisms.	• Numerous Ediacaran fossil localities
700				• Appearance of soft-bodied metazoan invertebrates ("Ediacaran biota").	
1400				• First eukaryotic cells.	• Bitter Springs Chert
2500				**Diverse cyanobacteria**	• Gunflint Chert
3500	Archean				• North Pole Barite
3960				? ? **Oldest fossil life, prokaryotic,**	
	Hadean			? **one-celled organisms** ? Biosynthesis event(s) ?	
4600–6000				**Origin of solar system, earth**	

1

Collecting Mazon Creek Fossils at Pit 11:

Amateur Perspective

Andrew A. Hay and Samuel Kruty

Prior to the early 1950s, Mazon Creek fossils were collected from river bank outcrops and shaft and small strip mine spoil heaps and were mainly plants, plant fragments, and rarely insects. When Pit 11 was opened, the word swiftly spread among amateur collectors that "Peabody" was generously allowing concretion collectors to hunt on the spoil piles of the active strip mine.

At Pit 11 the concretions were in general found to be smaller than those collected elsewhere. There were a great number of "duds"—that is, concretions that, after being hammered open, revealed nothing visible on the opened surface. Hay remembers hearing some of the first collectors of Pit 11 fossils recount their initial disappointment. Expecting large, well-preserved plants, they found unrecognizable or indistinct shapes that came to be known as "blobs." Most were thrown away. Years later these peculiar shapes were identified as fossil jellyfish! Shrimp, worms, fish, and occasional insects were the prizes to be found here. Enterprising souls brought their finds to the Field Museum, where they came to the attention of Dr. Richardson.

During the workweek, collectors entered the mine office or lab to ask for permission and to sign a release. Melbourne McKee, the mine chemist, was a great source of advice to amateurs and scientists alike. On Sundays, collectors went to the mine machine shop, winding their way between the giant machine tools and even larger pieces of mining equipment being repaired there, searching for the foreman. Permission was granted after the foreman glanced at those in the party. "Stay out of the way of the trucks! Don't park on the roads! Don't go into the working face! Good Luck!" With these admonitions collectors were off to take a look into the trench where the mine shovel was stripping 50 to 100 feet of sand, shale, and soil overburden off the coal. Underfoot the earth shook, arousing awe and admiration at the immense power of the machine. A few more bites and the podlike feet swung down to inch the shovel forward as if it were a mechanism in a science fiction thriller. Then the shovel, and in later years the "wheel," piled the stripped overburden in long hills adjacent to the open trench.

After parking their vehicles on the shoulder of the mine roads, collectors were up and over the 40 to 50 foot high spoil banks to climb and search the sides of the hills for concretions. It was soon learned that specific animals were to be found in distinct locations, not scattered equally over the entire area. Some hills bore numerous discoid concretions—those unwanted blobs. On other hills collectors might be more apt to find concentrations of shrimp, or chitons, or clams, or rarely Tully Monsters. Locations of favorite collecting areas were guarded with the paranoia of fishermen trying to keep the best fishing hole a secret. Hay once questioned Francis Tully as to where he found his best specimens. "I collect them from all over Pit 11," he responded cagily. In another instance, Hay attempted to follow a stooped, frail-looking collector who was noted for his collection of rare shrimp. Sensing a competitor, the shrimp collector was up, over, and down the nearest hill, disappearing into a several square mile wilderness of steep gullies, lakes, and hills. Secrecy was not the only challenge collectors faced. Summer meant pesky deer flies. And one has to experience the thrill of struggling up a steep hill, arriving at the summit, and finding oneself staring into the eyes of a four foot long blue racer snake. If it had rained recently, what had been silt from a Coal Age river would turn into a sticky mud that entrapped cars and the occasional group-rented bus. More than one collector found the mud a veritable quicksand into which it was easy to sink up to the waist. Smart collectors learned to get out of the mine if a thunderstorm threatened.

Holotypes of new species turned up every now and then, even as the private collections of the serious collectors began to grow. Dr. Eugene Stanley Richardson, Jr. ("Gene" to the collectors), realized that if a complete history and the paleoecology of the area were to be developed, a rapport with amateur as well as professional collectors would be very important. He realized that the possessive and cautious nature of many collectors could lead them to suppress their finds, so he gently coaxed them into sharing information and fossils. Gene Richardson had the capacity to make the questioning nonprofessional feel accepted and worthy of attention, of making basic inquiry welcomed. The strength of the bond established between Gene Richardson and the collectors cannot be overstated.

No university, museum, or research group could boast of an array of field assistants such as Gene had mustered. They were out in all weather, climbing, digging, hoping for "the big one"—something rare to bring in to him. Gene usually found time to look over what was presented for his inspection. Published papers gave credit to the finders of newly described creatures. Sometimes the discoverer's name was latinized to grace for all time a new genus or species—fuel enough to fire up any collector. Like Linnaeus, scientists Johnson, Richardson, Schram, and others had fun naming specimens like the eight-armed jellyfish *Octomedusa pieckorum* (after Helen and Ted Piecko). *Smithixerxes Juliarum* was named for Julia Smith, the six-year-old who found this tiny, rare shrimp. More than 70 Mazon Creek species were named after collectors. ESCONI, the Earth Science Club of Northern Illinois was immortalized through *Esconites zelus.* A favorite is *Escumasia roryi,* named after E. Scumas Rory, the author of *The Dancing Worm of Turkana.* Rory (actually E. S. Richardson) told the story of an elaborate hoax perpetrated by Bryan Patterson about a living Tully Monster.

An annual open house at the Field Museum became the forum for this amateur-professional exchange, with many collectors vying for Gene's attention. Many "oohs" and "aahs" could be heard coming from those gathered around the binocular microscope and attached Polaroid camera (allowing the collector to go home with a remembrance should he decide to give up his find). Samuel Kruty's youngest son, Peter, showed Gene a cockroach he had found at Benson's Farm—the beaming 11-year-old face unforgettable as he decided to donate the rare fossil to the museum upon seeing Gene's response.

About 20 years ago, Peabody sold Pit 11 to Commonwealth Edison Company for the construction of a nuclear power-generating station. Some collecting areas were lost. However, the area around the reactor cooling pond was leased to the State of Illinois to serve as a conservation area for fishing and fossil collecting. Rules have changed, and collecting has gotten harder as vegetation has slowly reclaimed the once actively eroding spoil heaps. This has not stopped the collectors, however, and periodically, wondrous things are brought back from Pit 11 to the open house now sponsored by Northeastern Illinois University.

2

History of Coal Mining in Northeastern Illinois

Christopher T. Ledvina

Mazon Creek fossils were collected long before coal was actively mined in the Mazon Creek area of northeastern Illinois. Concretions were found in limited exposures provided by streams and other natural topographic lows. The true magnitude of the fossil resource would never have been known, however, without the advent of an active coal industry in the area. The relationship between coal mining and fossil finds is so close that this chapter on the coal mining history of northeastern Illinois is essential to provide a thorough understanding of the fossil phenomenon.

Northeastern Illinois, the Discovery of Coal, and the Earliest Days

For the purposes of this chapter, northeastern Illinois consists of Will, Grundy, Livingston, Kankakee, and LaSalle Counties. This area holds an important place in the history of the coal industry and the development of our nation. In 1673, the first discovery of coal in the New World was made by the explorers Marquette and Joliet near what is now Utica in LaSalle County, Illinois (Bement, 1929). They noticed coal (referred to as *charbon de terre*) in the outcroppings along the Illinois River (Figure 2.1). LaSalle found coal again near the same place in 1679 (he made references to *"mines de charbon"* and *"cole mines"*), and Hennepin in 1682 (he mentioned "some pit coal in several places"). In contrast, coal was not discovered in Appalachia until 1736, by Benjamin Winslow. We know today that Illinois constitutes the world's single largest reserve of bituminous coal, with over 160 billion tons in the ground (Smith and Stall, 1975).

Though coal had been discovered, not until 1810 were significant quantities mined. The first mining and shipping of coal in North America took place in southwestern Illinois, from along the bluffs of the Big Muddy River in downstate Jackson County. This coal was sent by barge to New Orleans. Around this time, St. Louis was a rapidly growing frontier town, and it became a large consumer of the coal mined in southwestern Illinois. Over time, markets for coal from southwestern Illinois expanded and production increased. The area remains to this day an important coal producer.

Into the 1800s, Chicago too was growing rapidly. Wood was the primary fuel until about 1850, when the Chicago and Alton Railroad, now part of the Burlington Northern, first brought coal to Chicago from southwestern Illinois. With this market, and the St. Louis market, southwestern Illinois saw extensive coal mining more or less continuously. Northeastern Illinois coal remained mostly unexploited, however, except perhaps for some local trade and consumption from shallow drift mines along rivers. Though southwestern Illinois mines served Chicago, shipments carried via the Great Lakes from mines in Pennsylvania were the major source of coal for the burgeoning city.

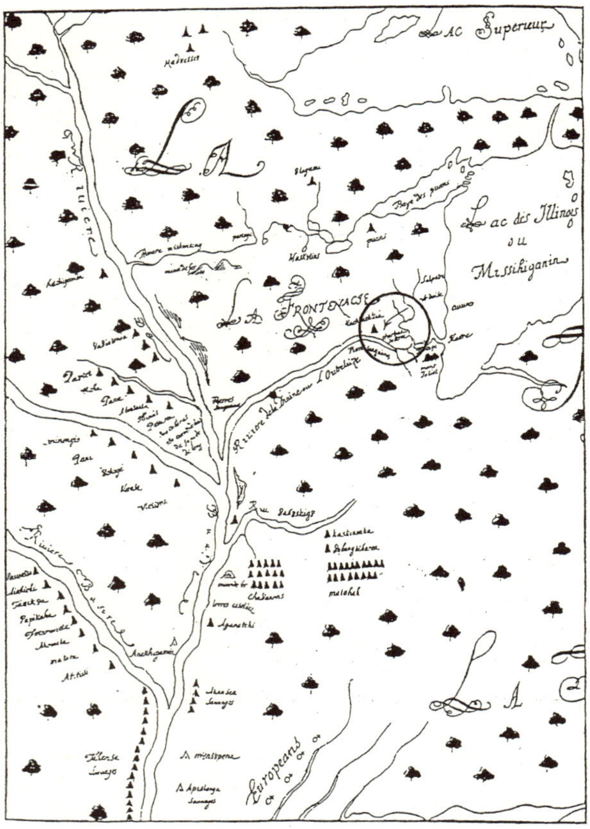

Figure 2.1. Early published map by Marquette and Joliet showing *carbon de terre.* Note errors in depicting the geography (from Bement, 1929).

Detail courtesy Illinois State Historical Library.

Mining in northeastern Illinois began along the route of the Illinois and Michigan Canal. The canal, opened in 1848, exposed coal during its excavation. Later, the canal served areas of LaSalle County (and later other counties of northeastern Illinois), where small amounts of coal were being mined along the Illinois River. Though the canal was initially built to transport goods to the Mississippi River via the Illinois River, it was soon transporting coal to Chicago from mines in northeastern Illinois. In 1849, 5,150 tons reached Chicago by this route. This tonnage represents the first known record of production in northeastern Illinois. However, this early mined coal was so high in sulfur that it eroded the grates of any stove or boiler it was used in. For this reason, northeastern Illinois coal was at first used almost exclusively as domestic hearth coal. It is almost certain that the coals mined were the Colchester (No. 2) and Summum (No. 4) Coals. Southwestern Illinois and Pennsylvania remained the primary sources for coal burned in Chicago. By 1856, three shaft mines were operating in LaSalle County, but other than to Chicago, no coal was being exported from the northeastern Illinois area.

Though no one at the time understood the geological reason, certain coals in northeastern Illinois were low in sulfur. We know now that only coals with a nonmarine deltaic roof (like the Francis Creek Shale) are low in sulfur—a relationship of great importance today (Ledvina, 1988). Early mining was done around outcrops or subcrops. It was suspected that coals continued inland from outcroppings, a suspicion frequently confirmed during the digging or drilling of water wells. For example, a water well sunk in Braidwood in 1865 struck a four foot seam of coal at 65 feet (the Colchester Coal) under a thick nonmarine roof (the Francis Creek Shale). This low sulfur coal was not only conveniently located but of surprisingly good quality.

Prime for a Boom

After the discovery of low sulfur coal, a concatenation of circumstances would soon put northeastern Illinois on the map as the greatest coal producing area in the country:

1) Northeastern Illinois now had high-quality coal at shallow depth, rarely more than 150 feet below the surface.
2) Chicago was larger and growing faster than St. Louis. It became the greatest coal market in the west.
3) The Illinois and Michigan Canal facilitated the transportation of coal north to Chicago and south via the Illinois and Mississippi Rivers. The Great Lakes provided additional shipping possibilities.
4) Northeastern Illinois's economic base was shifting from agriculture to industry because of Chicago's strategic location. The labor force was supplemented by a steady and increasing flow of immigrants, providing a pool of cheap labor.
5) In 1854 railroads became not only coal transporters but also coal consumers when Illinois coal was found to be a satisfactory fuel. Chicago was hub to most lines of the rapidly developing railroad industry.
6) Coal-fueled steamships were becoming more common on the Great Lakes and on the rivers of the Midwest.

Under these conditions, northeastern Illinois coal mining boomed. Although a few large companies were mining coal, most operations consisted of small, single-owner mines. By 1870, mining in northeastern Illinois had become a major industry. Over 50 mines were opened, with the coal produced going primarily to Chicago, but shipped to more than 15 states as well. Immigrant labor, mostly English, Scottish, Irish, Italian, and Eastern European, poured in from Chicago to fill the demand for workers. Northeastern Illinois had become the greatest coal mining area in North America.

Production Began from Underground Mines

All production until 1920 was from underground mines using the longwall mining system employed in England (Figure 2.2). In this

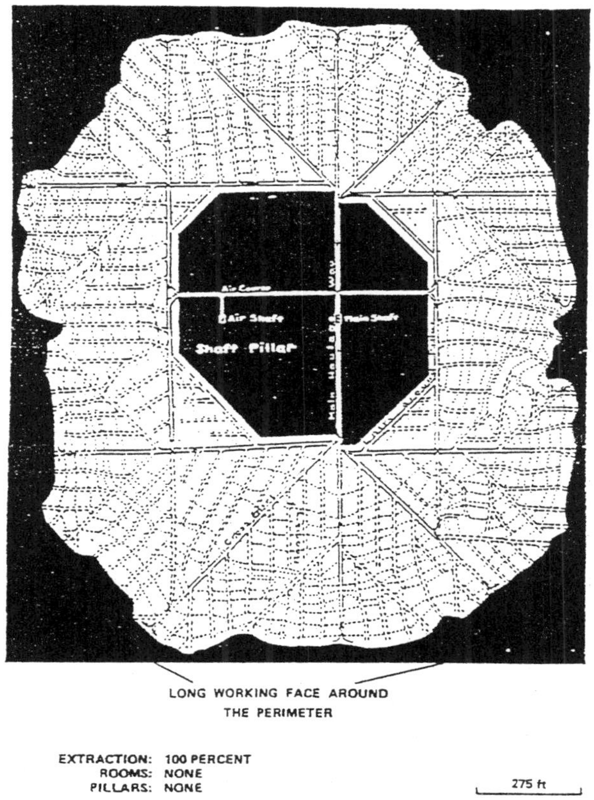

Figure 2.2. Map of a mine worked on the longwall system. Note the shafts and the pillar left for shaft support. All the coal was removed in an ever-widening circle away from the shaft, allowing the roof to cave in onto the mining surface. Entrances were kept open by rock packwalls (from Bement, 1929).

method, a shaft was sunk down to the coal, where it was surrounded by a large coal pillar. The coal was entirely removed in an increasing radius from the shaft, and the roof was allowed to cave in as mining progressed. Coal was hauled to the shaft bottom via mule, and the full cars were hoisted up the shaft. These mines could not reach a large size or operate for long because haul roads and ventilation became difficult to maintain with the rock falls created by the intentionally caved roof. In addition, the system produced a lot of waste rock. It is interesting to note that, in a very different modernized and mechanized form, the longwall system has recently been reintroduced to Illinois with great success. Modern longwall mining, however, takes advantage of the best characteristics of room and pillar mining and of historic longwall mining.

Northeastern Illinois was known as the longwall district because of the common use of the

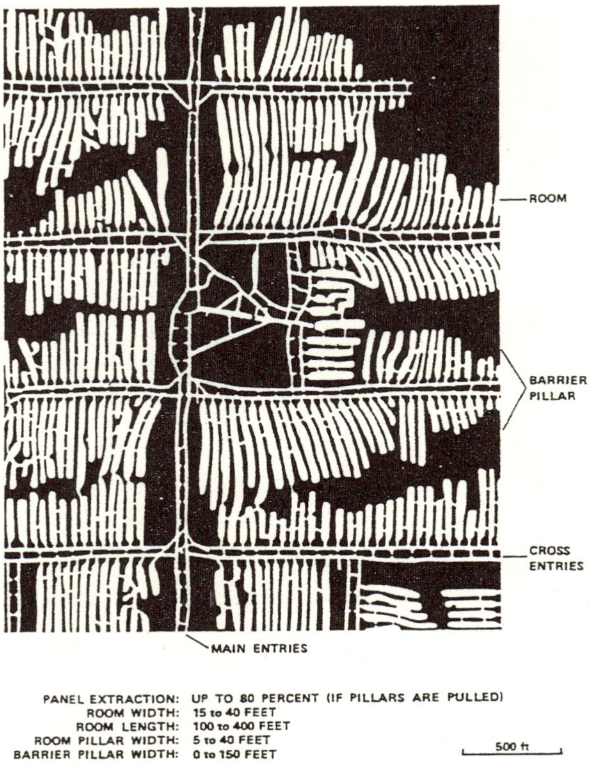

Figure 2.3. Map of a mine worked on the room and pillar system. Though half the coal is left unmined permanently, this system allowed greater areas to be mined, allowed individual mines to operate longer, and was conducive to mechanization (from Bauer et al., 1993).

Figure 2.4. Hand picking coal in the early 1900s. In those days, rock was removed by women, children, or men by visual inspection—difficult work at best. Modern mines use sophisticated machines to clean coal. Courtesy Freeman United Coal Mining Company.

Figure 2.5. Mine of the times showing conical spoil heap and track. Note how the shape of the heap is governed by the terminus of the track. Note also the tipple buildings. Courtesy John Taylor.

longwall mining method. Northeastern Illinois was also known as the third vein district because the most commonly mined coal (the Colchester [No. 2]) was the lowest of the three coals found in the area. The other two coals were the Summum (No. 4) and the Herrin (No. 6), which were mined locally but too thin and too high in sulfur to be in great demand. A few mines in the area used room and pillar mining (Figure 2.3). Here, pillars were left to support the roof. This method became more and more common into the late 1800s and early 1900s.

Typical mines of the time had a wooden headframe with a wooden tipple or building where the mine output was hoisted and dumped. In some cases, coal was shipped raw, and in others, the tipple had a crew of pickers separating out the rock (Figure 2.4). Some mines, near LaSalle and Peru, were drift mines: mines cut horizontally into a naturally occurring outcrop. Others were slope mines: mines using an angled tunnel for access. Most mines were shaft mines with vertical access. Waste rock was dumped on a cone-shaped spoil heap ("gob pile") served by a small waste car running up the heap on tracks (Figure 2.5). Remains of these spoil heaps dot the area today, although the remains of tipples and headframes are long gone (Figure 2.6). Each mine had a washhouse, mule stables, a machine shop and storage, a tipple, and a steam hoist house. A steam fan supplied air into the mine to dilute methane gas, and the majority of these mines had two separate shafts to provide intake and return airflow (Figures 2.7, 2.8).

In general, mines of northeastern Illinois

were developed under a cover of Francis Creek Shale up to 100 feet thick that was overlain by unconsolidated Pleistocene soils. The loose nature of these materials made sinking shafts difficult. In addition, the Pleistocene overburden was saturated with water, and most mines had to be pumped continuously. The caving effects of longwall mining, together with the friable nature of the Francis Creek Shale, meant that mines of the area had very poor roofs. Concretions in the mine roof, so prized for their fossils, also made the roof hazardous (Figure 2.9). In places where the roof was Oak Grove Limestone or Mecca Quarry Shale, roof conditions were probably better, but the associated coal was high in sulfur. Longwall mining caused complete and rapid caving clear to the surface. Cribs and packwalls kept haul roads open, but just barely.

Figure 2.6. Conical spoil heap from a longwall mine. The shape of the heap is a function of the way the gob was stacked. These heaps dot the countryside of northeastern Illinois. Though all other traces of mining are gone, the heap, and often the fossils it contains, remains. Courtesy Bob Heins.

Mines operated principally by hand loading. Coal was broken down by hand with picks (Figure 2.10). The pressure from the caving roof would cause the coal to crack and spall, making mining easier. In some cases, the coal would be undercut by hand, providing relief space so that hand-drilled holes could be loaded

Figure 2.7. Mine surface works in northeastern Illinois, circa 1911. This mine is the Big Four Eileen Mine in Coal City, Illinois. Note the main shaft headframe to the left, the tipple and coal hoisting headframe in the center, and boiler houses to the right. Except for the boiler houses, the entire surface works are of wooden construction. Courtesy Jane Campbell.

Figure 2.8. Mine surface works around the turn of the century. This photograph shows one of the larger mines operating in northeastern Illinois, Wilmington Star Mining Company's Mine No. 4. Note the wooden tipple and headframe. Courtesy Jane Campbell.

Figure 2.10. Work at the face of a longwall mine, circa 1890. Roof pressures cause the coal to crack and spall, reducing the work the miners had to do to break the coal down. The work was still tedious, however. Note the rock packwalls along the rib. In general, rock used for the packwalls came from the roof. Courtesy Illinois State Geological Survey.

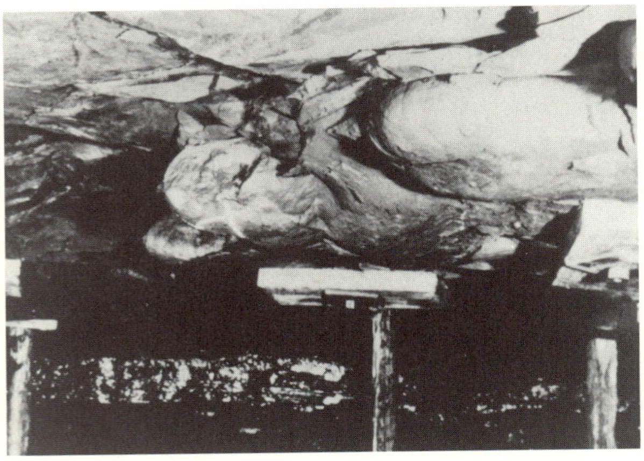

Figure 2.9. Concretions in the roof. Though prized by fossil collectors, concretions were hated by miners. They fell unpredictably, often with fatal results. Their presence generally weakened the roof even when they did not fall. Though this photograph shows large concretions in the roof of a modern mine, it conveys a sense of what early northeastern Illinois miners had to deal with. Courtesy Illinois State Geological Survey.

Figure 2.11. Undercutting by machine. The most difficult job miners had to do was to undercut or cut the kerf. This step was essential to breaking the coal down with explosives. Even in longwall mines, undercutting and blasting was often necessary. This photograph shows undercutting by a shortwall machine, circa 1910. Courtesy Zeigler Coal Holding Company.

with black powder and shot. With time, compressed air and, later, electric machines were used to undercut and drill the coal prior to shooting down with black powder (Figure 2.11). Permissible (flameless) high explosives were later used to reduce the possibility of an explosion. The coal was loaded by hand into pit cars and pulled out of the mine by mules or manpower (Figure 2.12). The cars were then hoisted, and the coal delivered into the tipple, where it could be sized, cleaned, and sorted (Figure 2.13). The rock cleaned from the coal usually contained considerable roof material, especially concretions. That is why the conical spoil heaps of underground mines make such good places to collect concretions today.

Though not known for great advances in coal mine mechanization, northeastern Illinois did benefit from the technological advances of the late 1800s and early 1900s. True mechanization was to occur in other mining districts to the south, which had thicker, better coals and better mining conditions. Thus underground mining in northeastern Illinois gave way to boom mining in central and southern Illinois, where seams were thicker and coal was of higher quality. This move in production was fostered in part by the development of railroads to the south.

Mining Was a Dangerous Business

Dangers to miners in northeastern Illinois were many. Numerous miners were taken out dead or almost so from roof falls. Wet conditions made work miserable on the one hand but prevented fires and explosions on the other. One of the biggest hazards in northeastern Illinois was flooding. Overlying Pleistocene materials were excellent aquifers full of buried valleys. Deluges of water were not uncommon and at their worst could lead to disaster, as at the Diamond Mine in Braidwood. The Diamond Mine Disaster, in which 69 men lost their lives, was the worst mining accident in the United States up to the time of its occurrence (Figure 2.14).

On 16 February 1883, warm rains had fallen in abundance for several days and melted the mantle of winter snow. As a consequence, the immediate area, over 10 square miles, was flooded to a depth of a foot or more. The Diamond Mine and several others in the area had problems enough dealing with water throughout the year. Overlying Pleistocene soils were saturated and riddled with aquifers. Because of the longwall mining system used and the breaks to the surface it provided, this water found its way into the workings of the Diamond Mine and other nearby mines on a continual basis. These mines had to be pumped of water every day of operation. The surface flood existing on 16 February, however, was especially likely to force its way into the workings 65 to 70 feet below.

Whether the flooded condition on the surface was unusual is hard to say. It is certain that similar quantities of water had lain there in springtime during other years. This year, however, the standing water would have fatal consequences. Superintendents of neighboring mines refused to let their men go below into the workings that morning, but this was not the case at the Diamond Mine owned by the Wilmington Coal Mining and Manufacturing Company (Illinois Department of Mines and Minerals, 1931).

Around midday, the cager noticed an unusual amount of water flowing to the sump at the shaft bottom. He sent word into the far areas of the mine, and men began to hurry to the shaft. The cager then rode the cage to the surface to ascertain the cause of the increased flow. Noticing nothing unusual, he went back below to find the water at such a depth that he had to pull a young trapper boy up out of the water and onto the cage. By now, those men who had heeded the previous warning were clambering out the second, or escapement, shaft (Ledvina, 1988, in preparation).

Figure 2.12. Mule haulage. Most mines of the area used mules to pull loaded cars to the shaft for hoisting. A mule's life of underground work was hard. The old-timers say, however, that they were lovingly cared for. Courtesy Southern Illinois University Coal Research Center.

Figure 2.13. Typical tipple of an underground mine. Technology at the time was primitive. Dumping of cars, sizing, screening, washing, and sorting was done in the tipple. At the site known as the preparation plant today, the same functions occur, only using highly mechanized, modern technologies. This view looks down the track of the spoil pile. Courtesy Jane Campbell.

Figure 2.14. The Diamond Mine after the disastrous flood of 16 February 1883. the mine was never recovered, and its remains are to this day the tomb of more than 30 miners and their mules. Courtesy Jane Campbell.

The mine was filling rapidly, and all those working in remote areas of the mine who had not gotten word or who had failed to heed the word were cut off from any hope of escape. Since both shafts were located at the lowest point of the fairly level coal seam, water accumulated in the shaft areas first. Anyone not at the shaft bottom ready to be hoisted was quickly stranded. The water was then left to reach the outer areas of the mine "at its own deadly leisure." A complicating factor was the presence of a dip or swail in the entry to the escapement shaft. Being a low place, this 30 foot section filled with water first and, except for a few brave souls who ventured to swim through it, barred the escape of dozens seeking the safety of the shaft.

The breach to the surface had occurred at an old air shaft. Water could be seen boiling its way down this opening within an hour of the initial alarm. This discovery let all know that any men remaining in the mine were surely dead. Men who had gotten out by the escapement shaft or up the main shaft were the only survivors. Two men went down the escapement shaft to lend assistance shortly after the breach. They joined the ranks of those left dead below.

Knowing that nothing could be done to save lives, it was decided by management to attempt to recover bodies. To do this, the mine had to be pumped dry. Pumping could not commence until a coffer dam had been built around the site of the breach. This task completed, all mines in the immediate vicinity suspended operations and provided the Diamond Mine with pumps. Water was pumped for 38 days before the mine was empty enough to enter. Upon entry, the mine was found to be in a deplorable condition. The roof had fallen everywhere, the air was bad, and most entries were blocked with silt, gravel, and debris carried in by the torrent. Because of the danger, only an incomplete job of recovery could be accomplished. Of the 69 drowned, approximately 30 bodies were recovered. The rest were left for the ages with a mine as their tomb.

The site of the Diamond Mine Disaster is today marked by an obelisk in the town of Diamond. The names of the dead are listed on this marker. Nearby is an Illinois Historical Society marker with a brief description of the unfortunate event.

The disaster taught numerous lessons that ultimately resulted in improvements in the mining industry as a whole. First, the Illinois Department of Mines and Minerals was founded to oversee the first systematic record keeping and inspection of mines. Second, the state legislature passed the first comprehensive mining act, which would become a model for other states. Finally, laws were emplaced that required every mine to be fitted with safe and secure escapement shafts and at least two methods of egress. Coal mining remained dangerous work, however. By 1926, nearly 6,000 Illinois coal miners had died in mining accidents.

The Birth of Two Coal Mining Giants

In 1866, Chicago and Boston investors formed the Chicago and Wilmington Coal Company. The company opened several mines near Braidwood and soon acquired properties in Danville in eastern Illinois. It then became the Chicago, Wilmington, and Vermilion Coal Company. This company later became one of the most important operators in Illinois upon the expansion of its mining operations into Franklin and Macoupin Counties. After this expansion, the company was known as the Chicago, Wilmington, and Franklin Coal Company. Following mergers with several other companies, portions of C. W. and F. survive today as Freeman United Coal Mining Company. Freeman United is now part of General Dynamics.

The Wilmington Star Mining Company started operations in northeastern Illinois in the late 1800s. This financially strapped company was taken over by the Buchanan family, who later expanded Wilmington Star to become the Ohio Valley Mining Company with the opening of mines in Franklin County. Ohio Valley was subsequently renamed Old Ben Coal Company. Old Ben became a division of the Standard Oil Company (Ohio) and is now a division of Zeigler Coal Holding Company. Old Ben and Freeman United are two of Illinois's and the nation's largest coal producers. The two companies have been responsible for some of the greatest advances in coal mining, both in safety and production. From their roots in northeastern Illinois they are thus tied to coal mines the world over.

The Production Peak and Decline

The year 1895 saw the greatest number of operating mines in northeastern Illinois, with over 200 hoisting. Coal production peaked in 1902 with over three million tons per year being mined (Illinois Department of Mines and Minerals, 1931). After 1902, production steadily declined despite a small rally during World War I. By 1929, production had dropped to one million tons. The Depression brought about a wholesale closing of mines. For all practical purposes, underground production was finished by 1930, with only a few small room and pillar mines continuing to operate into the 1950s.

Another contributor to the decline of mining was a change in coal consumption patterns. Until the early 1900s, coal was used domestically for heating and cooking and industrially for steam, rail transportation, and coking. The domestic and rail transportation market shrank into the 1900s and by the 1950s had dried up almost entirely. Industrial use continued, but changed from steam generation in factories to power generation by electrical utilities. Coinciding with these changes, underground mechanization of mines in thicker, higher quality coals and new, extensive railroads made mining in central and southern Illinois more attractive. The only viable option for northern Illinois was surface mining.

Surface Mining Dominates the Scene

In 1866, strip mining, with horse-drawn scrapers, was first tied successfully near Danville in eastern Illinois. This development would hold interesting consequences for the mining future of northeastern Illinois, where the shallow coal and the relatively soft overburden were perfectly suited to strip or surface mining. It was surface mining that was to dominate northeastern production from 1920 until the end of production in 1976.

In 1920, a new company, Northern Illinois Coal Company, drew up plans to adapt recently developed mechanized earthmoving equipment to the removal of overburden over the Colchester Coal in Grundy and Will Counties. By 1927, Northern Illinois Coal was operating the first large-scale strip mining operations in the country. The Pleistocene cover over the Francis Creek Shale was easily removed, and surface mining became very successful. Surface mining could produce coal at lower cost and consequently drove many underground operators out

Figure 2.15. Shovel used for overburden removal. Early shovels were not nearly as large as this 1940s vintage monster. The bucket capacity of this machine was over 20 cubic yards. These machines operated around the clock to expose the coal. Courtesy Illinois State Geological Survey.

Figure 2.16. Loading coal into trucks for transport to the plant. After overburden removal, smaller shovels excavated the coal and loaded it into large trucks. Courtesy J. Howe.

of business. Thus a few surface mines, usually owned by one operator, replaced the production—albeit at a lower level—of scores of operators.

In surface mining, a large shovel excavates a long linear pit. The shovel works its way into the pit, enlarging it into a trough several thousand feet long. The pit is then expanded laterally by excavating overburden from one side and placing it on the other side (Figure 2.15). The coal is thus uncovered, scooped up using smaller shovels, and transported to the tipple (or preparation plant in a modern mine) by railcar or truck (Figure 2.16). This entire mining system is known as "area strip mining" and proceeds along the plan shown in Figure 2.17. By its nature, area strip mining uncovers large volumes of roof shale, often containing fossil-bearing concretions.

As underground mining technology evolved, so did strip mining technology. Large shovels used for overburden removal gave way to draglines in the 1950s (Figure 2.18). Draglines gave way to the first bucket wheel excavator used in surface coal mining. While shovels and draglines could move up to 200 cubic yards of overburden every two minutes, they operated in a complex series of steps. Bucket wheels, however, removed overburden in an efficient continuous process (Figures 2.19, 2.20). The bucket wheel was the last of the huge overburden-removing machines used.

By law, modern surface-mined land must be reclaimed; the land must be returned to its original use and contour. All surface mining in northeastern Illinois ended before this law was enforceable, however; so no real reclamation occurred. This was great for fossil collecting as it left long rows of spoil and exposed concretions. It was the advent of strip mining that really started fossil collecting on a large scale, and Peabody Coal opened its mines to all who wished to collect.

An aspect of strip mining worthy of note is the numbering of the various pits. Pits are numbered when mining properties are allocated. A pit refers to the start, advancement, and recovery of a single area of surface mining. A single pit advances until the property is worked out. Equipment is then withdrawn, and a totally new advancing pit is started elsewhere. The new pit could be Pit 2 or Pit 20 depending on how the mining company numbers its properties.

Mining Ends in Northeastern Illinois

Northern Illinois Coal Company continued its strip mining operations until it was purchased by the Peabody Coal Companies in the late 1950s. Strip mining continued until 1976, when Peabody worked out its reserves and abandoned its last pit, Pit 11. The closing of Pit 11 marked the end of mining in the entire northeastern Illinois district. The area had given up nearly 120 million tons of coal.

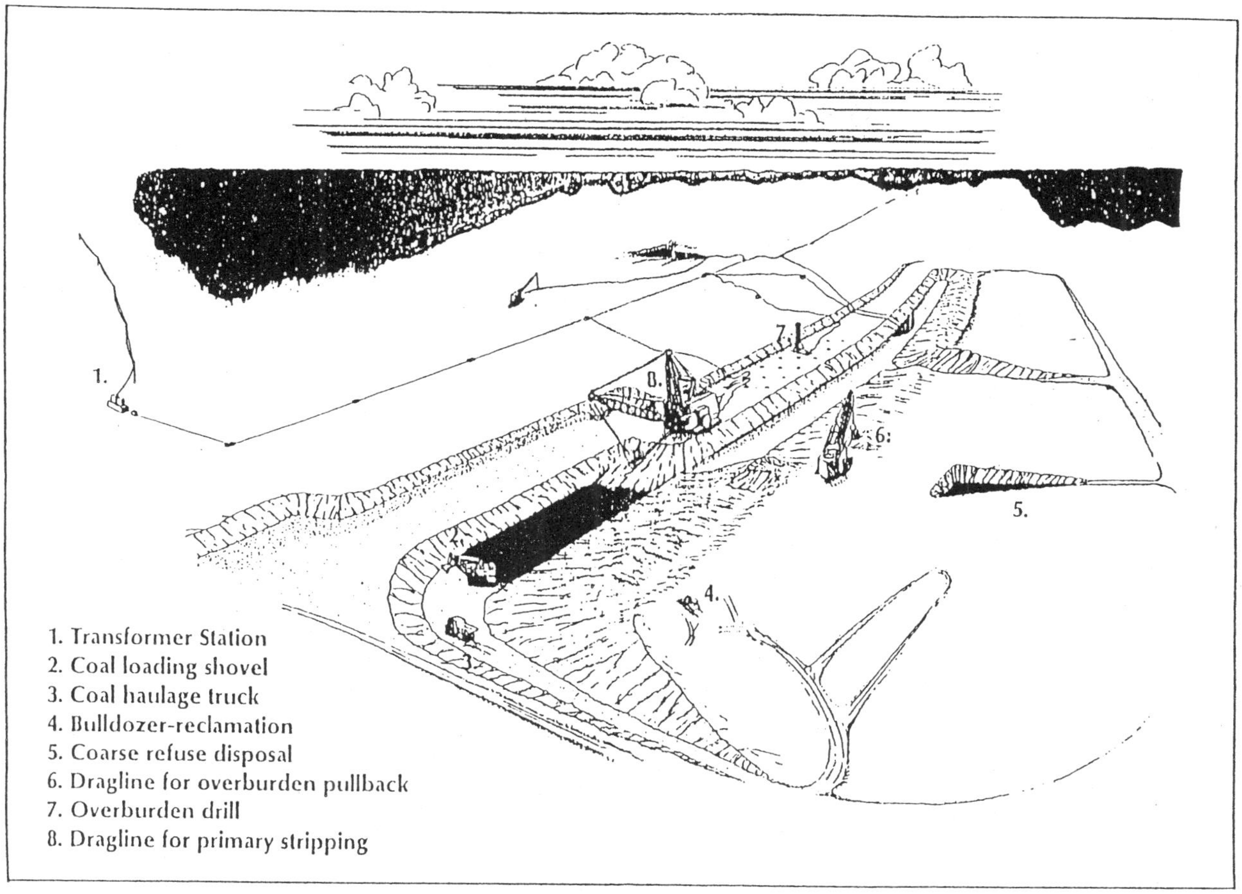

Figure 2.17. General plan of area strip mining. Though this view shows a totally modern operation, the sequence of steps and the types of machines used have changed little since the 1920s. Highwall blasting is a fairly recent introduction, however. Also, there has been a general trend toward ever larger machines. Courtesy Old Ben Coal Company.

Northeastern Illinois was indeed the first boom coal mining region in the country. It brought the Illinois coal industry into worldwide prominence. In addition, the area helped attract settlers into Illinois. Many of today's inhabitants trace their ancestry to the early coal miners of northeastern Illinois. The area fueled the growth of Chicago and its industry and thereby participated in the country's great industrial revolution. The area had all the romance and grandeur typical of the early days of industrialization. Northeastern Illinois's place in the history books is thus assured.

Commonwealth Edison, the Mazon Creek Project, and Fossil Collecting Today

After Peabody Coal ceased mining operations in northeastern Illinois, the strip-mined land, most of it unreclaimed, was to be put to yet another use. Commonwealth Edison, an electrical utility, purchased most of Peabody's unreclaimed strip-mined land and used some of this land for siting a nuclear power station. The lakes left between spoil piles over the surface-mined land were to be interconnected into a cooling lake. The Braidwood nuclear power station was thus designed, built, and commissioned. It went on-line in the late 1980s. Other portions of the strip-mined land that Commonwealth Edison acquired were leased to the Illinois Department of Conservation, sold to private developers, and either leased or sold to local sportsmen's clubs.

Various efforts were made locally and nationally to ensure continued access to the Mazon Creek fossil resource. Commonwealth Edison agreed to allow collecting under a pass holder system. This system was initially the responsibility of the Field Museum of Natural History in Chicago. The responsibility was transferred

Figure 2.18. Dragline, circa 1950. Though not very popular in northeastern Illinois, the dragline was a simpler and more versatile machine for overburden removal than the shovel. It worked well in soft overburden such as that of northeastern Illinois. Courtesy J. Howe.

Figure 2.19. Bucket wheel excavator. The bucket wheel excavator was first successfully applied in northeastern Illinois and was later used in other surface mines in Illinois and elsewhere. It was especially productive where highwall blasting was not required and made overburden removal a continuous process. Courtesy Illinois State Geological Survey.

Figure 2.20. Discharge of overburden from bucket wheel excavator. This large bucket wheel excavator made by Krupp of Germany was the world's first of its kind. Courtesy Illinois State Geological Survey.

to Northeastern Illinois University in 1984 and is currently administered by the Earth Science Department's Mazon Creek Project.

The Mazon Creek Project oversees fossil collecting on all Commonwealth Edison property regardless of the lessor. The project issues passes, develops collecting policies, encourages cooperation between amateur collectors and scientists, announces new fossil finds, and works as an advocate for the Mazon Creek fossil phenomenon.

Fossil collecting thus continues over the strip-mined land. There are a number of threats to collecting, however. One is the advance of vegetation over the spoil heaps. This growth has severely reduced the exposure necessary for good collecting. Other threats are liability and security issues. The project exists to deal with these problems through a collective effort. Thus collecting is, at present, assured. Nevertheless, the project requires continued support to maintain this assurance.

References

Bauer, R. A., Billy A. Trent, and Paul B. DuMontelle. 1993. Mine subsidence in Illinois: facts for homeowners. Illinois State Geol. Surv. Environ. Geol. Notes, 144, 16 pp.

Bement, A. 1929. Illinois coal: a non-technical account of its occurrence, production, and preparation. Illinois State Geol. Surv. Bull., 56, 123 pp.

Illinois Department of Mines and Minerals. 1931. A Compilation of Coal Reports of Illinois from the Earliest Records to 1930. Illinois Department of Mines and Minerals, Springfield, 177 pp.

Ledvina, C. T. 1988. The mining geology of the Herrin (No. 6) Coal of Illinois: discrete features. M.S. thesis, Northeastern Illinois University, Chicago, 670 pp.

———. In preparation. The Pictorial History of Coal Mining in Illinois. Illinois Mining Institute, Champaign, 486 pp.

Smith, W. H., and J. B. Stall. 1975. Coal and water resources for coal conversion in Illinois. Illinois State Geol. Surv. Coop. Rep., 4, 79 pp.

3

Geologic Setting of the Mazon Creek Area Fossil Deposit

Gordon C. Baird

Most of the fossils discussed here were collected within an area of northeastern Illinois encompassing portions of Will, Grundy, Kankakee, LaSalle, and Livingston Counties. This region is herein referred to as the Mazon Creek area (Figure 3.1). The Mazon Creek fossils occur in a sequence of coal-bearing layered sedimentary rock deposited over 280 million years ago during the Pennsylvanian Period, a geologic time division known informally as the "Coal Age."

On a regional scale, the rocks lie within a broad down-warped structural sag in the earth's crust termed the Illinois Basin (Figure 3.1C). These deposits typically display cyclic alternations of marine and nonmarine sedimentary strata, interpreted as advances and retreats of shallow inland seas across what is now the heartland of the United States.

The Mazon Creek fossils are found in siderite ("ironstone") concretions that occur within the Francis Creek Shale member of the Carbondale Formation. The Francis Creek Shale is a deposit of gray mudstone, muddy siltstone, and sandstone. This unit overlies the regionally extensive Colchester (No. 2) Coal member (Figure 3.2). Strip mining and deep mining of this coal unit account for the numerous spoil dumps of Francis Creek Shale overburden in the Mazon Creek area, which are the usual collecting sources of Mazon Creek fossils (Figure 3.1). The coal is usually about 1 meter or slightly less in thickness (Culver, 1923; Willman and Payne, 1942; Smith, 1970). Most scientists agree that the Colchester Coal is a carbonized (distilled) remnant of a considerably thicker peat deposit that represents a former coastal swamp forest.

The Francis Creek member is regionally quite variable in thickness. It locally exceeds 25 meters (80 ft.) in thickness to the east and southeast of Mazon Creek but is thinner or entirely absent in other parts of the state (Smith, 1970; Wright, 1965).

The sideritic concretions yielding the exceptionally preserved organisms are only found in Francis Creek deposits that exceed 15 meters in thickness. These concretions are found typically in the lowermost 3 to 5 meters of the unit (Figure 3.2.2). The concretions often occur in distinct bands in exposures along Mazon Creek and also at Pit 11, the principal localities for marine fossils south of Braidwood. Drill cores obtained near Pit 11 reveal that concretions apparently occur in silty mudstone, which is often marked by a distinctive cyclic banding (Baird et al., 1985a, 1986; see Chapter 5). Stratigraphically higher Francis Creek beds are generally silty to sandy, and they commonly display current ripples.

Thin Francis Creek deposits near the town of LaSalle and farther west yield sideritic concretions, but these are generally barren. Near Galesburg and Peoria, the thin Francis Creek Shale strata yield brachiopods, bivalves, and

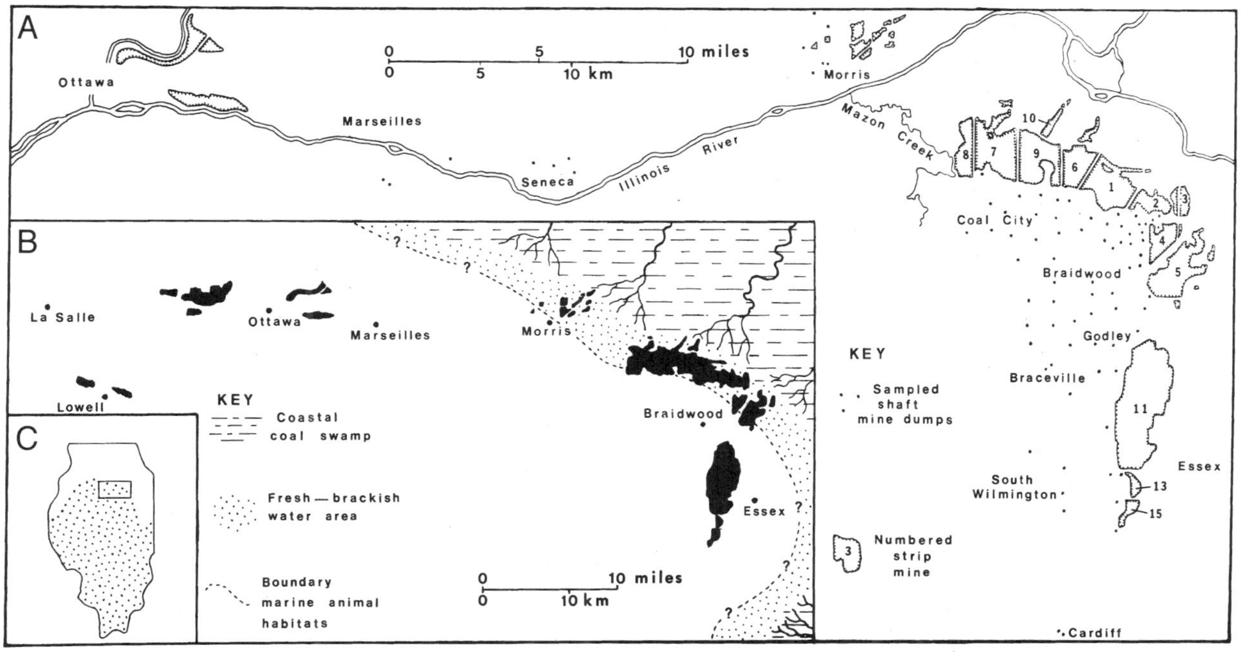

Figure 3.1. Mazon Creek area showing location of coal strip mines and underground mines. Dots denote sampling sources for data gathered as part of the Mazon Creek census project (from Baird et al., 1985a). Insert shows location of Mazon Creek area in the northeastern part of the Illinois Basin. The stippled area denotes the approximate inferred position of the Pennsylvanian sea. Note that most of the outcrop margin of the No. 2 Coal lies beneath Pleistocene glacial deposits (from Baird et al., 1986).

bryozoans, which usually occur in mudstone rather than within concretions (Shabica, 1971; Baird et al., 1985a, 1985b).

Thin Francis Creek deposits are overlain by another regionally widespread unit called the Mecca Quarry Shale member, which is about 0.5 meter in thickness in most areas (Figure 3.2.1). This shale is distinctly black in color, and it typically peels apart into platy or sheety layers (Smith et al., 1970; Zangerl, 1980). Like other Pennsylvanian sheety black shales, the Mecca Quarry Shale is rich in organic matter, often to the degree that it will burn if exposed to a flame. Such shales record deposition of mud and organic matter in anoxic (oxygen-starved) bottom settings, and they often yield unusual concentrations of trace metals (Coveney et al., 1983; Coveney, 1985).

The Mecca Quarry Shale is noteworthy because it locally yields diverse assemblages of superbly preserved marine vertebrates including sharks, extinct fish types related to sharks, and archaic bony fishes. Their preservation reflects mass mortality events associated with oxygen depletion. Gene Richardson worked extensively on this unit as a coworker and coauthor with Rainer Zangerl on the classic, 10-year study of the fossils and depositional setting recorded in the Logan Quarry and Mecca Quarry Shales in Indiana. The centerpiece publication of this project, Fieldiana Memoir Volume 4 (Zangerl and Richardson, 1963), ranks as one of the most detailed North American paleoecological studies known. The Mecca Quarry member is absent over the thick Francis Creek Shale deposits that contain Mazon Creek fossils and is not present in Grundy, Will, and Kankakee Counties (Smith, 1970; Baird, 1979).

Collecting Localities

The Mazon Creek area includes more than 100 specific collecting localities for Mazon Creek fossil organisms (Figure 3.1). The majority of collecting localities are spoil piles of abandoned coal mines (Figures 3.1, 3.3). Weathering of backpiled Francis Creek overburden from strip mines and older shaft mine spoil piles has produced a veneer of exposed concretions that can

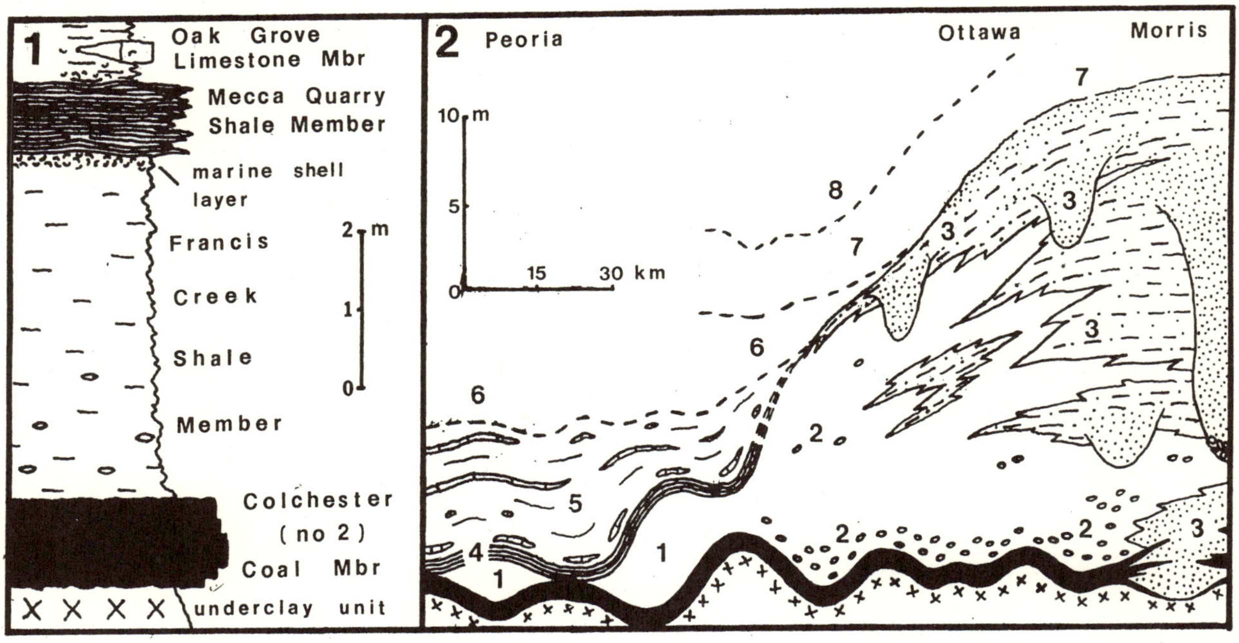

Figure 3.2. Francis Creek Shale member: stratigraphy and lithofacies. 1 Schematic vertical stratigraphic section of Francis Creek thin facies, Lowell, Illinois; 2 Cross section of Francis Creek deposit showing lateral gradation from thin facies in western Illinois to thick deposits near Mazon Creek. Units include: Underclay unit (XXX) below Colchester (No. 2) Coal member; *1,* Francis Creek burrowed mudstone lithofacies; *2,* Francis Creek laminar silty mudstone facies with sideritic concretions; *3,* Francis Creek siltstone-sandstone facies; *4,* Mecca Quarry Shale member; *5,* Marine strata above Mecca Quarry member; *6–8,* Younger cyclic marine and nonmarine units (from Baird et al., 1985).

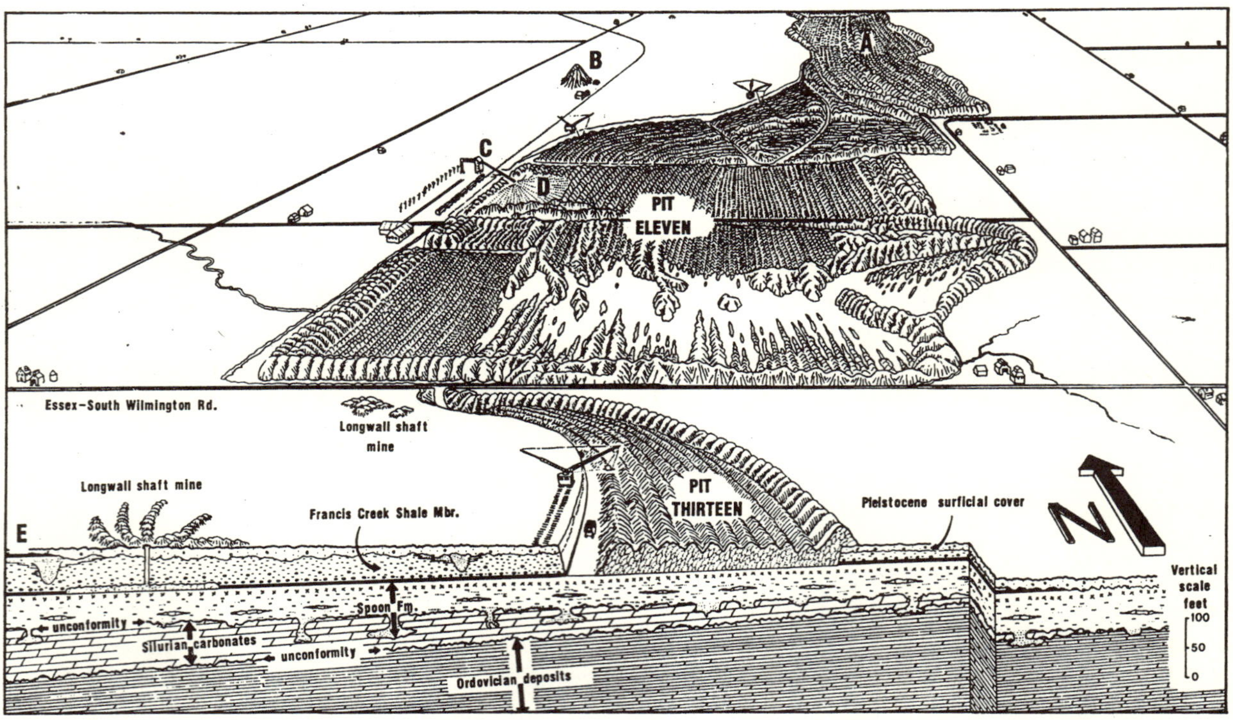

Figure 3.3. Diagrammatic cross section of Pit 13, adjacent to Pit 11, showing the mining process (ca. 1960) and generalized stratigraphy. Three major erosion surfaces (unconformities) occur beneath Silurian carbonates, beneath Pennsylvanian claystones and sandstones, and beneath Pleistocene ("ice age") surficial deposits. Strata dip generally toward the west-southwest. Both the Silurian and Pennsylvanian deposits terminate east of the strip mine due to the truncation beneath Pleistocene glacial deposits. X's beneath coal (*black band*) denote underclay (from Baird and Sroka, 1990).

be collected easily. Underground (shaft) mine spoil piles yielding Mazon Creek fossils numbered about 85 in the late 1970s. Some of these hills have been leveled in recent years through mine reclamation efforts, and more face this prospect as land use patterns change. Most of the shaft mine dumps yield marine fossils.

Although fewer in number as discrete localities, strip mines are most important as sources for fossils because they contain large areas of exposed dump surface for collecting. Consequently, the bulk of specimens described in this atlas come from strip mine areas. Famous strip mine localities for plants and nonmarine (Braidwood-type) animals include the long-abandoned Northern Illinois Coal Company Pits 1 and 6 northwest of Braidwood. Although these areas are now overgrown and closed to outsiders, they were once an important source for large and impressive plant fossils that grace museum exhibits all over the world.

The most important locality in the Mazon Creek area is the abandoned Peabody Coal Companies Northern Mine (Pit 11) south of Braidwood (Figures 3.1, 3.3). This strip mine and the adjacent abandoned McElvain strip mine bordering it on the east are the most renowned sources for marine (Essex) organisms. Perhaps as much as 80 percent of all the marine fossils ever collected from the Francis Creek Shale member have come from this large mined-out area. Pit 11 was the mine where the Essex fauna was first recognized as distinct from the nonmarine Braidwood fauna (see Richardson, 1956; Johnson and Richardson, 1966, 1970). Pit 11 served as the principal locality for joint Field Museum–University of Chicago paleoenvironmental studies in the 1960s and early 1970s when the mine was still active (see Johnson and Richardson, 1966; Richardson and Johnson, 1971; Shabica, 1970, 1971, 1979).

Intensive census sampling of numerous other strip mines during the late 1970s and early 1980s showed that the Essex fauna occurs in parts of the northern belt of strip mines north of Braidwood and Coal City, an area once thought to yield only Braidwood-type organisms (Figure 3.1). Essex fossils occur in strip mine areas as far west as Ottawa, LaSalle County (Baird, 1979; Baird et al., 1985a, 1986); however, the preservation of these organisms is generally poor in this area.

Outcrop localities where scientists and amateurs can collect fossiliferous concretions directly from shale are found almost solely along Mazon Creek southeast of Morris, Illinois. Long-known places such as Benson's Farm and Carr's Farm have yielded plants and nonmarine to minimally marine (Braidwood-type) animals. However, the bluffs adjacent to the junction of Claypool Creek by the White Tie Road overlook are now known to contain concretions yielding marine (Essex-type) animals, including the medusan *Essexella,* the echiuroid *Coprinoscolex,* as well as occasional polychaetes (Baird et al., 1985a, 1986; Baird and Sroka, 1990). At Morris, a partially submerged strip mine highwall yields abundant Braidwood-type bivalves and plants. Near Ottawa, LaSalle County, another partly submerged highwall yields concretions that contain marine (Essex) animals including the undescribed solemyid bivalve (see Chapter 8A). This clam is herein referred to as *"Edmondia"* or "common solemyid." Other marine animals found near Ottawa include the sea cucumber *Achistrum* and occasional polychaete worms (Baird et al., 1985a, 1986).

It should be noted that Francis Creek deposits in west-central (Fulton County) Illinois and Francis Creek–equivalent beds in Missouri and Oklahoma contain one or more of the key plant and animal associations found in Mazon Creek area localities. Although none of these localities is nearly as prolific as the region encompassing Morris, Coal City, Braidwood, and Essex, Illinois, a significant array of unusual and often rare soft-bodied animals have been recovered from them (Baird et al., 1985b; Sroka, 1996).

References

Baird, G. C. 1979. Lithology and fossil distribution, Francis Creek Shale in northeastern Illinois, pp. 41–67. *In* M. H. Nitecki (ed.), Mazon Creek Fossils. Academic Press, New York.

Baird, G. C., C. W. Shabica, J. L. Anderson, and E. S. Richardson, Jr. 1985a. Biota of a Pennsylvanian muddy coast: habits within the Mazonian Delta Complex, northeastern Illinois. Jour. Paleo., 59:253–281.

Baird, G. C., and S. D. Sroka. 1990. Geology and geohistory of Mazon Creek area fossil localities, Illinois, pp. C1–C70. *In* W. Hammer and D. F. Hess (eds.), Geology Field Guidebook. North-Central Section Geological Society of America, Macomb, Illinois.

Baird, G. C., S. D. Sroka, C. W. Shabica, and T. L. Beard. 1985b. Mazon Creek–type fossil assemblages in the U.S. Midcontinent Pennsylvanian: their recurrent character and paleoenvironmental significance. *In* H. B. Whittington and S. Conway-Morris (eds.), Extraordinary fossil biotas: their ecological and evolutionary significance. Phil. Trans. R. Soc. London B, 311:87–98.

Baird, G. C., S. D. Sroka, C. W. Shabica, and G. J. Kuecher. 1986. Taphonomy of Middle Pennsylvanian Mazon Creek area fossil localities, northeast Illinois: significance of exceptional fossil preservation in syngenetic concretions. Palaios, 1:271–285.

Coveney, R. M., Jr. 1985. Temporal and spatial variations in Pennsylvanian black shale geochemistry, pp. 247–266. *In* W. L. Watney, R. L. Kaesler, and K. D. Newell (eds.), Recent interpretations of Late Paleozoic cyclothems. Proc. 3rd Field Conf., Midcontinental Section, Soc. Econ. Paleo. Mineral., Lawrence, Kansas.

Coveney, R. M., Jr., B. G. Woodland, and R. Zangerl, 1983. Metalliferous shales of the Illinois Basin (field trip 10); pp. 147–171. *In* R. H. Shaver et al. (eds.), Field Trips in Midwestern Geology, volume 2. Indiana Geological Survey, Bloomington.

Culver, H. E. 1923. Geology and natural resources of the Morris Quadrangle. Illinois State Geol. Surv. Bull., 43:95–204.

Johnson, R. G., and E. S. Richardson, Jr. 1966. A remarkable Pennsylvanian fauna from the Mazon Creek area, Illinois Jour. Geol., 74:626–631.

———. 1970. Fauna of the Francis Creek Shale in the Wilmington area, pp. 53–60. *In* W. H. Smith, N. B. Nance, M. E. Hopkins, R. G. Johnson, and C. W. Shabica (eds.), Depositional environments in parts of the Carbondale Formation—western and northern Illinois. Illinois State Geol. Surv. Guideb. Ser., 8.

Richardson, E. S., Jr. 1956. Pennsylvanian invertebrates of the Mazon Creek area, Illinois. Fieldiana Geol., 12:3–76.

Shabica, C. W. 1970. Depositional environments in the Francis Creek Shale, pp. 42–52. *In* W. H. Smith, N. B. Nance, M. E. Hopkins, R. G. Johnson, and C. W. Shabica (eds.), Depositional environments in parts of the Carbondale Formation—western and northern Illinois. Illinois State Geol. Surv. Guideb. Ser., 8.

———. 1971. Depositional environments in the Francis Creek Shale and associated strata. Unpublished Ph.D. dissertation, University of Chicago, 206 pp.

———. 1979. Pennsylvanian sedimentation in northern Illinois: examination of delta models, pp. 13–40. *In* M. H. Nitecki (ed.), Mazon Creek Fossils. Academic Press, New York.

Smith, W. H. 1970. Lithology and distribution of the Francis Creek Shale in Illinois. *In* W. H. Smith, N. B. Nance, M. E. Hopkins, R. G. Johnson, and C. W. Shabica (eds.), Depositional environments in parts of the Carbondale Formation—western and northern Illinois. Illinois State Geol. Surv. Guideb. Ser., 8.

Smith, W. H., N. B. Nance, M. E. Hopkins, R. G. Johnson, and C. W. Shabica. 1970. Depositional environments in parts of the Carbondale Formation—western and northern Illinois (Field trip road log). Illinois State Geol. Surv. Guideb. Ser., 8:1–30.

Sroka, S. D. 1996. Synthesis of a Middle Pennsylvanian soft bodied fauna from Western Illinois. Unpublished Ph.D. dissertation, University of Illinois, Urbana, 197 pp.

Willman, H. B., and J. N. Payne. 1942. Geology and mineral resources of the Marseilles, Ottawa, and Streator Quadrangles. Illinois State Geol. Surv. Bull., 66, 388 pp.

Wright, C. R. 1965. Environmental mapping of the beds of the Liverpool cyclothem in the Illinois Basin and equivalent strata in the northern midcontinent region. Unpublished Ph.D. dissertation, University of Illinois, 100 pp.

Zangerl, R. 1980. The Pennsylvanian black, carbonaceous, sheety shales of the Illinois and Forest City Basins, pp. 225–238. *In* R. L. Langenheim, and C. J. Mann (eds.), Middle and Late Pennsylvanian Strata on Margin of Illinois Basin (Guidebook). 10th Field Conf., Great Lakes Section, Soc. Econ. Paleo. Mineral., Danville, Illinois.

Zangerl, R., and E. S. Richardson, Jr. 1963. The paleoecological history of two Pennsylvanian black shales. Fieldiana Geol. Mem., 4, 352 pp.

4A

Fossil Distribution and Fossil Associations

Gordon C. Baird

In any Mazon Creek locality, collectors nearly always find fossil plant fragments that were transported to the site of burial. In addition, these same localities yield either a sparse and low-diversity assemblage of nonmarine or minimally brackish animals (Braidwood fauna) or a much richer grouping of marine creatures (Essex fauna). Furthermore, localities yielding only Braidwood animals are characterized by abundant and often well-preserved plants, whereas localities characterized by marine animals yield a plant fraction that is usually distinctly lower in taxonomic diversity and preservation quality (Baird et al., 1985). This pattern is revealed through lateral shifts in organism dominance along a northeast-to-southwest sampling transect from areas yielding only Braidwood organisms to those yielding abundant marine taxa (Figure 4A.1).

The Braidwood fauna, other than the rare insects, arachnids, and myriapods, is a prosaic association of nonmarine or brackish animals that is locally dominated by small freshwater bivalves, especially in the case of the strip mines east of Morris (Figure 4A.2). The Braidwood nonbivalve component includes the xiphosuran (horseshoe crab) *Euproops danae,* the syncarid shrimps (*Palaeocaris typus* and *Acanthotelson stimpsoni*), diminutive bivalved crustaceans belonging to the class Branchiopoda, occasional sand-sized bivalved crustaceans belonging to the class Ostracoda, isolated lungfish scales, teeth of xenocanth fish, fish excrement (coprolites), a shark egg capsule (*Palaeoxyris*), and very rare amphibians (Figure 4A.2). Some of these organisms also occur in deposits yielding Essex animals, most notably some bivalves and the syncarid shrimp *Acanthotelson stimpsoni.* A few forms, including *Euproops, Palaeocaris* (Figures 4A.3A, 4A.3B), and xenocanth teeth, occur exclusively in localities lacking Essex animals as indicated by census sampling.

The Essex fauna contains a richer grouping of marine animals that were variably tolerant of changeable salinity and water turbidity (Figure 4A.2). It is the published descriptions of the numerous medusae, worms, shrimp, vertebrates, and problematica from the 1960s onward that have stimulated the surge of scientific interest in this deposit. Yet, like the Braidwood association, the Essex fossil samples are typically dominated by a few common forms (Figure 4A.2), most notably by the rhizostome medusan *Essexella asherae. Essexella* ("the blob") is the most common taxon in Essex samples. This form, like the plant fraction, may have been dispersed widely within the Mazon Creek area by waves and currents only to die and sink in fresher coastal waters. Depending on the sampling site, the next most frequently found Essex taxa are a common solemyid known as the "clam-clam" (*"Edmondia"* of earlier reports; see Figure 4A.3C), the eocarid shrimp *Belotelson magister,* a holothurian ("sea cucumber") currently being described (Sroka, 1987), and one or more trace fossils produced by animal activity.

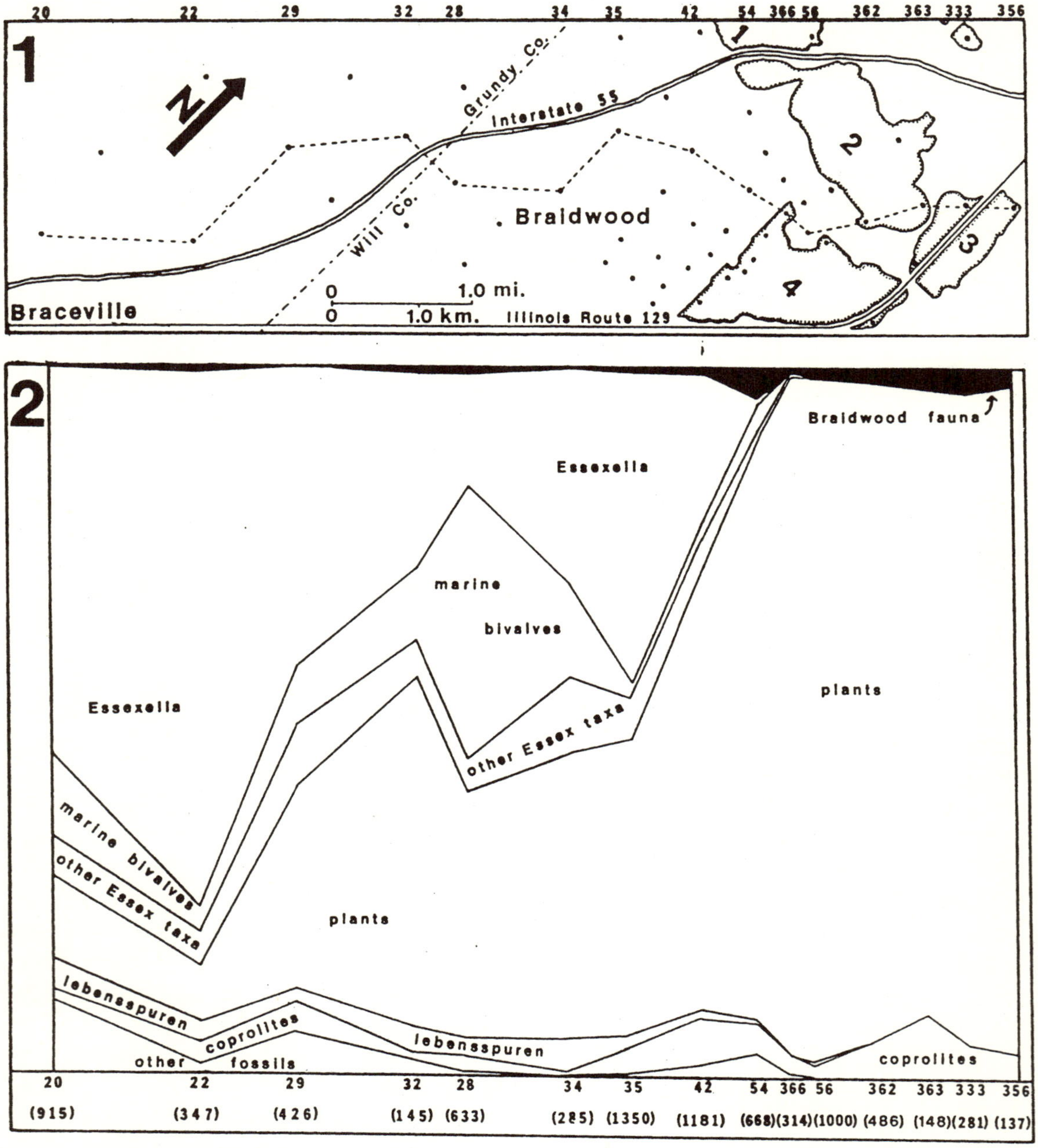

Figure 4A.1. 1 Change in the relative abundance of major organism groups along an inferred onshore-offshore transect across the Braidwood-Essex faunal boundary. Eight kilometer transect, subparallel to Interstate 55, connects 15 sampling localities, which include both strip and shaft mine nodule sources. Stippled borders denote boundaries of numbered strip mines. Dots indicate sampling localities within strip mines and on shaft mine dumps (dots outside of strip mine areas). Numbers at top correspond to numbered localities along transect. 2 Percentage variation of major fossil groups in transect samples. Quantity of specimens from each locality in parentheses. Note abrupt northeastward disappearance of Essex taxa between localities 35 and 56, with corresponding reduction in number of Braidwood animals in the southwestward direction (from Baird et al., 1985).

A major regional census collecting project across the entire Mazon Creek area (Baird et al., 1985) reveals the presence of several small, acre-sized areas in strip mines where rare or unusual organisms were dominant or important; these include "Jellyfish Hill" (*Octomedusa pieckorum* abundant), "Chiton Hills" in Pits 11 and 13 (*Glaphurochiton carbonarius* common), "H Hill" in Pit 11 (the possible hydrozoan *Etacystis communis* abundant), "*Lingula* Hill" in

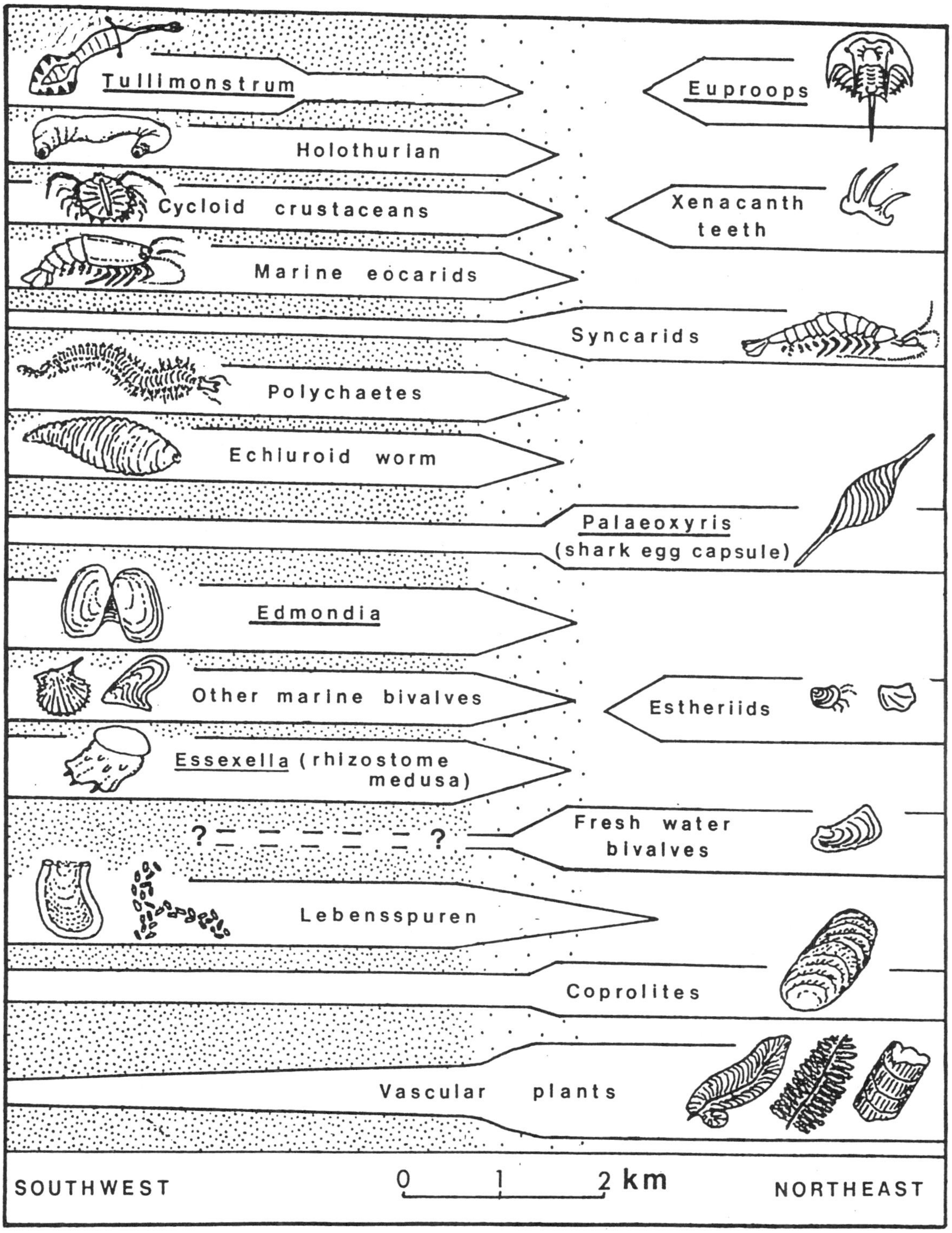

Figure 4A.2. Generalized onshore-offshore distributions of important fossil groups and key genera. Abundances are not shown to actual proportions but are generalized by relative width of the bars. Dense stippling denotes nearshore euryhaline marine (Essex) deposits, light stippling marks the narrow zone of Essex-to-Braidwood faunal change, and lack of stippling indicates nonmarine or minimally brackish (Braidwood) deposits. Note "one-way" mixing of Braidwood organisms into marine areas, with notable absence of landward mixing of Essex animals into Braidwood deposits (from Baird et al., 1986).

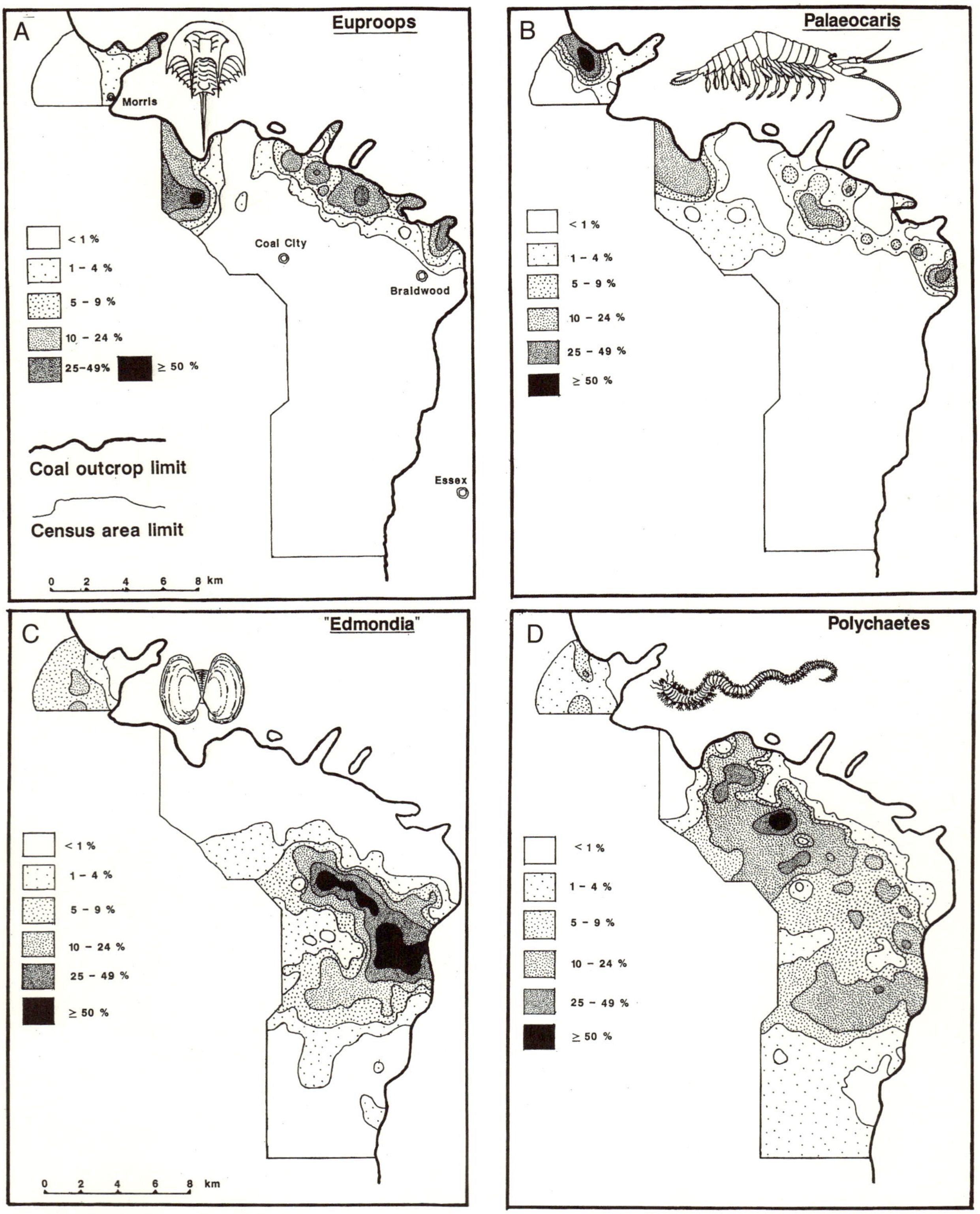

Figure 4A.3. Percentage abundance of four key Mazon Creek area taxa relative to total animal fraction in census samples (number of taxon/total number of animals, excluding lebensspuren and coprolites). Source of data includes sampling areas in strip mines and natural exposures (*dots*), as well as shaft mine dump piles shown in Figure 3.1. Computer-plotted percentages are shown as contoured values for A *Euproops danae;* B *Palaeocaris typus;* C *"Edmondia";* and D polychaetes (excluding *Spirorbis*) across the Mazon Creek area. Values below 1 percent for *Euproops danae* and *Palaeocaris typus* essentially denote absence of those taxa in samples. Note restriction of *Euproops* and *Palaeocaris* to northern sampling localities dominated by Braidwood-type (minimally marine to freshwater) taxa and general abundance of *"Edmondia"* and polychaetes in the region of Essex-type bottom animal occurrences southwest of Braidwood.

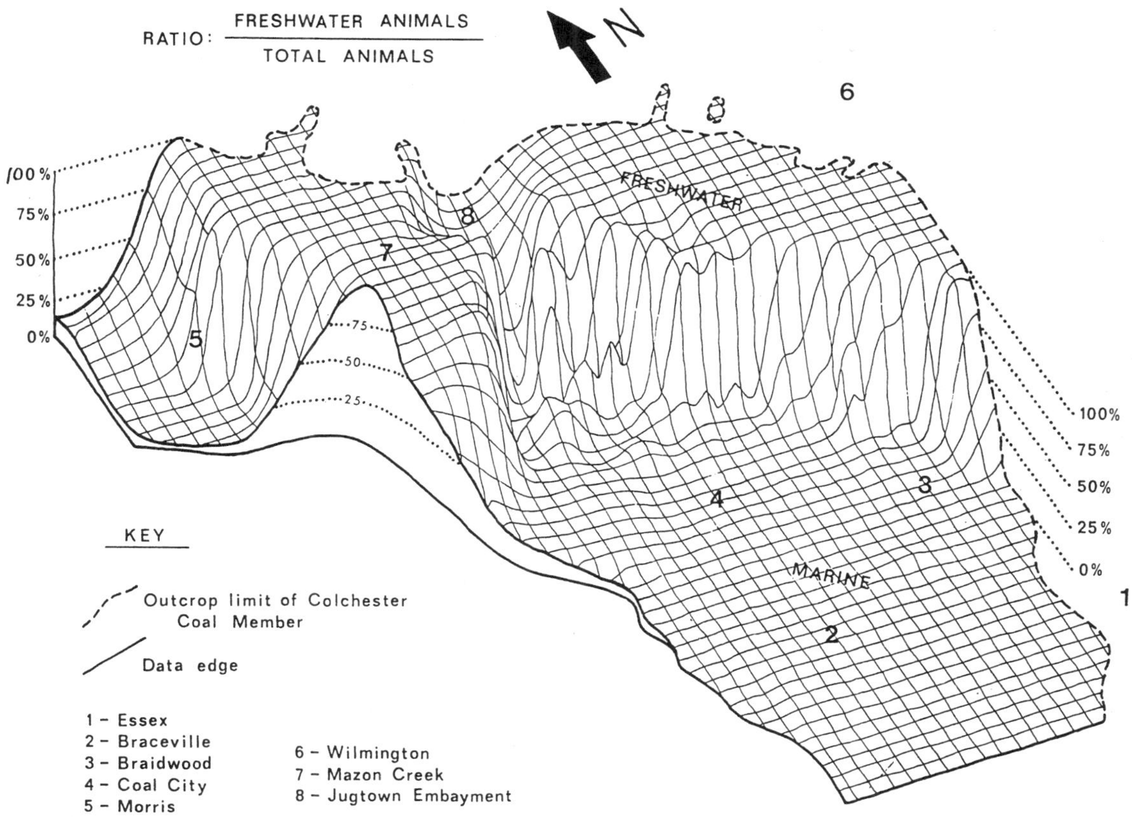

Figure 4A.4. Plot of ratio of Braidwood animals to total fauna for all samples. Visually improved, computer-generated, oblique three-dimensional projection of percentage data. Note prominent steep "escarpment" showing abrupt seaward decrease in percentage of Braidwood animals as the Braidwood-Essex faunal boundary is crossed, as well as conspicuous absence of Essex animals in samples from the "plateau" (from Baird et al., 1985).

Pit 11 (abundant inarticulate brachiopods), and "Fan Worm Hill," a shaft mine dump near Braceville (undescribed polychaete worm). Other locally common components of the Essex fauna include several polychaete worms, most notably *Esconites zelus* and *Didontogaster cordylina,* an echiuroid worm *Coprinoscolex ellogimus,* the cycloid crustacean *Cyclus americanus,* the hoplocarid shrimp *Kallidecthes richardsoni,* a bizarre species of the gastropod *Strobeus* (usually found attached to the jellyfish *Essexella*), the agnathan vertebrates *Esconichthys apopyris* ("the blade") and *Gilpichthys greenei,* together with the problematic animal *Tullimonstrum gregarium* (Figure 4A.2).

Two major subassociations are observed in areas of Essex animal occurrence. The first grouping, best observed in the Kankakee County part of Pit 11, in Pit 13, and on shaft mine dumps near South Wilmington, is dominated by swimming and floating animals. Most census samples from this region yield abundant blobs with lesser numbers of *Octomedusa, Etacystis, Belotelson, Kallidecthes, Strobeus* associated with *Essexella,* a small crustacean referred to as the "flea shrimp," and *Tullimonstrum.* Less common but still significant taxa in these samples include the cubomedusan *Anthracomedusa turnbulli,* the bivalve *Myalina* sp., the syncarid shrimp *Acanthotelson,* a palaeoniscoid fish *Esconichthys,* and the coelacanth fish *Rhabdoderma.* Except for the aforementioned localized occurrences of *Lingula, Coprinoscolex,* various polychaetes, *Glaphurochiton,* and the holothurian, the fauna in the samples is overwhelmingly composed of forms that swam or floated above the sea bottom.

The second Essex faunal subgrouping is best observed at localities such as the northern third of Pit 11, the abandoned Morris Coal Company strip mine ("Chowder Flats") west of Morris (now leveled and covered), the abandoned strip mines east of Ottawa, and numerous shaft mine dumps south of Braidwood, Eileen, and Coal City. Bottom-dwelling organisms, particularly the common solemyid bivalve *"Edmondia"* (Figure 4A.3C), are usually dominant at these localities. Polychaetes (Figure 4A.3D) and the holothurian are particularly common in these areas, as is also the echiuran *Coprinoscolex.* Abundant trace fossils representing within-sediment (infaunal) animal movement, characteristic of these localities, attest to the burrowing activity of worms, clams, and sea cucumbers.

The regional collecting census confirms the earlier recognition of a "nonmarine" (nonmarine or minimally brackish) Braidwood fauna and a saltwater (Essex) animal association in the Mazon Creek area (Johnson and Richardson, 1966; Richardson and Johnson, 1971). These animal groupings are discrete and usually separated regionally by a very narrow boundary zone where the associations intergrade, particularly in the vicinity of Braidwood (Figures 4A.1.1, 4A.2, 4A.4). It is significant that the northern to northeastward (landward) limit of Essex animal occurrence is quite abrupt while the Braidwood fauna and Mazon Creek flora are observed to decrease more gradually in abundance toward the southwest within the region of Essex animal abundance (Figures 4A.1, 4A.2, 4A.4). The southwestward (seaward) decrease in the quality of plant preservation across the marine part of our census area indicates that mixing of the two faunas was largely a one-way process; plants and many Braidwood animals were transported seaward into areas of saline water by outflow from delta distributaries or tidal channels, but there was no similarly effective mechanism to move marine animals into Braidwood animal habitats.

Finally, the sharp regional boundary between areas of Braidwood and Essex animal occurrence, the regional gradients in fossil preservation, and the presence of unusual localized animal associations (e.g., Chiton Hills and *Lingula* Hill) collectively indicate that most Essex animals were buried near the sites where they lived.

References

Baird, G. C., C. W. Shabica, J. L. Anderson, and E. S. Richardson, Jr. 1985. Biota of a Pennsylvanian muddy coast: habitats within the Mazonian Delta Complex, northeast Illinois. Jour. Paleo., 59:253–281.

Baird, G. C., S. D. Sroka, C. W. Shabica, and G. J. Kuecher. 1986. Taphonomy of Middle Pennsylvanian Mazon Creek area fossil localities, northeast Illinois: significance of exceptional fossil preservation in syngenetic concretions. Palaios, 1:271–285.

Johnson, R. G., and E. S. Richardson, Jr. 1966. A remarkable Pennsylvanian fauna from the Mazon Creek area, Illinois. Jour. Geol., 75:626–631.

Richardson, E. S., Jr., and R. G. Johnson. 1971. The Mazon Creek faunas, pp. 1222–1235. Proc. North American Paleo. Conv., 5–7 September 1969, Field Mus. Nat. Hist., vol. 1. Allen, Lawrence, Kansas.

Sroka, S. D. 1987. Preliminary studies on a complete fossil holothurian, from the Middle Pennsylvanian Francis Creek Shale of Illinois. Proc. 6th Int. Echinoderm Conf., Victoria, British Columbia.

4B

Relative Abundance of Different Mazon Creek Organisms:

A Closer Look at "Abundant," "Scarce," and "Rare"

Gordon C. Baird and John L. Anderson

Perseverance does not always pay off for every collector who stalks the coal company dumps or wades Mazon Creek. Although probabilities remain constant over time, chance sometimes matches a great fossil with the right collector on a day well remembered. Indeed, one hunter may find two lampreys in one day, while another collecting just as intensely finds none in a year. Gene Richardson recalled the first time that his Field Museum colleague Bill Turnbull was taken to the strip mines. Turnbull promptly secured a splendid specimen of the snakelike amphibian *Aornerpeton* (=*Phlegethontia*), a prize for which some veterans have searched in vain for decades.

But what controls the probability? Unless you are so foolish as to walk in the fresh bootprints of one or more zealous amateurs, your chances are influenced by the original absolute and relative abundances of organisms fossilized in concretions within the exposed and weathered rock. An approximate measure of the relative rarity or abundance of Mazon Creek organisms at various localities can be established through census collecting. However, the quality of such data depends on the amount of material collected, identified, and recorded. Roy L. Moodie tentatively determined that on average at least 100,000 concretions from Mazon Creek would have to be split to yield an amphibian fossil (Moodie, 1912, p. 279); this figure is little more than an educated guess based on extrapolation, however, because no one has ever collected half a million concretions from along the creek to firmly establish the odds.

What we do know is that good figures for relative and absolute abundance of common-to-scarce fossils now exist for both individual localities and for the Mazon Creek area overall as a result of the recent census project (see Baird et al., 1985a). As for truly rare organisms, one is obviously more at the mercy of chance discovery, as the Turnbull incident makes clear, and our relative abundance data for these organisms must be read with caution. One delightful aspect of the collecting of these fossils, however, is that the many categories of rare organisms make one's chances of finding a rare form during several days collecting quite high.

Tables 4B.1 and 4B.2 show the relative proportions of different categories obtained from the census for all localities yielding only minimally marine to nonmarine (Braidwood) animals (Table 4B.1) and for all localities yielding marine (Essex) organisms (Table 4B.2). These aggregate data sum numerous localities, so the variation between localities is averaged out and obscured. For example, in Braidwood animal collecting areas, bivalves display a patchy regional distribution, being either abundant or absent at each locality. Likewise, in marine (Essex) areas, most benthic organisms ("*Edmondia*," *Aviculopecten*, *Euphemites*, chitons, holothurians, and trace fossils) occur in great concentrations locally (Baird et al., 1985a, 1985b; see Chapter 3). However, the tables pro-

Table 4B.1
Relative Rarity of Selected Braidwood Fossils

Fossil	Number	Percentage*
Pecopteris	8,971	26
Plant stem	4,572	13
Epidophyllums	4,476	13
Macerated plant	4,233	12
Coprolite	2,670	7.8
Neuropteris	2,496	7.3
Annularia	1,430	4.2
Misc. plants	885	2.6
Bark	839	2.5
Calamites	770	2.3
Clam	632	1.8
Shrimp	162	0.5
Misc. molluscs	149	0.4
Horseshoe crab	98	0.3
Millipede	21	0.1
Fish scale	20	0.1
Insect	11	0.03
Fish	6	0.02
Spider	4	0.01
Centipede	1	0.003

Note: Relative rarity of key categories based on summed totals of all census categories collected from all minimally marine to nonmarine (Braidwood) localities (basis: total of sampling stations). Although some of these categories, most notably nonmarine bivalves, fluctuate in relative rarity among individual localities, the relative rarities listed are farily representative of most of these localities, particularly for the common plant taxa. Relative numerical and percentage of key Mazon Creek organisms computed by J. L. Anderson from Mazon Creek census data gathered by G. C. Baird.

*A total of 56,974 Braidwood concretions were collected. The percentage is based on 34,172 concretions (60 percent of total) that contained fossil remains and excludes 22,802 "dud" concretions (40 percent of total) that did not contain fossils.

Table 4B.2
Relative Rarity of Selected Essex Fossils

Fossil	Number	Percentage*
Blob (*Essexella*)	35,420	42
Plants	25,072	29
Burrow / trail	5,062	5.9
Clam (*"Edmondia"*)	4,673	5.5
Coprolite	4,070	4.8
Worms	2,409	2.8
Misc. molluscs	1,656	1.9
Shrimp (*Belotelson*)	1,497	1.8
Myaline	1,190	1.4
Misc. shrimp	446	0.5
Cyclus	412	0.5
Tully	318	0.4
Pecten	292	0.3
Octomedusa	262	0.3
Blade (*Esconichthys*)	95	0.11
Misc. fish	60	0.07
Insect	25	0.03
Millipede	12	0.014
Hydroid (*Drevotella*)	12	0.014
Horseshoe crab	9	0.011
Spider	9	0.011
Amphibian	3	0.004
Kottixerxes	2	0.002
Centipede	2	0.002

Note: Relative rarity of key categories based on summed totals of all census categories collected from all marine (Essex) localities (basis: total of sampling stations). Although the relative rarity of some of the major categories (e.g., *Essexella*) is usually consistent between most marine localities, relative proportions of some of these (clams, holothurians, burrows, and chitons) display a regionally patchy distribution. Relative numerical and percentage of key Mazon Creek organisms computed by J. L. Anderson from Mazon Creek census data gathered by G. C. Baird.

*A total of 229,979 Essex concretions were collected. The percentage value is based on 85,180 concretions (37 percent of total) that contained fossil remains and excludes 144,799 "dud" concretions (63 percent of total) that did not contain fossils.

vide an abundance ranking for taxa that encompasses the entire Mazon Creek area based on a total of 119,352 fossils sampled (excluding "duds"). They also provide a rough measure of the overall relative rarity of key organisms sought by amateurs and scientists.

One notable pattern that was clearly expected is that 15 to 20 fossil categories (excluding "duds"), out of a total of 350-plus possible taxa, overwhelmingly dominate census samples. Within Braidwood localities, the 10 most common categories constitute 91.7 percent of collected fossils, and the top 15 compose 97.8 percent of the take. A similar pattern is observed for the higher diversity Essex fauna, but dominance levels are somewhat lower; for estuarine localities, the top 10 include 83.3 percent of fossils collected, and the top 15 categories include 90.5 percent.

Typically, the mix of fossils that turns up in a day's collecting is a prosaic assemblage of plant parts, clams, and coprolites at Braidwood localities, and a somewhat more interesting mix of *Essexella, "Edmondia," Belotelson* molts, holothurians, coprolites, and trace fossils in marine areas. However, the sheer number of different categories of relatively rare taxa significantly increases the odds of really scoring on any particular day. Thus, the chance of finding a rare fossil is better than even at 250 fossils collected, and near certainty at 1000 fossils. This latter figure is what one person can collect in one day, assuming that he or she is the first to go to known choice areas in the early spring or after great storms. Seasoned and presevering veterans at "the Pit" know that, like a slot machine with numerous winning combinations, the area will continue to yield treasures. Moreover, unlike at a casino, there is more treasure because more untapped winning combinations surely await discovery. Let's all keep searching, collecting, and sharing our finds!

References

Baird, G. C., C. W. Shabica, J. L. Anderson, and E. S. Richardson, Jr. 1985a. Biota of a Pennsylvanian muddy coast: habitats within the Mazonian Delta Complex, northeast Illinois. Jour. Paleo., 59:253–281.

Baird, G. C., S. D. Sroka, C. W. Shabica, and G. J. Kuecher. 1985b. Taphonomy of Middle Pennsylvanian Mazon Creek area fossil localities, northeast Illinois: significance of exceptional fossil preservation in syngenetic concretions. Palaios, 1:271–285.

Moodie, R. L. 1912. The Pennsylvanian amphibia of the Mazon Creek Shale, Illinois. Kansas University Sci. Bull., 6:323–359.

4C

Francis Creek Diagenetic Events

Gordon C. Baird

Mazon Creek fossils occur within sideritic ("ironstone") concretions, which typically formed around buried organisms soon after they were entombed. These concretions are the result of localized chemical reactions, sometimes initiated by decay of the organism itself, that occurred at shallow depths beneath the sediment surface. Siderite precipitation occurred in the absence of free oxygen (as the iron in the carbonate is in the reduced state) through bacterially mediated reactions. Such reactions usually produced a hard cemented mud rind around buried organisms, often before significant animal or plant decay had commenced. Thus, fine exterior details of organisms are preserved as molds and composite impressions within concretions. Sometimes the actual soft tissue and cell structure of these organisms is also preserved.

Where three-dimensional hard objects such as woody stems and mollusc shells occur in concretions, they are preserved as molds. Moldic fossils are simply cavities in rock retaining the exterior or interior form of the organism that is now missing as a result of decay or dissolution. In the Mazon Creek area strip mines, one can easily find abundant glacially transported, buff-yellow boulders and cobbles of Silurian dolostone containing molds of trilobites and crinoid stems. These are visible as holes or pits in the rock. Details on these and on moldic Mazon Creek fossils are negatives of original interior or exterior surface detail. Even delicate plant foliage is preserved in the form of molds in some concretions (Schopf, 1979). This is why fern and seed fern pinnae often display veneers of sphalerite, pyrite, and white powdery kaolinite, as well as residual carbonaceous material. The narrow space, corresponding to the width of a few leaf cells, is where mineral salts precipitated after formation of the concretion and degradation of the plant material.

Fossils of medusae, hydrozoans, polychaetes, molted crustaceans, larval vertebrates, and *Tullimonstrum gregarium* are often preserved as composite mold impressions. In these instances, the soft tissue was compressed and degraded prior to cementation of enclosing muds. Many blobs and Tully Monsters display superposition of top- and bottom-side details. Anatomical features on smaller *Esconichthys* ("blades") and molted shrimp are often expressed only as ghostlike patterns of light coloration or as thin organic films.

Regional gradients in the quality of fossil preservation exist within the five-county study area. Animal preservation in concretions is often poorer where there is evidence of sediment disturbance by burrowing organisms. Blobs found in parts of Pit 11, in the southern part of Pit 13, and particularly in Pit 15 illustrate this point; they are often preserved as rectangular patches of light coloration with minimal detail and relief. Many of these surfaces show evidence of intense burrowing along the bedding plane on which the fossil lies. These localized burrow concentrations indicate that scavengers may have feasted on these remains prior to formation of the concretion (Baird et al., 1986). In areas where trace fossils are abundant, often in small, localized areas where the bivalve *"Edmondia"* and the holothurian are common, soft-bodied and lightly skeletonized animals are also poorly preserved. Frequent bottom scavenging and mud ingestion by "grave-robbing" scavengers and deposit feeders, both on and in bottom muds, may explain these small areas of poor fossil preservation (see observations and discussions in Schäfer, 1972; Plotnick, 1986). The poor fossil preservation and increased evidence of burrowing activity in thinner Francis Creek deposits marginal to the Mazon Creek area are due to slower rates of sediment accumulation and organism burial in settings farther offshore from distributaries (Baird et al., 1985a, 1986) because prolonged exposure of soft tissue on or just beneath the substrate favors its rapid decay through aerobic processes.

Preservation quality of Essex animals declines slightly to the south and west of Pit 11, and it decreases markedly to the west of Seneca, LaSalle County (Baird, 1979; Baird et al., 1985b, 1986). In the abandoned strip mines east of Ottawa, LaSalle County, soft-bodied organisms such as medusae are rare and usually poorly preserved. Lightly skeletonized animals such as holothurians are common, but only marginally preserved in this same area. West of Ottawa, only the bivalve *"Edmondia"* and occasional plant debris have been found in concretions. In western LaSalle County and in Bureau County still farther to the west, the Francis Creek Shale member thins to less than 5 meters in thickness in most places and is composed of very fine grained and typically burrowed mudstone. Sideritic concretions collected from shaft mine dumps in Bureau County display centers of pyrite and have not yielded fossils.

Concretions, particularly those from areas yielding Essex animals, display interior zones of pyrite cementation. These pyrite cores or bubblelike pyrite halos, often partly to completely enclosing fossils, are a nuisance to the collector because they impede proper splitting of nodules, whether by hammer blows or by the freeze-thawing technique. The pyrite is believed to have formed soon after the burial of these organisms. Bacterial reduction of seawater sulfate in oxygen-depleted, near-surface mud produces black monosulfide, which later converts to the more stable mineral pyrite (see Berner, 1971, for discussion of chemical models for pyrite formation).

In most areas of Braidwood and Essex animal occurrence, seawater sulfate was evidently in short supply. Under these conditions the anaerobic bacteria could have maintained sulfate reduction only so long as a sulfate supply could be obtained from the decaying organisms or from the overlying water. Once this supply was depleted, new reactions involving methanogenic bacteria would have commenced, leading to siderite precipitation (see Berner, 1971, 1981; Claypool and Kaplan, 1974). The abundance of iron in the estuarine environment, coupled with the weak supply of seawater sulfate, is believed to explain the occurrence of abundant siderite within the Francis Creek Shale (Woodland and Stenstrom, 1979).

Did buried organisms actively trigger concretion growth? For proper consideration, this question must be restated as two sequential questions. First, did buried organisms serve as passive nuclei for concretion growth in the manner that a string will serve as a surface for the nucleation of crystals? And second, did buried organisms serve as chemical triggers (catalysts) for siderite precipitation due to their decay within the sediment?

The first question can generally be answered in the affirmative given that Mazon Creek fossils not only occur in concretions but also closely govern nodule shape (as is usual with blob-, plant-, and *Tullimonstrum*-bearing concretions). The abundance of barren concretions in census counts leaves open the possibility that some nodules formed in the absence of nuclei. We suspect that "duds" are usually fossil-bearing concretions cored by unsplittable pyrite, excessively small fossils, irregular trace fossils, or diffuse, badly decayed organic material. In short, we cannot exclude the possibility that all of these nodules did, in fact, form around fossil material. However, most of the platy and irregular sideritic debris ignored by collectors shows little evidence of having formed around fossils.

The second question concerning the role of decay processes in triggering concretion growth, remains only partly answered to date. The occurrence of fossils as concretion nuclei indicates some relationship between buried organisms and diagenetic reactions, but the degree of interaction is unknown. A few nodules clearly formed in microenvironments of active bacterial decay (Woodland and Stenstrom, 1979). Such decay would have produced ammonia and alkaline conditions conducive to the production of siderite precipitation in the vicinity of the organism, as illustrated by the decay of "juicy" nuclei such as coprolites, which are often surrounded by robust subspherical concretions more massive than those around other fossils (Baird et al., 1986). However, for many small plant and animal fossils, the case for the nucleus's being a chemical catalyst is less firm. This problem is the subject of laboratory simulation of decay processes within muds (see Allison, 1986, 1988a, 1988b).

The rapidity of concretion growth is particularly illustrated by the fact that concretions contain up to 80 percent siderite cement (Woodland and Stenstrom, 1979). This implies that siderite precipitation occurred in water-rich surface mud that was sufficiently uncompacted to contain about 80 percent water by volume. A very early diagenetic origin for these concretions is also indicated by the fact that compacted, nonconcretionary mudstone is compressively draped around concretions that resisted compaction. An additional line of evidence is the presence of load-induced compactional faulting of the unconsolidated mudstone. This faulting is due to rapid deposition and is observed to cross-cut (fracture) concretions that had already formed (see Baird et al., 1986). Septarian (radiating) fracture systems within concretions, usually filled with sulfide mineral precipitates (pyrite, sphalerite) or with the problematic white powdery clay mineral kaolinite, formed due to dewatering of muds (syneresis) during nodule formation (Figure 4C.1). Precipitation of the sulfides and the kaolinite appears to be a later event that remains to be studied in detail.

Laboratory simulations of animal decay in sediments (Allison, 1988a, 1988b) indicate that bacterial sulfate reduction is more damaging to organic remains than are the siderite-producing methanogenic reactions. In the absence of diagenetic mineralization, shrimp and polychaete remains were usually destroyed in the sediment by anaerobic decay within a period of 25 weeks. However, decay is slower in the methanogenic interstitial environment, thus increasing the chances that the remains would still be present by the time siderite precipitation commences under natural conditions. In either case, preservation of soft-bodied and lightly sclerotized invertebrate remains requires early diagenetic mineralization, particularly when the precipitate is siderite as opposed to pyrite or calcium carbonate. This fact is important because good Mazon Creek fossil preservation involves enclosing concretionary siderite. Fossil nuclei in pyrite nodules have not been reported from Mazon Creek area localities, nor have any been found in marine limestone concretions within the Francis Creek deposits in western Illinois. Moreover, Francis Creek localities yielding marginally recognizable Essex animals in LaSalle County are always characterized by siderite concretions with abundant interior pyrite. These deposits are thought to record slower deposition farther from the delta source where marine conditions would have been more pervasive (Baird et al., 1985a). The abundance of pyrite in these concretions and its association with poorly preserved fossils may be further evidence for this.

Fossils in nodules often occur in the lower third or quarter part of concretions (Figure 4C.1). At the White Tie Road overlook along Mazon Creek, Essex animals (*Essexella, Coprinoscolex,* polychaetes) and plants occur in the lower part of all fossil-bearing concretions found in the mudstone bank (Baird et al., 1986). This positioning is due either to preferential upward growth of nodules into less compacted overlying mud layers or to the effect of an upward migration of chemical-rich waters around the growing concretion during mud compaction (Figure 4C.1).

These observations reveal the true "up-down" orientation of *Essexella* impressions, even for fossils collected on mine dumps! Both

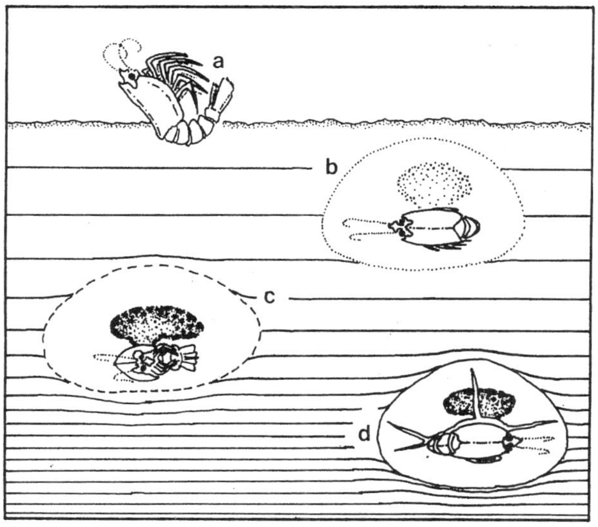

Figure 4C.1. Sequential events in the formation of sideritic concretions: *a,* Sedimentary record receives dead organism. *b,* Bacterial sulfate reduction proceeds in local decay microenvironment ("gas vent" zone) above organism, prior to depletion of interstitial sulfate. Subsequent methanogenic reactions produce siderite (*outline*). *c,* Continued siderite precipitation hardens protoconcretion, which resists compaction. *d,* Dewatering of surrounding muds causes laminae in unconsolidated sediments to deform (drape) around nodule. Syneresis (dewatering) within concretion produces septarial fractures that break (cross-cut) earlier formed pyrite (or iron monosulfide) halo around decay center. Note differential upward growth of sideritic concretion (from Baird et al., 1986).

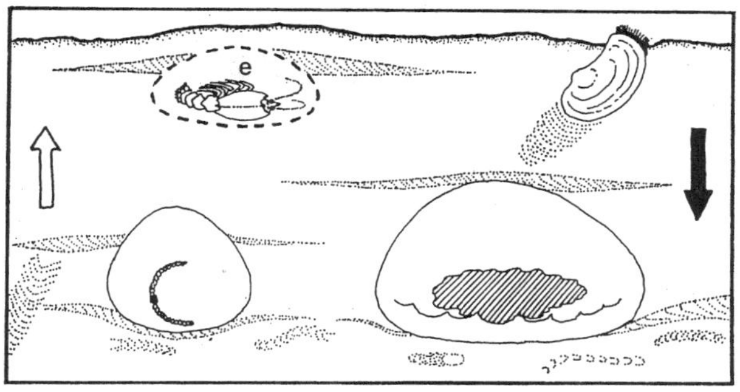

Figure 4C.2. (Also Figure 5.7B). Dark arrow (*right*) denotes compaction and dewatering of mud around growing nodules under thin sediment cover. White arrow (*left*) signifies upward migration of reaction ions owing to compactional loading of underlying peat and mud; this may explain differential upward (eccentric) growth of concretions. Dashed line around buried *Belotelson* (e) is protoconcretion in process of forming.

in the White Tie bank section and at Pit 11, blobs occur skewed to one side of the concretion; when split, the blob impression on the thicker concretion part has convex relief, and on the thinner counterpart, concave relief. Since *Essexella* ornamentation is concave (negative) on the lower counterpart, we can relate the ornamentation to the way the jellyfish originally settled on the bottom. The downward convexity of fold and pleat detail makes sense when one considers that this organism would have had some weight, particularly if loaded from above by rapidly deposited sediment (see Figures 4C.1, 4C.2 [5.7B]). Subsequent decay and collapse of the jellyfish would have allowed still-unconsolidated mud to fill in the underlying concave depression to produce a composite impression (Baird et al., 1986).

Seilacher (1970) classified localities yielding exceptional paleontological information into the categories of *Konzentrat Lagerstätten* (concentration deposits) and *Konservat Lagerstätten* (conservation deposits) characterized by exceptional, often soft-bodied, faunas. Recently, he further classified the Konservat Lagerstätten into bottom-smothered assemblages recording rapid burial and stagnation assemblages preserved due to an absence of oxygen (Seilacher et al., 1985). However, Allison (1988a, 1988b) showed that an absence of oxygen by itself will not produce soft-bodied fossil assemblages; he found that an absence of oxygen will only retard soft-part decay. Allison (1988a) further showed that soft-bodied organisms will be preserved only where remains have been buried rapidly and where within-sediment mineralization has occurred early enough to preserve impressions of the undecayed or partially decayed organism. Thus, the Mazon Creek fossil preservation story conforms to this last scenario involving rapid burial followed by extremely rapid siderite precipitation around organisms.

References

Allison, P. A. 1986. Soft-bodied animals in the fossil record: the role of decay in fragmentation during transport. Geology, 14:979–981.

———. 1988a. Konservat-Lagerstätten: cause and classification. Paleobiology, 14:331–344.

———. 1988b. The role of anoxia in the decay and mineralization of proteinaceous macrofossils. Paleobiology, 14:139–154.

Baird, G. C. 1979. Lithology and fossil distribution, Francis Creek Shale in northeastern Illinois, pp. 41–67. *In* M. H. Nitecki (ed.), Mazon Creek Fossils. Academic Press, New York.

Baird, G. C., C. W. Shabica, J. L. Anderson, and E. S.

Richardson, Jr. 1985a. Biota of a Pennsylvanian muddy coast: habitats within the Mazonian Delta Complex, northeast Illinois. Jour. Paleo., 59:253–281.

Baird, G. C., S. D. Sroka, C. W. Shabica, and T. L. Beard. 1985b. Mazon Creek–type fossil assemblages in the U.S. midcontinent Pennsylvanian: their recurrent character and paleoenvironmental significance. *In* H. B. Whittington and S. Conway-Morris (eds.), Extraordinary fossil biotas: their ecological and evolutionary significance. Phil. Trans. R. Soc. London B., 311:87–98.

Baird, G. C., S. D. Sroka, C. W. Shabica, and G. J. Kuecher. 1986. Taphonomy of Middle Pennsylvanian Mazon Creek area fossil localities, northeast Illinois: significance of exceptional fossil preservation in syngenetic concretions. Palaios, 1:271–285.

Berner, R. A. 1971. Principles of Chemical Sedimentology. McGraw-Hill, New York, 240 pp.

———. 1981. A new geochemical clarification of sedimentary environments. Jour. Sediment. Petrol., 51:359–365.

Claypool, G. E., and I. R. Kaplan. 1974. The origin and distribution of methane in marine sediments, pp. 99–139. *In* I. R. Kaplan (ed.), Natural Gases in Marine Sediments. Plenum, New York.

Plotnick, R. E. 1986. Taphonomy of a modern shrimp: implications for the arthropod fossil record. Palaios, 1:289–293.

Schäfer, W. 1972. Ecology and Paleoecology of Marine Environments. University of Chicago Press, Chicago, 568 pp.

Schopf, J. M. 1979. Evidence of soft-sediment cementation enclosing Mazon plant fossils, pp. 105–128. M. H. Nitecki (ed.), Mazon Creek Fossils. Academic Press, New York.

Seilacher, A. 1979. Begriff und bedeutung der fossil-Lagerstätten. Neues Jahrb. Geol. Paläo. Mh., 1970, pp. 34–39.

Seilacher, A., W. E. Reif, and F. Westphal. 1985. Sedimentological, ecological and temporal patterns of fossil-Lagerstätten. Phil. Trans. R. Soc. London B, 311:5–23.

Woodland, B. G., and R. C. Stenstrom. 1979. The occurrence and origin of siderite concretions in the Francis Creek Shale (Pennsylvanian) northeastern Illinois, pp. 69–104. *In* M. H. Nitecki (ed.), Mazon Creek Fossils. Academic Press, New York.

5A

Paleoenvironmental Setting of the Mazon Creek Biota

Gordon C. Baird

Inferred Paleogeography

During Pennsylvanian time, most of the ancient continents had converged to form a "supercontinent." This large landmass has been named Pangaea, which means "all Earth"(see Figure 5A.1). Prior to this, North America was part of a smaller supercontinent composed of continents and continental fragments called Laurasia. This entity collided with an even larger, sutured aggregate of continents now known as Gondwanaland to produce Pangaea. When this great collision occurred during the Middle Pennsylvanian period, the side of North America that faced the shrinking ocean between the two supercontinents was compressed, buckled, and uplifted. The northern parts of present-day Africa and South America converged with southeastern North America to produce a mountain range that extended in long, curving belts from what is now New England into Mexico (Rowley et al., 1985).

These mountains, visible today as the deeply eroded terrains of the Appalachian and Ouachita ranges, supplied large quantities of sediment to tropical river systems in lands on both sides of the mountains (Figure 5A.1). Flowing to shallow seas, these rivers created large deltas at their mouths and periodically inundated the inhabitants of the shoreline and shallow seas with vast quantities of mud, silt, and sand.

Paleomagnetic studies indicate that the earth's equator extended across what is now North America, with the position of present-day Chicago located near this paleoequator during Middle Pennsylvanian time (Scotese et al., 1979; Bambach et al., 1980; Rowley et al., 1985). Paleoclimatic studies confirm that tropical conditions prevailed within the Illinois Basin during the Middle Pennsylvanian. The rising mountains, at peak development during the Middle and Late Pennsylvanian, may have created conditions of atmospheric circulation that produced extensive rainfall in eastern North America (Phillips and Dimichele, 1981; Phillips et al., 1985). Periodic moist episodes in conjunction with rising sea level appear to explain the development of migrating coastal swamp forests that produced the extensive peat layers that would ultimately be converted to coal seams. The periods of deposition of the Colchester, Springfield, and Herrin peats coincided with the creation of a very wet and relatively nonseasonal climate, as indicated by the plant types, widespread swamps, and large deltas associated with the coal-bearing sequences (Phillips et al., 1985). In the Pangaea reconstruction (Figure 5A.1), most of the Pennsylvanian coal basins, extending from Oklahoma to European Russia, form a line subparallel to and nearly coincident with the inferred position of the ancient equator (Scotese et al., 1979; Rowley et al., 1985).

Cyclic repetitions of nonmarine sandstones, soils, coals, marine limestones, and black shale deposits occur in the Illinois Basin Pennsylva-

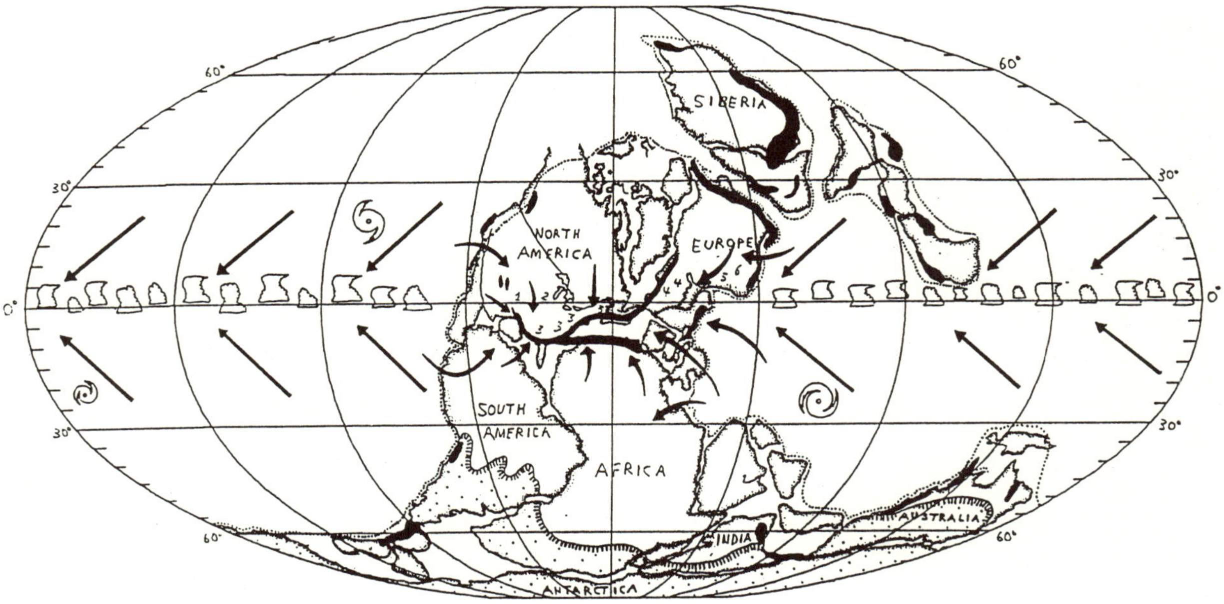

Figure 5A.1. Inferred Middle Pennsylvanian (Westphalian) global paleogeography showing linkage of present-day continents and microcontinents (*lightly stippled borders*) into an inferred supercontinent, Pangaea, outlined by a single heavy stippled line. This reconstruction is based on geomagnetic data (Scotese et al. 1979; Nie Shangyu et al., 1990), matches (jigsaw puzzle–like "fits") of concordant continent shapes and geological patterns, as well as paleoclimatic evidence. Collision-generated mountain ranges are indicated by broad black belts. Tropical wind and climate patterns over low-latitude parts of the circum-Pangaean (Panthalassa) Ocean are generalized from conditions on a modern rotating earth; vertical clouds mark the equatorial moist zone, and both northern and southern trade wind belts (*large arrows convergent on equator*) are shown. Wind patterns in equatorial Pangaea are generalized from discussions in Rowley et al. (1985); arrows converging toward equatorial Pangaea denote both moisture-laden, onshore (monsoonal) winds generated by seasonal differential heating of the land, and winds convergent on belts of equatorial collisional mountains (including Appalachian and Ouachita ranges in North America) due to differential heating of these elevated terrains. Both equatorial conditions plus convergent moisture-releasing air movement are believed to explain the distribution of the economically significant coal basins yielding the paleoequatorial Euramerican coal flora (Rowley et al., 1985). Numbered basins include *1,* Arkona Basin; *2,* Illinois Basin; *3,* Appalachian coal basins; *4,* Irish and British coal basins; *5,* Ruhr Basin (Germany); and *6,* Donetz Basin (C.I.S.). Inferred maximum extent of glacial and periglacial conditions (hatchured border and coarsely stippled area) associated with one or more large icecaps in southernmost Pangaea are also shown (see Veevers and Powell, 1987).

nian succession. These repeating sequences, termed "cyclothems" (Wanless and Weller, 1932), have been a long-standing subject of controversy. They have been interpreted as local depositional effects produced by delta growth and abandonment (see Ferm, 1970; Shabica, 1971; Merrill and Martin, 1976). The great extent of many cyclothems and their distinct succession of deposits suggests that they record global fluctuations in sea level as well as widespread warping of the continent due to collisional stresses (Wanless and Shepard, 1936; Heckel, 1986; Heckel et al., 1980; Klein and Kupperman, 1992). The argument for sea level oscillations is independently supported by evidence for extensive glaciation in the Pennsylvanian south polar region on Gondwana (Crowell, 1978; Veevers and Powell, 1987; see Figure 5A.1). According to this account, cyclic expansion and contraction of the ice cap on Gondwana caused repeated lowering and rising of sea level, similar to that produced during the recent Pleistocene "ice ages." Thus, a period of glacial melting in the Pennsylvanian south polar region of Gondwana may explain the transgressive drowning of the tropical Illinois seacoast that initiated the dramatic events at Mazon Creek.

Colchester Coal Swamp

The Pennsylvanian Period was notable for the development of swamp forests along coastal lowlands and on river floodplains (Figure 5A.2). Over long periods of time, the accumulated organic debris built up a thick peat body. Subsequent burial of this peat layer was followed by

Figure 5A.2. Some terrestrial and aquatic Mazon Creek animals as they might have appeared in life. Scene along a channel on the Mazonian Delta complex, winding through a reconstructed Pennsylvanian coal swamp: Amphibians shown are *Ptyonius (decomposing carcass)*, *Ophiderpeton* (*swimming*), *Brachydectes* (*on bank*). The giant dragonfly (710 mm wingspan) is described by Kukalova-Peck (see Chapter 14B). In the foreground are the centipede *Mazoscolopendra* and an insect larva.

its slow alteration to a seam of coal. Approximately 10 meters, and possibly more than 20 meters, of peat produced the 1 meter thick Colchester (No. 2) Coal seam (see Winston, 1986, 1988, for estimations of peat-to-coal compression ratios). Some workers now believe that ancient peats of this type may have accumulated vertically to as much as 8 meters above high tide under a saturated, raised swamp. In modern areas of high precipitation, swamps may build mud-free peat deposits above sea level (see McCabe, 1984, 1992; Pfefferkorn et al., 1971). The Colchester Coal is part of the most extensive coal seam known in the geologic record. When the equivalent Croweburg Coal is taken into account, a continuous Colchester-Croweburg seam is observed to run from western Indiana to Oklahoma. Eventually, the Colchester Coal swamp was inundated during marine transgression. By the time of Francis Creek mud deposition, much of the Mazon Creek area had been inundated by marine water.

The Colchester swamp forest, though luxuriant and tropical, differed fundamentally from any modern forest (Figure 5A.2). The angiosperms, or flowering plants, that dominate most present-day forests had not yet evolved, and the conifers, only beginning to appear in the Pennsylvanian, were not yet a component of coastal lowland settings. Large trees were

Figure 5A.3. Common plant genera found in the Francis Creek Shale member within the Mazon Creek area. **1** Pteridosperm leaf element *Neuropteris scheuchzeri.* **2a–d** Fern leaf element *Pecopteris* sp. showing variations in form of pinnae; **2b** shows spore tetrads on leaf pinnules. **3** Pteridosperm leaf element *Alethopteris searlii.* **4a,b** Needlelike *Lepidodendron* foliage element *Lepidophylloides.* **5** Modified leaf elements from cone organ *Lepidostrobus,* which belongs to the lycopod genus *Lepidodendron.* **6** Annulated bark organ *Calamites* sp. belonging to sphenopsid ("horsetail"). **7** Foliage element *Annularia* sp. belonging to sphenopsid. **8** Seed *Trigonocarpus* sp. belonging to a pteridosperm. **9** Sphenopsid foliage element *Asterophyllites* sp. **10** Sphenopsid foliage element *Sphenophyllum* sp. **11** Outer bark layer, *Lepidodendron* sp. showing rhombic sculptured surface of leaf scars. **12a,b** *Neuropteris* pinna which has been buried edgewise. **13** *Pecopteris* pinna showing evidence of current-induced flexing during burial.

present in the Colchester swamp, but these belonged to archaic plant groups that are either extinct or represented only by diminutive modern descendants (Figure 5A.2). The Colchester swamp was dominated by lycopods, most notably the "scale trees" *Lepidodendron* and *Lepidophloios* (Figures 5A.2, 5A.3.11), which probably formed the middle and upper stories of the swamp forest (see Pfefferkorn, 1979; Pfefferkorn et al., 1971; Phillips and Peppers, 1984; Phillips et al., 1985). Sphenopsids, of which modern *Equisetum* (the "horsetail") is a member, were represented as a minor component by *Calamites* (Figures 5A.2, 5A.3.6, 5A.3.7). Other subsidiary components included tree ferns, the source of the common leaf form *Pecopteris* (Figure 5A.3.2); pteridosperms ("seed ferns"), which produced the pinnule genera *Neuropteris, Cyclopteris,* and *Alethopteris* (Figure 5A.3.1, 5A.3.3, 5A.3.12); and cordaitales, an extinct plant group whose woody tree trunks are often found permineralized.

Plant parts observed in Francis Creek concretions represent a macrodetrital flora that was transported to the sea from terrestrial sources. Unlike the Colchester peat swamp, which was dominated by lycopods as was typical for coals of this age (Peppers and Pfefferkorn, 1970), the Francis Creek flora is dominated by fern, pteridosperm, and sphenopsid debris. This composition suggests that many of the plants found in the Francis Creek concretions had been transported considerable distances from upriver and upland sources well inland from the coastal coal swamp. Actualistic comparison of a modern coastal peat swamp flora in Alabama with the macrodetritus component in distributary sediments nearby suggests that the swamp contributes relatively little debris to active deltaic channels and that plant debris from inland (non-swamp) sources is dominant in the deltaic sediments (Gastaldo et al., 1987).

It was in these swamp and upland settings that many of the terrestrial animals of the Braidwood fauna presumably lived. Insects, including cockroaches and giant dragonfly-like forms (protodonata), inhabited the swamps, as did spiders, large millipedes, and centipedes (Figure 5A.2). The horseshoe crab *Euproops danae* probably crawled on the forest floor (Fisher, 1979), along with amphibians and even reptiles.

Mazonian Distributary Complex

The Francis Creek Shale member is a sequence of gray, silty mudstone produced by the deposition of mud in water along a river-influenced coast (Baird et al., 1985). Two major events are recorded in this unit: a relative rise in sea level and the outbuilding (progradation) of one or more river-fed distributaries that buried the already submerged peat layer under a mantle of deltaic and prodeltaic sediments (Figures 5A.4–5A.6). Based on the occurrence of marine (Essex) animals in the lower part of the Francis Creek Shale across most of the Mazon Creek area (Baird et al., 1985), it appears that the inundation of the swamp preceded the period of most riverine sediment discharge.

Delta progradation is suggested by evidence of upward-coarsening of Francis Creek sediments from silty mudstone at the base to sandstone facies with interspersed channels in the upper third of the unit (Figure 5A.4). This vertical profile can be seen both in sections along Mazon Creek and in drill core sections obtained near Pit 11 (Shabica, 1971; Kuecher, 1983; Baird et al., 1985). Numerous channels containing massive sandstone are well displayed along Mazon Creek and in strip mine headwalls. Moreover, numerous test bore holes and strip mine excavations show that several of the larger channels display a conspicuous northeast-southwest trend (Shabica, 1971; Baird and Sroka, 1990). Shabica (1971), Baird et al. (1985), and Baird and Sroka (1990) interpret the upward-shoaling change within the Francis Creek, as well as the alignment of the channels, as the expression of delta progradation into a coastal estuary from the east or northeast. A delta margin interpretation for the Francis Creek is also supported by isopach data that show that the Francis Creek is expressed as a major depocenter in northeast Illinois (Figure 5A.10, below).

The deltaic character of Francis Creek deposits is well displayed in the numerous exposed banks along Mazon Creek, in partly submerged highwalls in the adjacent strip mine (Pit 8), and at Morris. Most of these exposures contain variable proportions of siltstone and fine sandstone. The massive, locally developed sandstone units appear to be crevasse-splay channels associated with distributary deposits (Shabica, 1971; Baird et al., 1985; Baird and Sroka, 1990). Thin-bedded siltstone sequences containing climbing ripples, small channeloid sandstone lenses, and layers marked by load-induced sediment failure record the spilling of mud and debris-laden floodwaters over levees into adjacent bays and wetlands, as well as minor levee breaks (Shabica, 1971; Baird et al., 1985). Silty mudstone deposits, particularly those yielding sideritic concretions containing plants and nonmarine or minimally marine Braidwood animals, record the deposition of finer sediment under quieter conditions in bays bordering the delta. Finer mud containing buried marine Essex animals accumulated in a larger, open water, prodelta-estuary setting seaward of the distributaries (Figure 5A.5).

Possible modern analogues of the Mazonian deltaic complex are certain tropical deltas, including the Ganges-Brahmaputra, Mahakam, and Mekong, which are characterized by numerous active and abandoned distributary channels that are continually modified by tidal processes

(Coleman, 1969; Coleman and Wright, 1975; Allen et al., 1979). On such deltas, linear sand and mud ridges are observed to migrate laterally within distributary channels under the influence of tides (Figure 5A.5). Moreover, where mud supply and tidal activity are both high, the delta front region may be expressed as an ill-defined coastal region of migrating mud ridges, turbid channels, mudflats, and islands as on the Ganges-Brahmaputra Delta (Coleman, 1969).

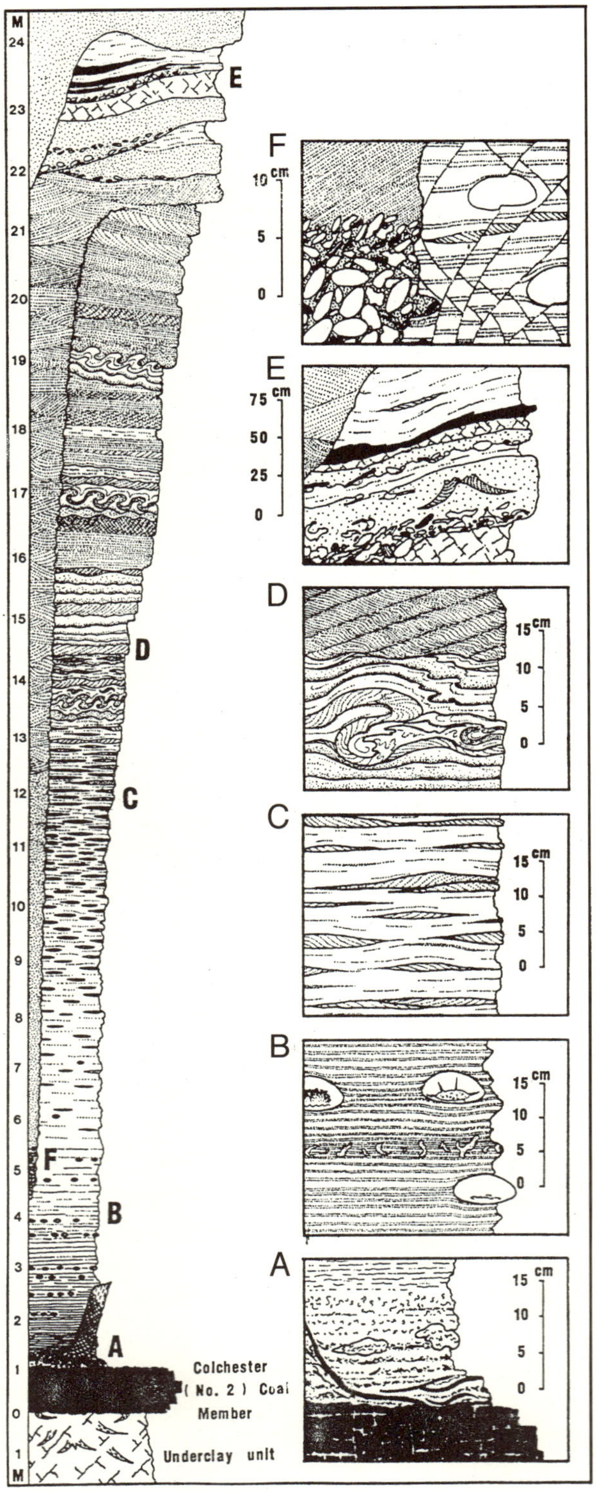

Figure 5A.4. Diagrammatic stratigraphic section through the Francis Creek Shale member in the vicinity of Godley, Will County, and Pit 11. Information based on work at Pit 11 (Shabica, 1971, 1979) and study of drill cores obtained in the Pit 11 vicinity (Baird et al., 1986; Baird and Sroka, 1990). Lettered divisions include **A** transgressive plant-debris-rich blocky mudstone above coal; **B** silty mudstone characterized by tidal "pinstripe" laminations and sideritic concretions yielding Essex fauna and plants; **C** silty mudstone with tidal lenticular bedding; **D** thin bedded siltstone and fine sandstone deposits displaying soft sediment failure and climbing ripples; **E** channeloid sandstone, underclay, and impure coal (delta plain or levee deposits); and **F** distributary channel bedload accumulation of reworked sideretic concretions, underclay clasts, vitrinite shreds, and fusain. Note early compactional microfaulting of mudstone near channel (from Baird and Sroka, 1990).

In the light of the above observations it is possible to interpret some laminated Francis Creek silty mudstone deposits as being sedimentary records of lateral accretion (and migration) of linear mud ridges or bars as opposed to strict vertical sediment accumulation. Conspicuous inclined bedding observed within Francis Creek silty mudstone facies along Mazon Creek near the White Tie Road overlook thus may record either the forward advance of a distributary mouth bar into the area, as interpreted by Baird et al. (1985), or the lateral migration of a muddy distributary ridge or bar under tidal influence.

Numerous sedimentological features are suggestive of conditions proximal to a delta: rapid mud and silt deposition indicated by climbing ripples; numerous escape burrows, sometimes containing the smothered animal (Figure 5A.7, 5A.8); engulfed upright trees; and, of course, the extraordinary preservation of Mazon Creek fossils (Shabica, 1971). The seaward flow of river water explains, in part, the abundance of land plant debris associated with aquatic Mazon Creek animals. This flow (although influenced by cycles of the tide) explains the dominantly one-way (seaward) mixing of plants and terrestrial animals in areas of marine animal occurrence.

Again, a delta-influenced setting for Francis Creek accumulation is further supported by

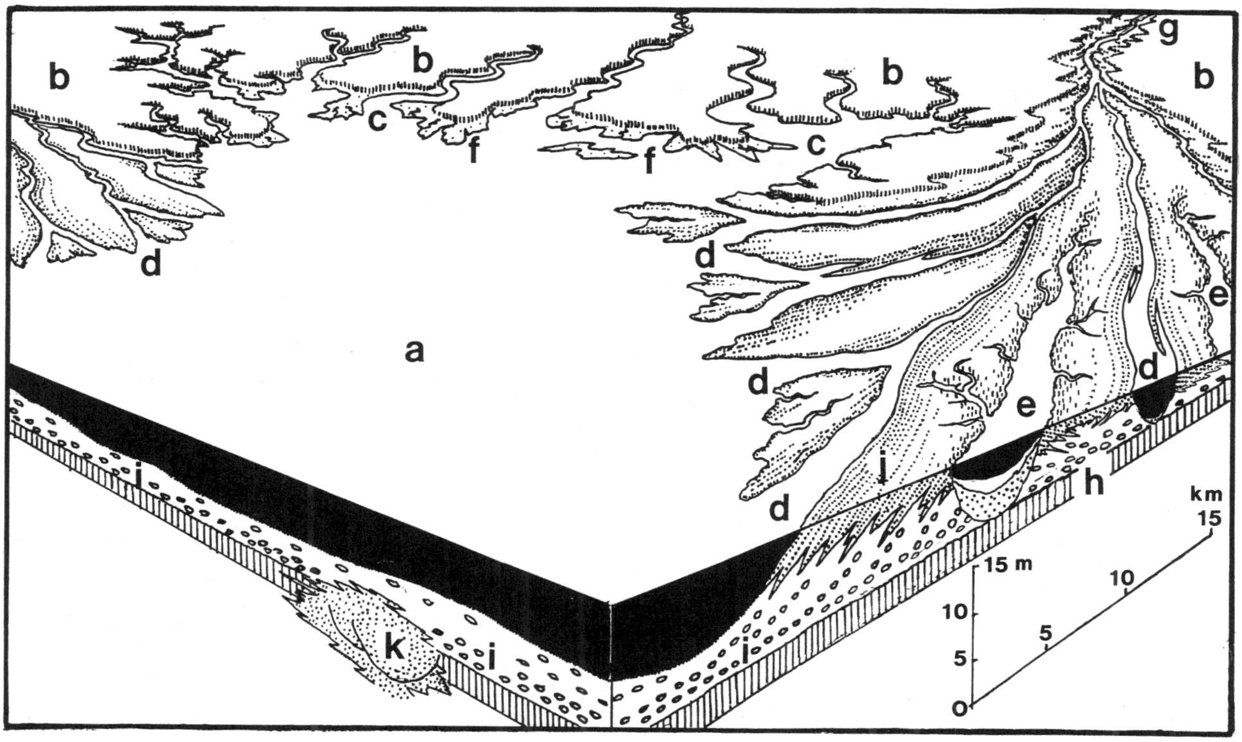

Figure 5A.5. River-influenced Mazonian estuary. Schematic reconstruction of Mazon Creek area aquatic paleohabitat as a turbid interdeltaic bay-estuary along a tropical mesotidal or macrotidal coast. In this model, delta progradation occurs during transgressive drowning of the Colchester peat swamp. The settings most conducive to burial and preservation of Mazon Creek organisms are areas of migrating muddy tidal channels and mud bars bordering the Colchester swamp (*f*) and regions near active distributaries (*d*) where progradation and lateral accretion of muddy bar fingers ensure rapid burial of organic remains. Estuarine bay setting (*a*) experiences less torrential sedimentation and supports the marginally marine (Essex) fauna, including some burrowing animals. Lettered features include *a,* interdeltaic estuarine bay; *b,* Colchester peat swamp; *c,* tidal channels or waterways into the peat swamp; *d,* active distributary channels; *e,* abandoned distributary channels bordered by swampy, subsiding delta plain; *f,* intertidal mudflats and migrating mud banks along the Colchester swamp coast; *g,* fluvial channel beyond the tidal reach; *h,* Colchester peat under sediment cover; *i,* Francis Creek mud deposit with early diagenetic sideritic concretions; *j,* lateral accretion of distributary channel margin under influence of tides; and *k,* abandoned and buried fluvial distributary channel that once extended through the coastal peat swamp prior to transgressive inundation of the swamp.

widespread evidence of rapid burial of organisms. Episodic bottom currents often flipped up fern and seed fern leaves during rapid sediment inundation so that these leaves occur upended ("on edge") within concretions (Figures 5A.3, 5A.7). Leaves buried this way are common in the strip mines to the east of Morris, and in Pits 7 to 10 north of Coal City. The common occurrence of *Edmondia* with the valves spread wide ("butterflied") is typical of rapidly buried clam populations. Essex animals including bivalves, chitons, polychaete worms, and the brachiopod *Lingula* occasionally show evidence of attempts to escape (Figures 5A.7, 5A.8), a clear indication of death by smothering (Shabica, 1971; Richardson, 1980; Baird et al., 1986). The aforementioned occurrence of bivalves at the end of escape burrows (Figures 5A.7, 5A.8) is particularly strong evidence for torrential sedimentation.

Mazonian Estuary

The occurrence of Mazon Creek animals over part, or all of five counties indicates that marginal marine Francis Creek habitats were widespread. Moreover, the dominance of mudstone in most Francis Creek sections yielding these organisms suggests that the Francis Creek Shale member records the filling of a coastal estuary with deltaic sediment. Transgression of the coast and deepening of the estuary are indicated further by evidence for a thick, unbroken record of sediment aggradation observed in drill cores through Francis Creek deposits (Kuecher, 1983; see Figure 5A.4). Rapid sea level rise and, to a lesser extent, compression of the underly-

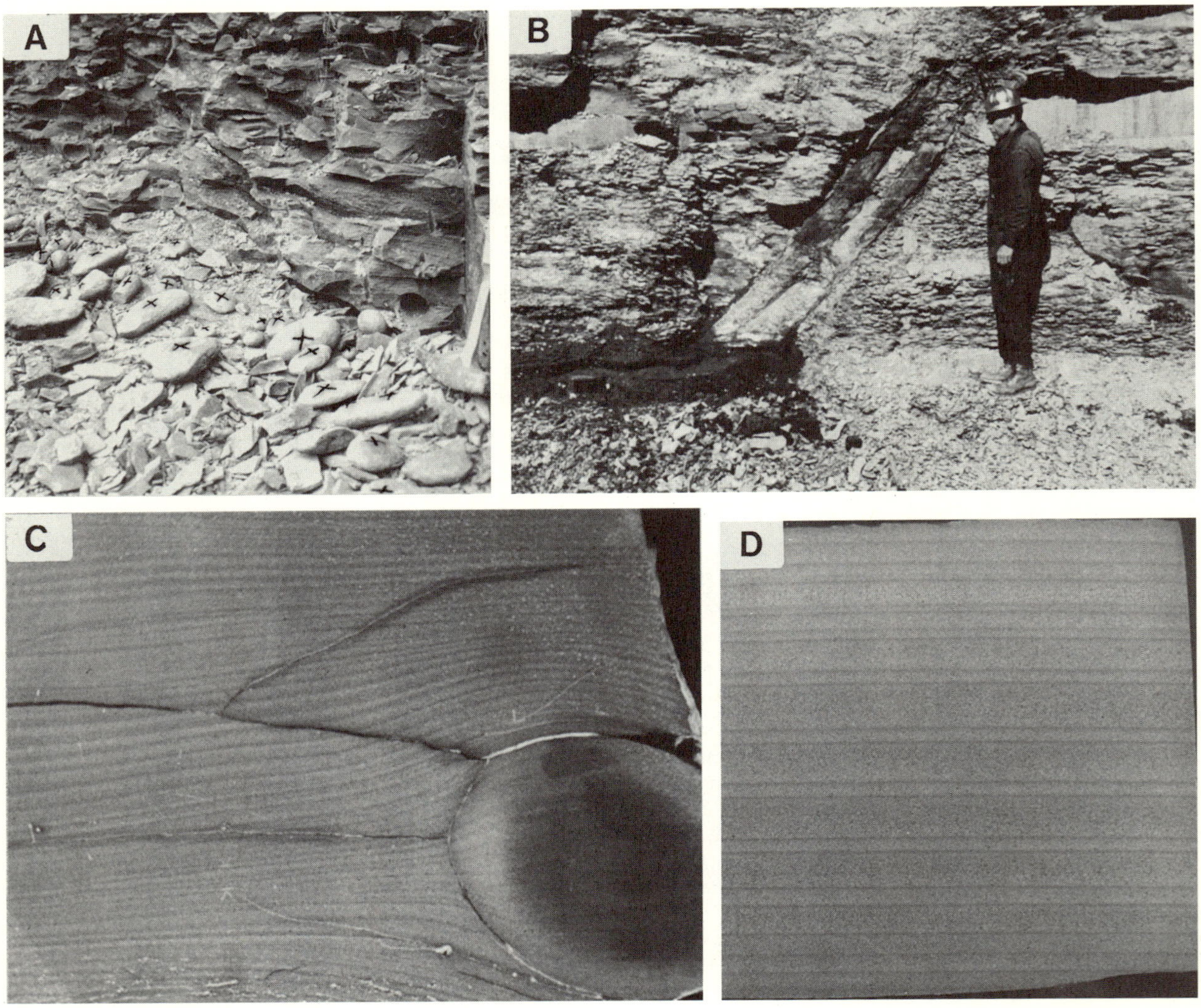

Figure 5A.6. Depositional characteristics of lower part of the Francis Creek Shale member. **A** Silty mudstone facies yielding fossiliferous sideritic concretions. Several nodules are marked × for up-orientation record. Headwall of abandoned strip mine at east edge of Morris, Grundy County. **B** Obliquely standing tree that was engulfed during deposition of basal Francis Creek mud. Headwall of now abandoned Peabody Coal Company Northern Mine (Pit 11) near Braidwood, Will County (from Shabica, 1971, 1979). **C** Vertical cross section of Francis Creek silty mudstone deposit showing in situ sideritic concretion. Note associated pinstripe lamination of tidal origin. Specimen collected by C. Shabica at Pit 11. **D** Pinstripe lamination recording twice-daily (semidiurnal) flood-ebb tide events in Mazonian estuary; thick couplets (top, bottom) record neap tide events (see also Figure 5.9 and discussion in text). Vertical section from Dames and Moore drill core obtained from near present location of Braidwood Nuclear Generating Station near Pit 11 (from Kuecher, 1983).

ing peat layer would have provided the necessary accommodation space for sediment accumulation in shallow water. Modern estuaries are embayments formed by the drowning of coasts by a rise in sea level. Since they are often associated with the overflow of freshwater from rivers, salinity usually drops up-estuary from normal marine (35 parts per thousand salt in modern oceans) at the estuary mouth to freshwater within the river system. Salinity and turbidity can vary daily and seasonally in response to tides and rainfall.

The most important sedimentological discovery at the Mazon Creek fossil beds has been the recognition of highly regular cyclic laminations ("pinstripe" lamination) in the lower Francis Creek section (Kuecher, 1983; Baird et al., 1985, 1986; Kuecher et al., 1990). Core sections of mudstone deposits taken at Pit 11 show remarkable evidence of flood-ebb tidal rhythmites indi-

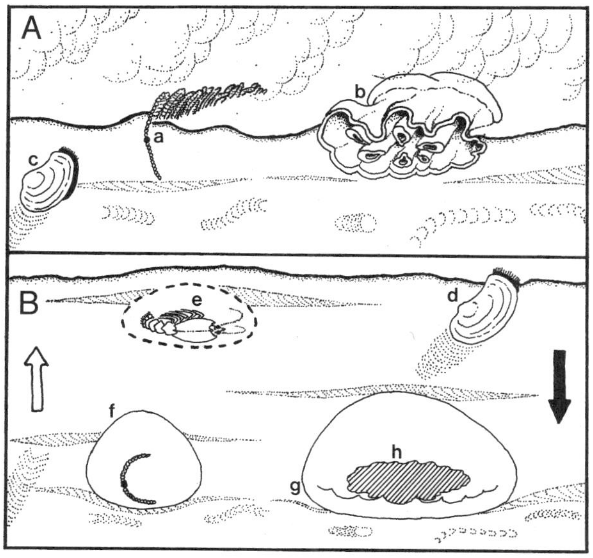

Figure 5A.7. Burial of Mazon Creek organisms. **A** Flood event showing death of organisms by bottom smothering and freshwater incursion (from Baird et al., 1986). **B** Subsequent burial of organisms in A with onset of very early diagenesis at shallow sediment depth. Organisms include *a,* upended pinna of *Pecopteris* sp.; *b,* partially covered medusoid *Essexella asherae; c, d,* bivalve *"Edmondia"* sp. making an escape attempt; *e, f, g,* sideritic concretion formation respectively around buried shrimp *Belotelson magister* (*e*), *Pecopteris* (*f*), and *Essexella* (*g*). The dark arrow (*right*) denotes compaction and dewatering of mud around growing nodules under thin sediment cover. The white arrow (*left*) signifies upward migration of reactive ions owing to compactional loading of underlying peat and mud; this may explain the differential upward (eccentric) growth of concretions discussed in Chapter 4C. Dashed line around buried *Belotelson* (*e*) is a protoconcretion in process of forming.

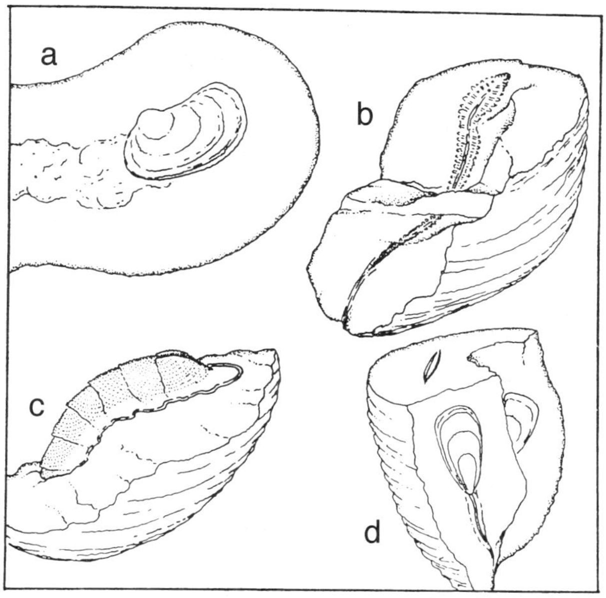

Figure 5A.8. Essex animals buried during lethal bottom-smothering events. Composite sketches from concretion specimens. **a** Bivalve *"Edmondia"* sp. at end of subhorizontal burrow; clam has rotated about 90° from life orientation possibly due to decay gases. **b** Polychaete annelid worm *Didontogaster cordylina* ("tummy tooth") with body oriented obliquely and extending through sediment layers. **c** Chiton *Glaphurochiton concinnus* with body flexed at angle to bedding. **d** Specimens of infaunal inarticulate brachiopod *Lingula* sp., which are vertically oriented in apparent life position within the sediment. These specimens display not only the trace of associated burrows but also a ghostlike trace of the muscular pedicle that produced them. Figures based on unoriented pickings from mine dump surfaces; worms and brachiopods are presumed to have been buried "head up" (from Baird et al., 1986).

cating rapid, but very regular, sediment accumulation under quiet conditions (Figures 5A.6C, 5A.6D, 5A.9). This significant vertical aggradational record of daily or even twice-daily (semidiurnal) deposition of mud from suspension during the alternating seaward and landward flows is rarely observed; it reflects an absence of disturbance and transport of coastal bottom mud by the scouring action of currents and by burrowing organisms. We recognize not only a twice-daily flood-ebb tidal cycle in these deposits but also a higher order, twice-monthly (fortnightly) variation in tidal current activity correlative to the neap-spring cycle (Kuecher, 1983; Baird et al., 1986).

Recent work by others on Pennsylvanian deposits shows that unbroken or minimally disturbed aggradational records of flood-ebb-generated pinstripe lamination are present in several other Late Carboniferous units (Kvale and Archer, 1990, 1991; Kvale et al., 1989; Feldman et al., 1994; Archer et al., 1991; Martino and Sanderson, 1993). Not only are twice-daily flood-ebb rhythmites observed in these localities, but the neap-spring cycle and higher order tidal cycles are recognized as well (Archer et al., 1991; Feldman et al., 1994). The most recent advances include recognition of a monthly lunar ellipticity (apogee-perigee) cycle and a six-month (solstice-equinox) cycle in Pennsylvanian deposits of Kentucky (Martino and Sanderson, 1993). A seasonal wet-dry cycle is recognized for a pinstripe-lamination-bearing interval at the late Pennsylvanian Hamilton fossil locality in Kansas, a locality which yields a biota similar to those of Mazon Creek localities

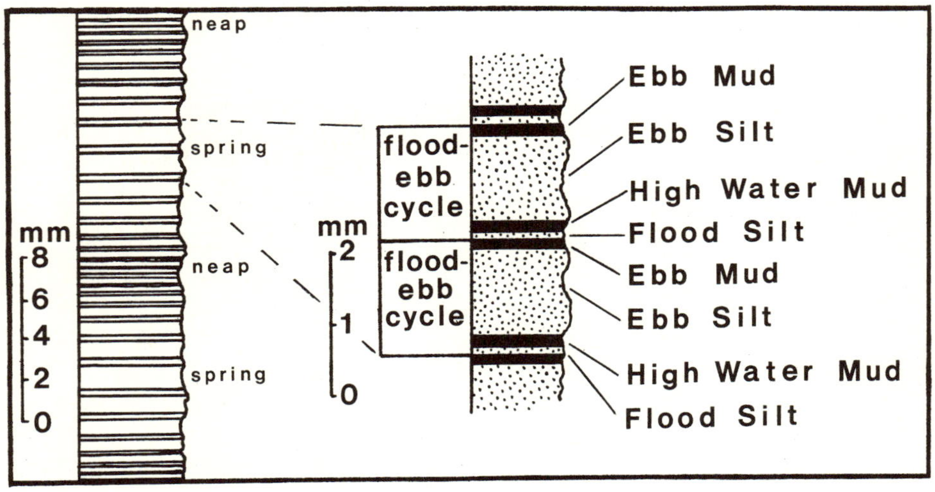

Figure 5A.9. Depositional model for semidiurnal tide origin of cyclic pinstripe lamination in lower part of Francis Creek Shale (see also Figure 5A.6C, 5A.6D). Four-part flood-ebb tidal cycles are shown at right. Higher order synodic neap-spring cycles are shown on left (from Baird et al., 1986).

(Schultze and Maples, 1992; Feldman et al., 1994). This ongoing research bodes well for those persons attempting to calculate precise rates of sedimentation for ancient deposits. Furthermore, the continuing search for new pinstripe-lamination-bearing units increases the potential for discovery of new "Mazon Creeks." Although flood-ebb-generated pinstripe lamination is known from deposits on modern temperate mesotidal and macrotidal coasts (Dalrymple et al., 1991; Tessler and Gigot, 1989; Tessler, 1993), more work is needed to see whether this bedding is a pervasive feature associated with modern tropical deltas and delta-influenced coastal settings.

Faunal Gradient within Estuary

A river-influenced estuary is indicated by the aquatic component of the Mazon Creek biota. Conditions in river-influenced estuaries exclude the more sensitive open marine organisms. Moreover, during periods of storms and sudden rainfall, mass mortality events occur, setting the stage for rapid burial and selective preservation of these taxa.

The Braidwood fauna (Figure 5A.10), which is characterized by a very low diversity aquatic assemblage of syncarid shrimps, a xiphosuran, diminutive bivalves, brachiopods, and ostracodes, has been designated "nonmarine" in most publications. This may be somewhat inaccurate in light of the fact that Francis Creek deposits at Benson's Farm on Mazon Creek, a locality yielding essentially no Essex forms, are now known to be characterized by the aforementioned tidal flood-ebb laminations (Kuecher et al., 1990). Rather they represent minimally brackish to near-freshwater conditions along the estuary margin and in channel waterways within the tidal reach. Similarly, Maples and Schultze (1988), Schultze and Maples (1992), and Feldman et al. (1994) argued that many Braidwood-type aquatic animals are brackish forms, based on comparisons of several Pennsylvanian soft-bodied animal assemblages from nearshore deposits (see discussion in Chapter 5B). A seafloor reconstruction of the Braidwood aquatic fauna, including both "rare" and "common" taxa is presented herein (Figure 5A.10).

The Essex fauna (Figures 5A.11, 5A.12), which as noted lacks the diverse shelly fauna of offshore Pennsylvanian environments, corresponds more closely to associations found in modern brackish water (Pearse and Gunter, 1957). Although referred to as "marine" both here and in earlier reports, the Essex fauna is not an assemblage typical of offshore marine Pennsylvanian deposits. There are no calcareous algae, no stony sponges or corals, no articulate brachiopods, cephalopods, or crinoids (Baird et al., 1985, 1986). Sensitive to fluctuations in water salinity and turbidity, these organisms are only found in offshore, open marine mudrock and limestone deposits. The presence of polychaete worms, small burrowing bivalves, shrimp, cycloid crustaceans, larval fish, and a holothurian confirms an estuarine interpretation for Essex organisms at most lo-

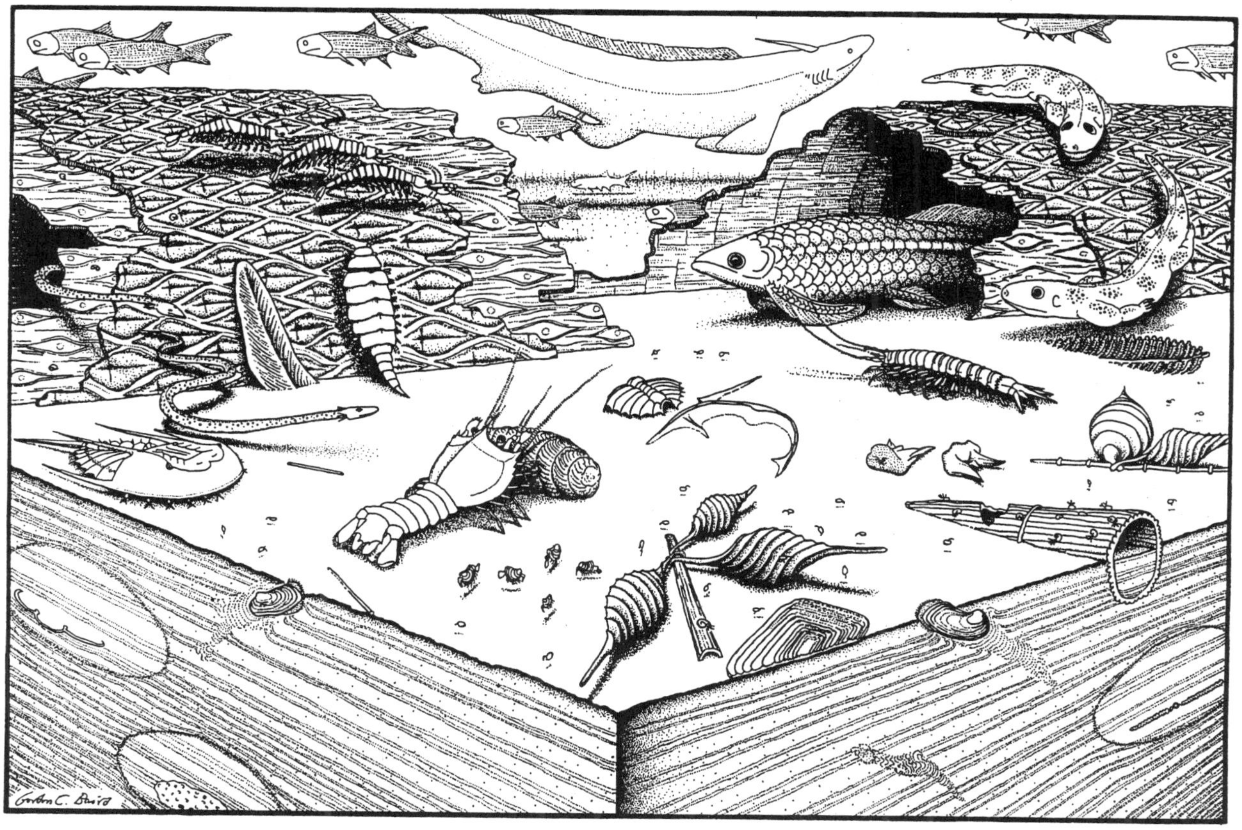

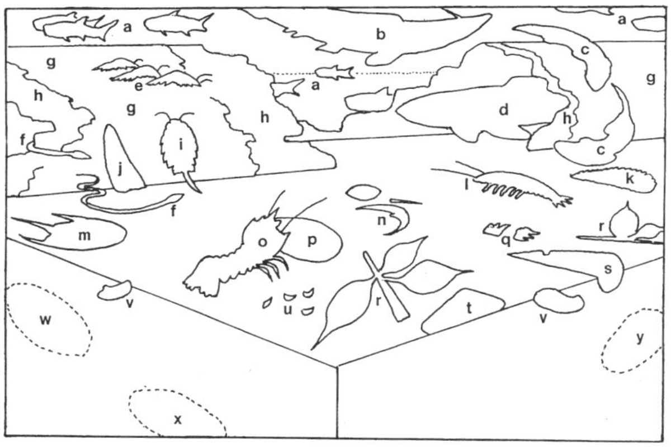

Figure 5A.10. Schematic reconstruction of brackish to freshwater (oligohaline-limnic) aquatic animal association constituting the Braidwood fauna. Animals are shown in association with transported terrestrial plant debris. Underlying sediments are characterized by cyclical flood-ebb tidal rhythmites, few trace fossils, and incipient formation of sideritic concretions at shallow depth. Lettered organisms include *a, Elonichthys peltigerus; b,* xenacanth shark; *c,* amphibian *Amphibamus grandiceps; d,* lungfish *Conchopoma; e,* syncarid shrimp *Palaeocaris typus; f,* snakelike amphibian *Aornerpeton* (formerly *Phlegethontia*); *g,* trunk of lycopod showing "scale bark" pattern characteristic of *Lepidodendron; h, Lepidodendron* underbark layer; *i,* euthycarcinoid arthropod; *j,* pteridosperm ("seed fern") pinnule of *Neuropteris scheuchzeri; k,* pinna (frond part) of fern *Pecopteris; l,* syncarid shrimp *Acanthotelson; m,* horseshoe crab *Euproops danae; n,* undisturbed molt ensemble of *Euproops; o,* crustacean *Anthracaris gracilis; p,* spiral coprolite (excrement) of shark; *q,* xenacanth shark teeth; *r,* probable chondrichthyan egg capsule *Palaeoxyris* attached to plant debris; *s,* sphenopsid ("horsetail") stem of *Calamites* encrusted by shelled polychaete worm *Spirorbis; t,* scale of lungfish *Ctenodus; u,* conchostracan (bivalved crustacean) *Pemphilimnadiopsis ortoni; v,* diminutive bivalve with associated burrowing trace; *w,* concretion forming around *Euproops; x,* concretion forming around coprolite; *y,* concretion forming around *Pecopteris* pinna. Unlettered swimming crustaceans are ostracodes. Rare aquatic animals not shown include a eurypterid, certain insect larvae, and possibly some arachnids. Reconstructions are based on numerous publications, and the input of systematists including R. Zangerl (chondrichthyans), F. R. Schram (crustaceans), and D. Bardack (fish).

calities (see Pearse and Gunter, 1957; Carriker, 1967; Perkins, 1974; Day, 1981). Significantly, the shoreward faunal transition from Braidwood to Essex biofacies is characterized by a sharp drop in animal diversity in areas of the nonmarine or minimally brackish Braidwood association (see Chapter 4A, Figures 4A.1–4A.4), which resembles biotic gradients observed in the upper reaches of many modern estuaries (Carriker, 1967).

It appears that the Essex-to-Braidwood faunal transition occurs within a strongly river-

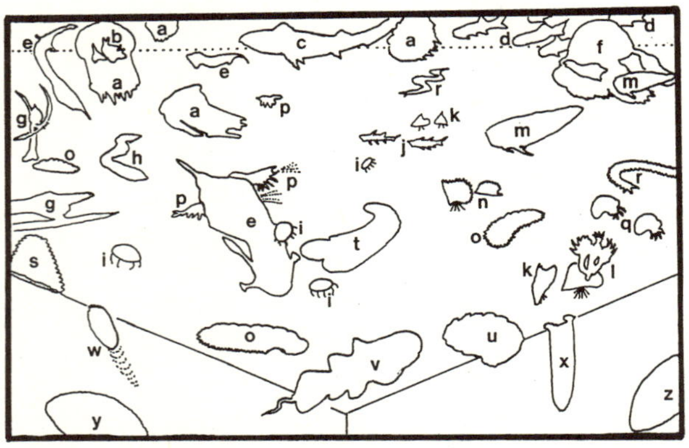

Figure 5A.11. Schematic reconstruction of nearshore estuarine marine habitat inhabited by organisms (Essex fauna) that could tolerate variations in salinity and turbidity. The sedimentation rate was high as indicated by cyclic, semidiurnal flood-ebb tidal rhythmites preserved in the sediment. Most bottom-dwelling organisms could not tolerate the high sedimentation rate, and trace fossils (burrowing and crawling marks) are relatively uncommon. Consequently, this subassociation of the Essex fauna is dominated by swimming and floating organisms. It includes forms usually found in the Kankakee County part of Pit 11, as well as on the mine dumps to the south and west of this area. Lettered taxa include *a,* rhizostome medusan *Essexella asherae* in different poses and conditions; *b,* gastropod *Strobeus* feeding on *Essexella; c,* generalized chondrichthyan responsible for spiral coprolites; *d,* generalized palaeoniscoid; *e, Tullimonstrum gregarium* in various poses; *f,* cubomedusan *Anthracomedusa turnbulli; g,* problematic organism *Etacystis communis* in two benthonic poses; *h,* agnathan chordate *Gilpichthys greenei; i,* medusan *Octomedusa pieckorum; j,* acanthodian fish *Acanthodes beecheri; k,* myalinid bivalve with attachment byssus; *l,* myalinid with attached gooseneck barnacles *Illilepas damrowi; m,* agnathan or possible lungfish *Esconichthys apopyris; n,* pectinid bivalve *Aviculopecten; o,* polyplacophoran *Glaphurochiton concinnus; p,* shrimp *Belotelson magister; q,* small swimming crustacean known as the "flea shrimp"; *r,* eunicid polychaete *Esconites zelus; s,* tree fern foliage fragment *Pecopteris* sp.; *t,* seed fern foliage fragment *Neuropteris scheuzeri; u,* spiral chondrichthyan coprolite; *v,* newborn coelacanth *Rhabdoderma* with yolk sac still present; *w,* solemyid bivalve *"Edmondia"; x,* U-shaped burrow *Diplocraterion* with wormlike inhabitant—high sedimentation rate during spring tide phase has forced inhabitant to migrate upward within its burrow trace; *y,* early diagenetic protoconcretion with coprolite nucleus; *z,* protoconcretion with arthropod nucleus and early diagenetic pyrite (or pyrite precursors) above nucleus. Common widespread fossils are *Essexella* (*a*), *Pecopteris* (*s*), *Neuropteris* (*t*), and coprolites (*u*). Locally common forms are *Strobeus* (*b*), myalinids (*k*), *Aviculopecten* (*n*), *"Edmondia"* (*w*), and *Diplocraterion* (*x*). Less frequent taxa include *Tullimonstrum* (*e*), *Etacystis* (*g*), *Octomedusa* (*i*), *Belotelson* (*p*), and *Esconites* (*r*). Scarce taxa include *Esconichthys* (*m*) and the "flea shrimp" (*q*). Rare taxa are *Anthracomedusa* (*f*), *Gilpichthys* (*h*), *Acanthodes* (*j*), *Illilepas* (*l*), *Glaphurochiton* (*o*), and *Rhabdoderma* (*v*). Very rare forms include articulated palaeoniscoids and sharks (shown), as well as tetrapods and allochthonous terrestrial arthropods (not shown). This reconstruction benefited from critical review and advice from Merrill W. Foster (Bradley University), Frederick R. Schram (Universiteit van Amsterdam), Rainer Zangerl (Rockville, Indiana), and David Bardack (University of Illinois, Chicago) during its preparation.

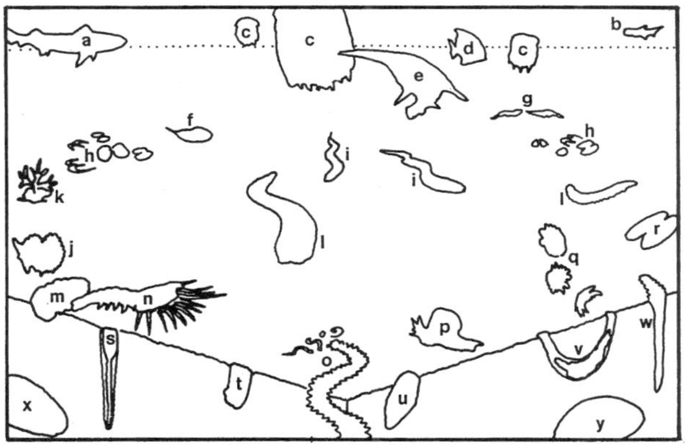

Figure 5A.12. Reconstruction of Mazonian estuary seascape showing the typical association of animals comprising the Essex fauna that is observed in the northern (Will County) part of Pit 11, at Chowder Flats west of Morris, and on numerous shaft mine dumps near Braceville, Coal City, Eileen, and Braidwood. This faunal association is rich in benthonic animals including several shallow burrowers as well as pelagic forms dominated by the jellyfish *Essexella.* Shallow burrowing locally disrupts the conspicuous pinstripe lamination of the Francis Creek deposit that was generated by neap-spring tidal cycles (see text). Animals in the foreground are shown in relative actual size. Taxa include *a,* generalized chondrichthyan responsible for numerous spiral coprolites (*m*) in assemblages; *b,* palaeoniscoid fish; *c,* rhizostome medusoid *Essexella asherae; d,* deep-bodied "sunfish" palaeoniscoid *Platysomus; e,* small chondrichthyan *Bandringa rayi; f,* xiphosuran *Palaeolimulus; g,* hoplocarid shrimp *Kallidecthes richardsoni; h,* solemyid bivalve *"Edmondia"* with splayed "butterflied" valves; *i,* polychaete worm *Astreptoscolex; j,* pectinid bivalve *Aviculopecten* with attachment byssus; *k, Aviculopecten* valve with attached hydroid *Drevotella proteana; l,* undescribed holothurian; *m,* spiral chondrichthyan coprolite; *n,* paleocarid shrimp *Belotelson magister; o,* eunicid polychaete *Esconites zelus* devouring small wormlike animal; *p,* bellerophontid gastropod *Euphemites; q,* cycloid crustacean *Cyclus americanus; r,* splayed empty shell of *Solemya radiata; s,* inarticulate brachiopod *Lingula* in burrow; *t,* bivalve *Schizodus* in burrow; *u,* bivalve *"Edmondia"* in burrow; *v,* echiuroid *Coprinoscolex* in burrow; *w,* polychaete *Didontogaster ellogimus* in burrow; *x,* protoconcretion containing *Essexella* (wavy cross section) with overlying pyrite halo. Abundant fossils figured are *Essexella* (*c,y*), *"Edmondia"* (*h,u,x*), the holothurian (*l*), coprolites (*m*), and burrow traces. Common forms include *Aviculopectin* (*j*), *Astreptoscolex* (*i*), *Esconites* (*o*), *Didontogaster* (*w*), *Coprinoscolex* (*v*), *Belotelson* (*h*), *Kallidecthes* (*g*), and *Cyclus* (*q*). Uncommon taxa include *Drevotella* (*k*), *Lingula* (*s*), *Schizodus* (*t*), *Euphemites* (*p*), and *Solemya* (*r*). Rare forms include *Palaeolimulus* (*f*), palaeoniscoids (*b,d*), and chondrichthyans (*a,e*). The fish appear to be wide ranging and are found in both the Braidwood and Essex localities. This reconstruction benefited from critical review and advice from Merrill W. Foster (Bradley University), Frederick R. Schram (Universiteit van Amsterdam), Rainer Zangerl (Rockville, Indiana), and David Bardack (University of Illinois, Chicago) during its preparation.

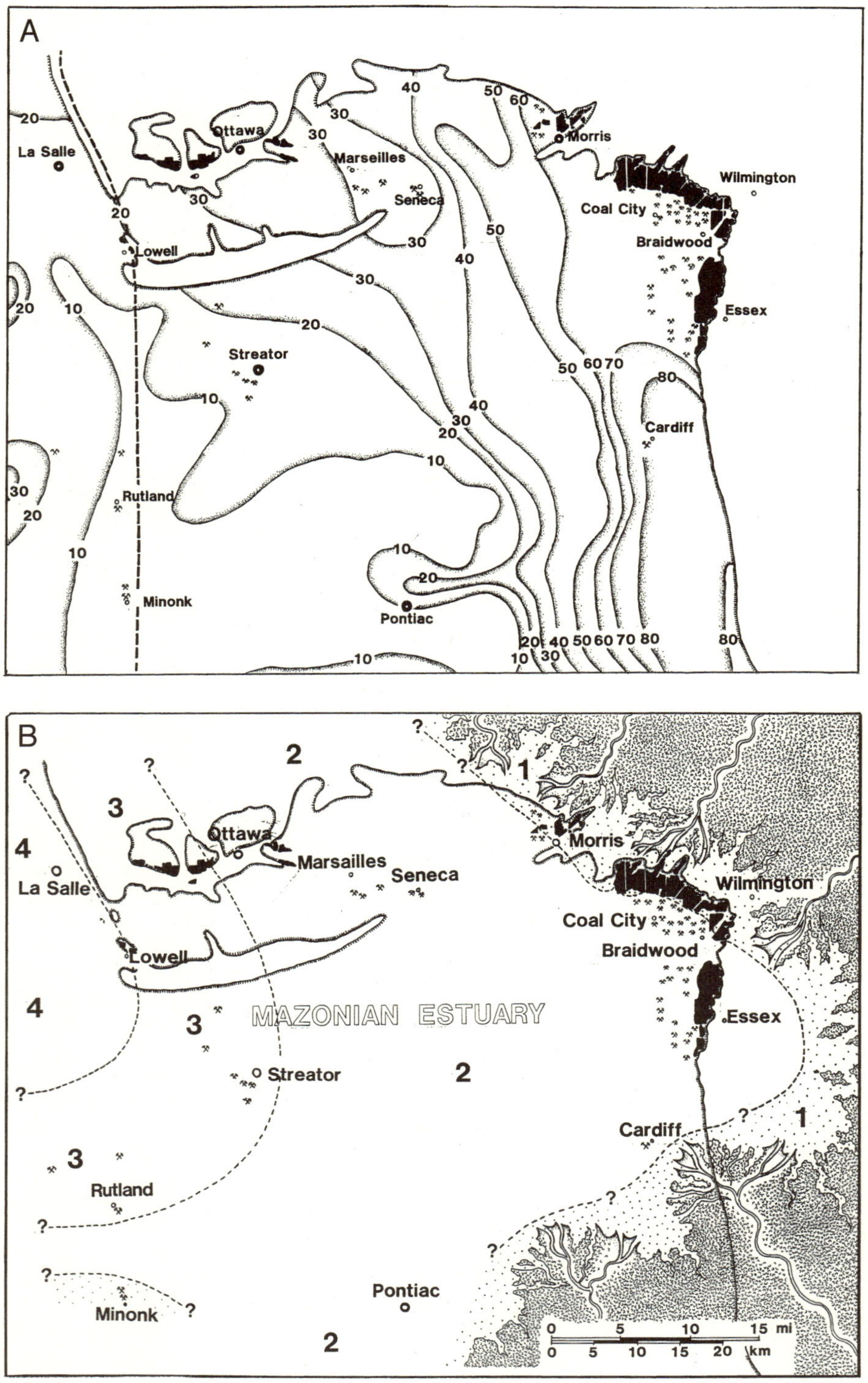

Figure 5A.13. **A** Isopach map showing thickness of Francis Creek Shale. Dashed line shows location of LaSalle Arch. The ⚒ symbols show shaft mines; black areas show strip mines. **B** Paleogeographic reconstruction showing *1,* Braidwood areas; *2,* Essex areas; *3,* marginal Essex areas; and *4,* normal marine areas.

influenced body of water corresponding to a *positive* (also called *normal*) estuary regime with good development of a halocline gradient (Knox, 1986). The Essex fauna clearly falls along a community gradient (coenocline) between the Braidwood association and the offshore stenohaline biotas on the shelf. However, it is not possible to establish average salinity levels of Essex paleohabitats. At best, we can place the subassociations within the Essex fauna as fitting somewhere within a broad range of possible salinities falling between the oligohaline and middle polyhaline zones (from 5 up to 18–25 parts per thousand salt) using the

Venice system of brackish water classification (see Carriker, 1967).

Estuary Dimensions

The dimensions and boundaries of the inferred estuarine embayment remain poorly delimited (Figure 5A.13). The direction of open sea apparently was to the west of the present Mazon Creek area, near Peoria and Galesburg (Shabica, 1971; Baird et al., 1985, 1986). Until recently, however, too little was known about fossil distribution to the south and southwest to allow reconstruction of the proposed partially enclosed embayment. Recent fossil samplings from Minonk in Woodford County, about 70 kilometers southwest of Morris, show that the nonmarine Braidwood fauna and associated flora are abundant in this area while Essex organisms are absent (Figure 5A.13B). This discovery, together with the occurrence of abundant plant fossils on shaft mine dumps at Cardiff, 5 miles south of Peabody Coal's Pit 11, suggests that the Essex localities were bounded to the south by nonmarine habitats (Baird and Sroka, 1990). The area where Essex animals are found thus appears to have been a semienclosed estuarine embayment (Figure 5A.13B).

Modern Mazon Creek Analogue

Tidally dominated, fast-growing, tropical deltas within modern estuarine embayments may prove to be only partial analogues for the Francis Creek deposit. Francis Creek mudstone displays a "poured in" quality over an unusually large area. Moreover, the absence or near absence of burrowing in the Francis Creek gives it more the aspect of a restricted lagoon than the variably bioturbated prodelta deposits of many modern deltas. An intuitive assessment of the Francis Creek deposit advanced here is that it records a, non-steady-state coastal event that was unusual for its rapidity and extent during the Pennsylvanian. It appears that transgressive inundation of the coastal coal swamp was followed by a sudden bay-filling event that produced extreme turbidity, excluded most burrowing organisms, and resulted in near-torrential sediment aggradation. It is possible that the only good modern analogue for this event is when anthropogenic deforestation or some other major disruption around a modern estuary has caused catastrophic sediment influx into the sea. If such disturbed places are the modern Francis Creek analogue, then we must locate and investigate them.

References

Allen, G. P., D. Laurier, and J. Thouvenin. 1979. Etude sedimentologique du delta de la Mahakam. Notes et Memoires, vol. 15. Compagnie Francais des Petroles, Paris.

Archer, A. W., E. P. Kvale, and H. R. Johnson. 1991. Analysis of modern equatorial tidal periodicities as a test of information encoded in ancient tidal rhythmites, pp. 189–196. *In* D. G. Smith, G. E. Reinson, B. A. Zeitlin, and R. R. Rahmani (eds.), Clastic Tidal Sedimentology. Canadian Soc. Petrol. Geol. Mem., 16.

Baird, G. C., C. W. Shabica, J. L. Anderson, and E. S. Richardson, Jr. 1985. Biota of a Pennsylvanian muddy coast: habitats within the Mazonian Delta Complex, northeastern Illinois. Jour. Paleo., 59:253–281.

Baird, G. C., and S. D. Sroka. 1990. Geology and geohistory of Mazon Creek area fossil localities, Illinois, pp. c1–c70. *In* W. R. Hammer, and D. F. Hess (eds.), Geology Field Guidebook. 24th annual meeting of the North-Central Section, Geol. Soc. Amer., 24–30 April 1990, Macomb, Illinois.

Baird, G. C., S. D. Sroka, C. W. Shabica, and G. J. Kuecher. 1986. Taphonomy of Middle Pennsylvanian Mazon Creek area fossil localities, northeast Illinois: significance of exceptional fossil preservation in syngenetic concretions. Palaios, 1:271–285.

Bambach, R. K., C. R. Scotese, and A. M. Ziegler. 1980. Before Pangaea: the geographies of the Paleozoic world. Amer. Sci., 16(1):26–38.

Carriker, M. R. 1967. Ecology of estuarine benthic invertebrates: a perspective, pp. 442–487. *In* G. H. Lauff (ed.), Estuaries. Amer. Assoc. Adv. Sci. Publ., 83.

Coleman, J. M. 1969. Brahmaputra River: channel processes and sedimentation. Sediment. Geol., 3:129–239.

Coleman, J. M., and L. D. Wright, 1975. Modern river deltas: variability of processes and sand bodies, pp. 99–149. *In* M. L. Broussard (ed.), Deltas: models for exploration. Houston Geological Society, Houston.

Crowell, J. C. 1978. Gondwanan glaciation, cyclothems, continental positioning, and climatic change. Amer. Jour. Sci., 278:1345–1372.

Dalrymple, R. W., Y. Makino, and B. A. Zeitlin. 1991. Temporal and spatial patterns of rhythmite deposition on mudflats in the macrotidal Cobequid Bay–Salmon River Estuary, Bay of Fundy, Canada, pp. 137–160. *In* D. G. Smith, G. E. Reinson, B. A. Zaitlin, and R. A. Rahmani (eds.), Clastic Tidal Sedimentology. Canadian Soc. Petrol. Geol. Mem., 16.

Feldman, H. R., A. W. Archer, E. P. Kvale, C. R. Cunningham, C. G. Maples, and R. R. West. 1993. A tidal model of Carboniferous Konservat Lagerstätten formation. Palaios, 8:485–498.

Ferm, J. C. 1970. Allegheny deltaic deposits. Soc. Econ. Paleo. Mineral. Spec. Publ., 15:246–255.

Fisher, D.C. 1979. Evidence for subaerial activity of *Euproops danae* (Merostomata, Xiphosurida), pp. 379–448. *In* M. H. Nitecki (ed.), Mazon Creek Fossils. Academic Press, New York.

Gastaldo, R. A., F. P. Douglass, and S. M. McCarroll. 1987. Origin, characteristics, and provenance of plant macrodetritus in a holocene crevasse splay, Mobile Delta, Alabama. Palaios, 2:229–240.

Heckel, P. H. 1986. Sea-level curve for Pennsylvanian eustatic marine transgressive-regressive depositional cycles along midcontinent outcrop belt, North America. Geology, 14:330–334.

Heckel, P. H., R. C. Price, and S. R. Schutter. 1980. Glacial-eustatic model for black shale-bearing Pennsylvanian cyclothems of midcontinent and Illinois, pp. 139–158. *In* R. L. Langenheim and C. J. Mann (eds.), Middle and Late Pennsylvanian Strata on Margin of Illinois Basin (Guidebook). 10th Field Conf., Great Lakes Section, Econ. Paleo. Mineral., Danville, Illinois.

Klein, G. de V., and J. B. Kupperman. 1992. Pennsylvanian cyclothems: methods of distinguishing tectonically induced changes in sea level from climatically induced changes. Geol. Soc. Amer. Bull., 104(2):166–175.

Knox, G. A. 1986. Estuarine Ecosystems: A Systems Approach, Vol. 1. 289 pp.

Kuecher, G. J. 1983. Rhythmic sedimentation and stratigraphy of the Middle Pennsylvanian Francis Creek Shale near Braidwood, Illinois. Unpublished M. S. thesis, Northeastern Illinois University, Chicago, 143 pp.

Kuecher, G. J., B. G. Woodland, and F. M. Broadhurst. 1990. Evidence of deposition from individual tides and of tidal cycles from the Francis Creek Shale (host rock to the Mazon Creek biota), Westphalian D (Pennsylvanian), northeast Illinois. Sediment. Geol., 68:211–221.

Kvale, E. P., and A. W. Archer. 1990. Tidal deposits associated with low-sulfur coals, Brazil Fm. (Lower Pennsylvanian), Indiana. Jour. Sediment. Petrol., 60:563–574.

———. 1991. Characteristics of two, Pennsylvanian-age, semidiurnal tidal deposits in the Illinois Basin, pp. 179–188. *In* D. G. Smith, G. E. Reinson, B. A. Zeitlin, and R. A. Rahmani (eds.), Clastic Tidal Sedimentology. Canadian Soc. Petrol. Geol. Mem., 16.

Kvale, E. P., A. W. Archer, and H. R. Johnson, 1989. Daily, monthly, and yearly tidal cycles within laminated siltstones of the Mansfield Formation (Pennsylvanian) of Indiana. Geology, 17:365–368.

Maples, C. G., and H. P. Schultze, 1988. Preliminary comparison of Pennsylvanian assemblage of Hamilton, Kansas, with marine and nonmarine contemporaneous assemblages, pp. 253–273. *In* G. Mapes and R. H. Mapes (eds.), Regional geology and paleontology of Upper Paleozoic Hamilton Quarry area in southeastern Kansas. Kansas Geol. Surv. Guideb. Ser., 6.

Martino, R. L., and D. D. Sanderson. 1993. Fourier and autocorrelation analysis of estuarine tidal rhythmites, lower Breathitt Formation (Pennsylvanian), eastern Kentucky, USA. Jour. Sediment. Petrol., 63:105–119.

McCabe, P. J. 1984. Depositional environments of coal and coal-bearing strata, pp. 13–42. *In* R. A. Rahmani and R. M. Flores (eds.), Sedimentology of Coal and Coal-Bearing Sequences. Spec. Publ. Int. Assoc. Sediment., 7(1).

———. 1992. Geology of coal: environments of deposition, pp. 469–482. *In* H. J. Gluskoter, D. D. Rice, and R. B. Taylor (eds.), Economic Geology, U.S. Geology of North America Series. Geological Society of America, Boulder, Colorado.

Merrill, G. K., and M. D. Martin. 1976. Environmental control of conodont distribution in the Bond and Mattoon formations (Pennsylvanian, Missourian), northern Illinois, pp. 243–271. *In* C. R. Barnes (ed.), Conodont Paleoecology. Geol. Assoc. Canada Spec. Paper, 15.

Nie Shangyu, D. B. Rowley, and A. M. Ziegler. 1990. Constraints on the location of Asian microcontinents in Palaeo-Tethys during the Late Paleozoic, pp. 397–410. *In* W. S. McKerrow and C. R. Scotese (eds.), Paleozoic Palaeogeography and Biogeography. Geol. Soc. London Mem., 12.

Pearse, A. S., and G. Gunter. 1957. Salinity, pp. 129–158. *In* J. W. Hedgpeth (ed.), Treatise on Marine Ecology and Paleoecology: Ecology. Geol. Soc. Amer. Mem., 67.

Peppers, R. A., and W. H. Pfefferkorn. 1970. A comparison of the floras of the Colchester (No. 2) Coal and the Francis Creek Shale, pp. 61–72. *In* W. H. Smith, N. B. Nance, M. E. Hopkins, R. G., Johnson, and C. W. Shabica (eds.), Depositional environments in parts of the Carbondale Formation—western and northern Illinois. Illinois State Geol. Surv. Guideb. Ser., 8.

Perkins, E. J. 1974. The Biology of Estuaries and Coastal Waters. Academic Press, London, 678 pp.

Pfefferkorn, W. H. 1979. High diversity and stratigraphic age of the Mazon Creek flora, pp. 129–142. *In* M. H. Nitecki (ed.), Mazon Creek Fossils. Academic Press, New York.

Pfefferkorn, W. H., R. A. Peppers, and T. L. Phillips. 1971. Some fern-like fructifications and their spores from the Mazon Creek compression flora of Illinois (Pennsylvanian). Illinois State Geol. Surv. 468:1–29.

Phillips, T. L., and A. T. Cross. 1991. Paleobotany and paleoecology of coal. Plate 4, Reconstruction of a Pennsylvanian coal swamp. *In* H. J. Gluskoter, D. D. Rice, and R. B. Taylor (eds.), The Geology of North America. Vol. p-2, Economic Geology, U.S. Geological Society of America, Boulder, Colorado, 622 pp.

Phillips, T. L., and W. A. Dimichele. 1981. Paleoecology of Middle Pennsylvanian age coal swamps in southern Illinois/Herrin Coal member at Sahara Mine No. 6, pp. 231–284. *In* K. J. Niklas (ed.), Paleobotany, Paleoecology, and Evolution, vol. 1. Praeger, New York.

Phillips, T. L., and R. A. Peppers. 1984. Changing patterns of Pennsylvanian coal swamp vegetation and implications of climatic control on coal occurrence. Int. Jour. Coal Geol., 3:205–255.

Phillips, T. L., R. A. Peppers, and W. A. Dimichele. 1985. Stratigraphic and interregional changes in Pennsylvanian coal-swamp vegetation: environmental inferences. Int. Jour. Coal Geol., 5:43–109.

Richardson, E. S., Jr. 1980. Life at Mazon Creek, pp. 217–224. *In* R. L. Langenheim and C. J. Mann (eds.), Middle and Late Pennsylvanian Strata on Margin of Illinois Basin (Guidebook). 10th Field Conf., Great Lakes Section, Soc. Econ. Paleo. Mineral., Danville, Illinois.

Rowley, D. R., A. Raymond, J. T. Parrish, A. L. Lottes, C. R. Scotese, and A. M. Ziegler. 1985. Carboniferous paleogeographic, phytogeographic and paleoclimatic reconstruction, pp. 17–42. *In* T. L. Phillips and C. B. Cecil (eds.), Paleoclimatic Controls on Coal Resources of the Pennsylvanian System of North America. Int. Jour. Coal Geol., 5.

Schultze, H. P., and Maples, C. G. 1992. Comparison of the late Pennsylvanian faunal assemblage of Kinney Brick Company Quarry, New Mexico, with other late Pennsylvanian Lagerstätten. New Mexico Bur. Mines Min. Res. Bull., 138:231–242.

Scotese, C. R., et al. 1979. Paleozoic base maps. Jour. Geol., 87:217–277.

Shabica, C. W. 1971. Depositional environments in the Francis Creek Shale and associated strata. Unpublished Ph.D. dissertation, University of Chicago, 206 pp.

———. 1979. Pennsylvanian sedimentation in northern Illinois: examination of delta models, pp. 13–40. *In* M. H. Nitecki (ed.), Mazon Creek Fossils. Academic Press, New York.

Tessler, B. 1993. Upper intertidal rhythmites in the Mont-Saint-Michel Bay (N.W. France): perspectives for paleoreconstruction: Marine Geol., 110:355–367.

Tessler, B., and P. Gigot, 1989. A vertical record of different tidal cyclicities: an example from the Miocene marine Molasse of Digne (Haute Provence, France). Sedimentology, 36:767–776.

Veevers, J. J., and C. McA. Powell, 1987. Late Paleozoic glacial episodes in Gondwanaland reflected in transgressive-regressive depositional sequences in Euramerica. Geol. Soc. Amer. Bull., 98:475–487.

Wanless, H. R., and F. P. Sheppard. 1936. Sea level and climatic changes related to Late Paleozoic cycles. Geol. Soc. Amer. Bull., 47:1177–1206.

Wanless, H. R., and J. M. Weller. 1932. Correlation and extent of Pennsylvanian cyclothems. Geol. Soc. Amer. Bull., 43:1003–1016.

Winston, R. B. 1986. Characteristic features and compaction of plant tissues traced from permineralized peat to coal in Pennsylvanian coals (Desmoinesian) from the Illinois Basin. Int. Jour. Coal Geol., 6:21–41.

———. 1988. Paleoecology of Middle Pennsylvanian-age peat-swamp plants in Herrin Coal, Kentucky, U.S.A. Int. Jour. Coal Geol., 10:203–238.

5B

Regional, Temporal, and Evolutionary Significance of the Mazon Creek Biota

Gordon C. Baird and Christopher G. Maples

Mazon Creek localities in northeastern Illinois are an important example of a Konservat Lagerstätte ("preservational gold mine") that yields important and unusual paleontological information (see Seilacher, 1970; Seilacher et al., 1985). Other well-known invertebrate-dominated Lagerstätten, yielding soft-bodied and lightly sclerotized organisms, include the Middle Cambrian Burgess Shale locality, the Lower Devonian Hunsrückschiefer in Germany, the Jurassic Solnhofen locality in Germany. Although these Lagerstätten have yielded fossils of remarkable evolutionary significance, assemblages at these localities are sparse and less diverse when compared to the large numbers of jellyfish, worms, shrimp, and other exotica obtained from the Mazon Creek area. Moreover, localities yielding Mazon Creek–type fossils may be much more common and widespread than previously thought (Baird et al., 1985).

Calver (1968) described marine communities from the Middle Pennsylvanian (Westphalian Stage) of northern England. He recognized a predictable onshore-offshore sequence of communities at several stratigraphic levels; these associations were, in respective seaward order: estheriid, *Planolites*, foraminiferal, *Lingula*, myalinid, and productid associations, which together represent a common assemblage spectrum from nonmarine to fully marine. One grouping (estheriid) is largely nonmarine, and it roughly corresponds to part of what is called the Braidwood fauna in Illinois. Calver's (1968) myalinid association, yielding *Edmondia*, *Myalina*, and small gastropods, may be the hard-part record of assemblages comparable to the Essex fauna. His recognition of the myalinid association at several stratigraphic levels, albeit lacking soft-bodied taxa, suggested that Essex-type biotas may have been pervasive in Westphalian nearshore environments.

Discovery of additional Mazon Creek–type localities has confirmed this prediction. Concretionary Braidwood-type nonmarine fossil assemblages dominated by diminutive bivalves, syncarid shrimp, *Palaeoxyris*, estheriids, and xiphosurans in association with a major detrital plant component are known from several stratigraphic levels in Illinois and Indiana (see Canfield, 1959; Boneham, 1975; Gastaldo, 1977; Baird et al., 1985). Recently, additional occurrences of concretionary Braidwood animals, in association with abundant detrital plants, have been discovered at stratigraphic levels of Westphalian C age (below the level of the Francis Creek Shale) in strip mines in eastern Oklahoma near Sallisaw, Sequoyah County (Baird et al., 1985). Braidwood-type plants and animals were discovered in strip mines near the towns of Henryetta and Morris in eastern Oklahoma (Baird et al., 1985). Similarly, fossils of this type were collected from various mine localities in the British Coal Measures. A siderite-concretion-enclosed fauna, rather similar to the Braidwood animal association, but yielding a few rare problematic taxa of a more marine character, is described from Upper Pennsylvanian (Stephanian) deposits at Montceau-les-Mines in eastern France (Pacaud et al., 1981; Heyler, 1985). Siderite-concretion-enclosed faunas of a Braidwood type also are known from younger Pennsylvanian (Virgilian) units in Kansas (e.g., the Lawrence Shale, Wood Siding Formation). That numerous localities yield the Braidwood association suggests that this grouping represents one or more important Pennsylvanian nearshore-to-onshore associations.

The Essex fauna may also be a widespread faunal association. Essex animals were discovered at the Sunspot Strip Mine in Astoria, Illinois, in the 1960s (Richardson and Johnson, 1971; Shabica, 1971; Sroka, 1996). More significantly, both Braidwood and diverse Essex taxa were obtained from strip mines developed into the Croweburg Coal near the western Missouri town of Windsor in Henry County. Mines west of Windsor have yielded key Essex animals such as *Essexella,* polychaetes, *Edmondia,* chitons, the shrimp *Belotelson,* a holothurian, and fish (Baird et al., 1985). Recently, Baird discovered Essex taxa (*Essexella, Belotelson,* and *Tyrannophontes*) in concretions on shaft mine dumps near the northern Missouri towns of Brookfield and Marceline in Linn County. The recurrence at these localities of well-defined Braidwood and Essex biotas as discrete associations within estuarine-deltaic facies is striking. It indicates that the Braidwood and Essex associations were spatially widespread biofacies associated with river-influenced Pennsylvanian coastal waters.

The Mazon Creek localities are but one of several Carboniferous Lagerstätten yielding superbly preserved aquatic and terrestrial animals. Important nonconcretionary Lagerstätten (and their most important or most common biotas) include the Mississippian Glencartholm locality of Scotland (arthropods, vertebrates), the Mississippian Granton shrimp beds of Scotland (worms, arthropods, chordates, the conodont animal), the latest Mississippian Bear Gulch locality in the Heath Formation in Montana (worms, arthropods, chordates, problematica), the Middle Pennsylvanian Logan and Mecca Quarry Shales of Indiana (arthropods, vertebrates), the Middle Pennsylvanian Linton locality in Ohio (arthropods, vertebrates), the Middle Pennsylvanian Nýřany locality in the Czech Republic (arthropods, vertebrates), the Late Pennsylvanian Garnett and Hamilton Lagerstätten of Kansas (plants, arthropods, vertebrates), the Late Pennsylvanian Robinson Lagerstätte of Kansas (vertebrates), and the Late Pennsylvanian Kinney Brick Quarry in New Mexico (arthropods, bivalves, vertebrates). All of these localities, with the exception of Bear Gulch and the possible exceptions of the Logan and Mecca Quarry Shales, represent variably restricted nearshore settings as is discussed below (see summaries in Maples and Schultze, 1989; Schultze and Maples, 1992; Cunningham et al., 1993; Schultze et al., 1994; Feldman et al., 1994).

Comparisons of the biotas and inferred paleoenvironments of these localities are difficult to make. Maples and Schultze (1989) attempted qualitative placement of seven Pennsylvanian Lagerstätten (Nýřany, Linton, Mazon Creek [Braidwood fauna], Mazon Creek [Essex fauna], Montceau-les-Mines, Garnett, Hamilton, Robinson) along a hypothetical onshore-offshore paleoenvironmental continuum using biota composition and geologic context as a guide. The Mazon Creek biota (Braidwood-Essex faunas) bridged the midpoint of this spectrum. Localities (Robinson, Garnett, Hamilton) yielding a problematic mixture of eurytopic parallic and terrestrial organisms in association with stenotopic marine forms (foraminifera, bryozoans, articulate brachiopods) plotted in a seaward position relative to the Essex fauna. Lagerstätten (Nýřany, Linton, Montceau-les-Mines) dominated by nonmarine and terrestrial taxa and a near absence of amphidromus organisms (organisms that live most of their lives in nonmarine settings but occasionally migrate into marine conditions) and anadromous organisms (organisms that live most of their lives in marine settings but occasionally migrate into nonmarine conditions) plotted in a landward position relative to the Braidwood fauna. Subsequent quantitative comparison of eight Pennsylvanian Lagerstätten (the above localities, plus the Kinney Brick locality) revealed difficulties of habitat comparison owing to the dominance in faunal lists of rare taxa found only at a few of the total localities (Schultze and Maples, 1992). Furthermore, dependence on long lists of rare terrestrial arthropods and mobile vertebrates renders difficult the comparison of these various biotas. Of particular significance has been the lack of attention (lack of collecting?) given to common, mundane forms at many of these localities, particularly aquatic invertebrates, which have vital paleoenvironmental significance; sketchy lists of aquatic invertebrates (cnidarians, worms, crustaceans, molluscs) from several localities renders existing lists problematic (Schultze and Maples, 1992). Comparison of various Lagerstätten with the Essex and Braidwood faunas is further hampered by the lack of faunal differentiation between them in published lists. Ongoing analysis of a larger census database for Mazon Creek localities will serve to quantitatively define and subdivide these associations. Such work should provide insight as to probable faunal composition of the invertebrate fraction preserved at other Pennsylvanian Lagerstätten.

Schram (1979) argued convincingly, from systematic study of biotas from the Mississippian Glencartholm and Bear Gulch localities, from the Mazon Creek Lagerstätte, and from younger occurrences, that Mazon Creek organisms, particularly those of the Essex fauna, constituted stable associations that persisted through the Carboniferous. This assertion suggests that there is a predictive temporal phylogenetic and ecological continuum into which Mazon Creek–type communities can be placed. This is confirmed by discovery of a diverse Mazon Creek–type fauna in the medial Triassic Grés-à-Voltzia deltaic deposits in the Alsace region of France (see Gall, 1971, 1983).

The Grés-à-Voltzia biota includes a rich terrestrial association of plants, insects, diplopods, arachnids, and tetrapods. Autochthonous marine animals include medusae, the brachiopod *Lingula,* eunicid polychaetes, decapod and isopod crustaceans, estheriids, euthycarcinoid arthropods, limulids, and small bivalves. Additional forms include vertebrates and the problematic form *Palaeoxyris,* which may be a chondrichthyan egg capsule (Zidek, 1976; McGhee, 1982). Although fossils are preserved as impressions in shale rather than as molds in sideritic concretions, the biota closely resembles the Mazon Creek fauna in having components from terrestrial and schizohaline sources. Moreover, the Grés-à-Voltzia deposit is interpreted to record brackish to hypersaline estuarine conditions on a subtropical delta; seasonal mass mortality in restricted ponds and waterways is believed to have been due to rapid salinity changes as well as to seasonal drying of ephemeral water habitats (Gall, 1983).

The association of both the Mazon Creek and Grés-à-Voltzia faunas with similar inferred paleoenvironments shows that the resemblances between these faunas is no coincidence. The Grés-à-Voltzia fauna indicates that Essex or very similar associations apparently span the Late Paleozoic and Early Mesozoic without dramatic changes. This is particularly striking in that this association spans the greatest of all Phanerozoic mass extinctions at the Permo-Triassic boundary. We believe that the 75-million-year longevity of the Essex fauna reflects the inherent adaptation of these organisms to high-stress, marginal marine environments. Hence, this adaptation may have conferred a survival edge to these taxa during the extinction event. It should be noted that our generalized comparison of these faunas is at familial and higher classificatory scales and may represent more a recurrence of bauplans and trophic strategies than actual persistence of the same or related biotas.

As a final observation, it is particularly noteworthy that the Essex aquatic invertebrate fauna, which lacks rugosans, trilobites, articulate brachiopods, bryozoans, and pelmatozoans, is dominated by bivalves and crustaceans. Whether this reflects the inability of sessile benthic organisms to cope with high rates of sedimentation in these settings (see discussion in Feldman et al., 1993) or larger scale macroevolutionary patterns remains to be tested. Whatever the ultimate cause, the bivalve and crustacean associations, though commonly linked with restricted estaurine environments in the Late Paleozoic and Triassic, presage the spread of mollusc- and crustacean-dominated invertebrate communities in widespread shelf settings in the later Mesozoic and Cenozoic. This appears to be another potential example of the onshore-to-offshore migration of new evolutionary faunas documented by Sepkoski and Miller (1985), Jablonski and Bottjer (1983), and Jablonski et al. (1983) for Paleozoic and Mesozoic communities. As such, Late Paleozoic estuarine settings may have served as evolutionary crucibles for initial development of many modern high-diversity, open-shelf mollusc- and crustacean-dominated communities.

Acknowledgments

C. G. M. acknowledges valuable discussions about Lagerstätten and marginal marine systems with A. W. Archer, C. R. Cunningham, H. R. Feldman, H.-P. Schultze, and R. R. West, all of whom have been supported in Lagerstätten research by either NSF grant EAR 8903792, or NSF grant EAR 9018079, or both. H. R. Feldman and E. C. Brett reviewed the manuscript and improved the final version.

References

Baird, G. C., S. D. Sroka, C. W. Shabica, and T. L. Beard. 1985. Mazon Creek–type fossil assemblages in the U.S. Midcontinent Pennsylvanian: their recurrent character and paleoenvironmental significance. *In* H. B. Whittington and S. Conway-Morris (eds.), Extraordinary fossil biotas: their ecological and evolutionary significance. Phil. Trans. R. Soc. London B, 311:87–98.

Boneham, R. F. 1975. Chieftain No. 20 flora (Middle Pennsylvanian) of Vigo County, Indiana. Proc. Indiana Acad. Sci., 84:89–113.

Calver, M. A. 1968. Distribution of Westphalian marine faunas in northern England and adjacent areas. Proc. Yorkshire Geol. Soc., 37:1–72.

Canfield, J. E. 1959. Fossil plants of Indiana. Indiana Geol. Surv. Rep. Prog., 14, 45 pp.

Cunningham, C. R., H. R. Feldman, E. K. Franseen, R. Gastaldo, G. K. Mapes, C. G. Maples, and H.-P. Schultze. 1993. The Upper Carboniferous (Stephanian) Hamilton fossil-Lagerstätte (Kansas, U.S.A.): a valley-fill, tidally influenced depositional model. Lethaia, 26:225–236.

Feldman, H. R., A. W. Archer, E. P. Kvale, C. R. Cunningham, C. G. Maples, and R. R. West. 1993. A tidal model of Carboniferous Konservat-Lagerstätten formation. Palaios, 8:485–498.

Feldman, H. R., R. Lund, C. G. Maples, and A. W. Archer. 1994. Taphonomy and origin of the Bear Gulch Limestone (Namurian, Montana, U.S.A.). Geobios. M.S., 16:283–291.

Gall, J. D. 1971. Faunes et paysages du Grés-à-Voltzia du nord des Vosges: essai paléoecologique sur le Buntsandstein Superieur. Mem. Serv. Carte Geol. Alsace Lorraine, 34, 217 pp.

———. 1983. Ancient Sedimentary Environments and the Habitats of Living Organisms. Springer, New York, 219 pp.

Gastaldo, R. A. 1977. A Middle Pennsylvanian nodule flora from Carterville, Illinois, pp. 133–156. *In* R. C. Romans (ed.), Geobotany. Plenum, New York.

Heyler, D. 1985. Introduction, pp. 3–6. *In* D. Heyler and S. Secrétan (eds.), Table 2 Ronde Internationale du C.N.R.5: Sur le Gisement Stéphanien de Montceau-les-Mines. Bull. Trim. Soc. d'Hist. Nat. Amis Mus. d'Autun, 116.

Jablonski, D., and D. J. Bottjer. 1983. Soft-bottom epifaunal suspension-feeding assemblages in the Late Cretaceous: implications for the evolution of benthic paleocommunities, pp. 747–812. *In* M. J. S. Tevesz and P. L. McCall (eds.), Biotic Interactions in Recent and Fossil Benthic Communities. Plenum, New York.

Jablonski, D., J. J. Sepkoski, Jr., D. J. Bottjer, and P. M. Sheehan. 1983. Onshore-offshore patterns in the evolution of Phanerozoic shelf communities. Science, 222:1123–1125.

Maples, C. G., and H.-P. Schultze. 1989. A preliminary comparison of the Pennsylvanian faunal assemblage of Hamilton, Kansas, with marine and nonmarine contemporaneous assemblages, pp. 253–273. *In* G. Mapes and R. H. Mapes (eds.), Geology and paleontology of the Hamilton Quarry area in southeastern Kansas. Kansas Geol. Surv. Guideb. Ser., 6.

McGhee, G. R., Jr. 1982. The problematica *Palaeoxyris, Vetacapsula* and *Fayolia:* a morphological comparison with recent chondrichthyan egg cases. Proc. No. Amer. Paleo. Conv., 3:365–369.

Pacaud, G., W. D. I. Rolfe, F. Schram, S. Secrétan, and D. Sotty. 1981. Quelques invertébrés nonveaux du Stéphanian de Montceau-les-Mines. Bull. Soc. d'Hist. Nat. Autun, 97:37–43.

Richardson, E. S., Jr., and R. G. Johnson. 1971. The Mazon Creek faunas, pp. 1222–1235. *In* Proc. North American Paleo. Conv., 5–7 September 1969, Field Mus. Nat. Hist., vol. 1. Allen, Lawrence, Kansas.

Schram, F. R. 1979. The Mazon Creek biotas in the context of a Carboniferous faunal continuum, pp. 159–190. *In* M. H. Nitecki (ed.), Mazon Creek Fossils. Academic Press, New York.

Schultze, H.-P., and C. G. Maples. 1992. Comparison of the Late Pennsylvanian faunal assemblage of Kinney Brick Company Quarry, New Mexico, with other Late Pennsylvanian Lagerstätten. New Mexico Bur. Mines Min. Res. Bull., 138:231–242.

Schultze, H.-P., C. G. Maples, and C. R. Cunningham. 1994. The Hamilton Konservat-Lagerstätte: Stephanian terrestrial biota in a marginal-marine setting: Trans. R. Soc. Edinburgh Earth Sci., 84:443–451.

Seilacher, A. 1970. Begriff and Bedeutung der Fossil-Lagerstätten. Neues Jahrb. Geol. Paläo. Mh., 1970, pp. 34–39.

Seilacher, A., W. E. Reif, and F. Westphal. 1985. Sedimentological, ecological and temporal patterns of fossil Lagerstätten, pp. 5–23. *In* H. B. Whittington and S. Conway Morris (eds.), Extraordinary fossil biotas: their ecological and evolutionary significance. Phil. Trans. R. Soc. London B, 311:3–4.

Sepkoski, J. J., Jr., and A. I. Miller. 1985. Evolutionary faunas and the distribution of paleozoic marine communities in space and time, pp. 153–190. *In* J. W. Valentine (ed.), Phanerozoic Diversity Patterns: Profiles in Macroevolution. Princeton University Press, Princeton, New Jersey.

Shabica, C. W. 1971. Depositional environments in the Francis Creek Shale and associated strata. Unpublished Ph.D. dissertation, University of Chicago, 206 pp.

Sroka, S. D. 1996. Synthesis of a Middle Pennsylvanian soft-bodied fauna from western Illinois. Unpublished Ph.D. thesis, University of Illinois, Urbana, 197 pp.

Zidek, J. 1976. A new shark egg capsule from the Pennsylvanian of Oklahoma, and remarks on the Chondrichthyan egg capsules in general. Jour. Paleo., 50:907–915.

6

Cnidaria

Steven D. Sroka

Phylum Cnidaria (from the Greek word *knide* meaning "nettle" and the Latin *aria* meaning "like or connected with") includes the familiar jellyfish (Scyphozoa and Cubozoa), hydralike animals (Hydrozoa), anemones and corals (Anthozoa), and a few less familiar fossil groups. These are most of the animals once referred to as the "Coelenterata." Most, but not all, of the members of this phylum are characterized by radial or biradial body symmetry arranged around a longitudinal axis, which accounts for the "spokes of the wheel" pattern on the skeletons of colonial reef or solitary corals. The organisms in this phylum possess two thin layers of body wall tissue with a thin to thick gelatinous connective tissue sandwiched in between. Basically, cnidarians are a watery sac with a single opening that serves as both anus and mouth. A distinguishing characteristic unique to Cnidaria are cells termed cnidocysts, which contain specialized stinging organelles (nematocysts). Nematocysts are abundant on tentacles, where they are used to capture and paralyze prey and discourage predators. All cnidarians are carnivorous.

Only two body forms are found among cnidarians: polyps, which are more or less cylindrical in outline and adapted to a sessile mode of life; and medusae or jellyfish, which are adapted for weak swimming or floating.

Four classes are recognized within the phylum Cnidaria: Anthozoa, Hydrozoa, Scyphozoa, and Cubozoa. Cubozoa was until recently considered an order (Cubomedusae) of Scyphozoa. Hydrozoa includes both medusae and solitary and colonial polyps. A typical colonial hydrozoan is attached to the substrate by means of a rootlike feature called the stolon. Attached to the many stalks that can arise from the stolon are individual feeding zooids (animals) or hydranths. The hydranths, which can vary in shape, all have a terminal mouth with tentacles arranged in a small circle. Hydranths supply nutrients for the colony by capturing and ingesting small prey. Medusae (when present) have a velum. The velum, which is used in swimming, is a shelflike feature produced by the inward projection of the bell margin. It partially encloses the open side of the bell. Hydrozoans are found in marine and some freshwater environments.

Members of the class Scyphozoa are generally solitary free-swimming or floating forms with the polyp stage reduced or absent. Medusae are bell-shaped but do not contain a velum. The margin of the bell is typically scalloped, with notches containing sense organs. Scyphozoans are all marine. Cubozoans are bell-shaped medusoid cnidarians having four flattened sides, with a tentacle or group of tentacles hanging from each corner of the square. At the base of each group of tentacles (or tentacle) is a tough, flat bladelike structure called the pedalium. The bell is not scalloped but contains a velum that aids in swimming. Cubozoans were separated from scyphozoans based on their bell shape and nature of the bell margins, their nematocysts, their possession of a velum, and to some extent their lifestyle (Werner, 1975; Dunn, 1982; Brusca and Brusca, 1990). Cubozoans are active predators in tropical and subtropical oceans. Only a single order, Cubomedusae, is known for this class.

Anthozoans are marine, flowerlike solitary or colonial polyps. This group includes anemones, corals, sea fans, and sea pens. The digestive cavity is subdivided by radial divisions (mesenteries) bearing nematocysts. Many members of this class secrete a calcareous skeleton.

Cnidaria is one of the oldest known phyla. Soft-bodied cnidarians, notably medusae, appeared 620 to 570 million years ago in late Precambrian time, prior to the first appearance of shelled organisms. Despite the longevity of this group, the fossil record of the soft-bodied cnidarians is spotty and poor. The stony corals alone are well represented in the rock record.

The Mazon Creek cnidarian biota therefore is especially significant, as it provides a rich sample of groups that are found at only a few other fossil localities in the world. The number of soft-bodied cnidarians in the Essex fauna exceeds that of any other fossil locality, and the diversity (numbers of different species) of these forms rivals the diversity of the considerably younger Jurassic Solnhofen locality in Germany (see Foster, 1979). These species can be placed into groups above the family level using modern classification schemes. Although the Mazon Creek jellyfish can be placed within these higher level groups, many lack features necessary for this and their placement is problematic at its best.

The scyphozoan medusa *Essexella asherae* is the most abundant organism in the Essex fauna, composing more than 50 percent of all fossils collected from the Francis Creek Shale at Pit 11. Other less commonly found scyphozoans include *Octomedusa pieckorum, Reticulomedusa greenei,* and the rare *Lascoa mesostaurata.* Cubozoa is represented is by *Anthracomedusa turnbulli.* Possible hydrozoans include *Drevotella proteana, Mazohydra megabertha,* and the unusual *Etacystis communis.* Several additional undescribed or recently discovered species are part of the Mazon Creek fauna (Foster, 1979; Sroka, 1996) but are not included in this paper. Many of the forms described here are also known from western Illinois (Sroka, 1996). No anthozoans have been found in the Mazon Creek fauna.

Figure 6.1. *Anthracomedusa turnbulli,* NEIU MCP 210, Pit 11, Kankakee County; top view showing septa and stomach in upper right center of specimen. Scale bar = 10 mm.

Class Cubozoa
Anthracomedusa turnbulli
Johnson and Richardson, 1968
Figures 6.1, 6.2

This cubomedusan has a cube-shaped bell bearing four flattened sides. The bell widths vary from 2 to 10 cm. Each bell corner appears to bear a very short, thick, pedalium that branches into smaller pedalia with tentacles. Four septa appear to project into the corners of the stomach. Tentacles are sometimes crinkled, as they are in living cubomedusae.

Anthracomedusa is similar to the members of the modern family Chirodropidae, the sea

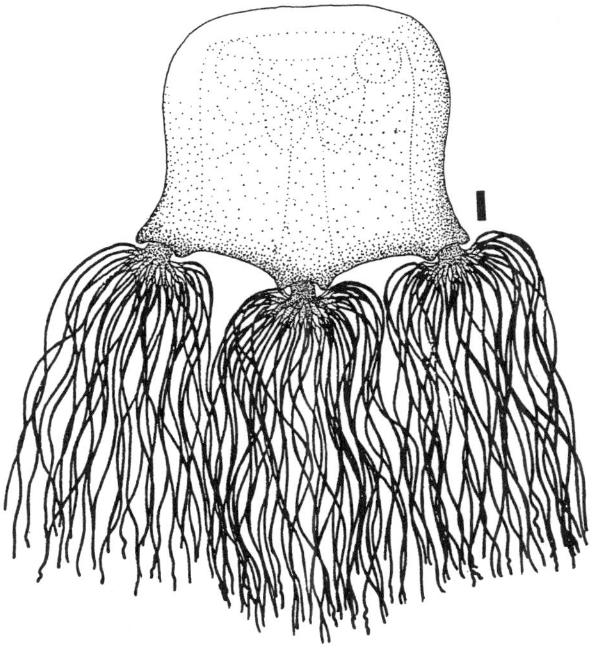

Figure 6.2. *Anthracomedusa turnbulli,* modified from Foster (1979); side view. Scale bar = 10 mm.

Figure 6.3. *Essexella asherae,* NEIU MCP 606, Pit 11; side view. Scale bar = 10 mm.

wasps. This family includes some of the most dangerous living invertebrates. It also includes the fastest swimming coelenterates. Members of this family live in shallow tropical waters and are efficient predators.

Anthracomedusa fossils are found in the southernmost part of Pit 11 and have been found at "Chowder Flats," just west of Morris, and in western Illinois.

Class Scyphozoa
Essexella asherae Foster, 1979
Figures 6.3–6.5

For many years collectors referred to these common ill-defined mushroom-shaped, circular or elongate forms as "blobs" and discarded them. Short statements on blobs can be found in some of the general works on the Essex fauna, but the first published work devoted solely to their description and discussion was by Nadasdy (1976). Foster (1979) provided a more complete description and discussion of the blobs and proposed the name *Essexella asherae,* which has been generally accepted.

The morphological features within this species vary greatly from specimen to specimen and may be due to preservational history rather than original parts.

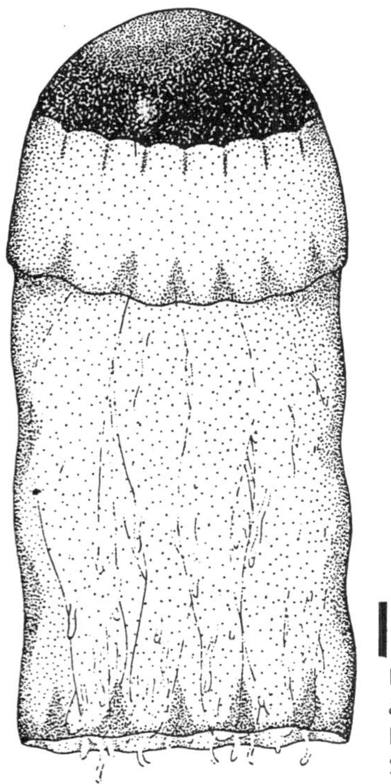

Figure 6.4. *Essexella asherae,* modified from Foster (1979). Scale bar = 10 mm.

Figure 6.5. "Multiple blobs," *Essexella asherae,* NEIU MCP 206, Pit 11. Scale bar = 10 mm.

Essexella asherae is a medium-sized (1–15 cm in length) scyphomedusan with a manubrium (a tubelike structure) extending into eight complex oral lobes with tentacle-like structures below the bell. Both structures are almost enclosed within a membrane skirt hanging down from the bell. The majority have been preserved lying on their sides, with the skirt masking internal structures (Foster, 1979).

Blobs have been found in the same concretions with plants, crustacea, and polychaetes and also as "multiple blobs" (Figure 6.5). Individual blobs on concretions containing multiple individuals are usually small (a few centimeters in length) with smooth skirts. Also found on the skirts of numerous specimens of *Essexella* is the small snail *Strobeus.* These thin-shelled snails, in apparent life positions, may have been pelagic predators (Foster, 1979) or scavengers feeding on dead medusae.

This jellyfish is usually the most abundant taxon in Essex localities and occurs throughout the spatial range of the Essex fauna in the Mazon Creek area, as well as near Astoria, Illinois. *Essexella asherae,* like other Essex fauna cnidarians, was a predator. Foster (1979) suggested that the configuration of skirt and oral lobes made an excellent trap, preventing prey from moving or escaping.

Lascoa mesostaurata Foster, 1979
Figures 6.6, 6.7

A rare scyphozoan described from a single specimen from Astoria, Illinois. According to Foster

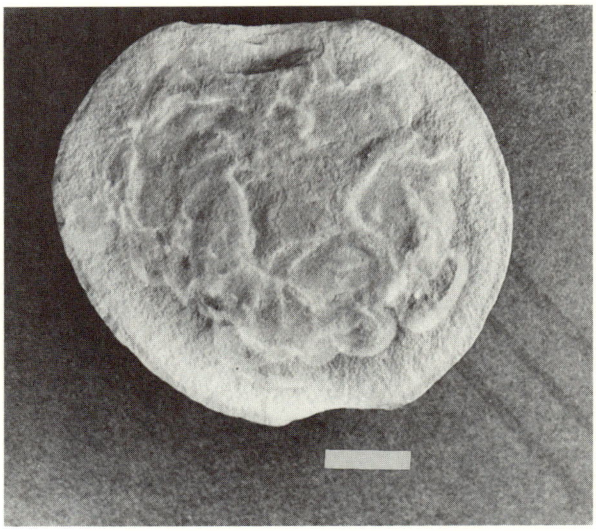

Figure 6.6. *Lascoa mesostaurata,* holotype, Lasco collection; top view. Scale bar = 10 mm.

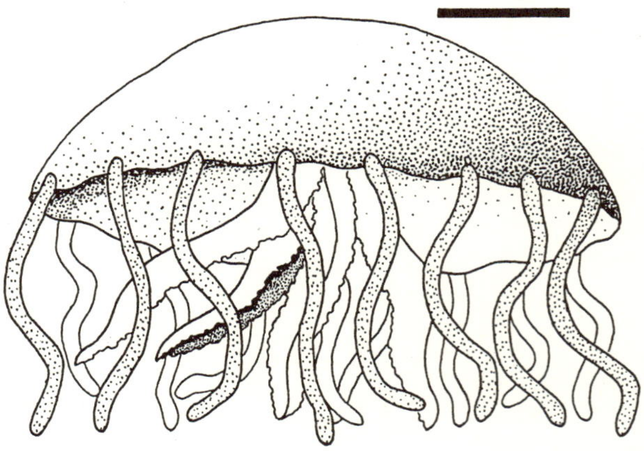

Figure 6.7. *Lascoa mesostaurata,* modified from Foster (1979); side view. Scale bar = 10 mm.

(1979, pp. 233–234) *Lascoa* is a medium-sized medusa, 4 cm in diameter, circular in outline, with a mouth that is drawn into four lobes. The oral lobes extend slightly beyond the rim of the bell. Near the edge of the bell there are approximately 16 relatively thick tentacles.

Foster noted that *Lascoa* bears a superficial resemblance to living members of the order Semaestomae. He compared *Lascoa* to the extant genus *Aurelia,* with the only obvious difference being the presence of much thicker, longer, and less numerous marginal tentacles in the former.

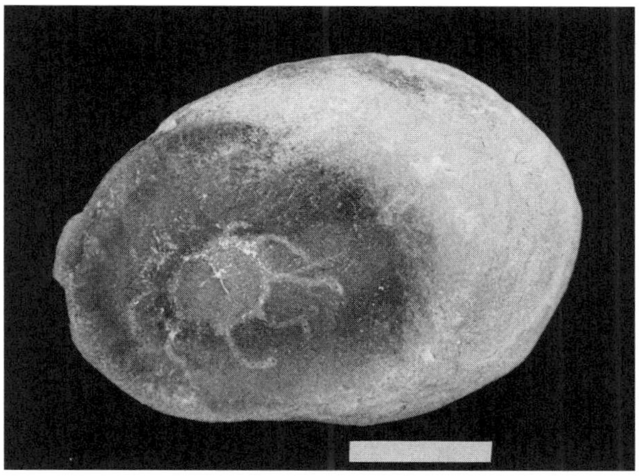

Figure 6.8. *Octomedusa pieckorum,* NEIU MCP 406, Pit 11; mouth impression in center of specimen. Scale bar = 10 mm (estimate).

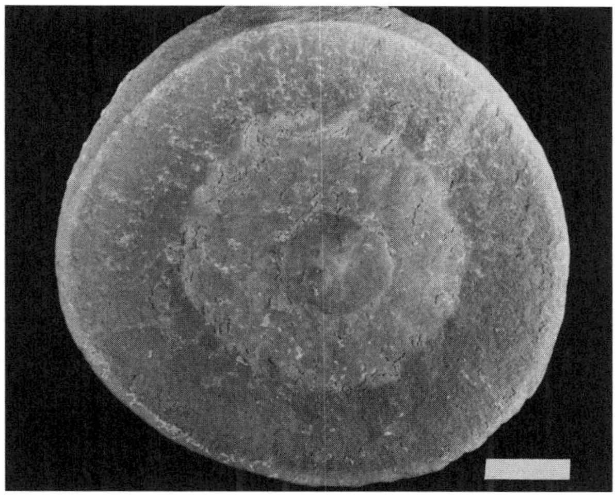

Figure 6.10. *Reticulomedusa greenei,* NEIU MCP 444, Pit 11. Scale bar = 10 mm.

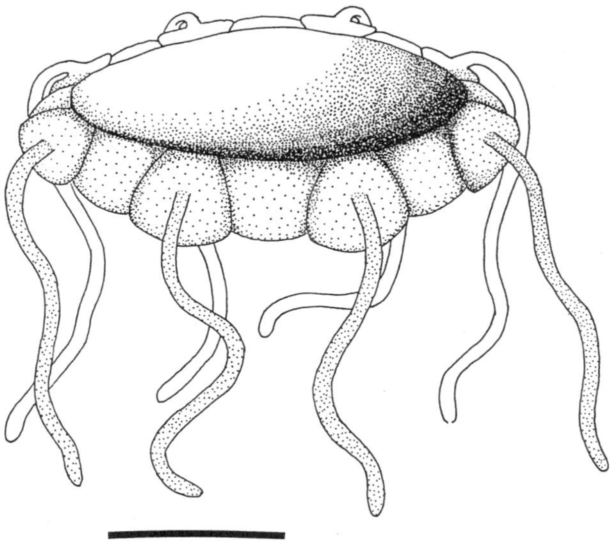

Figure 6.9. *Octomedusa pieckorum,* modified from Foster (1979). Scale bar = 10 mm.

Octomedusa pieckorum Johnson and Richardson, 1968 Figures 6.8, 6.9

Octomedusa pieckorum is a small octagonal medusa with eight tentacles whose length ranges from a few nubs to slightly greater than the diameter of the bell. Foster (1979) observed that the bell diameter ranges from approximately 3 to 21 mm. In the center of well-preserved specimens there is either a light-colored subquadrangular region representing the cross section of the manubrium or furrows in a cruciform pattern representing the mouth impression (Foster, 1979).

Concretions with multiple individuals of *Octomedusa* are fairly common. *Octomedusa pieckorum* has been found in concretions with *Belotelson magister, Essexella asherae, Etacystis communis,* polychaete worms, and *Tullimonstrum gregarium.*

Foster (1979) placed *O. pieckorum,* with its scalloped margin, large quadriparite mouth, and marginal coronal furrow, in the scyphozoan order Coronate. This order presently contains the smallest known living scyphozoan medusa.

Octomedusa occurs at most places in Pit 11 but is only abundant in the southern part of Pit 11 and is the overwhelmingly dominant taxon (64 percent) at the locality known as "Jellyfish Hill" (see Foster, 1979, p. 196).

Reticulomedusa greenei Foster, 1979 Figure 6.10

Foster described *R. greenei* as a medium-sized (4.5 cm) scyphomedusan with a distinct gastrovascular network in the bell and, in well-preserved specimens, a raised cross probably corresponding to the quadripartite mouth.

Reticulomedusa is subcircular in outline with a small crescent-shaped lobe on one side. The margin of the bell is slightly scalloped. The circular central portion, which is raised, is concave. Within the concave area, slightly off

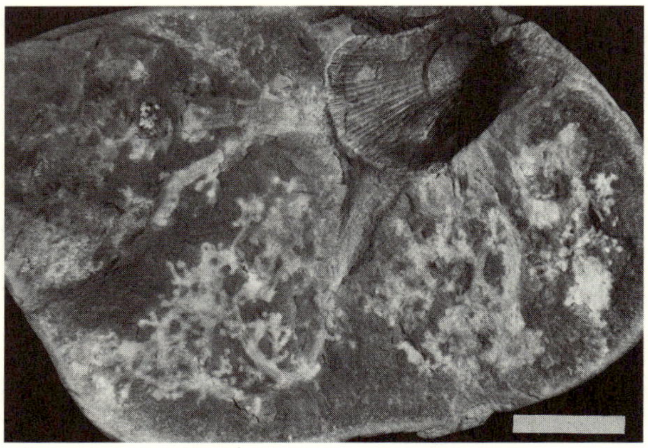

Figure 6.11. *Drevotella proteana*, NEIU MCP 439, Pit 11 (attached to *Aviculopecten mazonensis*). Scale bar = 10 mm.

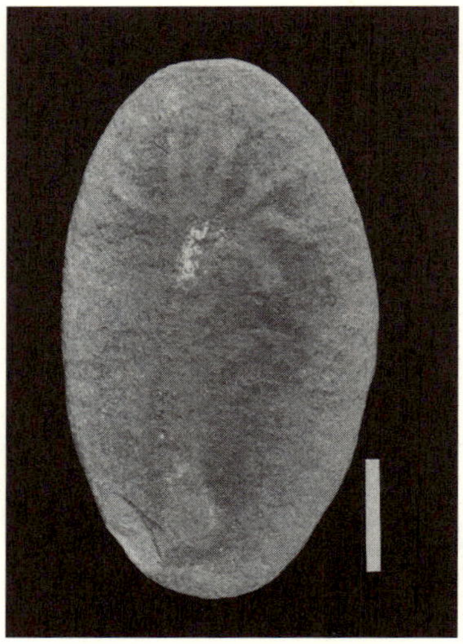

Figure 6.12. *Mazohydra megabertha*, holotype, FMNH P 32074. Scale bar = 5 mm.

center, is a flat and smooth circular area with a distinct raised cross. Between the concave area and the rest of the specimen there is a distinct break, which forms a rim. The surface of specimens is covered by fine raised ridges forming a reticulate pattern. Foster interpreted the reticulate pattern to correspond to a gastrovascular network similar to ones in modern rhizotome scyphozoans. In the holotype this pattern is obvious in the central concave area (Foster, 1979).

Very few specimens of *R. greenei* have been found to date.

Class Hydrozoa
Drevotella proteana
Nitecki and Richardson, 1972
Figure 6.11

Drevotella are benthic treelike sessile colonies of varying size (maximum height approximately 10 cm) and shape with two or more main stems of varying thickness. Stems show irregular dichotomous branching with smaller branches along their lengths. Swellings exist at the ends of the finest branches and probably represent hydranths (Nitecki and Richardson, 1972). The hydranths are broad and rounded compared to their stalks. Several specimens of *D. proteana* have been found attached to bivalves (*Aviculopecten mazonensis* and a myalinid), indicating that the shells were used as a supporting base. Nitecki and Richardson (1972) judged *Drevotella* to be a fossil hydrozoan.

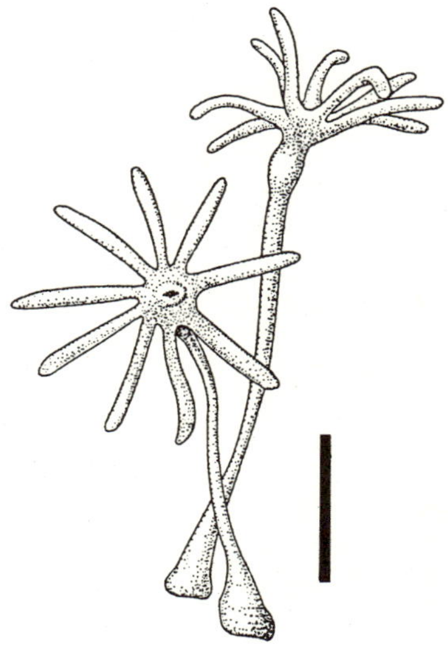

Figure 6.13. *Mazohydra megabertha*, modified from Schram and Nitecki (1975). Scale bar = 5 mm.

Mazohydra megabertha
Schram and Nitecki, 1975
Figures 6.12, 6.13

A large solitary, athecate, hydralike fossil with an elongate stalk. The body length of the holotype is approximately 15 mm. The gastral region is long and slender with a terminal pedal disk. The oral area is surrounded by a ring of eight or nine tentacles approximately one-quarter the body length. Schram and Nitecki (1975) indicated that *Mazohydra* is very much like the living *Hydra* except for its larger size and shorter, more numerous tentacles. It is considered to be the earliest known fossil hydra. *M. megabertha* is very rare in the Essex fauna, with only a few specimens having been found.

Figure 6.14. "The H," *Etacystis communis,* FMNH PP 16766. Scale bar = 10 mm.

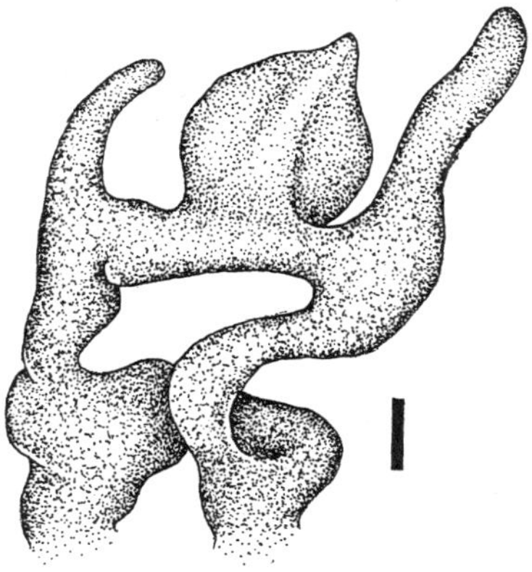

Figure 6.15. *Etacystis communis,* modified from Nitecki and Schram (1976) and Foster (1979). Scale bar = 10 mm.

Etacystis communis
Nitecki and Schram, 1976
Figures 6.14, 6.15

Known to collectors as "the H," *Etacystis* is a relatively large unsymmetrical H-shaped fossil animal or part of an animal with two unequal arms and a sac connected to a crossbar (peduncle). The longest leg of the H, termed the stolon (Nitecki and Schram, 1976), was apparently flexible. This structure, which varies in length, almost always extends off the edge of the concretion. Within some specimens there is a canallike area down the middle of the stolon (Nitecki and Schram, 1976; Foster, 1979). The stolon may have been both an anchoring organ and a device to connect adjoining animals, or a site of asexual budding (Nitecki and Schram, 1976). There is a heart-shaped saclike structure connected to the peduncle by a stalk with a small aperture located at the point where the stalk joins the sac. This aperture may represent the mouth (Nitecki and Schram, 1976).

Numerous specimens are found with papillations on the arms, peduncle, and sac. These papillations might be suckers, sticky pads, clumps of cnidocytes, remnants of lost tentacles, or even diagenetic alterations (Nitecki and Schram, 1976).

The natural affinities of *E. communis* have been debated since it was first described. Nitecki and Schram (1976), with some uncertainty, considered it to be similar to pterobranch hemichordates such as *Rhabdopleura,* which contains a pair of tentaculate lophophore arms. Foster (1979, p. 219) thought it possible that *Etacystis* is not related to hemichordates. He assigned *Etacystis* to the Hydrozoa (order Siphonophora) because of its planktic nature, its morphology, and its great range in size without appreciable change in form of the peduncle, sac, and arms.

References

Brusca, R. C., and G. J. Brusca. 1990. Phylum Cnidaria, pp. 211–261. *In* Invertebrates. Sinauer, Sunderland, Massachusetts.

Dunn, D. F. 1982. Cnidaria, pp. 669–705. *In* S. P. Parker (ed.), Synopsis and Classification of Living Organisms. McGraw-Hill, New York.

Foster, M. W. 1979. Soft-bodied coelenterates in the Pennsylvanian of Illinois, pp. 191–268. *In* M. H. Nitecki (ed.), Mazon Creek Fossils. Academic Press, New York.

Johnson, R. G., and E. S. Richardson, Jr. 1968. Pennsylvanian invertebrates of the Mazon Creek area, Illinois: the Essex fauna and medusae. Fieldiana Geol., 12(7):109–115.

Nadasdy, G. S., 1976. Preliminary investigations of the blobs, Essex fauna from Pit Eleven. Proc. Peoria Acad. Sci., 9:1–22.

Nitecki, M. H., and E. S. Richardson, Jr. 1972. A new hydrozoan from the Pennsylvanian of Illinois. Fieldiana Geol., 30(1):1–7.

Nitecki, M. H., and F. R. Schram. 1976. *Etacystis communis,* a fossil of uncertain affinities from the Mazon Creek fauna (Pennsylvanian of Illinois). Jour. Paleo., 50:1157–1161.

Schram, F. R., and M. H. Nitecki. 1975. Hydra from the Illinois Pennsylvanian. Jour. Paleo., 49:549–551.

Sroka, S. D. 1996. Synthesis of a Middle Pennsylvanian soft-bodied fauna from western Illinois. Unpublished Ph.D. dissertation, University of Illinois, Urbana, 197 pp.

Werner, B. 1975. Bau und Lebesgeschichte des Polypen von *Tripedalia cystophora.* (Cubozoa, class nov., Carybdeidae) und seine Bedeutung für die Evolution der Cnidaria. Helgoländer Wiss. Meers., 27:461–504.

7A

Polychaete Worms

Kirk Fitzhugh, Steven D. Sroka, Samuel Kruty, Michael D. Henderson, and Andrew A. Hay

The traditional concept of "worms" is derived from Linnaeus's (1758) grouping of all elongate, bilaterally symmetrical invertebrates, exclusive of insects, into the taxon Vermes (Greek, "worms"). Today, as many as 14 "worm" or vermiform phyla are recognized (Howell, 1962; Robinson, 1987; Brusca and Brusca, 1990), but only one, the Annelida (Greek, *annulatus* "annulated" or "ringed"), has a diverse fossil record in the Mazon Creek area (cf. Chapter 7B). The annelid body is composed of segments, with which is often associated an external repetition of body parts. This segmentation is also imposed on internal features such as circulatory, reproductive, nervous, and muscle systems.

Between two and four annelid classes have been recognized in the past, with most current classifications utilizing only three: the Polychaeta (marine bristleworms), Oligochaeta (earthworms and related aquatic forms), and Hirudinea (leeches). Polychaeta is the only annelid class represented in the Mazon Creek Essex fauna.

Morphology

Polychaetes, as well as all other annelids, have a general body plan that can be divided into four distinct regions, proceeding from anterior to posterior: prostomium, peristomium, trunk, and pygidium. General reviews of polychaete external morphology can be found, for example, in Fauchald (1977), George and Hartmann-Schröder (1985), and Brusca and Brusca (1990).

The prostomium and peristomium constitute the anterior or head region (Figure 7A.1A, 7A.1C). The polychaete prostomium displays a variety of shapes but is usually roughly conical to quadrangular and commonly bears sensory structures, such as eyes and antennae. Some taxa also have an additional sensory structure known as a caruncle, a posterodorsal extension of the prostomium or peristomium (Figure 7A.1C).

The peristomium usually appears as a ring or collar behind the prostomium and in polychaetes might bear one or two pairs of peristomial cirri (Figure 7A.1A, 7A.1C). A structure seen only in certain polychaete taxa is paired palps, located at the junction between pro- and peristomium, and ranging in shape from short and conical to elongate and cirriform. Short, conical palps are situated ventral to the prostomium (Figures 7A.1A, 7A.1C) and are usually only sensory in function. Elongate, cirriform palps are generally attached dorsally and are used as feeding or sensory devices. The mouth is ventrally located and surrounded by the peristomium.

In a number of polychaete taxa, the prostomium and peristomium are fused to some degree with one another as well as with the first few anterior segments of the trunk. The parapodia of these fused segments are often only repre-

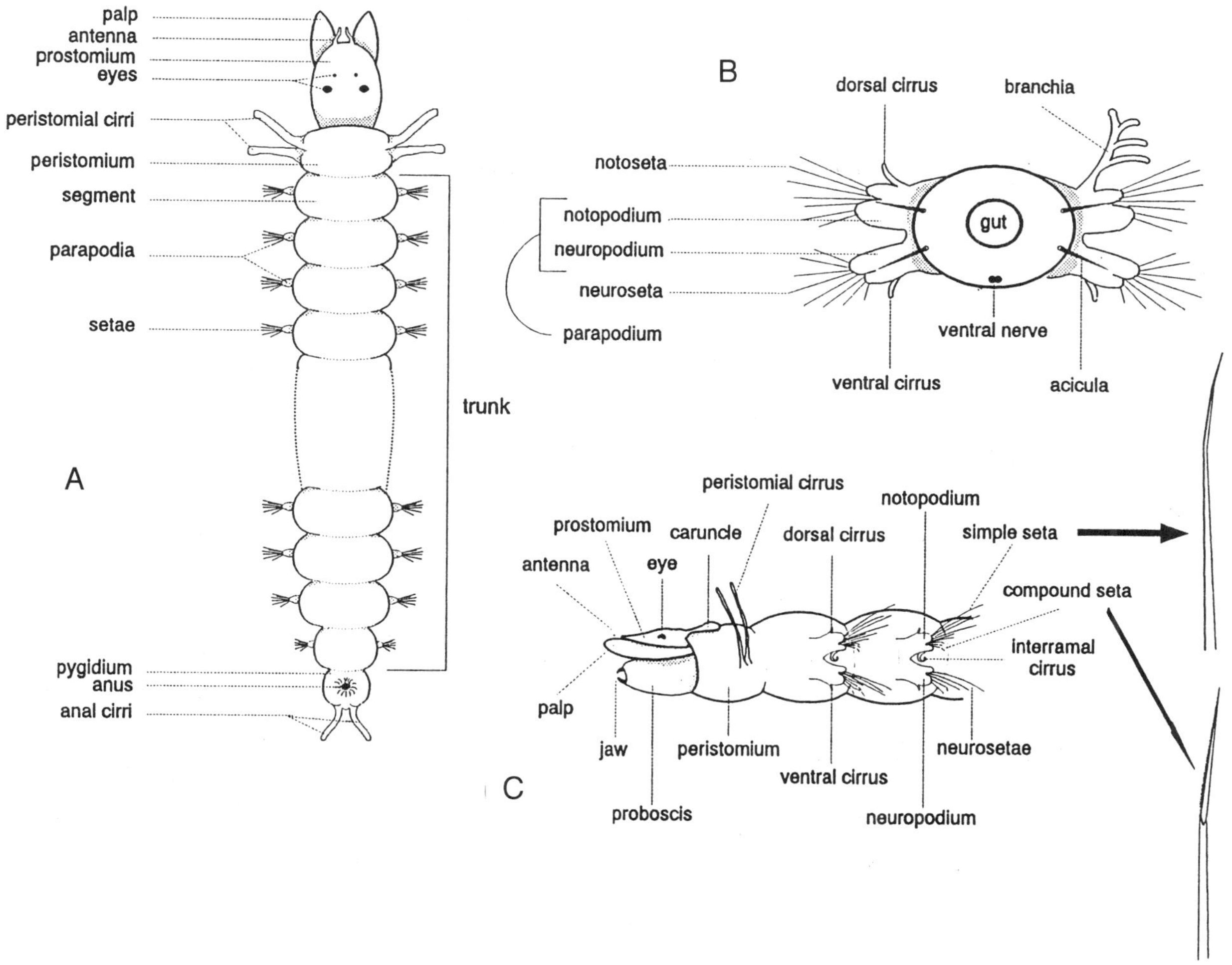

Figure 7A.1. Schematic representation of polychaete morphology. **A** Entire worm, dorsal view. **B** Cross section through a segment, showing parapodial structures in lateral view. **C** Lateral view of anterior end.

sented by pairs of cirri, closely resembling peristomial cirri, and called tentacular cirri.

The annelid trunk, which lies between the anterior presegmental prostomium and peristomium, and posterior postsegmental pygidium, is the body region composed of segments (Figure 7A.1A). Unlike oligochaetes or leeches, the segments of polychaetes often have paired, lateral, fleshy appendages or lobes called parapodia, which usually function as locomotory or respiratory organs. In most cases, a parapodium consists of two rami or lobes: an upper or dorsal ramus called a notopodium and a lower or ventral ramus called a neuropodium. A parapodium is said to be biramous when noto- and neuropodia are present and uniramous when only the neuropodium is present. Parapodial projections, such as dorsal and ventral cirri or branchiae, might be present. The parapodial rami of some taxa are supported by internal chitinous rods, called aciculae (sing., acicula). Parapodia usually bear numerous elongate bristles or setae (or chaetae) that extend from the body in a fanlike bundle called a fascicle, hence the name polychaete (Greek, *poly* "many" and *chaitē* "long hair"). The number, position, shape, and size of setae are important characters in polychaete classification and identification. For purposes of the present review, two setal forms will be recognized: simple and compound (Figure 7A.1C). A simple seta is constructed as a single, unjointed shaft, whereas a compound seta has a joint or articulation at some point along the shaft.

In most polychaetes, the anterior region of the gut is modified into an eversible pharynx or proboscis (Figure 7A.1C). Associated with the proboscis of some taxa are two or more jaws, usually located at the anterior end of the everted proboscis, and commonly used by carnivorous polychaetes to grasp prey items. Fossilized polychaete jaws, or jaw parts, are usually found in separate pieces and are commonly referred to as scolecodonts in the paleontological literature. The jaw pieces of most polychaetes consist of a pair or two pairs (cf. Figure 7A.9). Members of the superfamily Eunicea are unique, however, in that the jaw apparatus consists of a paired series of upper maxillae and a pair of lower mandibles (e.g., Figure 7A.4B; cf. George and Hartmann-Schröder, 1985; Fauchald, 1992). Of the Mazon Creek polychaetes, this elaborate jaw apparatus is exemplified by *Esconites zelus* Thompson and Johnson (Figures 7A.4, 7A.5). Mineralogically, polychaete jaws are composed of either scleroprotein, aragonite, or calcite (Colbath, 1986, 1989).

The posteriormost part of the body is terminated by a pygidium on which is located the anus and sometimes cirriform or digitate structures called anal cirri (Figure 7A.1A).

Ecology. The majority of polychaetes are free living and marine, though some species do occur in freshwater or semiterrestrial environments. Polychaetes are found in all marine habitats, at all depths, and are especially common in soft sediments. While polychaete external morphology is extremely diverse, several basic body plans are often identified to infer general habitat requirements, degree of motility, diet, and feeding behavior (e.g., Woodin, 1987). Keep in mind, however, that no generalization about polychaete ecology can be completely accurate; ecological diversification can be just as great as the array of morphological variations among polychaete species.

Polychaetes characterized by numerous, similar body segments and a muscular, eversible proboscis, as well as jaws, are usually inferred to be motile, epi- or infaunal predators, though there are numerous exceptions, whereas taxa with highly differentiated body regions are usually presumed to be sedentary, infaunal detritivores, with motility limited to burrowing activity, or else nonmotile tube dwellers. Similarly, tubicolous polychaetes tend to be either infaunal detritivores or epifaunal suspension feeders. Tubes are most often constructed of sand or mud, cemented together with mucus secreted by epithelial gland cells. A notable exception are members of the family Serpulidae or feather-duster worms, described later, which produce calcareous tubes through the secretion of calcium carbonate.

An additional relationship between worm and environment is often seen in the degree of parapodial development and general body shape (Woodin, 1987). For example, errant taxa tend to have relatively short bodies, well-developed parapodia and setal fascicles (e.g., Figure 7A.1B, 7A.1C), and flattened to cylindrical bodies. More sedentary taxa, such as burrowers, usually have a more elongate body, small parapodia, large or small setal fascicles, and cylindrical bodies. Tube-dwelling polychaetes usually have elongate bodies composed of few to many segments, poorly developed parapodia, small setal fascicles, and cylindrical bodies. It is important to note, however, that numerous exceptions to these generalizations occur—a testament to the adaptability and evolutionary success of polychaetes as a group.

According to Thompson (1979), the Mazon Creek polychaetes appear to have been primarily motile, epifaunal predators. Exceptions to this generalization appear to include *Didontogaster cordylina* Thompson (family Nephtyidae) and *Pieckonia helenae* Thompson (family Goniadidae), both predators hypothesized to have been infaunal. Thompson (1979) inferred habits of Mazon Creek polychaetes on the basis of what is known about extant taxa, in conjunction with fossil positions relative to bedding planes. The presumed dominance of errant epifaunal taxa over infaunal forms has yet to be adequately explained.

Classification

Since the work of Audouin and Milne Edwards (1834), polychaete families have most commonly been divided into two broad subclasses or orders, the Errantia and Sedentaria. The goal of this system has been to integrate gross structural morphology and life habits. While this classification scheme conveniently divides polychaete families into two relatively large, equitable groups, the Errantia and Sedentaria are not amenable to any sort of rigorous definition, nor can they accurately reflect hypotheses of phylogenetic relationships among families.

In response to these deficiencies, several attempts have been made recently to delineate orders, or at least groups of families, on the basis of particular suites of characters, such as proboscis and nephridia structure (Dales, 1962), musculature (Storch, 1968), or coelomic and segment development (Clark, 1969). In an effort to synthesize these studies, Fauchald (1977; see also Fauchald, 1974) attempted to subdivide the more than 80 polychaete families into 17 clearly defined orders using as much of the available external morphological data as possible. While there has been general recognition or acceptance of these recent endeavors to denote relationships among polychaete families (e.g., Pettibone, 1982; Brusca and Brusca, 1990), all of these schemes have distinct limitations. Thus, substantial revisions of order and family concepts, as well as hypotheses of evolution, will likely ensue.

Fossils

The generally poor and often questionable fossil record of polychaetes has effectively precluded the identification of a narrow geologic period during which the group arose. The variety of fossil trails, burrows, and tubes at least indicate the existence of wormlike taxa during the Proterozoic (Howell, 1962; Robinson, 1987). Most research on polychaete fossils has centered on jaw pieces, or scolecodonts, of a few families (e.g., Kielan-Jawarowska, 1966). On occasion, however, whole body remains of polychaetes have been found throughout the Paleozoic (Fauchald, 1974; Conway Morris et al., 1982; Briggs and Kear, 1993), and entire faunas are preserved in the famous Middle Cambrian Burgess Shale (Conway Morris, 1979; Butterfield, 1990), the Mississippian Bear Gulch Limestone (Schram, 1979), and the Pennsylvanian Mazon Creek area (Thompson, 1979).

To date, only two papers have dealt extensively with Mazon Creek polychaetes: Thompson and Johnson (1977) described the eunicid *Esconites zelus*, and Thompson (1979) described 10 species in seven families and noted the occurrence of four additional undescribed species. Thompson (1979) also erected a new family, Fossundecimidae, as well as the new subfamily Hystriciolinae in the Aphroditidae. An additional species, *Howellitubus whitfieldorum* Richardson (1956), originally described as a worm tube, has been shown to be a bone fragment (Richardson, 1980). Our present account briefly summarizes information on the 11 species described among Thompson and Johnson (1977) and Thompson (1979). Two additional species are included, assignable to the families Serpulidae and Flabelligeridae, respectively: *Spirorbis* sp. and Flabelligeridae sp. We also comment on the systematic position of *Palaeocampa anthrax* Meek and Worthen (1865).

The excellent preservation of Mazon Creek polychaetes commonly allows for ready identification of branchiae, cirri, and parapodia. While photographs of most species are provided, line drawings of taxonomically important features are also included, most of which are derived from the original publications. Readers should, however, be aware that no consensus of opinion exists regarding interpretations of structures or reconstructions based on fossil specimens. Species presented here are arranged by taxonomic order and family as these are currently recognized by most polychaete systematists (cf. Fauchald, 1977; Pettibone, 1982; George and Hartmann-Schröder, 1985). This system differs from that used by Thompson (1979), which was patterned after Day (1967), especially with regard to the composition of the families Aphroditidae and Glyceridae. Brief diagnoses, primarily derived from Fauchald (1977), are provided for each family and are intended as general guides to the placement of species in a family. By necessity, family diagnoses are largely based on extant taxa. Table 7A.1 summarizes past and present ordinal and familial distributions for Mazon Creek polychaetes; taxonomic changes shown in the table are discussed in the Remarks sections below.

Table 7A.1
Ordinal and Familial Distributions for Mazon Creek Polychaetes

Order	Current Family Designation	Previous Family Designation (if different)	Species
Amphinomida	Amphinomidae		*Rhaphidiophorus hystrix*
Eunicida	Eunicidae		*Esconites zelus*
Flabelligerida	Flabelligeridae		Flabelligeridae sp.
Phyllodocida	Polynoidae	Aphroditidae	*Fastuoscolex gemmatus*
	Polynoidae	Aphroditidae	*Hystriciola delicatula*
	Sigalionidae	Aphroditidae	*Dryptoscolex matthiesae*
	Nereididae	Fossundecimidae	*Fossundecima konecniorum*
	Goniadidae	Glyceridae	*Pieckonia helenae*
	Hesionidae		*Rutellifrons wolfforum*
	Nephtyidae		*Astreptoscolex anasillosus*
	Nephtyidae		*Didontogaster cordylina*
	Phyllodocidae		*Levisettius campylonectus*
Sabellida	Serpulidae		*Spirorbis* sp.
Incertae sedis	Incertae sedis		*Palaeocampa anthrax*

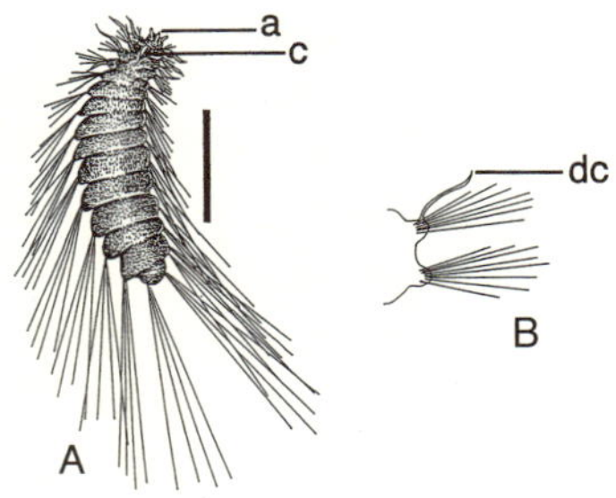

Figure 7A.2. *Rhaphidiophorus hystrix,* modified from Thompson (1979, fig. 10).
A Dorsal view.
Scale bar = 10 mm.
B Anterior parapodium.
Abbreviations: a, antenna; c, presumptive caruncle; dc, dorsal cirrus.

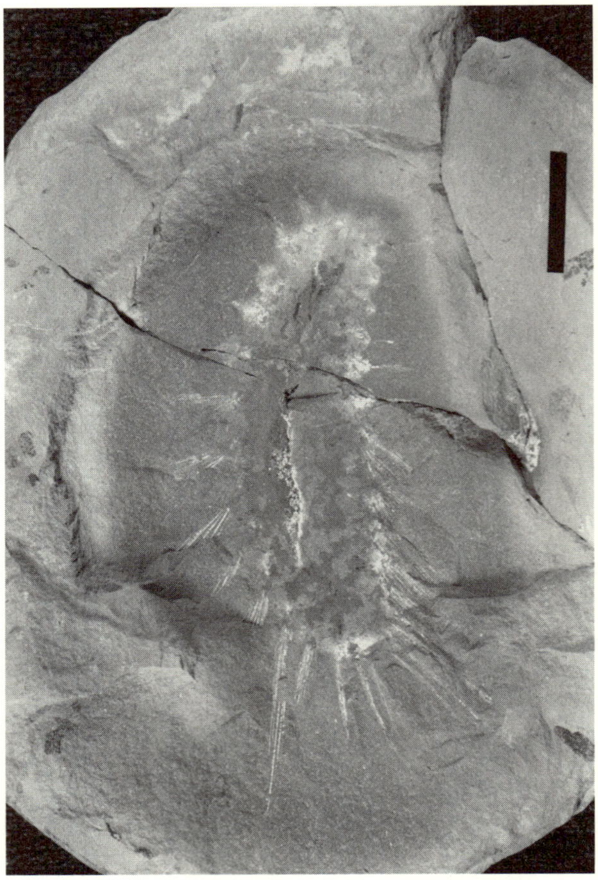

Figure 7A.3. *Rhaphidiophorus hystrix* Thompson, NEIU MCP 608. Scale bar = 10 mm.

Systematic Paleontology

Order Amphinomida *sensu* Fauchald, 1977
Family Amphinomidae Lamarck, 1818

Body elongate and cylindrical, or ovate and flattened. One to three antennae and paired, ventral, cirriform palps present. Caruncle usually present. Prostomium small, usually retracted between anterior segments. Proboscis eversible, jaws absent. Peristomium fused to anterior segments. Parapodia biramous. Noto- and neurosetae simple, calcareous, hollow, in distinct tufts; notosetae often spinose. Aciculae absent. Parapodial branchiae present as branched tufts.

Rhaphidiophorus hystrix Thompson, 1979
Figures 7A.2, 7A.3

Description. Body short, broad; total length 10–35 mm, with 9 to 16 segments (Figures

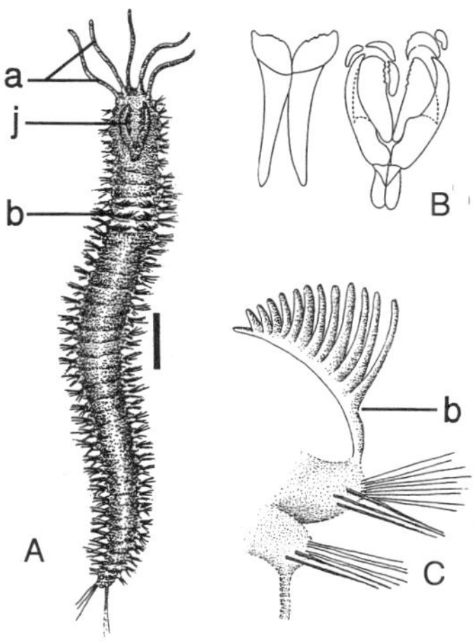

Figure 7A.4. *Esconites zelus,* reconstruction by Donald Auler from several sources.
A Dorsal view. Scale bar = 10 mm.
B Mandibles (*left*) and maxillae (*right*).
C Branchiate parapodium.
Abbreviations: a, antennae; b, branchia; j, jaws.

7A.2A, 7A.3). Anterior end usually poorly preserved; well-preserved specimens with anterior prostomium margin rounded, bilobed; three short antennae present. Anterior parapodial dorsal cirri present in some specimens. Parapodia biramous (Figure 7A.2B). Noto- and neurosetae stout; short in anterior segments, becoming very long in posterior segments; setae hollow, outer surfaces ornamented with fine, longitudinal ridges. Anal cirri not observed.

Remarks. Thompson (1979, pl. 6, fig. 7) described a small, teardrop-shaped depression or elevation in the anterior end of some specimens that she thought might be a caruncle. The caruncle is a sensory lobe behind the prostomium, characteristic in extant amphinomids (e.g., Figures 7A.1C, 7A.2A). In addition to the three "antennae" Thompson (1979, p. 191) also observed in four specimens "[s]omewhat longer projections . . . located below the outline of the prostomium," which she suggested were "peristomial or parapodial cirri." Since extant amphinomids do not have peristomial cirri, it is possible that these structures are palps.

Thompson (1979) hypothesized that *R. hystrix* was an epifaunal scavenger or possibly pelagic. Extant amphinomids are carnivores (Fauchald and Jumars, 1979) and typically epifaunal.

Order Eunicida *sensu* Fauchald, 1977
Family Eunicidae Lamarck, 1818

Body elongate. One to five occipital antennae and paired ventral palps present. Proboscis eversible, jaws present as paired mandibles and maxillae. Peristomium distinct. Notopodia represented only by branchiae and dorsal cirri, notoaciculae sometimes present; neurosetae simple and compound; neuroaciculae present.

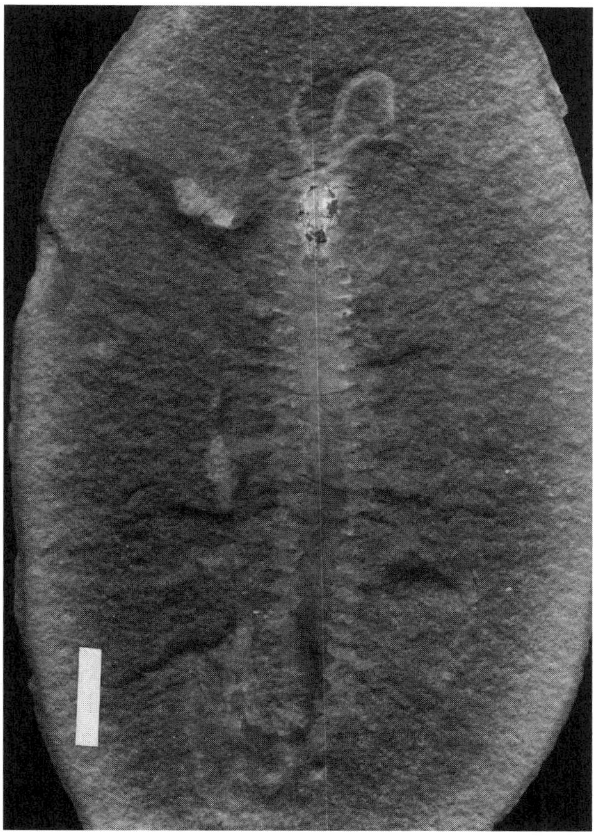

Figure 7A.5. *Esconites zelus* Thompson and Johnson, NEIU MCP 305. Scale bar = 10 mm.

Esconites zelus Thompson and Johnson, 1977
Figures 7A.4, 7A.5

Description. Total body length 3–11 cm, 28 to 80 segments (Figures 7A.4A, 7A.5). Anterior end usually poorly preserved; well-preserved specimens with five prostomial antennae and paired palps. Mandibles well developed, winglike, with short, broad carriers (Figure 7A.4B); maxillae total five on left side, four on right, first maxilla without teeth, all other dentate. Parapodia biramous (Figure 7A.4C); notopodia with two to four aciculae (2 mm in length) and long, slender setae; neuropodia with one to three aciculae and short, slender setae. Pectinate branchiae present on some parapodia (Fig-

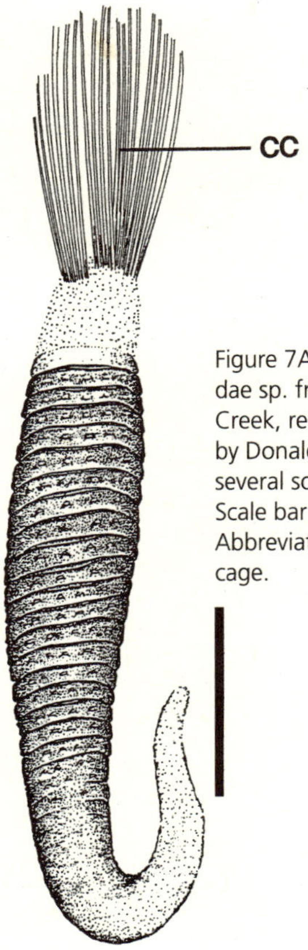

Figure 7A.6. Flabelligeridae sp. from Mazon Creek, reconstruction by Donald Auler from several sources. Scale bar = 10 mm. Abbreviation: cc, cephalic cage.

retractable between anterior segments. Palps present. Jaws absent. Notosetae simple, neurosetae simple or compound; aciculae absent; parapodia poorly developed.

Flabelligeridae sp.
Figures 7A.6, 7A.7

Description. Body anteriorly inflated, tapering posteriorly (Figures 7A.6, 7A.7A). Pro- and peristomium not visible. Number of segments indeterminable. Papillae present over entire body surface in well-preserved specimens. Arrangement of parapodia unknown. Setae of anterior segments very long, slender, directed anteriorly to form "cephalic cage" around anterior end (Figures 7A.6, 7A.7).

Remarks. The combination of distinct body surface papillae and modification of anterior setae into a cephalic cage is very characteristic of some extant flabelligerid genera (cf. Fauchald, 1977). As with extant taxa, the present species was probably a deposit feeder. A formal description of this species is being prepared by Hay.

Order Phyllodocida *sensu* Fauchald, 1977

ure 7A.4A, 7A.4C), exact distribution along body unknown. Pygidium with two long anal cirri.

Remarks. Esconites zelus is easily distinguished by its jaw apparatus. The species is very similar in appearance to such extant euniceans as *Eunice* Cuvier and *Marphysa* Quatrefages. Thompson and Johnson (1977) noted that the mandibles and maxillae are usually preserved close to one another, making it difficult to distinguish individual elements. According to these authors, the outlines of parapodia are only occasionally preserved, but parapodial shapes could be inferred from impressions left by aciculae and setae, though aciculae and setae are not always preserved together. We have not investigated the accuracy of the observation that notosetae are present.

Order Flabelligerida *sensu* Fauchald, 1977
Family Flabelligeridae Saint-Joseph, 1894

Body fusiform or cylindrical. Body surface usually covered by papillae. Pro- and peristomium

Remarks. The Phyllodocida currently contains about 30 extant families, most of which have been traditionally assigned to the old grouping Errantia. Thompson (1979) described nine Phyllodocida species in six families from the Essex fauna (Table 7A.1), of which three species were placed in the superfamily Aphroditacea, family Aphroditidae. The latter taxon, however, requires clarification before reviewing species in this order.

The Aphroditacea, collectively known as scaleworms, are a speciose group, currently with six extant families (Fauchald, 1977; Pettibone, 1982). The term "scaleworm" is derived from the modification of some dorsal cirri into scalelike structures called elytra (Figure 7A.8). Elytra at least partially cover the dorsum of the body and usually alternate along the body with unmodified dorsal cirri. The dorsal surfaces of elytra are smooth, or there can be protuberances called tubercles (Figure 7A.8B), which range in size from microscopic (microtubercles) to large (macrotubercles), with shapes varying

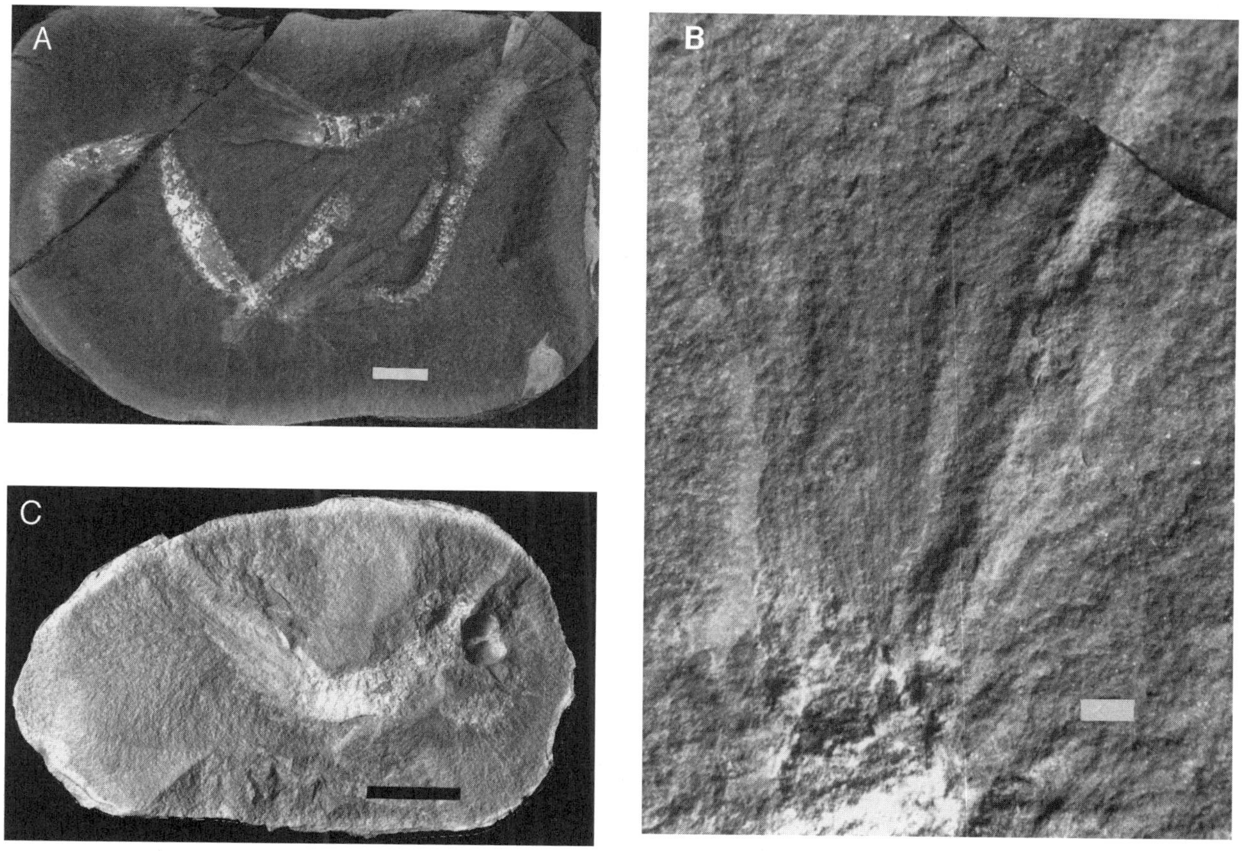

Figure 7A.7. Flabelligeridae sp. **A** Group of entire animals, NEIU MCP 605 (donated by John Prinos). Scale bar = 10 mm. Photograph Andrew Hay. **B** Anterior end of one NEIU MCP 605 specimen, showing cephalic cage. Scale bar = 1 mm. Photograph Andrew Hay. **C** NEIU MCP 607 (donated by Keith Holm). Scale bar = 10 mm.

from low and papilliform to erect and highly ornate.

The aphroditiform species described by Thompson (1979) were *Dryptoscolex matthiesae, Fastuoscolex gemmatus,* and *Hystriciola delicatula,* all of which were assigned to the Aphroditidae *sensu lato,* following Day's (1967) view that all scaleworms should be in a single family, relegating to subfamilies those taxa more traditionally regarded as families (cf. Table 7A.1). To accommodate Mazon Creek species, Thompson considered it necessary to erect the new subfamily Hystriciolinae, defined by the putative absence of elytra. For the purposes of our review, this taxon will be referred to as the "family" Hystriciolidae to be consistent with current systematic arrangements of Aphroditacea families (e.g., Fauchald, 1977; Pet-

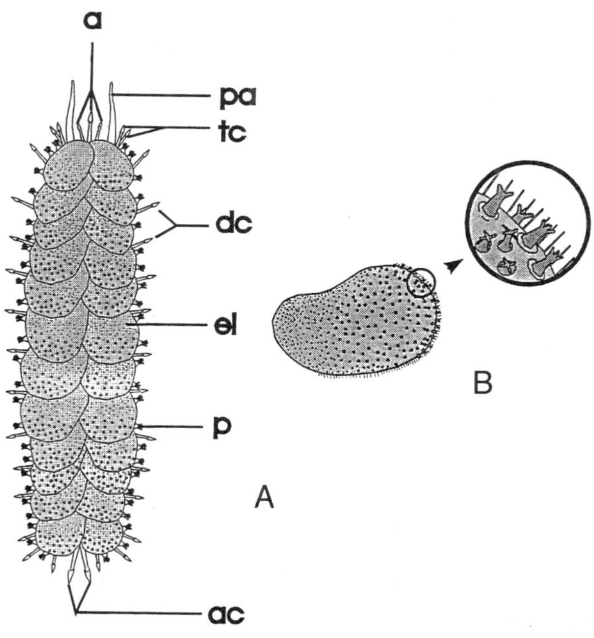

Figure 7A.8. Scaleworm morphology. **A** *Lepidonotus squamatus,* an extant species of the family Polynoidae, dorsal view, modified from Pettibone (1963, fig. 3a). **B** Dorsal view of a polynoid elytron; note distribution of tubercules over surface, modified from Pettibone (1963, fig. 5b). Abbreviations: a, antennae; ac, anal cirri; dc, dorsal cirrus; el, elytron; p, parapodium; pa, palp; tc, tentacular cirri.

tibone, 1982; Table 7A.1). Thompson (1979, p. 175) gave two reasons for establishing the Hystriciolidae. First, the absence of elytra was presumed to be real and not a preservation artifact; the latter condition being "unlikely considering the large number of specimens available for study and the extraordinary preservation of other soft parts, for example the papillae on *Fastuoscolex gemmatus*."

Thompson's interpretation of body surface papillae in *F. gemmatus* is, however, in error. These "papillae" are in fact macrotubercles of elytra. Further, we predict that the "absence" of elytra in *D. matthiesae* and *H. delicatula* is a taphonomic artifact. While the preservation of Mazon Creek polychaetes is certainly exceptional, the chances of discerning fine details of the body surface of scaleworm specimens are very limited. Given the potential for variability in elytra thickness, size, and tubercle development in extant taxa, it comes as little surprise that evidence of elytra has been found only in some (not all) specimens of *F. gemmatus*.

Thompson's (1979, figs. 1–3) reconstructions of *F. gemmatus, H. delicatula,* and *D. matthiesae* (Figures 7A.9, 7A.11, 7A.13, respectively) show dorsal cirri on all segments, a condition that, if correct, would preclude the presence of elytra (cf. Figure 7A.8A). Of the Mazon Creek scaleworms we have examined, none suggest the unequivocal distribution of dorsal cirri implied by Thompson. Indeed, what Thompson interpreted as dorsal cirri might in some cases be artifacts, leaving open the possibility that elytra were, in fact, present.

Thompson's (1979, p. 175) second reason for placing Mazon Creek scaleworms in the Hystriciolidae was that these species "do not fit into any one subfamily [= family] better than into another." This statement is inaccurate, however, since it implies that no distinctions exist between extant Aphroditacea *sensu stricto* families (= Aphroditidae subfamilies *sensu* Day, 1967). Indeed, even if the presence or absence of elytra is discounted, there is little difficulty in placing Mazon Creek scaleworms into extant families on the basis of jaw arrangements and setal forms. Thus, we do not recognize the Hystriciolidae and instead suggest that Mazon Creek scaleworms can readily be placed among at least two extant families: the Polynoidae and Sigalionidae.

Family Polynoidae Kinberg, 1858

Body dorsoventrally flattened, short to long. Prostomium usually with three short antennae. Peristomium fused to first segment. Paired palps long, ventral. Eversible proboscis with four jaws. Dorsal surfaces of elytra smooth or with micro- or macrotubercles. Parapodia biramous; noto- and neurosetae simple, margins often spinose; aciculae present. Dorsal and ventral body surfaces usually smooth; antennae, palps, tentacular cirri, and dorsal cirri sometimes with minute papillae.

Fastuoscolex gemmatus Thompson, 1979
Figures 7A.9, 7A.10

Description. Large-bodied species, total length 27–60 mm; body oval, flattened; mature specimens with 24 segments (Figures 7A.9, 7A.10). Prostomial appendages usually undiscernible; condition of palps unknown. Four jaws present, usually only two visible (Figure 7A.9A); inner jaw margin with a least four teeth and small hook on distal outer margin (Figure 7A.9B); proximal margin broadly rounded, with shallow medial notch. Presence of elytra over most of body indicated by low, conical macrotubercles, especially well developed posteriorly (cf. Thompson, 1979, pl. 2, fig. 5; pl. 3, fig. 1); sizes, shapes, and boundaries of elytra not discernable. Parapodia biramous, noto- and neurosetae present (Figure 7A.9C), setal margins hirsute, spinose, or serrated. Parapodial (presumed dorsal) and anal cirri long.

Remarks. Our placement of *F. gemmatus* in the Polynoidae is based on the presence of simple, serrated, or spinose setae and well-developed macrotubercles, the latter clearly indicating the presence of elytra. The elytra of other aphroditiform taxa lack macrotubercles. As we noted earlier, Thompson (1979) thought *F. gemmatus* lacked elytra and thus interpreted the macrotubercles as papillae distributed over the body surface. Similarly, the presence of elytra necessitates a reinterpretation of Thomp-

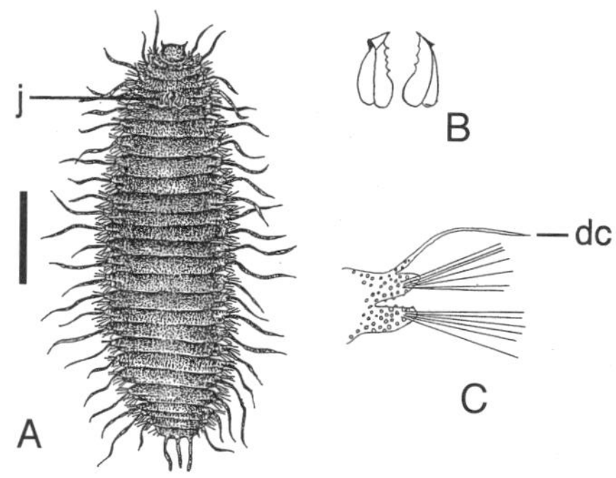

Figure 7A.9. *Fastuoscolex gemmatus,* modified from Thompson (1979, fig. 3). **A** Dorsal view. Scale bar = 10 mm. **B** Two of the four jaws. **C** Parapodium from median segment. Abbreviations: dc, dorsal cirrus; j, jaws.

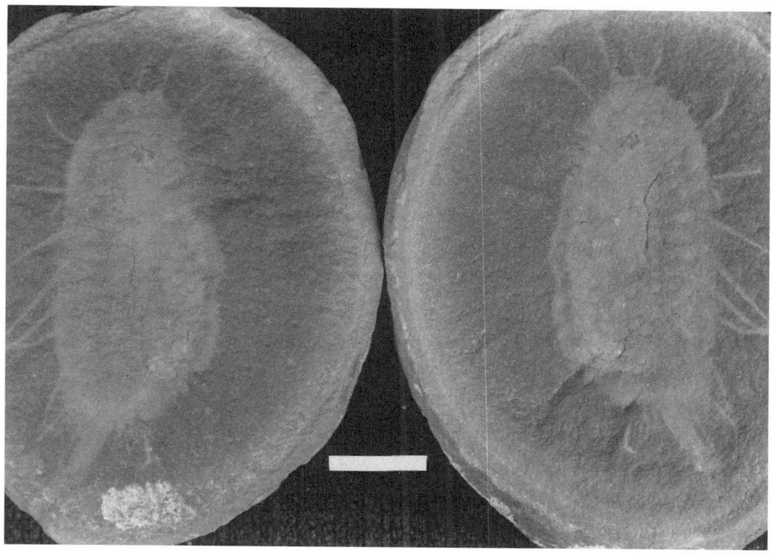

Figure 7A.10. *Fastuoscolex gemmatus* Thompson, NEIU MCP 395. Scale bar = 10 mm.

son's reconstruction of the species (Figure 7A.9) with dorsal cirri along the entire body.

According to Thompson (1979), and based on our own observations, the anterior end is very poorly preserved. Thompson did, however, observe what might be outlines of two frontal antennae on the prostomium. It is also possible that these features represent the palps.

Thompson (1979) hypothesized that *F. gemmatus* was an epifaunal predator, which is consistent with the habits seen in most extant polynoids (Fauchald and Jumars, 1979).

Hystriciola delicatula Thompson, 1979
Figures 7A.11, 7A.12

Description. Small-bodied species, oval in shape (Figures 7A.11A, 7A.12). Prostomium appears to have two short antennae. Eversible proboscis with four roughly triangular jaws (only two usually visible; Figures 7A.11B, 7A.12); proximal jaw margin deeply incised, forming two distinctly rounded lobes, about five teeth along inner lateral margin and blunt distal spur. Elytra or macrotubercles not observed. Parapodia biramous (Figure 7A.11C), one acicula present in each ramus. Noto- and neurosetae simple, slender; fine structures unknown. Long parapodial (presumed dorsal) and one pair of anal cirri present.

Remarks. As with *Fastuoscolex gemmatus, H. delicatula* appears to be a member of the Polynoidae. It is conceivable, however, that *H. delicatula* could be placed in one of several other aphroditiform families that have well-developed jaws and only simple neurosetae, for example, the Acoetidae (= Polyodontidae) or Eulepethidae.

As mentioned above for *Fastuoscolex gemmatus,* Thompson's (1979) description of a pair of prostomial antennae might actually be remnants of the palps. Thompson considered *H. delicatula* to have been an epifaunal predator.

Family Sigalionidae Kinberg, 1858

Body quadrangular or dorsoventrally flattened, usually long, slender. Prostomium with one to three antennae. Peristomium fused to first segment. Paired ventral palps long, slender. Eversible proboscis with four jaws. Dorsal surfaces of elytra smooth or with microtubercles. Parapodia biramous; notosetae simple, neurosetae compound; aciculae present. Papillae sometimes present on parapodia, remainder of body surface smooth.

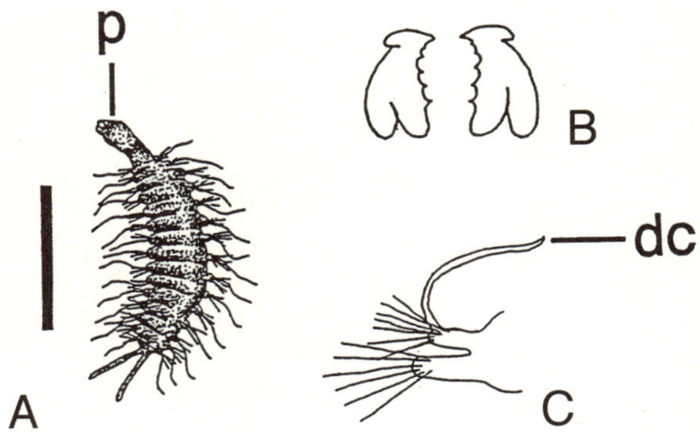

Figure 7A.11. *Hystriciola delicatula,* modified from Thompson (1979). **A** Dorsal view, note everted proboscis. Scale bar = 10 mm. **B** Two of the four jaws. **C** Parapodium. Abbreviations: dc, dorsal cirrus; p, proboscis

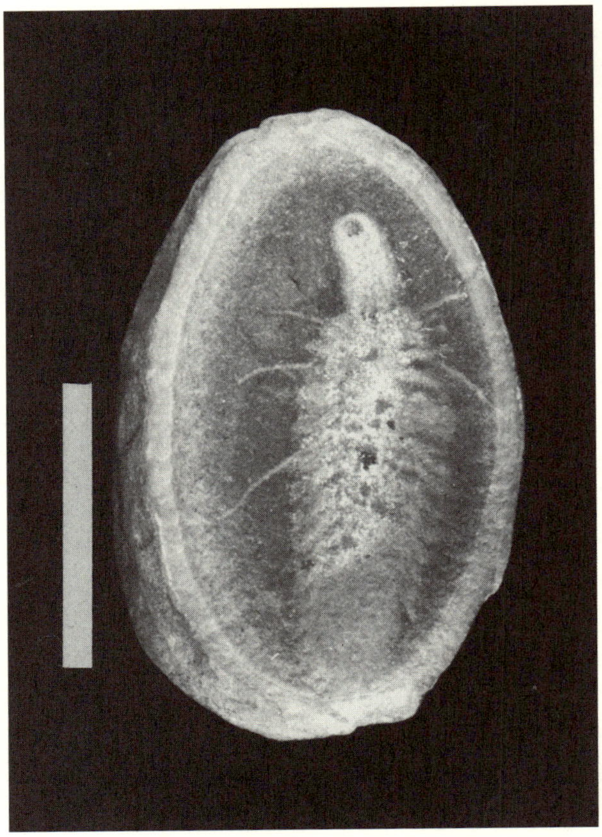

Figure 7A.12. *Hystriciola delicatula* Thompson, NEIU MCP 157. Scale bar = 10 mm.

Dryptoscolex matthiesae Thompson, 1979
Figures 7A.13, 7A.14

Description. Elongate, large-bodied species (Figure 7A.14), total length 2–10 cm, 30 to 52 segments. Prostomium large, rounded, with two antennae (Figure 7A.13A). Proboscis eversible; four stout, triangular-shaped jaws present (usually only two visible; Figure 7A.13B), each with six to eight teeth along inner margin and hook near distal outer margin, proximal margin with two equal-sized lobes. In specimens with proboscis inverted (typical condition), jaws 2–4 mm from anterior end. Parapodia biramous (Figure 7A.13C); noto- and neurosetae long, fine, in dense bundles; notosetae simple, presumed compound neurosetae observed in several specimens. Aciculae rarely preserved. Some specimens with parapodial (presumed dorsal) and two anal cirri.

Remarks. Our placement of *D. matthiesae* in the Sigalionidae is tentative, owing to the questionable presence of compound neurosetae observed in two specimens by Thompson (1979). None of the specimens we have examined provide evidence that neurosetae were compound. The most substantial hindrance is the coarse-grained nature of the rock matrix, which easily obscures any potential sign of an articulation.

The only other aphroditiform family with compound neurosetae is the Pholoidae Kinberg. Extant members of this family and sigalionids tend, however, to have distinct differences in body dimensions. Pholoidids are short and ovate, whereas sigalionids are usually elongate and slender. The body shape of *D. matthiesae* accords well with the general dimensions of sigalionids. Thompson's (1979) interpretation of the prostomium as bearing a pair of antennae should also be considered tentative; the palps are also located in this area and are usually considerably larger than antennae.

Thompson (1979) suggested that *D. matthiesae* might have been an epifaunal predator.

Family Nereididae Johnston, 1845

Body elongate. Prostomium with two antennae and paired ventral palps. Proboscis eversible, one pair of jaws present. Peristomium fused with anterior segments; two or four pairs of tentacular cirri present. Parapodia usually biramous, with dorsal and ventral cirri; notoacicu-

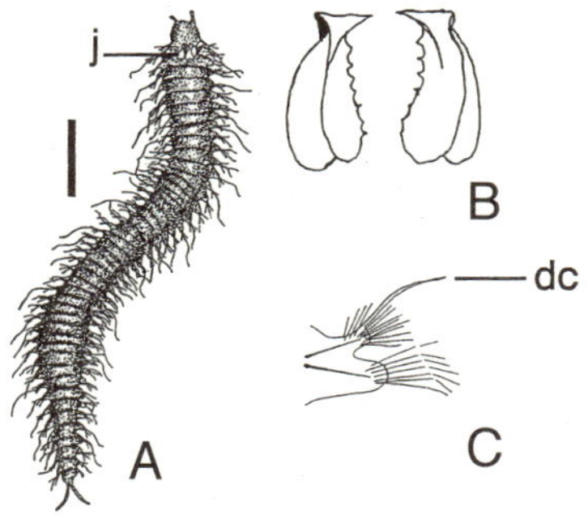

Figure 7A.13. *Dryptoscolex matthiesae,* modified from Thompson (1979, fig. 2). A Dorsal view. Scale bar = 10 mm. B Two of the four jaws. C Parapodium. Abbreviations: dc, dorsal cirrus; j, jaws.

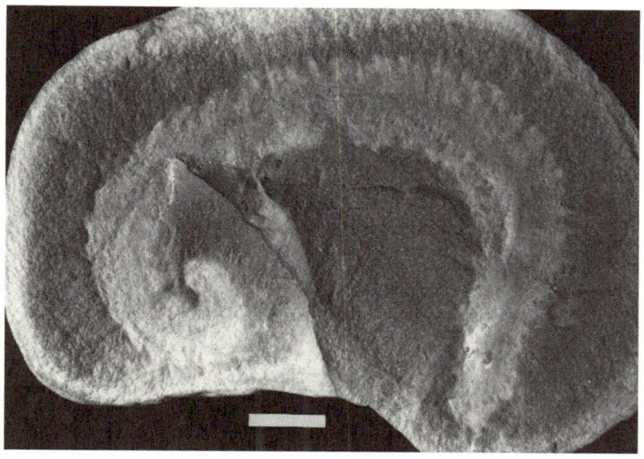

Figure 7A.14. *Dryptoscolex matthiesae* Thompson, NEIU MCP 91. Scale bar = 10 mm.

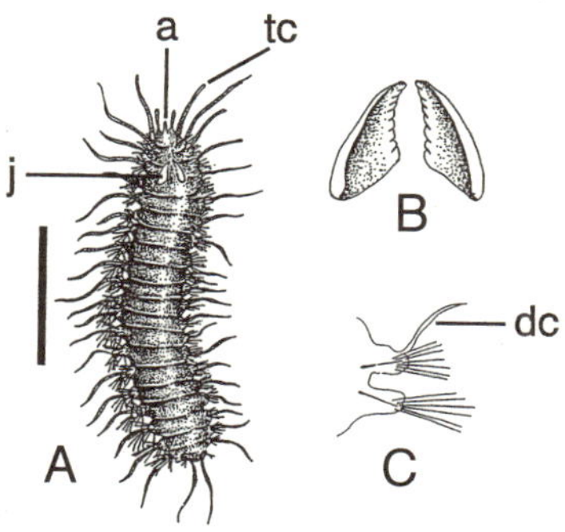

Figure 7A.15. *Fossundecima konecnyi,* modified from Thompson (1979, fig. 9). A Dorsal view. Scale bar = 10 mm. B Jaws. C Parapodium. Abbreviations: a, antennae; dc, dorsal cirrus; j, jaws; tc, tentacular cirrus.

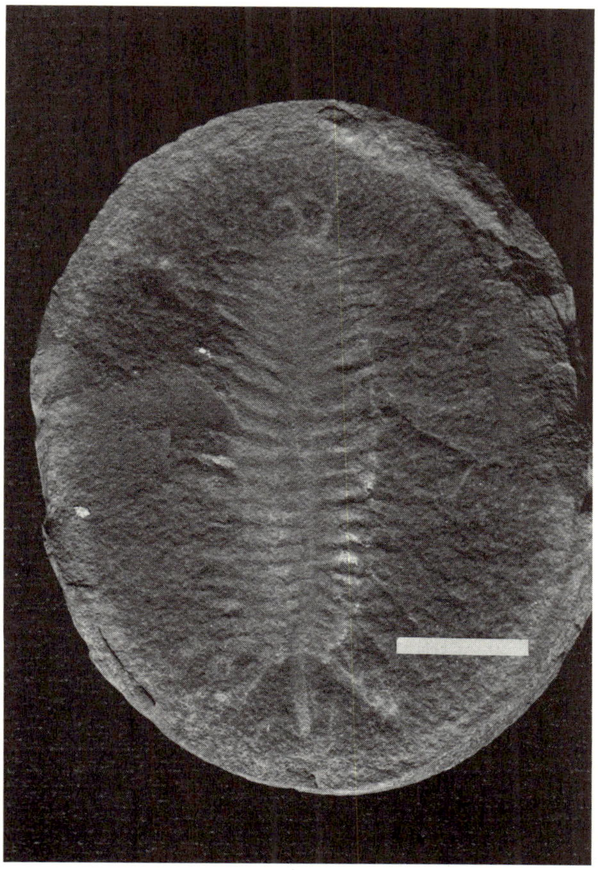

Figure 7A.16. *Fossundecima konecnyi* Thompson, NEIU MCP 203. Scale bar = 10 mm.

lae usually present, neuroaciculae present; noto- and neurosetae usually compound.

Fossundecima konecniorum Thompson, 1979
Figures 7A.15, 7A.16

Description. Medium-sized, broad-bodied species, total length 9–46 mm, with 12 to 31 segments. Prostomium with two or three short antennae; peristomium with two pairs of tentacular cirri. Proboscis eversible but usually inverted (Figures 7A.15A, 7A.16). One pair of large, triangular jaws present, each with about 10 blunt teeth along inner margin (Figures 7A.15B, 7A.16). Parapodia biramous (Figure 7A.15C); long cirri (presumed dorsal) on first few segments, becoming shorter on median and posterior segments. Noto- and neurosetae stout, forming short, fanlike bundles. Pygidium with two long anal cirri.

Remarks. In the original description, Thompson (1979) tentatively placed *F. konecniorum* in its own family, the Fossundecimidae, though she did suggest that the species might be a member of either the Aphroditidae subfamily Hystriciolinae (discussed above) or the Nereididae. Membership in the Hystriciolinae was dis-

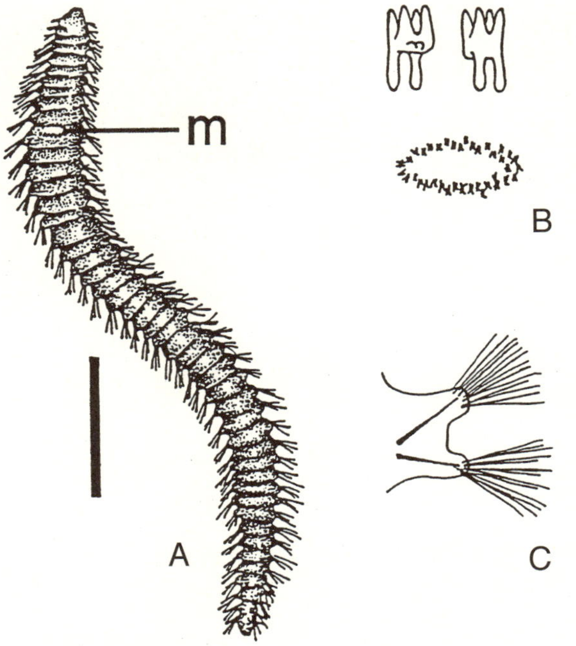

Figure 7A.17. *Pieckonia helenae,* modified from Thompson (1979, fig. 8). A Entire worm. Scale bar = 10 mm. B Individual micrognaths (*top*) and orientation on proboscis (*bottom*). C Parapodium. Abbreviations: m, micrognaths.

Figure 7A.18. *Pieckonia helenae* Thompson, NEIU MCP 376. Scale bar = 10 mm.

counted since there were two rather than four jaws, as well as differences in jaw shape. Thompson (1979, p. 190) then argued that while *F. konecniorum* might be closely related to the Nereididae, the jaws of this species "are considerably broader and flatter." Based on Thompson's illustrations and our own observations, we find no outstanding differences between the jaws of *F. konecniorum* and those of extant nereidids. As well, the general arrangement of what appear to be tentacular cirri and antennae strongly suggests that this species is a member of the Nereididae. The most striking difference between most extant nereidids and *F. konecniorum* is the latter species' short, very robust body and small number of segments.

Assuming *F. konecniorum* was a nereidid, we predict that at least the neurosetae were compound as opposed to simple as described by Thompson (1979). As we noted in the remarks for *Dryptoscolex matthiesae,* the inability to discern articulations in setae might be a taphonomic artifact.

Thompson (1979) hypothesized that *F. konecniorum* was an epifaunal predator. While most extant nereidids are thought to be carnivorous (Fauchald and Jumars, 1979), it is not at all uncommon to find them as members of the infauna.

Family Goniadidae Kinberg, 1866

Body long, slender. Prostomium conical, with two pairs of antennae. Peristomium distinct. Palps and peristomial cirri absent. Proboscis eversible, with terminal circlet of small (micrognaths) and large (macrognaths) jaw pieces. Parapodia anteriorly uniramous, posteriorly biramous. Notosetae simple, neurosetae compound; aciculae present.

Pieckonia helenae Thompson, 1979
Figures 7A.17, 7A.18

Description. Body long, narrow; total length 22–75 mm, with 24 to 64 segments (Figures 7A.17A, 7A.18). Anterior end slightly tapered,

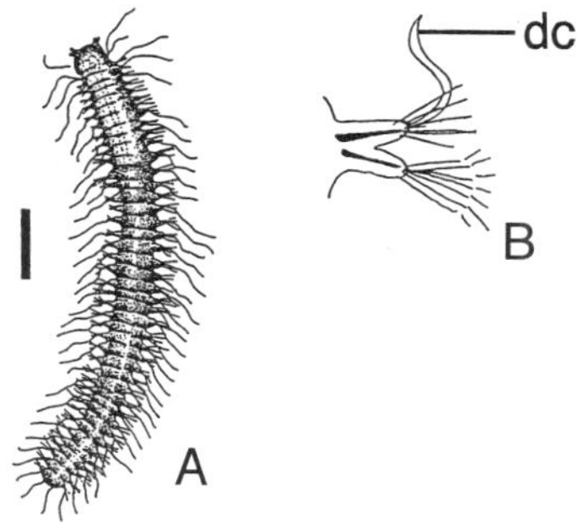

Figure 7A.19. *Rutellifrons wolfforum,* modified from Thompson (1979, fig. 5). A Dorsal view. Scale bar = 10 mm. B Parapodium. Abbreviation: dc, dorsal cirrus.

antennae not observed. Proboscis usually preserved inverted, with circlet of chair-shaped micrognath jaws (Figure 7A.17B). Parapodia biramous, with bundles of slender noto- and neurosetae (at least 20 setae per bundle); parapodial cirri not observed (Figure 7A.17C).

Remarks. Thompson (1979) apparently followed Day (1967) in placing *P. helenae* in the family Glyceridae Grube and the subfamily Goniadinae. Most current classifications (e.g., Fauchald, 1977), however, recognize the Glyceridae subfamilies Glyceridae and Goniadinae as separate families. This latter arrangement is followed here (see Table 7A.1).

Extant goniadids have four very small antennae at the distal end of a distinctly conical prostomium. Antennae were probably present in *P. helenae* but were not preserved well enough to be observed. Thompson (1979) described the prostomium as being unusually short, a condition which might be a preservation artifact. Similarly, we have not confirmed Thompson's observation that all parapodia are biramous, which is not the case in extant goniadids. *Pieckonia helenae* appears to especially differ from extant goniadids in that macrognath jaws are absent.

Thompson (1979) considered *P. helenae* to be a rapid burrower, citing as evidence several specimens preserved oblique to the bedding plane. Based on extant taxa, she suggested that it was also a predator.

Family Hesionidae Grube, 1850

Body short. Two or three antennae usually present. Ventral palps usually present. Peristomium fused with anterior segments; two to eight pairs of tentacular cirri. One pair of jaws sometimes present. Parapodia uni- or biramous; aciculae present; dorsal cirri long, slender. Notosetae, if present, simple; neurosetae compound.

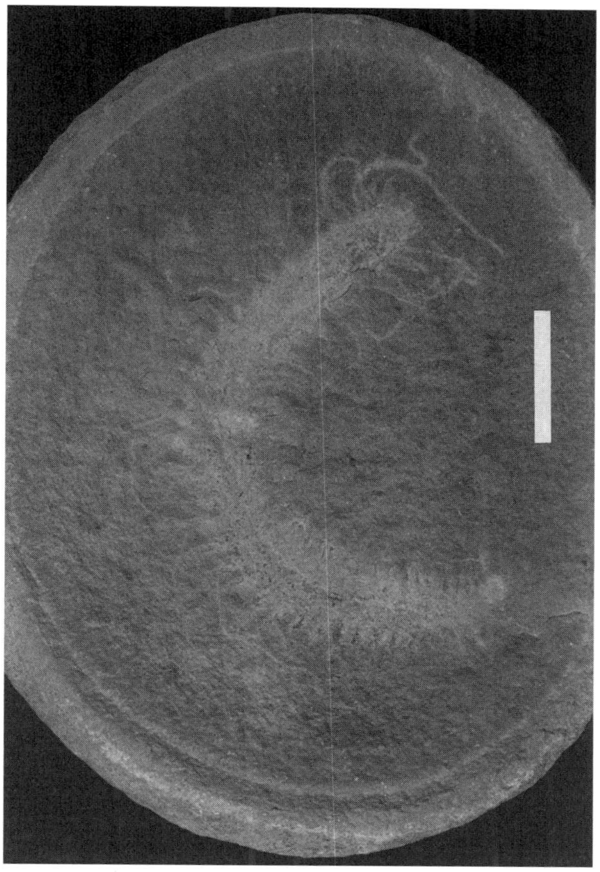

Figure 7A.20. *Rutellifrons wolfforum* Thompson, NEIU MCP 375. Scale bar = 10 mm.

Rutellifrons wolfforum Thompson, 1979
Figures 7A.19, 7A.20

Description. Body large, stout, with 34 to 40 segments; total body length 47–75 mm (Figures 7A.19A, 7A.20). Anterior prostomium margin blunt, rounded; demarcation between prostomium and tentacular region indistinct. Prostomial appendages or palps not clearly visible. Presumed tentacular cirri present on some specimens, total number unknown. Proboscis eversible, no jaws observed. Parapodia biramous, with cirri (presumed dorsal) in some specimens; notosetae short, simple; neurosetae longer, compound (Figure 7A.19B). One pair of anal cirri sometimes present.

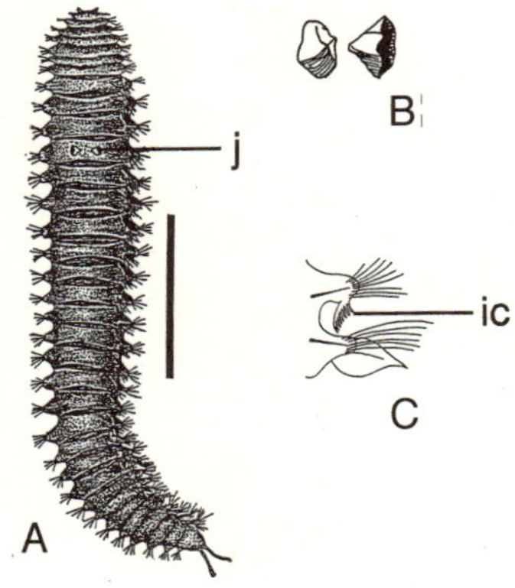

Figure 7A.21. *Astreptoscolex anasillosus,* modified from Thompson (1979, fig. 6). A Entire worm. Scale bar = 10 mm. B Jaws. C Median parapodium. Abbreviations: ic, interramal cirrus; j, jaw.

Figure 7A.22. *Astreptoscolex anasillosus* Thompson, NEIU MCP 433. Scale bar = 10 mm.

Remarks. As a group, the Hesionidae is a poorly defined family. We agree with Thompson's (1979) identification of *R. wolfforum* as a hesionid by the combination of an eversible pharynx, presence of long tentacular and parapodial cirri, and presence of simple notosetae and compound neurosetae. *Rutellifrons wolfforum* might have had antennae and palps, both of which tend to be small in hesionids; antennae are often easily detached. Thompson (1979) speculated that *R. wolfforum* was an epifaunal predator.

Family Nephtyidae Grube, 1850

Body long, slender. Two pairs of small antennae. Peristomium fused to anterior segments. Eversible proboscis with one pair of small, subterminal jaws. Parapodia biramous, well developed; aciculae present; noto- and neurosetae simple; interramal respiratory cirri usually present.

Astreptoscolex anasillosus Thompson, 1979
Figures 7A.21, 7A.22

Description. Body medium sized, with 29 to 42 segments; total length 16–55 mm (Figures 7A.21A, 7A.22). Anterior end blunt, rounded; posterior end slightly tapered. Prostomium small, poorly preserved; no appendages observed. Proboscis sometimes everted, two small conical jaws present at about midlength (Figure 7A.21A, 7A.21B); in specimens with proboscis inverted, jaws lie between segments 8 and 13. Bases of jaws deeply concave (Figure 7A.21B). Parapodia biramous, with lamellae in anterior segments (Figure 7A.21C); notosetae very slender, at least 30 per fascicle, shorter than neurosetae; neurosetae 2–4 mm long. Noto- and neuropodial aciculae present, though not always preserved. Interramal cirri sometimes present (Figure 7A.21C).

Remarks. The presence of a pair of small, conical jaws medial on the proboscis, and interramal cirri are characteristics of extant nephtyids. As well, nephtyids have two pairs of very small antennae along the anterior margin of the quadrangular-shaped prostomium. These structures probably have not been preserved well enough to be seen in *A. anasillosus.*

Although extant nephtyids are usually considered motile, epi- or infaunal predators, feeding on small invertebrates, some species are deposit feeders (Fauchald and Jumars, 1979). Thompson (1979) believed *A. anasillosus* to have been an epifaunal carnivore or omnivore since no specimens observed contained sediment in their guts.

Didontogaster cordylina Thompson, 1979
Figures 7A.23, 7A.24

Description. Medium-sized species with 29 to 68 segments; total length 12–87 mm. Anterior end slightly inflated, tapering to narrow posterior end (Figures 7A.23A, 7A.24). Prostomium anteriorly rounded, no appendages observed. Proboscis present, inverted in most specimens. Two conical jaws present (Figure 7A.23A, 7A.23B); each jaw base with three shal-

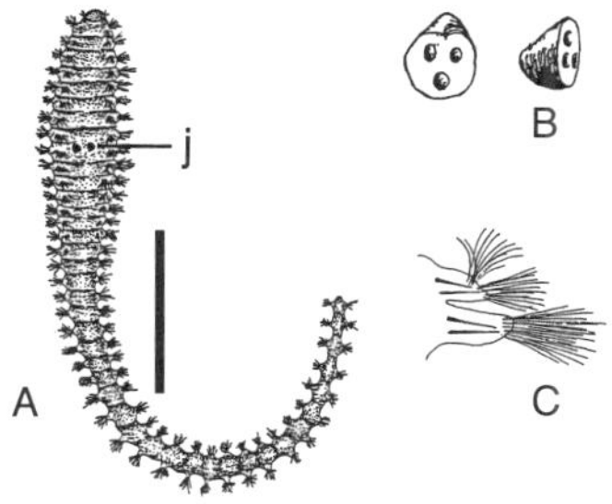

Figure 7A.23. *Didontogaster cordylina,* modified from Thompson (1979, fig. 7).
A Entire worm. Scale bar = 10 mm.
B Jaws.
C Anterior parapodium.
Abbreviation: j, jaw.

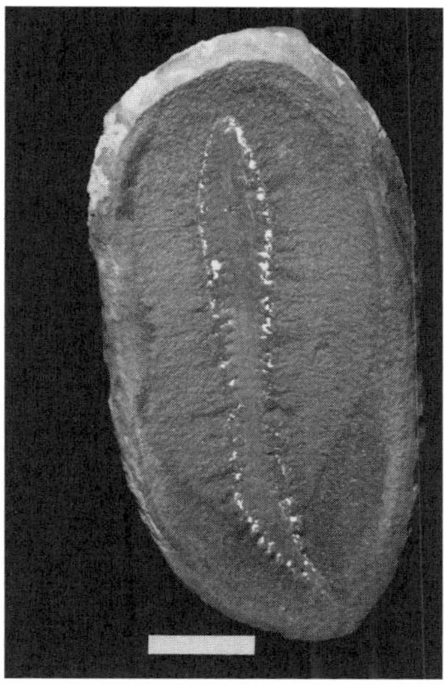

Figure 7A.24. *Didontogaster cordylina* Thompson, NEIU MCP 354. Scale bar = 10 mm.

low pits. Parapodia biramous; each ramus with two aciculae (Figure 7A.23C). Setal fascicles with short (less than 2 mm) and long (2 mm) setae; notopodia of well-preserved specimens with two distinct setal fascicles, one oriented dorsoventral to the other. Interramal and anal cirri not observed.

Remarks. This is the most abundant polychaete species in the Mazon Creek fauna. Thompson (1979) speculated that the three shallow pits located in the base of each jaw served as points of muscle attachment to the proboscis. Based on positions of some specimens relative to bedding planes, Thompson (1979) suggested that *D. cordylina* was a burrowing omnivore.

Family Phyllodocidae Williams, 1851

Body long, slender. Prostomium with four to five antennae; eyes sometimes present. Peristomium fused with anterior segments; two to four pairs of tentacular cirri. Palps absent. Eversible proboscis without jaws. Parapodia biramous; notopodia reduced, represented by foliose dorsal cirri; notoaciculae and notosetae usually absent. Neurosetae compound; notosetae, when present, simple.

Levisettius campylonectus Thompson, 1979
Figures 7A.25, 7A.26

Description. Body long, slender; segments narrow, numerous. Total length 48–93 mm. Anterior prostomium margin blunt, rounded (Figures 7A.25A, 7A.26); well-preserved specimens with eyes and what appear to be short prostomial antennae. Presumed tentacular cirri present in some specimens, total number unknown. Proboscis sometimes everted, jaws absent. Parapodia biramous; noto- and neurosetal fascicles with very thin, long setae; foliose dorsal cirri occasionally evident (Figure 7A.25B).

Remarks. Specimens of *L. campylonectus* are found infrequently. Thompson (1979) tentatively placed this species in the Phyllodocidae on the basis of general body shape, presence of eyes, tentacular cirri, and leaflike dorsal cirri. If *L. campylonectus* is a phyllodocid, we predict

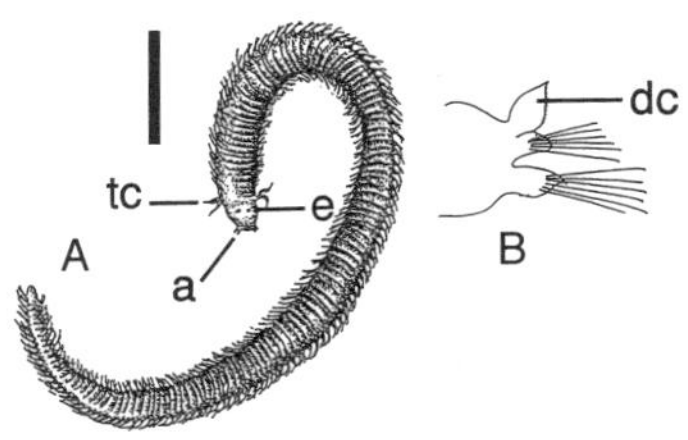

Figure 7A.25. *Levisettius campylonectus,* modified from Thompson (1979, fig. 4).
A Dorsal view. Scale bar = 10 mm.
B Parapodium. Abbreviation: dc, dorsal cirrus.

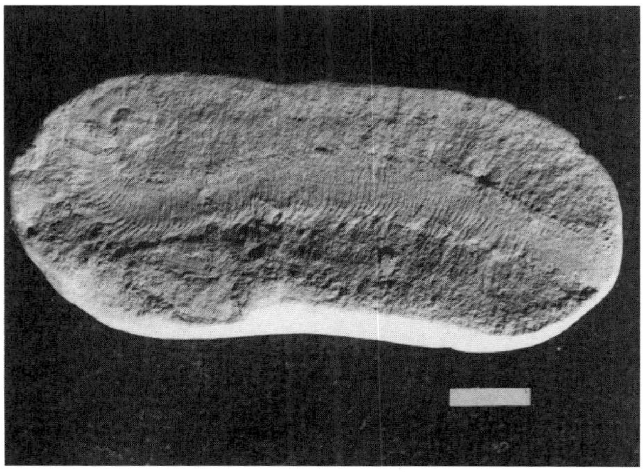

Figure 7A.26. *Levisettius campylonectus* Thompson, FMNH MCP 611. Scale bar = 10 mm.

Figure 7A.27. Extant *Spirorbis* sp., modified from several sources. Note branchial crown extending out of tube opening. Scale bar = 1 mm.

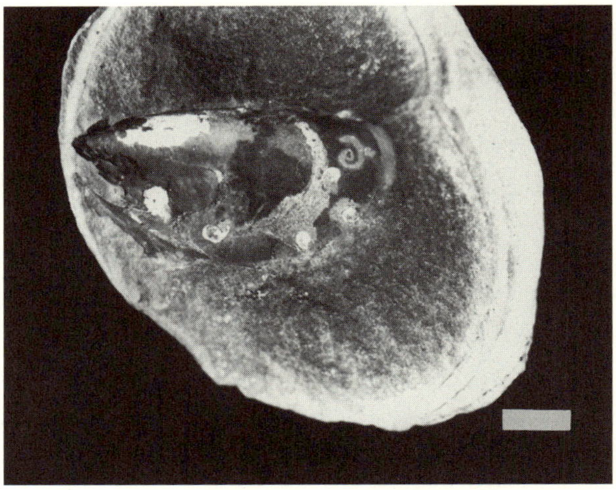

Figure 7A.28. Fossil tubes of *Spirorbis* sp., on *Septimyalina* sp., NEIU MCP 195. Scale bar = 5 mm.

that better preserved specimens would display at least two pairs of prostomial antennae, as implied by Thompson (1979), as well as aciculae and compound neurosetae. Based on extant phyllodocids, Thompson (1979) hypothesized that *L. campylonectus* was an active predator.

Order Sabellida *sensu* Fitzhugh, 1989
Family Serpulidae Rafinesque, 1815

Prostomium reduced, fused to peristomium. Palps modified into branchial crown. Proboscis and jaws absent. Parapodia biramous, very reduced; aciculae absent; noto- and neurosetae simple. Calcareous tube secreted by body wall; tubes straight, irregular, or tightly coiled.

Spirorbis sp.
Figures 7A.27, 7A.28

Remarks. Spirorbids are very small-bodied members of a polychaete group collectively known as feather-duster worms. Systematically, spirorbids have usually been regarded as a subfamily of the Serpulidae or separated into their own family, the Spirorbidae, as suggested by Pillai (1970). Fitzhugh (1989) reviewed problems with the placement of spirorbids in their own family, pointing out the benefits of maintaining these taxa in the Serpulidae.

One of the defining characters of the Serpulidae is the ability of its members to secrete a calcareous tube, a feature only seen in several other, disparate polychaete species. As a consequence, serpulids are generally well represented as fossils. Typically, serpulid tubes are attached to hard substrates, such as rocks or shells. In addition to these materials, spirorbids are also commonly found on algae and sea grasses. Serpulids are filter feeders, extending the ciliated branchial crown out of the tube and into the water (Figure 7A.27). Spirorbid tubes are very distinct in that they are tightly coiled (Figures 7A.27, 7A.28), either sinistrally or dextrally. This is in contrast to other serpulids, which produce straight or irregular tubes. Distinguishing fossil spirorbid species depends entirely on external tube sculpturing or ornamentation.

Dawson (1881) described *S. carbonarius,* collected from the Mazon River. The specimen was, however, so poorly preserved that Dawson regarded his identification as tentative. Thompson (1979) stated that Mazon Creek spirorbids are *S. pusillus* (Martin). Unfortunately, neither Dawson nor Thompson conducted a comprehensive study of spirorbids, leaving open the identification of Mazon Creek specimens.

Order incertae sedis
Family incertae sedis
Palaeocampa anthrax Meek and Worthen, 1865
Figures 7A.29, 7A.30

Description. All specimens incomplete; total body length 27–43 mm, rectangular in cross section, with 10 to 11 segments (Figures 7A.29A, 7A.29C, 7A.30A). Prostomium rarely preserved, appears to be partially retracted into anterior end; anterior prostomial margin of some specimens with what appear to be two short anten-

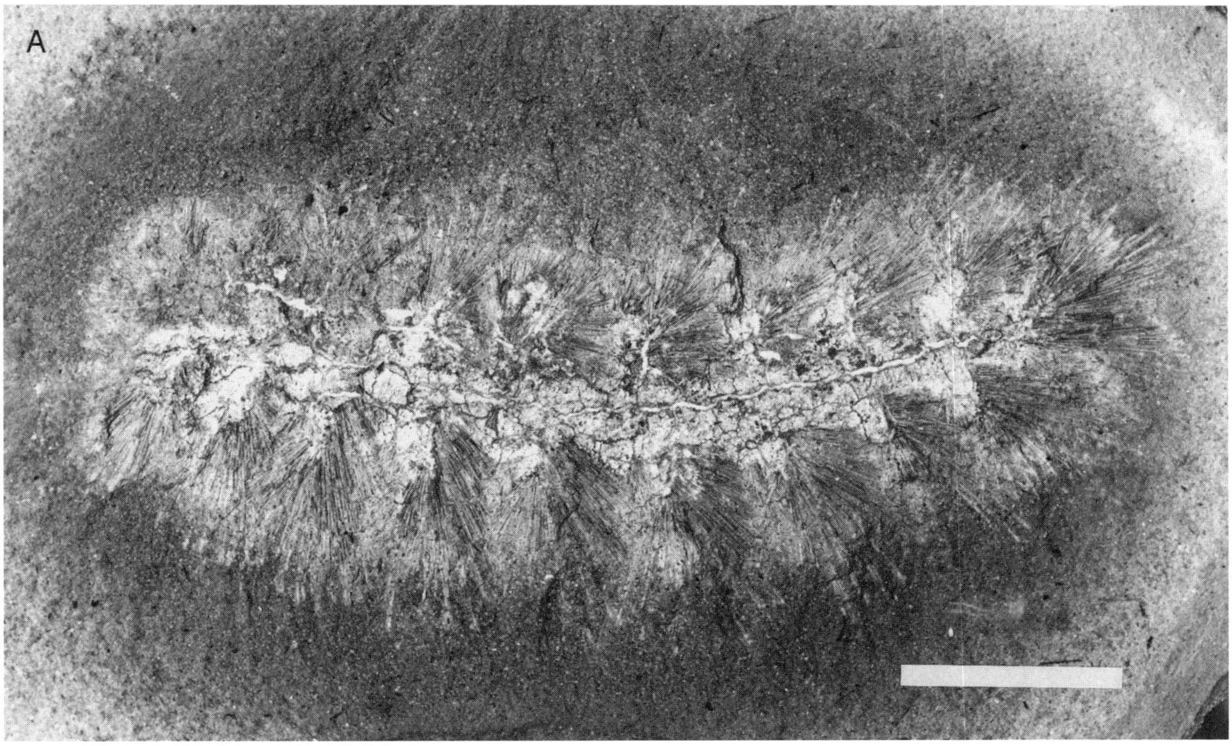

Figure 7A.30. *Palaeocampa anthrax* Meek and Worthen, USNM 38032. **A** Entire animal, FMNH MCP 549. Scale bar = 10 mm. **B** SEM micrograph of setae, FMNH PE 29384; setae are 120–130 μm (0.0048–0.0052 in.) in width. Photograph Mike Henderson.

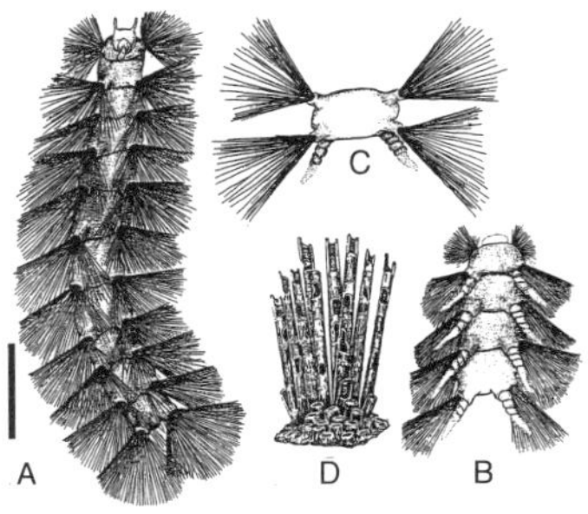

Figure 7A.29. *Palaeocampa anthrax* Meek and Worthen, reconstruction by Donald Auler. **A** Dorsal view. Scale bar = 10 mm. **B** Median segments, ventral view. **C** Segment in anterior view. **D** Setal fascicle drawn in situ; note that the setae are hollow.

nae or palps. Caruncle-like structure (median antenna?) present at posterior prostomium margin (Figure 7A.29A). Eversible proboscis not observed, jaws absent. Parapodia biramous (Figure 7A.29C), aciculae absent. Notopodia situated along dorsolateral surface (Figure 7A.29A), spirally twisted cirri (presumed ventral) sometimes preserved in association with neuropodia (Figure 7A.29B, 7A.29C). Noto- and neurosetae very stout, about 1.5 times longer than body width, hollow and appear calcareous (Figures 7A.29D, 7A.30B); setal surfaces ornamented with fine, longitudinal ridges.

Remarks. Palaeocampa anthrax, originally described as a caterpillar by Meek and Worthen (1865), then considered a myriapod by Scudder (1882, 1884), is now believed to be a polychaete. While the reconstructions in Figure 7A.29 are still preliminary, the species appears to be a member of the family Amphinomidae, to which *Rhaphidiophorus hystrix* Thompson belongs (see above). In fact, Rolfe et al. (1982) regarded *Palaeocampa* as a senior synonym of *Rhaphidiophorus* Thompson. A formal redescription of *P. anthrax* is being prepared by Henderson.

References

Audouin, J. V., and H. Milne Edwards. 1834. Recherches pour servir à l'histoire naturelle du littoral de la France, ou receuil de mémoires sur l'anatomie, la physiologie, la classification et les moeurs des animaux de nos côtes; ouvrage accompagné des planches faites de après nature. 2, Annélides, pt. 1. Crochard Libraire, Paris, 290 pp.

Briggs, D. E. G., and A. J. Kear. 1993. Decay and preservation of polychaetes: taphonomic thresholds in soft-bodied organisms. Paleobiology, 19:107–135.

Brusca, R. C., and G. J. Brusca. 1990. Invertebrates. Sinauer, Sunderland, Massachusetts.

Butterfield, N.J. 1990. A reassessment of the enigmatic Burgess Shale fossil *Wiwaxia corrugata* (Matthew) and its relationship to the polychaete *Canadia spinosa* Walcott. Paleobiology, 16:287–303.

Clark, R. B. 1969. Systematics and phylogeny: Annelida, Echiura, Sipuncula, pp. 1–68. *In* M. Florkin and B. Scheer (eds.), Chemical Zoology, vol. 4. Academic Press, New York.

Colbath, G. K. 1986. Jaw mineralogy in eunicean polychaetes. Micropaleontology, 32:186–189.

———. 1989. Taphonomy of recent polychaete jaws from Florida and Belize. Micropaleontology, 34:83–89.

Conway Morris, S. 1979. Middle Cambrian polychaetes from the Burgess Shale of British Columbia. Phil. Trans. R. Soc. London B, 285:227–274.

Conway Morris, S., R. K. Pickerill, and T. L. Harland. 1982. A possible annelid from the Trenton Limestone (Ordovician) of Quebec, with a review of oligochaetes and other annulate worms. Canadian Jour. Earth Sci., 19:2150–2157.

Dales, R. P. 1962. The polychaete stomodeum and the interrelationships of the families of Polychaeta. Proc. Zool. Soc. London, 139:389–428.

Dawson, J. W. 1881. Note on *Spirorbis* contained in an ironstone nodule from Mazon Creek, with millipede. Proc. Boston Soc. Nat. Hist., 21:157–158.

Day, J. H. 1967. A monograph on the Polychaeta of southern Africa. Part 1, Errantia. British Museum (Natural History), London.

Fauchald, K. 1974. Polychaete phylogeny: a problem in protostome evolution. Syst. Zool. 23:493–506.

———. 1977. The polychaete worms: definitions and keys to the orders, families, and genera. Nat. Hist. Mus. Los Angeles County Sci. Ser., 28:1–188.

———. 1992. A review of the genus *Eunice* (Polychaeta: Eunicidae) based upon type material. Smithsonian Contrib. Zool., 523, 422 pp.

Fauchald, K., and P. Jumars. 1979. The diet of worms: a study of polychaete feeding guilds. Oceanog. Marine Biol. Ann. Rev., 17:193–284.

Fitzhugh, K. 1989. A systematic revision of the Sabellidae-Caobangiidae-Sabellongidae complex (Annelida: Polychaeta). Bull. Amer. Mus. Nat. Hist., 192, 104 pp.

George, J. D., and G. Hartmann-Schröder. 1985. Polychaetes: British Amphinomida, Spintherida and Eunicida. *In* D. M. Kermack and R. S. K. Barnes (eds.), Synopses of the British Fauna (New Series), 32, 221 pp.

Grube, A. E. 1850. Die Familien der Anneliden. Arch. Naturgesch. Berlin, 16:249–364.

Howell, B. F. 1962. Worms, pp. W144–W177. *In* R. C. Moore (ed.), Treatise on Invertebrate Paleontology. Part W, Miscellanea. Geological Society of America and University of Kansas Press, Lawrence.

Kielan-Jawarowska, Z. 1966. Polychaete jaw apparatuses from the Ordovician and Silurian of Poland and a comparison with modern forms. Palaeo. Polonica, 16, 152 pp.

Kinberg, J. G. H. 1858. Kongliga Svenska Fregatten *Eugenies* Resa omking jorden under befäl af C. A. Virgin åren 1851–1853. Vetenskapliga Iakttagelser på Konung Oscar den Förstes befallning utgifna af K. Svenska Vetenskapsakademien. Andra Delen. Zool. 3rd Ann. Almqvist & Wiksells, Uppsala and Stockholm, 32 pp.

———. 1866. Annulata nova. Öfv. Svenska Vetensk. Akad. Förh., 22:239–258.

Lamarck, J.-B. 1818. Histoire naturelle des animaux sans vertèbres, vol. 5. Deterville, Verdiere, Paris.

Linnaeus, C. 1758. Systema naturae per règne tria naturae, secundum classes, ordines, genera, species, cum characteribus, differentiis, synonymis locis, vol. 1, 10th ed., rev. Laurentii Salvii, Stockholm, 823 pp. (Facsimile copy issued by the British Museum [Natural History] in 1939.)

Meek, F. B., and A. H. Worthen. 1865. Notice of some new types of organic remains from the Coal Measures of Illinois. Acad. Nat. Sci. Philadelphia, 17:41–53.

Pettibone, M. H. 1963. Marine polychaete worms of the New England region. 1, Aphroditidae through Trochochaetidae. U.S. Nat. Mus. Bull., 227:1–356.

———. 1982. Annelida, pp. 1–43. *In* S. P. Parker (ed.), Synopsis and Classification of Living Organisms, vol. 2. McGraw-Hill, New York.

Pillai, T. G. 1970. Studies on a collection of spirorbids from Ceylon, together with a critical review and revision of spirorbid systematics; an account of their phylogeny and zoogeography. Ceylon Jour. Sci. (Biol. Sci.), 8:100–172.

Rafinesque, C. S. 1815. Analyse de la Nature ou Tableau de l'Universe et des Corps organisés, etc. Palermo, 224 pp.

Richardson, Jr., E. S. 1956. Pennsylvanian invertebrates of the Mazon Creek area, Illinois, Marine Fauna. Fieldiana Geol., 12:59–67.

———. 1980. *Howellitubus* is not a worm tube. Jour. Paleo., 54:587.

Robinson, R. A. 1987. Annelida, pp. 194–204. *In* R. S. Boardman, A. H. Cheetham, and A. J. Rowell (eds.), Fossil Invertebrates. Blackwell, Palo Alto, CA.

Rolfe, W. D. I., F. R. Schram, G. Pacaud, D. Sotty, and S. Secretan. 1982. A remarkable Stephanian biota from Montceau-les-Mines, France. Jour. Paleo. 56:426–428.

Saint-Joseph, A. de. 1894. Les Annélides Polychètes des côtes de Dinard. Troisième partie. Ann. Sci. Nat. Zool., Paris, 7th ser., 17:1–395.

Schram, F. R. 1979. Worms of the Mississippian Bear Gulch Limestone of central Montana, U.S.A. Trans. San Diego Soc. Nat. Hist., 19:107–120.
Scudder, S. H. 1882. The affinities of *Palaeocampa* Meek and Worthen as evidence of the wide diversity of type in the earliest known myriapods. Amer. Jour. Sci., 24:161–170.
———. 1884. Two new and diverse types of Carboniferous myriapods. Boston Soc. Nat. Hist. Mem., 3(9):283–297.
Storch, V. 1968. Zur vergleichenden Anatomie der segmentalen Muskelsysteme und zu Verwandschaft der Polychaeten-Familien. Zeit. Morph Tiere, 63:251–342.
Thompson, I. 1979. Errant polychaetes (Annelida) from the Pennsylvanian Essex fauna of northern Illinois. Paleontographica. A, 163:23–199.
Thompson, I., and R. G. Johnson. 1977. New fossil polychaete from Essex, Illinois. Fieldiana Geol., 33(25):471–486.
Williams, T. 1851. Report on the British Annelida. Rep. British Assoc. Adv. Sci. London, 21:159–272.
Woodin, S. A. 1987. External morphology of the Polychaeta: design constraints by life habit? Bull. Biol. Soc. Washington, 7:295–309.

7B

Nonannelid Worms

Kirk Fitzhugh and Steven D. Sroka

In addition to members of the class Polychaeta (phylum Annelida; Chapter 7A), the following five extant phyla are each represented by a single species in the Mazon Creek Essex fauna: Nemertea, Nematoda, Priapulida, Echiura, and Chaetognatha. Extensive descriptions of these phyla can be found in invertebrate zoology texts (e.g., Brusca and Brusca, 1990).

Phylum Nemertea

The Nemertea (Greek, "a sea nymph") or Rhynchocoela (Greek, *rhynchos* "snout" and *coel* "cavity"), commonly called ribbon worms, are elongate, unsegmented, bilaterally symmetrical, and dorsoventrally flattened. Indeed, the general appearance and movement of nemerteans is quite reminiscent of free-living flatworms, to which the group is usually considered to be closely related.

Externally, the body is smooth, though it might sometimes have annulations due to contraction. Locomotory appendages are absent, and movement is accomplished by epidermal cilia or muscular contractions. The anterior end usually contains chemosensory slits or grooves, and simple eye spots. The mouth is a small, anteroventral opening. As a group, the Nemertea are unique in their possession of an eversible, proboscis-like structure (Figure 7B.1A), quite unlike the proboscis described in Chapter 7A for the Polychaeta (e.g., Figure 7A.1C). The nemertean proboscis is a muscular, eversible, blind tube surrounded by a fluid-filled cavity called the rhynchocoel. The rhynchocoel lies dorsal to, and is usually completely separate from, the gut. As a result of muscle contractions around the rhynchocoel, the proboscis is forcefully everted to the outside through a proboscis pore located dorsal to the mouth. Most nemerteans being predators, the proboscis is usually used to grasp prey, which are usually small invertebrates.

Most nemerteans are free living and marine, burrowing in soft mud and sand, or living in algal mats or crevices, or beneath rocks and shells. Several species are symbiotic with other invertebrates, and a few occur in estuarine, freshwater, and terrestrial environments. About 900 extant species have been described, ranging in length from less than 1 cm to several meters (Brusca and Brusca, 1990).

Archisymplectes rhothon Schram, 1973
Figures 7B.1A, 7B.4A

Description. Body slender, 5–12 cm long, distinctly longer than wide (Figure 7B.4A). Body usually convoluted (Figure 7B.1A) and only indicated by color differences in concretion. Proboscis everted in some specimens.

Remarks. Other than the outline of the body and everted proboscis, there are no distinct features identifying this species.

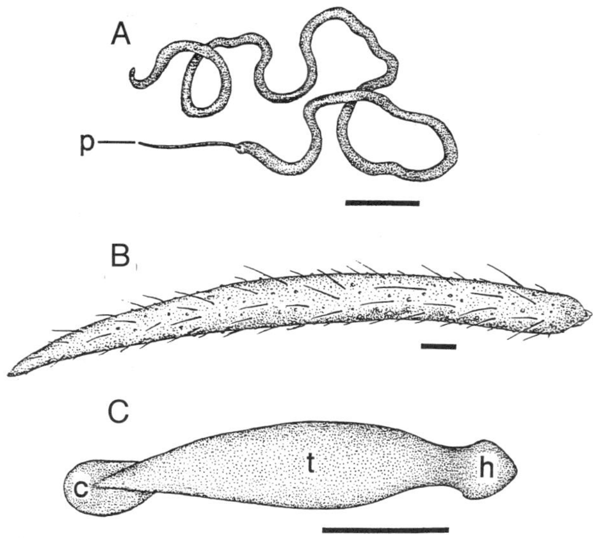

Figure 7B.1. Reconstructions of selected Mazon Creek nonannelid worms. A *Archisymplectes rhothon,* modified from Schram (1973, fig. 1D). B *Nemavermes mackeei,* modified from Schram (1973, fig. 1B). C *Paucijaculum samamithion,* modified from Schram (1973, fig. 1C). All scale bars = 5 mm. Abbreviations: c, caudal fin; h, head; p, proboscis; t, trunk.

Phylum Nematoda

Nematodes (Greek, *nema* "thread"), or roundworms, are elongate, bilaterally symmetrical, and unsegmented. The body is usually round in cross section, quite slender, with anterior and posterior ends slightly tapered. The body surface is covered by a layered cuticle of scleroprotein and can range from smooth to highly sculptured with spines and bumps. The mouth is located at the anteriormost part of the body and is often surrounded by spines, flaps, or papillae.

Most free-living nematodes are very small to microscopic, whereas parasitic species are often quite large. Extant species inhabit virtually every marine, freshwater, and terrestrial habitat. Free-living species are commonly detritivores, scavengers, or predators.

Nemavermes mackeei Schram, 1973
Figures 7B.1B, 7B.4B

Description. Body moderate to large, stout, up to 12.5 cm long (Figures 7B.1B, 7B.4B). Few external details present except for fine, hairlike spines preserved in some specimens as negative impressions over body surface. Anterior end of some specimens with short, fleshy, digitiform processes. Detritus occasionally present in gut.

Remarks. Schram (1973) suggested that the processes seen in the mouth region of some specimens might represent oral papillae. The species is not very common.

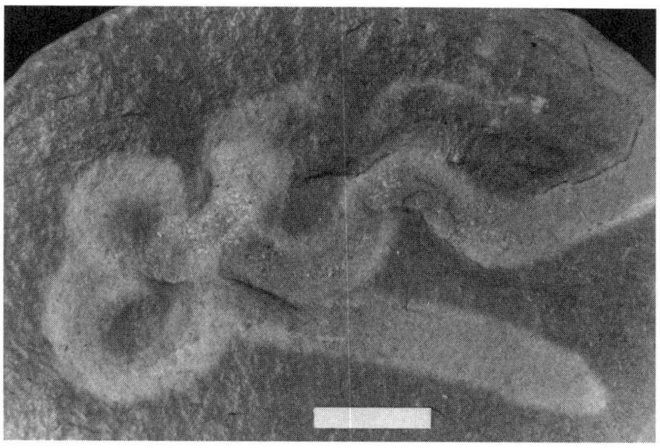

Figure 7B.4A. *Archisymplectes rhothon* Schram, NEIU MCP 437. Scale bar = 5 mm.

Figure 7B.4B. *Nemavermes mackeei* Schram, NEIU MCP 445. Scale bar = 10 mm.

Phylum Priapulida

The Priapulida (from *Priapos,* the Greek god of reproduction, symbolized by the penis) contain about 15 extant and six fossil species (Brusca and Brusca, 1990; Conway Morris, 1977). The group is entirely marine and benthic, with most species burrowing in soft sediments and feeding on small invertebrates. Priapulids are cylindrical, longer than wide, from 0.55 to 20 cm in length, and unsegmented. The body is

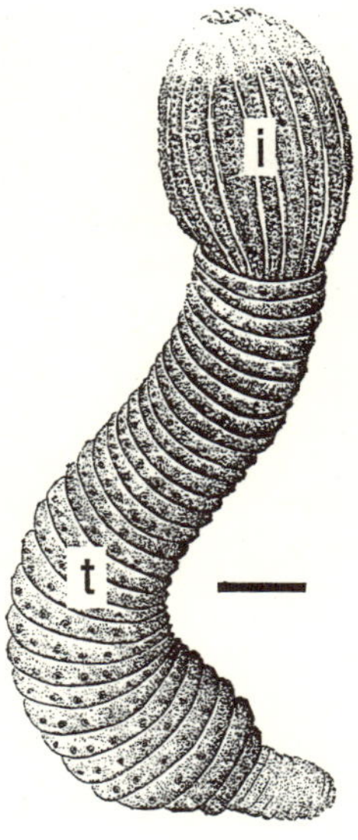

Figure 7B.2. *Priapulites konecniorum,* reconstruction modified from several sources. Scale bar = 5 mm. Abbreviations: i, introvert; t, trunk.

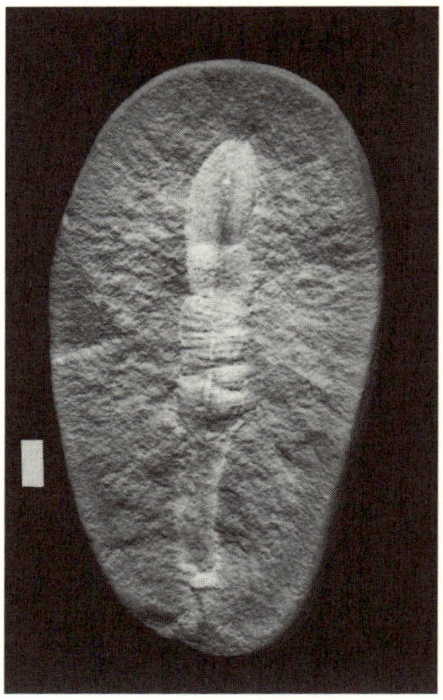

Figure 7B.4C. *Priapulities konecniorum* Schram, NEIU MCP 86. Scale bar = 5 mm.

composed of two or three distinct regions: an introvert (or proboscis), a trunk, and, when present, a caudal appendage. The introvert and trunk are covered by a thin cuticle, which usually has cuticular extensions such as spines and tubercles. As well, the trunk cuticle is often annulated, giving the impression of segments. The mouth is located at the anterior end of the introvert. A cuticle-lined, eversible pharynx is used to grasp prey items. The caudal appendage varies in shape between species and in some forms may function in gas exchange. The relationship of priapulids to other invertebrate phyla is still unresolved, though the group is sometimes considered to be related to the Nematoda (but see Eernisse et al., 1992).

Priapulites konecniorum Schram, 1973
Figures 7B.2, 7B.4C

Description. Body 4–6 cm long, composed of introvert, trunk, and caudal appendage (Figures 7B.2, 7B.4C). Introvert slightly inflated, with 16 to 20 papillate, longitudinal ridges; little detail visible in anteriormost region. Trunk of even width; cuticle with 30 to 40 distinct annulations. Caudal appendage poorly preserved, appearing paired in some specimens.

Remarks. The species is rarely found. *Priapulites konecniorum* closely resembles the extant species, *Priapulus caudatus* Lamarck.

Phylum Echiura

The Echiura (Greek, *echis* "serpentlike") are a relatively small group of unsegmented, bilaterally symmetrical worms. All are marine and benthic, and most occupy a burrow and are detritivores. The body is composed of two regions: a soft, saclike, cylindrical trunk with a voluminous body cavity and a proboscis at the anterior end of the trunk. The mouth is located on the trunk, near the base of the proboscis. The echiuran proboscis is a nonretractile, preoral structure and is actually homologous to the annelid prostomium (see Chapter 7A). The proboscis is elongate, muscular, dorsoventrally flattened, and extensile; the ventrum is concave, forming a longitudinal food groove. During feeding, the proboscis is extended over the sediment surface. Mucus secreted by the proboscis accumulates sediment particles, which are conveyed to the mouth by cilia on the proboscis groove.

The trunk surface is smooth to lightly papillate. Most echiurans have a pair of small, curved setae on the trunk ventrum, just posterior to the mouth. Several extant genera (*Echiurus* Guérin-Méneville and *Urechis* Seitz) also have one or two rings of setae encircling the posterior end of the trunk. The presence of setae, regarded as homologous to annelid setae, is usually considered important evidence for a close relationship between the Echiura and Annelida.

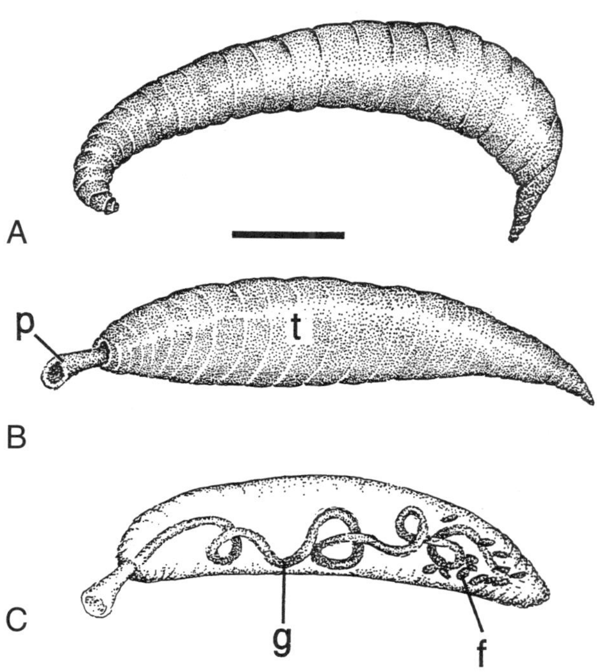

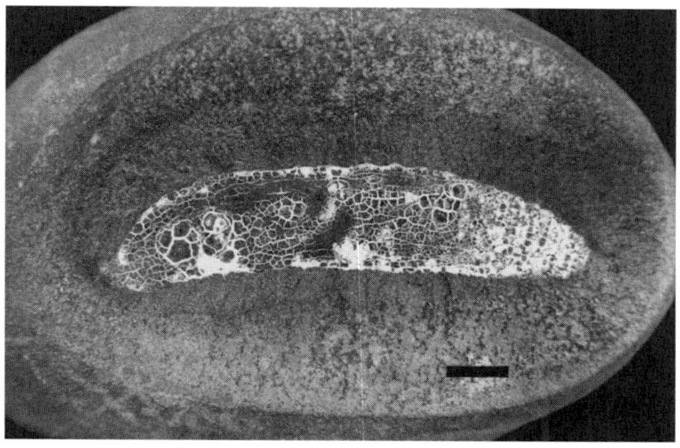

Figure 7B.3 (*Left*). *Coprinoscolex ellogimus,* reconstructions modified from Jones and Thompson (1977, fig. 3). **A** Trunk only, proboscis missing; anterior end to the left. **B** Proboscis and trunk. **C** Outline of coiled gut and fecal pellets beneath body wall. Scale bar = 5 mm. Abbreviations: f, fecal pellets; g, gut; p, proboscis; t, trunk.

Figure 7B.4D (*Above*). *Coprinoscolex ellogimus* Jones and Thompson, NEIU MCP 181. Proboscis is absent. Scale bar = 10 mm.

Coprinoscolex ellogimus
Jones and Thompson, 1977
Figures 7B.3, 7B.4D

Description. Body elongate, flattened (Figures 7B.3, 7B.4D); trunk cigar-shaped, 2–7.5 cm long. Proboscis flat, widest along anterior margin, 5–10 mm long (Figure 7B.3B). Anterior and posterior regions of trunk surface sometimes with prominent papillae. Setae not observed. Convoluted gut sometimes visible, 2 to 3 times longer than trunk (Figure 7B.3C). Outline of gut occasionally indicated by numerous rice-shaped fecal pellets, each 2 mm long.

Remarks. Coprinoscolex ellogimus is a common species of the Mazon Creek fauna. Prior to being described, it was identified as a leech by amateur collectors. Jones and Thompson (1977) noted that specimens commonly lack a proboscis. This observation is consistent with what is seen in extant echiurans, which are capable of detaching their proboscis from the trunk when disturbed (Stephen and Edmonds, 1972; K. Fitzhugh, personal observation). Jones and Thompson (1977) considered *C. ellogimus* to be an epifaunal deposit feeder.

Phylum Chaetognatha

Extant chaetognaths (Greek, *chaitē* "hair" and *gnathos* "jaw"), or arrow worms, are bilaterally symmetrical, elongate, streamlined, and unsegmented, though the coelomic cavity is divided into three compartments. Body regions consist of the head, trunk, and tail. The head contains dorsal eyes and a ventral mouth. Lateral to the mouth are a pair of large grasping spines, and above the mouth a set of smaller teeth. Paired, lateral fins are present on the trunk, and the tail has a single caudal fin. The body surface is smooth except for minute sensory bristles. Chaetognaths range in size from 0.5 to 12 cm and are found in all oceans and some estuarine environments. Most species are planktonic, but several are benthic. Most planktonic species have transparent bodies associated with their lifestyle. Chaetognaths are predatory carnivores, planktonic species feeding on crustaceans, small fish, and other chaetognaths. There are about 100 extant species (Brusca and Brusca, 1990).

Paucijaculum samamithion Schram, 1973
Figures 7B.1C, 7B.4E

Description. Body length 1.5–3 cm. Head region poorly preserved (Figure 7B.1C); no grasp-

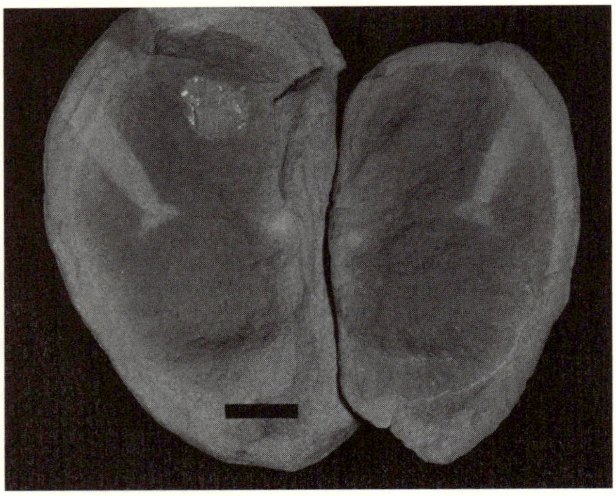

Figure 7B.4E. *Paucijaculum samamithion* Schram, NEIU MCP 70. Only posterior half of animal shown. Scale bar = 10 mm.

ing spines observed. Posterior trunk region of some specimens with what appear to be poorly developed lateral fins. Tail with distinct caudal fin, posterior margin rounded (Figures 7B.1C, 7B.4E).

Remarks. Schram (1973) noted that this species is often found in association with other marine invertebrates, such as shrimp.

References

Brusca, R. C., and G. J. Brusca. 1990. Invertebrates. Sinauer, Sunderland, Massachusetts.

Conway Morris, S. 1977. Fossil priapulid worms. Spec. Paper Palaeo. Assoc. London, 20, 95 pp.

Eernisse, D. J., J. S. Albert, and F. E. Anderson. 1992. Annelida and Arthropoda are not sister taxa: a phylogenetic analysis of spiralian metazoan morphology. Syst. Biol., 41:305–330.

Jones, D., and I. Thompson. 1977. Echiura from the Pennsylvanian Essex fauna of northern Illinois. Lethaia, 10:317–325.

Schram, F. R. 1973. Pseudocoelomates and a nemertine from the Illinois Pennsylvanian. Jour. Paleo., 47:985–989.

Stephen, A. C., and S. J. Edmonds. 1972. The Phyla Sipuncula and Echiura. British Museum (Natural History), London.

7C

Wormlike Animals:

Enteropneusta (Acorn Worms)

David Bardack

Phylum Hemichordata
Class Enteropneusta
Genus *Mazoglossus* new genus
Mazoglossus ramsdelli new genus and species
Figures 7C.1–7C.8

"Diagnosis." Small (specimens less than 10 cm total length) enteropneusts recognizable by outline shape of proboscis, collar, trunk, and small, circular branchial pores(?).

Type species. *Mazoglossus ramsdelli* new species.

Etymology. Named in reference to the Mazon Creek biota and glossus, a common ending for genera of acorn worms. The specific name is for Karlene Ramsdell, an indefatigable collector.

Holotype. FNMH PE 45207 part and counterpart, Pit 11.

Referred specimens. In excess of 50 cataloged specimens at the Field Museum (including FMNH PE 39786, PE 45206, PE 45207, and PE 45756) all from Pit 11. There are two dozen concretions from the Royal Ontario Museum (including ROM 44438, 44439, 44440, 44441, 49829, 49830, 49831, 49832, and 49833) all from Pit 11. Numerous other specimens are in private collections. One specimen in the T. Testa collection is from Henry County, Missouri.

Description. The fossils reported here belong to the extant group of "acorn worms" or enteropneusts that are included in the phylum Hemichordata along with the Pterobranchia. Living species are differentiated largely on the basis of internal, soft anatomical features (Hyman, 1959; Jefferies, 1986). Association of *Mazoglossus* with enteropneusts is based primarily on the three-part body outline comprising a proboscis, a sometimes visible proboscis stalk, a collar and an elongate trunk. A few other external anatomical features are described below. Essentially, *Mazoglossus* is a "form" genus.

Specimens range in total length from 3 to 8 cm and from 0.5 to 1.1 cm in width just behind the collar. Most concretions show one individual that, if long, may have the posterior half of the trunk curled toward the head. Some concretions show two or more individuals, a few of which lie parallel, head to head, but most are randomly clustered and often overlap each other. While body outlines of the fossils are smooth and intact in most cases, in some these outlines are irregular and suggest decay processes have begun. The specimens appear to have been preserved and fossilized outside of burrows as the concretions show no evidence of trails or differential fillings. However, some living species do not burrow, and these ontogenetically small fossil forms may not yet have begun to burrow. Most modern genera are confined to shallow waters, although a few occur at greater depths (Hyman, 1959).

Preserved shape of the proboscis is quite variable. This structure may appear long and slender as in living *Saccoglossus* or short and narrow as in *Glandiceps* or *Ptychodera.* The

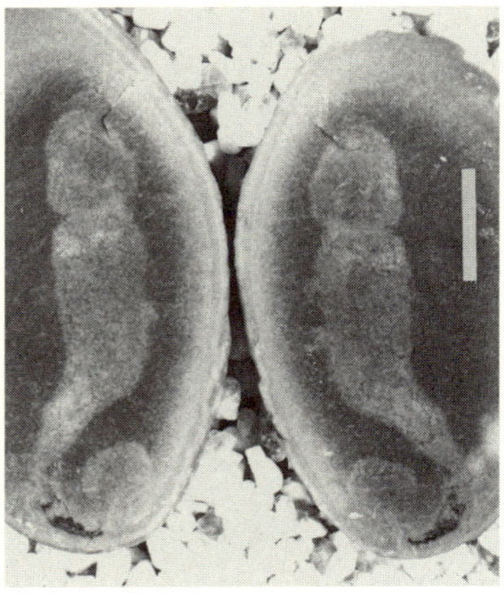

Figure 7C.1. *Mazoglossus ramsdelli,* ROM 49828. Whole animal in part and counterpart. Scale bar = 10 mm.

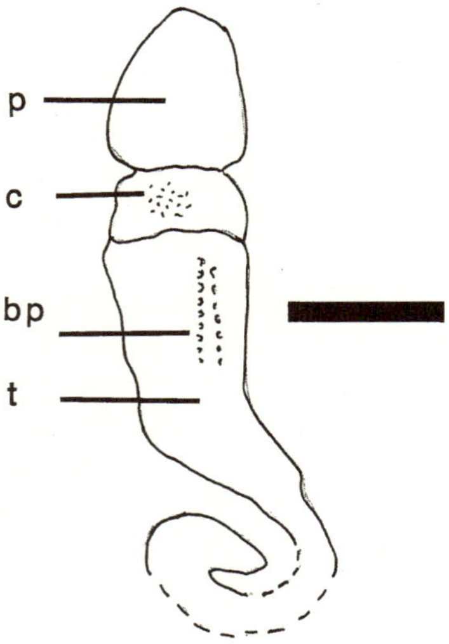

Figure 7C.2. *Mazoglossus ramsdelli,* drawing of ROM 49828. Scale bar = 10 mm. Abbreviations: bp, branchial pores; c, collar; p, proboscis; t, trunk.

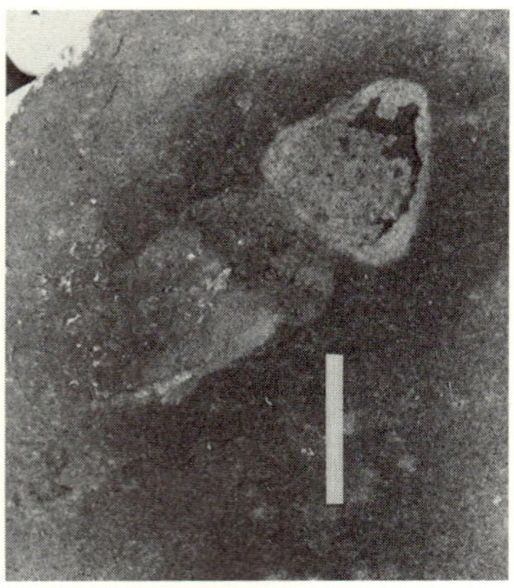

Figure 7C.3. *Mazoglossus ramsdelli,* ROM 49832; head and anterior part of trunk. Scale bar = 10 mm.

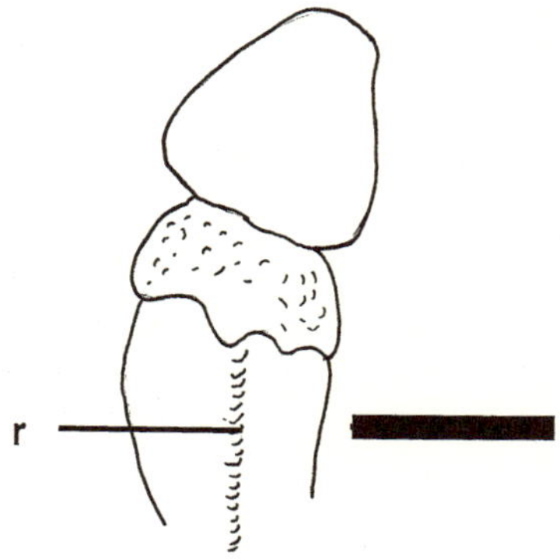

Figure 7C.4. *Mazoglossus ramsdelli,* drawing of ROM 49832. Scale bar = 10 mm. Abbreviation: r, midline ridge.

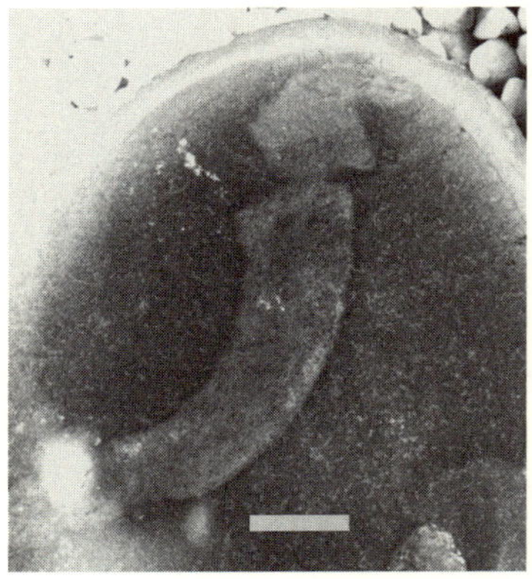

Figure 7C.5. *Mazoglossus ramsdelli,* ROM 49833; whole animal. Scale bar = 10 mm.

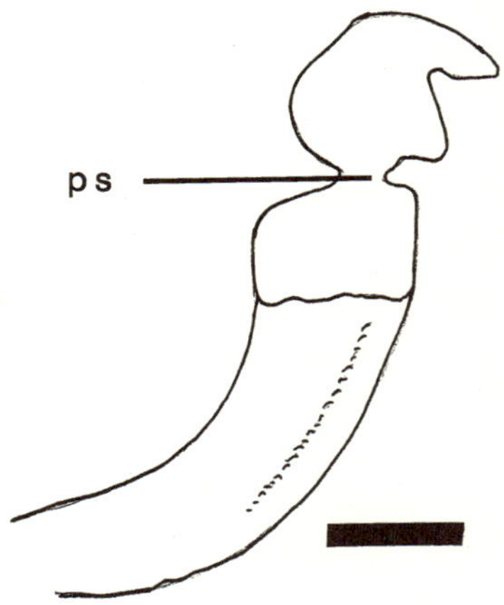

Figure 7C.6. *Mazoglossus ramsdelli,* drawing of ROM 49833. Scale bar = 10 mm. Abbreviation: ps, proboscis stalk.

proboscis ranges from 0.4 to 1.6 cm in greatest width and from 0.6 to 1.6 cm in greatest length. Varying shapes in the fossils probably reflect preservational differences of this muscular structure. A low correlation coefficient (r = 0.44 for 27 specimens) of length versus width supports the suggestion that proboscis shape is flexible—assuming that the specimens belong to one taxon. In some specimens with a slender, elongate proboscis, the front half is twisted sideways.

A constriction or gap separates the proboscis from the collar. In some individuals, a short proboscis stalk is visible between these structures. No evidence of a mouth is seen. In living

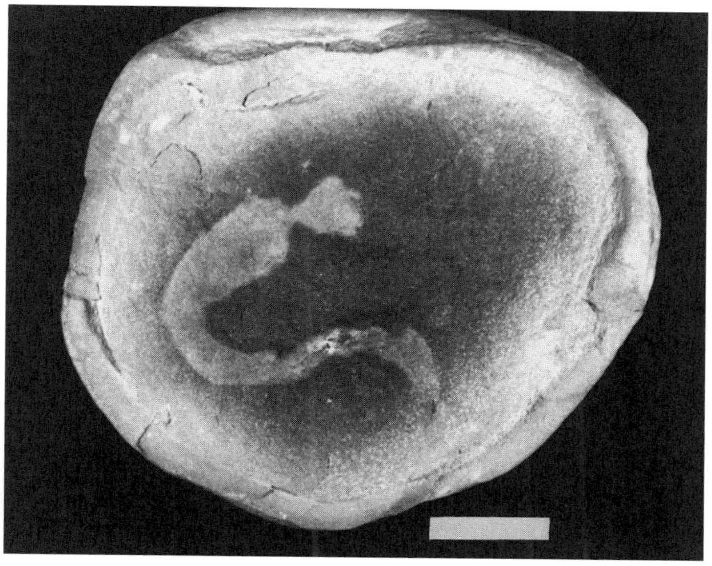

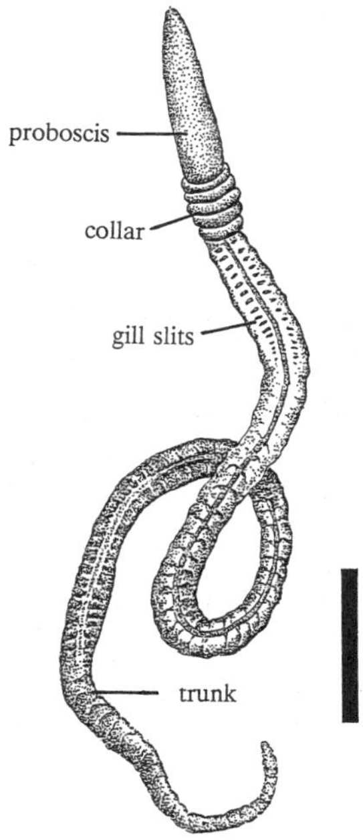

Figure 7C.7 (*Far left*). *Mazoglossus ramsdelli,* NEIU MCP 75; whole animal. Scale bar = 10 mm.

Figure 7C.8 (*Left*). *Dolichoglossus,* extant enteropneust, modified from Broadman (1987). Scale bar = 10 mm.

forms, a proboscis skeleton developed from basal epidermal and coelomic mesoderm forms within the proboscis stalk above the buccal area and extends into the proboscis (Hyman, 1959, fig. 23). Some fossils show faint outlines near the proboscis base that may be evidence of this "skeleton." The collar is particularly distinct on many specimens, extending beyond the width of the trunk and enfolding the base of the proboscis as a thick scarf. Its surface is often grooved or pitted.

The trunk forms most of the animal's length. It tapers gradually posteriorly, but in many concretions, the distal end is not preserved. On some specimens, the trunk shows a more or less regular, transverse pattern of epidermal grooves. In living acorn worms, the trunk bears a middorsal and midventral ridge. These are not definitively or consistently preserved on the Mazon Creek specimens.

Prominent external features on living acorn worms include the gonads, which may occupy anteroposteriorly elongate genital wings, and branchial pores sunk into depressions or along a flattened surface. There is no evidence of gonadal structures on *Mazoglossus,* but branchial pores(?) seem to be present on a few specimens when carefully illuminated from the side at a low angle. These pores appear as a series of small openings or circular incisions in an anteroposterior linear pattern. They seem to be limited to the anterior 25 percent of the trunk. Such pores, although prominent features of living forms, are difficult to identify on these fossils for several possible reasons. First, these are rather small acorn worms. In contrast, some living species reach over 1.5 m in length and many are 10–50 cm long. Second, the grain surface of the concretion has about the same dimension as the probable gill pores. Third, as individual acorn worms grow, the branchial sac is divided by a medial tongue bar and the pore becomes an elongate, larger, and more distinct structure. Again, as with so many Mazon Creek fossils, these small individuals show early growth stages and not the array of adult characters typical of the descriptions of living acorn worms. The circular form of the branchial pores supports this. Among living genera, in *Balanoglossus* and *Ptychodera,* gonads arise from ventrolateral wings of the body and the branchial pores are partially enclosed by the wings. In *Saccoglossus,* gill pores appear as an elongate series of small openings along a smooth body surface. The latter living form is most like that of *Mazoglossus.*

Discussion. One other fossil enteropneust, *Megaderaion sinemuriense* (Arduini et al., 1981) has been described. It is based on a single specimen from the lower Jurassic (Sinemurian)

Osteno deposit along the east bank of Lake Lugano in northern Italy. This specimen differs from *Mazoglossus* in a relatively longer collar and a pair of short, ovate structures termed "external gonads," which are probably equivalent to the genital wings described by Hyman (1959). In addition to the whole specimens, some dubious trace activities such as burrow and trail marks of acorn worms have been identified (see references in Arduini et al., 1981). As expected, Mazon Creek enteropneusts are confined to the Essex fauna.

References

Arduini, P., G. Pinna, and G. Teruzzi. 1981. *Megaderaion sinemuriense* n.g. n.sp., a new fossil enteropneust of the sinemurian of Osteno in Lombardy. Atti. Soc. Ital. Sci. Nat. Mus. Civ. Stor. Nat. Milano, 122(1–2):104–108.

Broadman, R. S., A. H. Cheetham, and A. J. Rowell. 1987. Fossil Invertebrates. Blackwell, Palo Alto, California.

Hyman, L. H. 1959. The Invertebrates: Smaller Coelomate Groups, vol. 5. McGraw-Hill, New York, 383 pp.

Jefferies, R. P. S. 1986. The Ancestry of the Vertebrates. Cambridge University Press, Cambridge, 376 pp.

8A

Bivalvia

Including Descriptions of Two New Species

Jack Bowman Bailey and Steven D. Sroka

The class Bivalvia (Pelecypoda) is the group of aquatic molluscs containing the clams, scallops, mussels, and oysters. With an estimated 15,000 living species, they are among the most successful of the molluscs, having colonized both marine and freshwater environments at all latitudes and temperatures and in water depths ranging from intertidal to abyssal. Their success seems odd in view of the fact that they, unlike their relatives the cephalopods and gastropods, have entirely lost much of the molluscan anterior anatomy, including the head, along with the brain and anterior sense organs, and the "toothed tongue" (radula) that other molluscs rely on for feeding.

The body of a bivalve is enclosed by a shell consisting of left and right calcareous valves that dorsally attach to one another by means of an elastic ligament, a horny springlike structure that is responsible for opening the valves. Attaching to the inner surfaces of the valves are one or two large adductor muscles that work against the ligament's elasticity to close the valves. The valves rotate to the opened and closed positions along a hinge line, an axis that passes through or near the ligament. Often, alignment of the closed valves is maintained by a series of hinge teeth (denticles) and sockets, located on a shelly thickening along the hinge known as the hinge plate.

Beneath each valve is a fleshy lining or lobe. The two lobes form the mantle, a soft envelope that secretes the valves and encloses the remaining soft anatomy. The mantle lobes merge dorsally; between them lies the mantle cavity, which contains the visceral mass, flanked on each side by a set of gills. The mouth is positioned at the forward end of the visceral mass; below lies the muscular foot, a protrusible organ used by many bivalves as a locomotion and digging tool. Primitively, margins of the mantle lobes were separate from one another. By Ordovician times, a few bivalves fused and modified the two lobes posteriorly to form extensible siphons to separate and direct the flow of water into and out of the mantle cavity. But mantle fusion and development of siphons is most notable in post-Paleozoic taxa. Oxygen and nutrients enter the inhalant siphon, and wastes are flushed out the exhalant siphon, conveniently located near the anus.

Bivalves have adopted a wide variety of lifestyles. Some are mobile, actively burrowing suspension feeders or deposit feeders. Deeper burrowers show considerable elongation of the siphons that extend to the entrance of the burrow. Sedentary bivalves are exclusively suspension feeders, and most lack siphons. Many of these, including mussels, certain scallops, and pearl oysters, attach themselves to hard substrata by a tuft of organic threads called the byssus. Byssally attached clams that adopt a partially buried lifestyle are termed endobyssate; those living as exposed surface dwellers are called epibyssate. Some sedentary bivalves lack a byssus. Oysters, for example, recline on one valve that becomes permanently cemented to an exposed hard surface.

Other suspension-feeding bivalves include nestlers and borers. Nestlers (e.g., some arks) are byssate bivalves that occupy preexisting nooks and hollows. Borers, such as phlolads and certain mussels, use the shell as a rasping tool to carve permanent domiciles in wood or reef rock. A few bivalves, notably some scallops, file shells, and awning shells, can even swim short distances by energetically clapping the valves together.

Mode of life can often be judged from morphology. Free-living burrowers are equivalved, more or less homomyarian (i.e., show equal development of the two adductor muscles), with a large foot and variable siphon development. Attached surface dwellers show considerable reduction of the anterior anatomy. Byssally attached and cemented taxa show either a marked reduction of the anterior adductor muscle (anisomyarian condition) or its total loss (monomyarian condition), and the foot is either reduced or absent. Reclining taxa are often inequivalved, having lost bilateral valve symmetry.

Shell Morphology

Shell morphology (Figure 8A.1) is important in determining both taxonomy and mode of life. The bivalve shell is fundamentally a spiral with a high rate of whorl expansion and moderate translation. The earliest growth stage of each valve is located at a pointed terminus called the beak. Covered by a thin outer horny layer called the periostracum, the shell grows through secretion of a series of calcareous laminae by the mantle; these appear on the shell surface as comarginal growth lines whose center of curvature is the beak. The region of the valve around the point of maximum curvature and extending to the beak is termed the umbo. The shell may be marked by radial ribs visible only on the exterior (costae) or by ribs discernible on both exterior and interior surfaces (plicae). Bivalves in which the beak is centrally placed are termed equilateral; those in which the beak is placed near either the posterior or anterior valve extremity are called inequilateral. In some bivalve shells, dorsal margins may extend away from the umbonal region of the shell to form auricles. Small auricles are sometimes called ears, while large auricles may be referred to as wings.

The valve interior usually has a dorsal thickening along the hinge line called the hinge plate; here the dentition (hinge teeth and sockets) and ligament control the articulation of the two valves. Other features that mark the interior surfaces of the valves are the adductor muscle scars, usually connected by an arcuate groove running parallel to the ventral curvature of the valve. Termed the pallial line, this groove is produced by the attachment of a closely set series of small mantle muscles. The line may either describe a simple arc or may show an embayment (pallial sinus) posteriorly to accomodate retraction of siphons.

There are several distinct grades of both dentition and ligament and numerous variations of each. For a more complete account of these and other aspects of bivalve morphology, see the *Treatise* volumes (Moore and Teichert, 1969–1971) and Pojeta (1987). The dentition fundamentally consists of two kinds of teeth: those situated beneath the beak (cardinal teeth) and those located on one or both sides of the beak (lateral teeth). Cardinals and laterals are separated from each other by a toothless portion of the hinge plate termed an edentulous space. In cases where the edentulous space is absent, the teeth are referred to as pseudocardinals and pseudolaterals (Pojeta, 1987). The most basic grades of dentition are of four kinds, but several other types have been described. The basic four are 1) the edentulous grade (no teeth), 2) the heterodont grade (cardinal and lateral teeth separated by an edentulous space), 3) the actinodont grade (pseudocardinals and pseudolaterals without an edentulous space), and 4) the taxodont grade (numerous teeth and sockets arranged in a closely spaced repeating series).

The ligament, usually located near the posterior portion of the hinge line, may either be external or internal. External ligaments may attach to the shell exterior along parallel grooves or may attach to small interior shelves (nymphs). If internal, the ligament may be located in a small pit called the resilifer (chondrophore) that takes the form either of a depres-

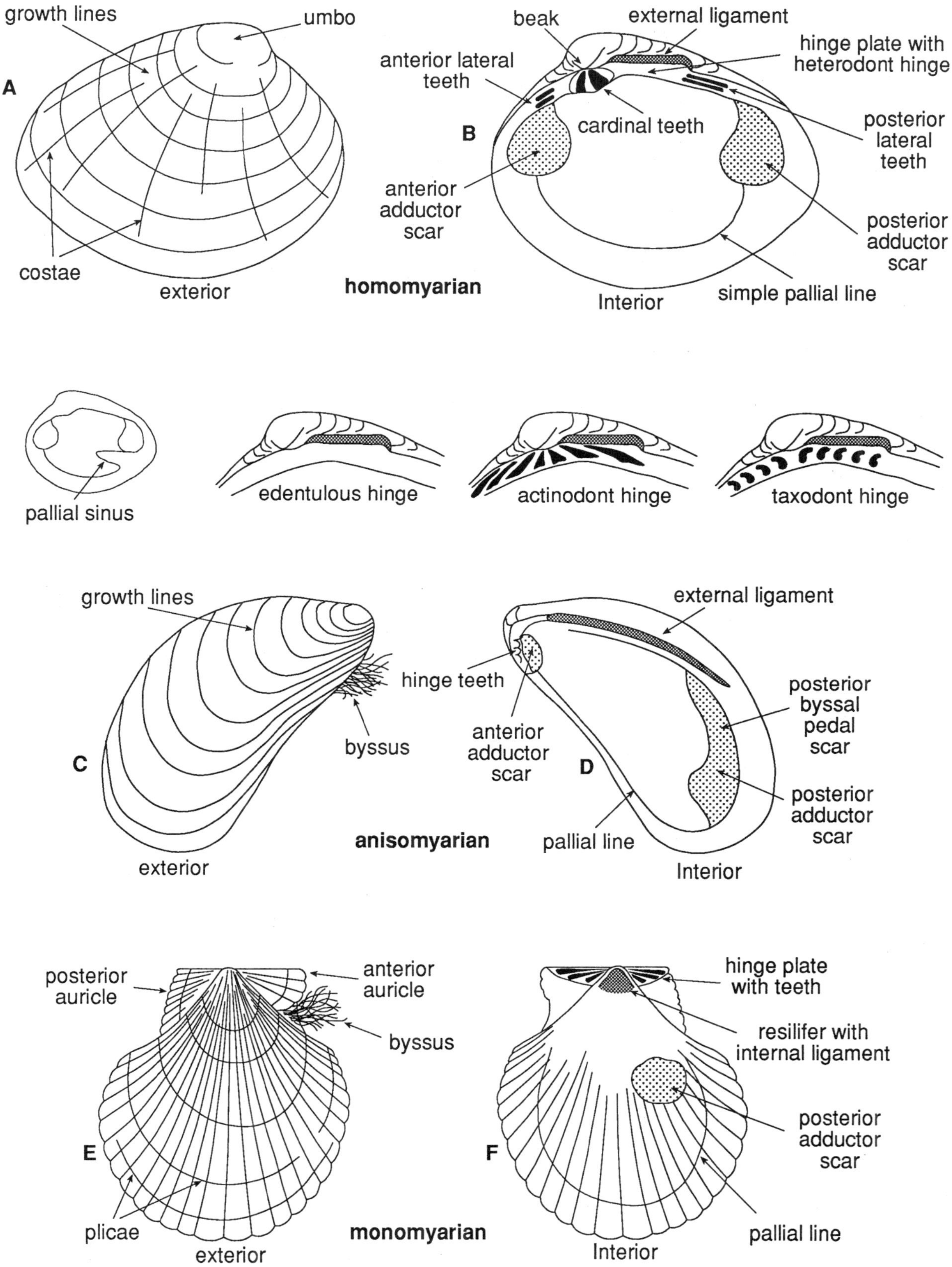

Figure 8A.1. Generalized bivalve shell morphology. **A, B** Homomyarian bivalve: **A** exterior view of a right valve; **B** interior view of the same valve. Alternative hinges and the form of a pallial sinus are shown in the figure subset. **C, D** Anisomyarian bivalve: **C** exterior view of a right valve; **D** interior view of same valve. **E, F** Monomyarian bivalve: **E** exterior view of a right valve; **F** interior view of same valve.

sion in the hinge plate or a spoon-shaped structure extending from the hinge plate.

Fossil Record and Classification

The history of the Bivalvia is one of increase in diversity with relatively few setbacks (Newell and Boyd, 1978). Bivalve anatomy is in many respects simpler or more specialized than that of other molluscan classes. Consequently, bivalves have remained a conservative group, and similar evolutionary themes and morphologies keep reemerging throughout their history, often in separate lineages. The resulting proliferation of both related and unrelated look-alikes (homeomorphs) makes classification difficult. As a result, most major taxonomic groups of bivalves defy simple morphologic definitions. They can only be understood within the context of their evolutionary history.

Generic and specific identification of fossil bivalves relies entirely on shell morphology, emphasizing ligamental attachment areas and interior features such as dentition, muscle scars, and pallial line. Because the Bivalvia are notoriously homeomorphic, classification is difficult enough where such data are plentiful. Without them, identification may be reduced to little more than best estimates based on shell outline and ornament.

Adding to this problem is polymorphism, a high degree of shell variability, which can occur within a given species owing either to genetic variation or differences in ecologic station. Early authors often failed to take variation into account when working with bivalves, seeing two or more species (or even genera!) where there is only one. Another common error was allowing stratigraphic or biogeographic occurrence to unduly bias identification of taxa. Such practices have resulted in the proliferation of superfluous generic and specific names applied in confusing ways. Not only have lists of synonyms become cumbersome (e.g., see the huge synonymies given in Bailey, 1983), but they mask our understanding of the conservative and geologically enduring nature of many of the taxa.

Bivalves show a spotty distribution among Paleozoic strata. They are frequently absent or rare among many Paleozoic faunas; elsewhere they are abundant and in a few cases predominant. Perhaps because of competition from Paleozoic communities dominated by brachiopods and bryozoans, bivalves were largely restricted to nearshore sands and muds and a few carbonate shoals (see Bretsky, 1968; Newell and Boyd, 1978).

Pojeta and Runnegar (1985) have shown that bivalves evolved from an archaic two-shelled molluscan class called the Rostroconchia, now extinct. Rostroconchs themselves were derived from still more primitive molluscs, the Monoplacophora, a class ancestral to gastropods and cephalopods.

The earliest bivalves occur in Early Cambrian strata in the form of a single group of very small burrowing marine bivalves, the order Fordilloida. Two genera are known, *Fordilla* and *Pojetaia.* Runnegar and Pojeta (1992) indicated that other coeval genera such as *Tuarangia* and *Pseudomyona* are not true bivalves. Waller (1990) considered the fordilloids to be an extinct group that split off early from the stem group leading to the class Bivalvia. But Runnegar and Pojeta (1992) argued that the fordilloids are either forerunners or primitive members of the bivalve subclass Palaeotaxodonta.

Bivalves underwent an important adaptive pulse in the Ordovician. During this time they significantly increased their range of lifestyles. Some were free-burrowing deposit feeders and suspension feeders. Others were sedentary suspension feeders, either partially buried or fully exposed surface dwellers. Still other suspension feeders adopted a nestling or boring mode of life. Along with the increase in lifestyles came a dramatic increase in diversity, and the bivalves split into five distinct subclasses:

Palaeotaxodonta. Cambro-Ordovician–Recent. These anatomically primitive bivalves represent the stem group of the class, probably descending from the Cambrian fordilloids (Runnegar and Pojeta, 1992). They are small burrowing inhabitants of marine and brackish waters, mostly deposit feeders. Examples: nut clams and awning shells.

Pteriormorphia. Ordovician–Recent. Sedentary and free-living surface dwellers. Many are attached by a byssus; others are cemented. All are suspension feeders. A few are occasional swimmers. Examples: arks, pearl oysters, pen shells, file shells, scallops, and some mussels.

Heteroconchia. Ordovician–Recent. The most diverse modern group. These are free-burrowing and boring suspension feeders with well-developed siphons in many post-Paleozoic genera. They inhabit both marine and freshwater environments. Examples: steamer clams, coquinas, tellins, geoducks, giant clams, and freshwater mussels.

Anomalodesmata. Ordovician–Recent. Important in the Paleozoic; uncommon today. Included are burrowing, nestling, partially buried, and a few cemented genera. All are suspension feeders, marine only. Examples: pandoras.

Isofilibranchia. Ordovician–Recent. These are sedentary suspension feeders attaching themselves to exposed surfaces. Others are borers. They inhabit marine, brackish, and freshwaters. Examples: certain mussels.

Silurian bivalve faunas are in general not as well known as those of the Ordovician, but they contain the earliest scallops. The earliest giant bivalves occur in the Silurian; some reached the size of cantaloupes. By Devonian times bivalves became common in marine and brackish water deltaic communities, and a few genera invaded freshwater habitats. The earliest cemented genera occur in the Early Carboniferous. Elsewhere during that time, bivalves continued to be important among marine, brackish, and freshwater deltaic communities. Nonmarine bivalves are much more common in the Carboniferous than the Devonian and are particularly abundant in association with coal seams.

Significant trends in life habits among middle and late Paleozoic bivalves are noted by Stanley (1977). Among sedentary taxa, these include major increases among the epibyssate clams and major decreases among the endobyssate species. Among the free burrowers, suspension feeders increased at the expense of deposit feeders.

Perhaps because of the extinction of the brachiopod–bryozoan community at the end of the Permian, the bivalves underwent another major adaptive pulse beginning in the Triassic. Like the Ordovician one before it, the Triassic pulse involved both a proliferation of new genera and a major shift in lifestyles. Stanley (1968, 1977) has shown that the greatest post-Paleozoic increase in bivalve diversity was among the siphonate burrowers (Heteroconchia). Although siphons first developed among Paleozoic bivalves by fusion of the mantle lobes, siphon development is not a main theme of bivalve evolution until the Mesozoic. Mantle fusion not only increased respiration and suspension-feeding efficiency through siphon development but also provided a tight seal for the mantle cavity during burrowing activities. Such adaptations are perhaps linked to deeper burrowing lifestyles in order to escape predation (J. Pojeta, personal communication).

Bivalves became significantly more abundant and widespread among marine faunas of the Mesozoic and Cenozoic. During these times, both the geographic range and the ecologic distribution of bivalves expanded considerably, and most modern genera first evolved. The Mesozoic is notable for the proliferation of borers as well as cemented surface dwellers such as the earliest oysters and the extinct rudists, which were bizarrely shaped reef builders. After a modest setback, roughly coinciding with the Cretaceous–Tertiary mass extinction event, the bivalves further increased their diversity throughout the Cenozoic.

Mazon Creek Bivalve Paleoecology

Although Mazon Creek fossils have been studied for more than a century, no formal papers have been published dealing solely with the bivalves. Much of the existing data is in the form of faunal lists (Schram, 1979), occasional mention in general articles, and two unpublished theses (Dickson, 1974; Sroka, 1984).

Mazon Creek bivalve faunas are unique in certain respects. Preservation is remarkable; specimens occasionally show rarely seen traces of the original soft parts such as the ligament, pallial musculature, byssal threads, the periostracum, and preserved burrows with the clams inside (Bailey, 1992; in press). The community is unusual on several counts: 1) The assemblage of taxa is distinctive. 2) Most specimens are small; some taxa seem stunted when compared with their counterparts in other coeval faunas. 3) The shells in most taxa are very thin and poorly calcified. 4) Most taxa are either edentulous or have only rudimentary dentition, suggesting that they relied largely on the organic ligament to maintain articulation of the valves. Such would be the case where shells are poorly calcified.

Past studies have divided Mazon Creek bivalves into two faunas, the Braidwood and the Essex. The two faunas appear to be relatively distinct, but to what extent the two may overlap has not been determined. More study is needed.

The Braidwood is a low-diversity assemblage dominated by freshwater mussels such as *Anthraconaia* and, to a lesser extent, *Anthraconauta.* These genera are well known outside the Mazon Creek, especially among the Upper Carboniferous coal-bearing strata of Europe and eastern North America. The remaining Braidwood bivalves are small thin-shelled myalinids, such as *Myalinella.* They are euryhaline (i.e., show broad tolerances to salinity variations) and elsewhere have been reported from marine, brackish, and freshwater deposits (Newell, 1942).

Baird and Sroka (1990, p. C40) postulated the Essex to be a possible estuarine environment. Prior to their study, the Essex community had long been thought to be marine. Essex bivalves are more diverse than Braidwood bivalves, and the freshwater genera are evidently lacking. Aside from euryhaline taxa such as the myalinids and solemyids, the marine character previously was supported by a seeming domination by edmondiids and scallops, and to a lesser extent heteroconchs, and perhaps a few isolated nuculoids. However, our study of the bivalve content of the Essex suggests that it bears little resemblance to true marine faunas in that the preponderance of Essex bivalves were adapted to a high-stress environment marked by fluctuating salinities and poorly oxygenated sediments.

The Essex contains mostly euryhaline brackish water bivalves with thin shells, along with a small admixture of stunted, marginal marine genera. As shown in Table 8A.1, the general makeup of the solemyid/myalinid-dominated Essex muds contrasts with offshore true marine bivalve mud communities, which are often monopolized by nuculoids and pholadomyaceans. For example, in Fulton County, the overlying Oak Grove Limestone Member (this is in the offlapping marine prodelta mud phase of the same Liverpool Cyclothem; the Francis Creek Member beneath it probably represents the gradual drowning of the forests of the Colchester No. 2 Coal, which is found at the base of the Francis Creek) is a thicker-shelled bivalve community dominated by nuculanaceans and, to a lesser extent, by both aviculopectinaceans and pholadomyaceans (Jackson, 1982); solemyids and myalinids are lacking entirely. In this respect other Carboniferous bivalve mud faunas associated with such formations as the Imo (Late Mississippian, Arkansas) and the Wewoka (Pennsylvanian, Oklahoma) more nearly resemble the Oak Grove except that within the Imo and Wewoka, aviculopectinaceans are rare and there is a richer diversity of larger and thicker-shelled nuculoids and pholadomyaceans. Of the faunas depicted in Table 8A.1, the Essex bivalves most nearly resemble those of the pyritic concretionary fissile shales above the Springfield No. 5 Coal (member 98 of Wanless, 1958) of the St. David Cyclothem. These also lack nuculoids but are dominated by solemyids, aviculopectinaceans, and pholadomyaceans; there are important differences, however—the taxa are different, consisting of larger thick-shelled forms. One other marine bivalve fauna, from the Shumway Limestone (Mattoon Formation, Late Pennsylvanian, Shumway, Illinois), shares similar limid, trigoniid, and edmondiid taxa with the Essex, but it shows its essentially marine character by the presence of nuculoids (*Nuculopsis girtyi*) and large thick-shelled myalinids (*Septimyalina perattenuata*) and the lack of solemyids. The robustness of the myalinids suggests adaptation to a firm supportive substrate; the Shumway probably represents a limey marine shoal, well offshore.

Because Essex bivalve fauna is so inconsistent with a normal marine assemblage, we interpret the Essex as onshore delta platform muds of a transitional environment such as a salt marsh, a marshy estuary, or a tidal flat. This conclusion is based on several essential points:

Table 8A.1
Midwestern Carboniferous Bivalve Faunas: Abundance and Diversity

Superfamily	Essex	Springfield	Oak Grove	Shumway	Wewoka	Imo
Nuculacea	O	O	?	C D1 S	A D4 M	A+ D4 M/L
Nuculanacea	O	O	A S	C D1 S	A+ D4 L	A+ D4 M
Solemyacea	A+ D3 M/L	A D2 M/VL	O	O	O	O
Ambonychiacea	A D3 S/M	R S	O	A D2 VL	C L	O
Mytilacea	O	O	O	C/R	C M	R S/M
Aviculopectinacea	A D4 S/M	C D2 S/M	C S	O	R VL	R S
Pteriacea	R D1 S	O	O	O	O	O
Limacea	R S	O	O	C D1 S	O	O
Pholadomyacea	R D4 S/M	C D2 L/VL	C S/M	C D1 M/L	C D3 L/VL	A D4 L/VL
Crassatellacea	?	R S	C/R D1 S	O	C/R D1 M	R L/VL
Kalenteracea	R D2 S	O	?	O	R M	R/C S
Trigoniacea	R D2 S/L	R D2 S/M	C S/M	C D1 S/M	R D1 S	A D2 M/L

Legend: A = abundant; C = common; R = rare; O = absent; D4 = high diversity; D3 = medium diversity; D2 = low diversity; D1 = very low diversity; S = small average size; M = medium average size; L = large average size; VL = very large average size

Note: The six nonsynchronous faunas of Carboniferous bivalves above are placed left to right in order of estimated proximity to shoreline: Essex (*extreme left*) = onshore; Imo (*extreme right*) = offshore. Normal marine assemblages (Oak Grove, Wewoka, and Imo) are dominated by thick-shelled taxa: nuculanaceans, nuculaceans, and pholadomyaceans; solemyaceans are noticeably absent. The brackish Essex is dominated by thin-shelled solemyaceans and ambonychiaceans (myalinids). The Springfield is basically similar to the Essex, but the presence of large pholadomyaceans, the lack of small myalinids, and the overall thicker shells (including the solemyaceans) of the Springfield taxa suggest that it is transitional between Essex-like brackish communities and normal marine communities such as the Oak Grove. The Shumway, which probably represents a firm substrate (offshore shoal), is similar to the Oak Grove but contains large, robust myalinids. Note that aviculopectinaceans occur in all environments, but smaller taxa tend to occur more onshore. The table is based on the following materials: Essex—FMNH, WIU, and NEIU MCP collections. Concretionary shale member (member 98 of Wanless, 1958) at the top of the Springfield No. 5 Coal (Carbondale Formation), Wyoming, Illinois—WIU collections. Oak Grove Limestone (Carbondale Formation), Wolf Covered Bridge, Illinois—Knox College (Galesburg, Illinois) collections. Shumway Limestone (Mattoon Formation), Shumway, Illinois—WIU collections. Wewoka Formation (Pennsylvanian of Oklahoma)—WIU collections. Imo Formation (Late Mississippian), Peyton Creek, Arkansas—WIU collections.

1) Evidence suggests that the preponderance of Essex bivalves were adapted to a high-stress environment marked by fluctuating salinities. Taxa are thin-shelled, mostly edentulous forms relying on ligament and periostracum for articulation instead of shelly hinge teeth, suggesting that the ability to secrete the calcium carbonate shell was limited.
2) The Solemyidae dominate all other families of Essex bivalves. As shown by one of us (Bailey, 1992, in press) the largest and most abundant Essex bivalve, the "clam-clam," is not an edmondiid as earlier authors had supposed but rather an undescribed member of the Solemyidae. Besides the "clam-clam," at least three other solemyid species occur in the Essex. Solemyids are noted for their tolerance to salinity variations and nutritional dependence on reducing environments—that is, sulfide-rich muds similar to those of pyrite and sphalerite-bearing shales in which the Essex fauna occurs. Modern solemyids thrive in organic, sulfide-rich sediments of salt marshes (see Reid, 1980; Reid and Bernard, 1980; Cavanaugh, 1983; Pojeta, 1988).
3) Myalinid bivalves are found in deposits of varying salinities. In contrast with Essex myalinids, normal marine myalinids like those of the Shumway Limestone attain significantly larger size, often with thick, robust shells. This suggests that Essex myalinids were somehow impaired in their ability to grow and calcify the shell.
4) Reports of nuculoid bivalves (Johnson and Richardson, 1970) in the Essex are spurious. In offshore marine muds such as those of the Oak Grove, Imo, and Wewoka (Table 8A.1) the nuculanaceans and nuculaceans ecologically replace the solemyids.
5) Although scallops are regarded as marine,

Essex genera are unusual. *Dunbarella, Euchondria,* and *Aviculopecten* are commonly found in dark and carbonaceous pyritic shales (Newell, 1937), often in association with both inarticulate and productid brachiopods. *Aviculopecten* seems to have been especially well adapted to a wide variety of environments. It has been reported in sediments ranging from open marine sediments, such as cross-bedded oolitic limestones, to restricted transitional marine deposits, including black pyritic limestones and carbonaceous shales. In offshore environments like the Imo and Wewoka, scallops tend to be larger, thicker shelled, and less common.

6) Although the Essex contains a variety of marine genera including limids, heteroconchs, edmondiids (i.e., *true* edmondiids), and related anomalodesmatans, all are rare, in some instances based on as few as one or two specimens. Perhaps they were carried in by tides. Furthermore, Essex representatives of these taxa appear smaller and thinner shelled than normal, suggesting some degree of stunting, as would be expected in a stressful transitional environment of salinity mixing.

Systematic Paleontology

Class Bivalvia Linné
Subclass Palaeotaxodonta Korobkov
(Subclass Protobranchia Pelseneer)
Order Solemyoida Dall
Superfamily Solemyacea Adams and Adams
Family Solemyidae Adams and Adams
Undescribed genus and species. Commonly known as the "clam-clam."
Figure 8A.2A, 8A.2B

The "clam-clam" is the most abundant of the Mazon Creek bivalves, and indeed one of the most common invertebrates in the fauna, accounting for as much as 70 percent of the total sample in a few locations. The shells are anteriorly elongated and shaped like smoothly rounded ovals, with low umbos and prominent comarginal growth lines. The valves are medium to large in size ranging from roughly 9 to 41 mm in length with a height/length ratio of about 0. 8. The paired valves nearly always remain joined together but widely open—that is, "butterflied"—in natural death pose (hence the collector's term, "clam-clam") and diverging from just behind the small beaks. Although the interior features of the shell are weakly impressed, specimens commonly show traces of the original ligament and periostracum and are occasionally found in association with burrows or trails. Early collectors and authors were nearly correct in their identification of this clam as *Solemya.* More recent authors have followed Dickson (1974), who misidentified it as *Edmondia* based on the shell outline, grooved edentulous hinge, and small internal ridges. One of us (Bailey, 1992, in press) has identified the primary ligament and hinge axis on the short side of the beaks; hence, the short end of the shell is posterior, and the long end is anterior. Anterior elongation of the shell is a diagnostic trait among solemyoids and nuculoids; whereas true examples of *Edmondia* are posteriorly elongate.

"Clam-clam" specimens frequently show a second portion of the ligament (Figure 8A.2A) extending between the butterflied valves in front of the beaks. This feature is not the primary ligament, but the anterior extension of the outer ligamental layer, a feature commonly seen in modern solemyids such as *Solemya* and *Acharax.* The specifics of this and other solemyids bivalves from the Essex fauna are published elsewhere (Bailey, in press).

Modern solemyids are euryhaline free burrowers common in sulfide-rich, reducing muds of salt marshes. Because the gut is either reduced or absent (Reid, 1980; Reid and Bernard, 1980), they are nutritionally dependent on intracellular symbionts in the form of chemosynthetic bacteria which oxidize sulfides (Cavanaugh, 1983). Concentrations of sulfide minerals such as sphalerite, pyrite, or pyrite pseudomorphs within the valves of the "clam-clam" suggest a possible feeding mode in which symbiotic bacteria served a similar role.

The "clam-clam" is restricted to the Essex community in a region including the Will County half of Pit 11 but less commonly in the Kankakee portion of the same pit. It is also known from shaft mines near Braceville and

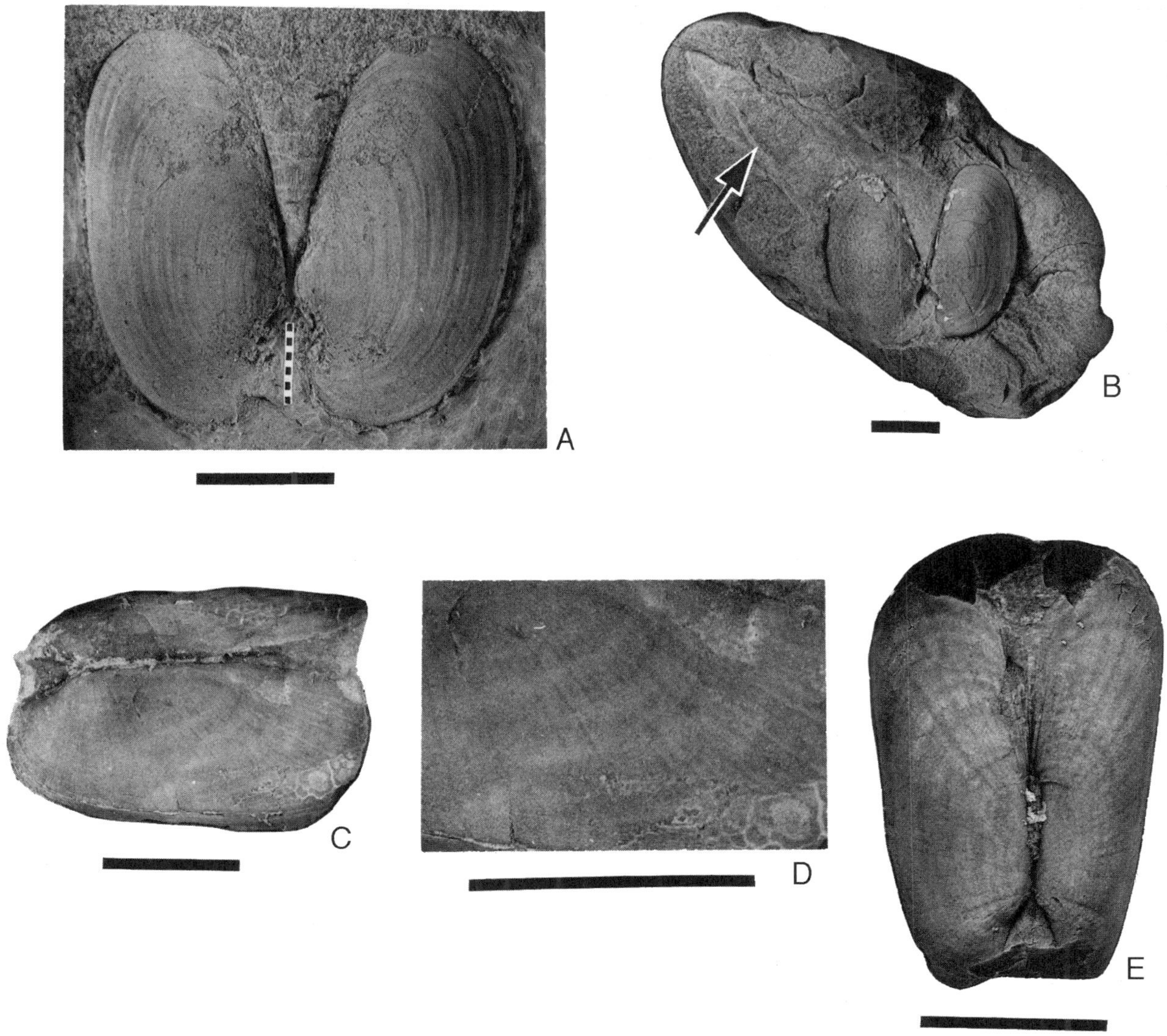

Figure 8A.2. Mazon Creek solemyid bivalves: Francis Creek Shale, Carbondale Formation, Pennsylvanian of Illinois. **A, B** Undescribed solemyid bivalve, commonly known as the "clam-clam": **A** Butterflied articulated specimen showing primary hinge axis (*dashed line*), FMNH PE 18152, Pit 11, Will-Kankakee Counties; this is the same specimen labeled "undetermined bivalve" by Johnson and Richardson (1970, pl. 5, fig. 6). **B** Butterflied articulated specimen with preserved burrow or trail (*arrow*), NEIU MCP 163, Pit 11. **C–E** *"Solemya" radiata:* **C** right dorsolateral view of articulated composite mold, FMNH PE 29875, Pit 11, Will County; **D** enlargement of same specimen showing radial plicae; **E** dorsal view of composite mold of articulated specimen, FMNH PE 52187, Pit 11, Will-Kankakee Counties.
All scale bars = 10 mm.

Braidwood, as well as mines at Ottawa and west of Morris (Chowder Flats). One of us (Bailey) has collected several examples from Essex locations near Astoria.

Genus *Solemya* Lamarck, 1818
"Solemya" radiata Meek and Worthen, 1860
Figure 8A.2C–8A.2E

"Solemya" (misspelled as *Solenomya* by early authors) *radiata* is confined to the Essex community but is much less common than its unnamed cousin the "clam-clam." Modern *Solemya* and its close relative *Acharax,* the "awning shells," are living fossils; Mazon Creek examples closely resemble modern specimens in every detail. Earlier examples are known from rocks as old as Devonian (Prosh, 1988; Bailey and Prosh, 1990). *"Solemya" radiata* is sometimes found within or near its own burrow; in rare cases, traces of the "awning," an organic frill of periostracum extending beyond the shell margins, are also preserved. The shells

are anteriorly elongated, shaped like smoothly rounded ovals, and marked by weak ribs (Figure 8A.2D) radiating from the low beaks. It is easily distinguished from the "clam-clam" by its lower, weaker umbos, radial ornament, and weaker comarginal growth lines.

There are at least three species closely resembling *Solemya* in the Essex community. *"Solemya" radiata* is the most common of these. Although it has been called *Solemya* by past authors (Richardson and Johnson, 1970; Johnson and Richardson, 1970), the ligament is not identical to *Solemya* in the strict sense but closely resembles that of a sister genus, *Acharax.* The morphological basis for transferring *"S." radiata* to *Acharax* is given elsewhere (Bailey, in press). Specimen sizes range from 12 to about 26 mm in length with a height/length ratio of about 0. 4. This species is relatively uncommon, restricted to a few locations in the Will County part of Pit 11. Additionally, four specimens have been found at shaft mines in Will County (G. C. Baird, personal communication, 1983).

Another rare *"Solemya"* is the larger species formerly called *Solemya trapezoides.* However, Carter (1990) argued that it should be treated as a separate subgenus (*Nacrosolemya*) of *Acharax.* Outside the Mazon Creek, shells referrable to this species are relatively common at several locations within the Carbondale Formation. It ranges from 24 to 32+ mm in length with a height/length ratio similar to that of *"S." radiata,* but with a more subangular outline.

The mode of feeding in *Solemya* and *Acharax* is similar to that of the "clam-clam" described above. Concentrations of sulfide minerals within the valves of *"S." radiata* suggest possible presence of chemosynthetic bacteria.

Subclass Pteriormorphia Beurlen
Order Pterioida Newell
Superfamily Ambonychiacea Miller
Family Myalinidae Frech

Myalinids are second only to the solemyids in abundance. The shells of Mazon Creek species are among the thinnest in the fauna. Their fragility is such that they are normally crushed flat by sedimentary compaction, which, judging from the preservation of other species in the fauna, was relatively slight. Myalinids normally show broad variation in shell form within each species (Newell, 1942). To this must be added the additional variation imposed by the compaction. As a result, the exact number of myalinid species in the fauna is difficult to determine and awaits further study. Past investigations of the myalinids (Dickson, 1974; Sroka, 1984) suggesting the presence of two larger, thick-shelled marine species, *Septimyalina perattenuata* and *Myalina wyomingensis,* are not supported because the crushed shell shapes in samples of identified specimens broadly overlap and the inner features, including the diagnostically important "umbonal deck" (septum), which distinguishes the two species, are as yet unknown among the available fossils. If the Mazon Creek examples belong to these two species, they must be viewed as stunted populations. Identification of the myalinids here is largely based on Newell's (1942) monograph on North American genera and species.

Genus *Myalinella* Newell, 1942
Myalinella meeki (Dunbar, 1924)
Figure 8A.3A–8A.3E

Most so-called *Septimyalina perattenuata* and *Myalina wyomingensis* from the Mazon Creek are probably the small widespread and thin-shelled Pennsylvanian–Permian genus *Myalinella.* Newell (1942) used size and thickness of the shell, the angle between the hinge line and umbonal ridge (α), and relative development of the umbonal septum and anterior auricles as aids in identifying myalinids. In these respects, the shells are in clear agreement with *Myalinella,* especially *M. meeki* as figured by Newell (1942, pl. 14, figs. 7–10, 12–14, esp. fig. 10) from the Carlton Limestone of Kansas. Composite molds lack the umbonal septum of *Septimyalina,* and the shells are so thin that they often appear wrinkled or even folded by sedimentary compaction (Figure 8A.3E). This suggests the shells were not only paper thin but also poorly calcified. The valves are triangular to subquadrate in outline and lack an anterior auricle.

However, the shell outline of *M. meeki* superficially resembles several thicker-shelled myalinids from the Pennsylvanian including *Septimyalina orthonota* (Mather), Hale Formation, Fayetteville, Arkansas, in Newell (1942, pl. 13,

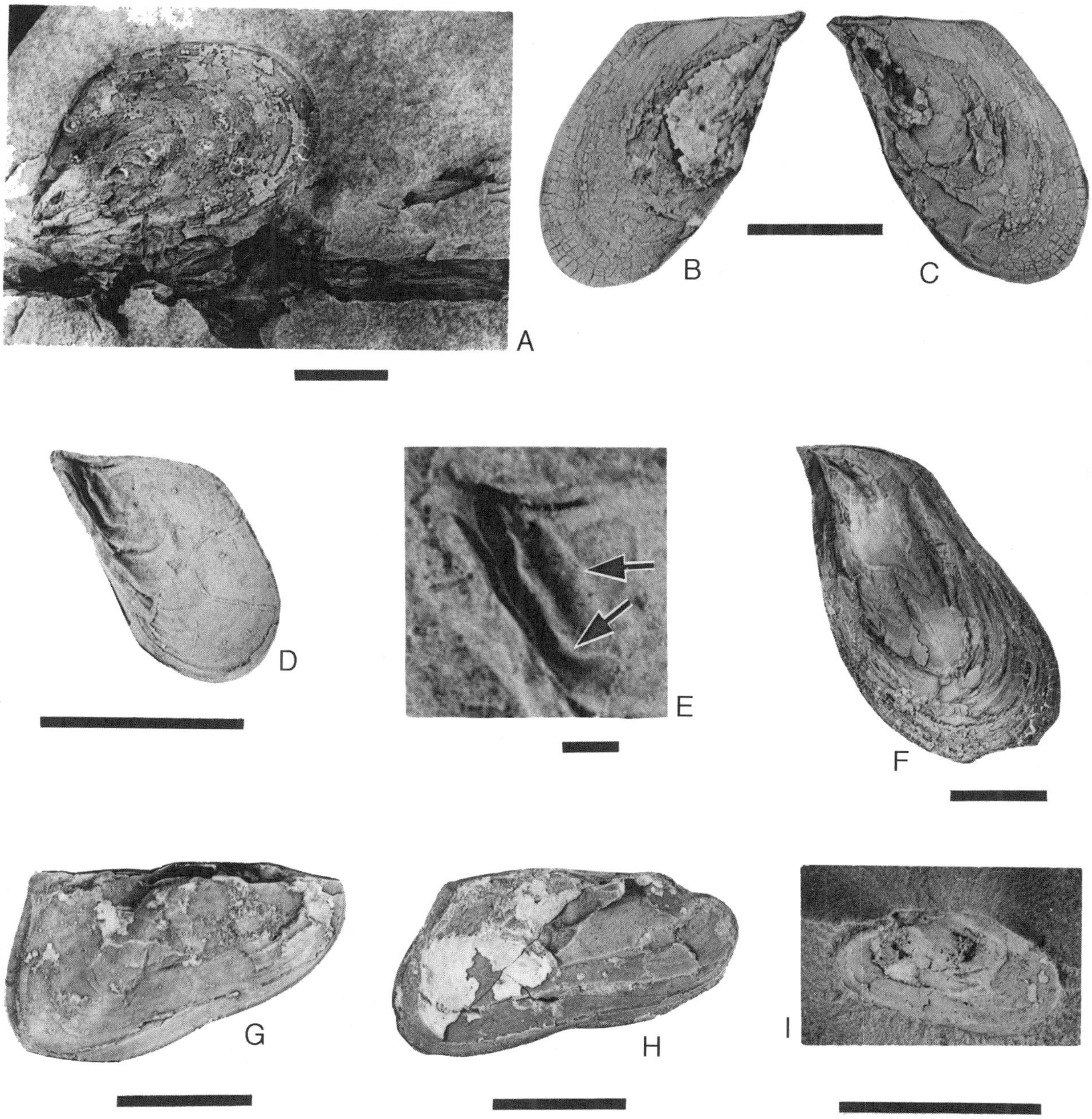

Figure 8A.3. Mazon Creek myalinid bivalves: Francis Creek Shale, Carbondale Formation, Pennsylvanian of Illinois. **A–E** *Myalinella meeki:* **A** Left lateral view of articulated specimen attached to a twig, Testa collection, NEIU MCP TVT 3409, Pit 11. **B, C** Articulated specimen, NEIU MCP 112, Pit 11, Kankakee County; right lateral view (B) and left lateral view (C). **D, E** WIU MC 1101; left lateral view (D) and detail of umbonal area showing lack of umbonal septum and rounded wrinkles (*arrows*) due to flattening of shells (E). **F** *Myalinella* sp., left lateral view, Testa collection, NEIU MCP TVT 3394. **G, H** *Anthraconaia ohioensis:* **G** right lateral view of articulated specimen, WIU MC 1102; **H** right lateral view of articulated specimen, WIU MC 1103. **I** *Anthraconauta* sp., left lateral view of WIU MC 1104. Scale bars: E = 1 mm; all others = 10 mm.

fig. 2); *S. perattenuata,* "Virgilian" of Nebraska, in Newell (1942, pl. 13, figs. 3, 5); *S. perattenuata,* Magoffin Member, Breathitt Formation, eastern Kentucky, in Carter (1990, fig. 29A); and *Myalina pernaformis,* Pottsville Group, Ohio, in Morningstar (1922, pl. 11, figs. 5–9).

The outline is also like *S. perattenuata* from the Shumway Limestone Member (Mattoon Formation, Effingham, Illinois) collected by Bailey. But here again, the valves of *S. perattenuata* are large, robust, and very thick, especially along the hinge line. However, it is possible that many of Newell's myalinellas are nothing more than stunted ecomorphs of larger myalinids.

Myalinella meeki is distinguishable from other Mazon Creek clams by the edentulous, mussellike valves, which are more or less triangular and spatulate in outline with mostly weak growth lines. The beaks are narrowly pointed, low or reclining, and situated at the anterior limit of the hinge. In some specimens they project forward slightly. The shells are small, less than 20 mm in maximum dimension.

Myalinella meeki was recognized by both Dunbar (1924, p. 201) and Newell (1942, p. 60) as having broad tolerances to salinity variations. Elsewhere it is frequently found in freshwater to brackish water deposits in association with nonmarine flora and fauna. It was a sedentary surface dweller and can occasionally be seen still attached to bits of wood (Fig. 8A.3A) with traces of the organic byssal threads still preserved (Bailey, 1992). In addition, shells are commonly encrusted with barnacles, *Spirorbis*, or unidentified plants. *Myalinella* evidently was gregarious. It is locally abundant at many Essex sites, especially the Will County portion of Pit 11 and among gob piles of several shaft mines. It also occurs in a band roughly trending northwest-southeast, parallel but seaward of the paleoshore position of Baird and Shabica (1980).

Myalinella sp.
Figure 8A.3F

This rare Essex bivalve is known from a single specimen donated by Thomas V. Testa. It is much larger than the other myalinids, measuring about 41 mm in maximum dimension (i.e., 30 mm in true length; 36 mm in height). In addition to its large size, it is distinguished from other myalinids by the relatively short hinge line and great posteroventral extension of the shell, where the growth lines are also better marked and more rugose than in other Mazon Creek myalinids. The beaks are small and narrow and project horizontally forward.

Assignment here to *Myalinella* is far from certain because internal shell features are unknown. The Testa specimen has the outline of *Myalina pernoides* from the Carboniferous of Britain as figured by Hind (1897, pl. 4, figs. 1, 1a). Newell (1942) believed this species should be placed in *Septimyalina*. But *Septimyalina*, in the strict sense, is a larger, robust-shelled genus; whereas the Testa specimen, like all Mazon Creek myalinids, has thin shells. The specimen also resembles one example of *Septimyalina perattenuata* figured by Newell (1942, pl. 13, fig. 11a, 11b), but here again, *S. perattenuata* is demonstratably a thick-shelled species. Furthermore, microscopic examination of the umbonal region of the Testa specimen (a composite mold) shows no indication of the umbonal septum associated with *Septimyalina*. Moreover, the size of the Testa specimen is small for a mature *S. perattenuata*.

Alternatively, the Testa specimen may perhaps be a senile *Myalinella meeki*. In many respects it resembles *M. meeki* except for the larger size and ventral elongation of the shell, which recurves backward in later stages of growth, a shell trait that Newell (1942) considered to be gerontic.

Freshwater Myalinidae

Roof shales of Pennyslvanian coal beds in Europe and eastern North America are noted for rich, varied faunas of freshwater and estuarine molluscs. The bivalves among these communities, which fall into two main groups, unionaceans and myalinids, have been well studied since the late 1800s by Hind, Weir, Eagar, and several others. Identification is often imprecise because most of the genera, unionacean and myalinid alike, are highly variable if not polymorphic. That is, the shell form of a given genus or species is likely to consist of more than one mode. The modes represent extremes linked to each other by an intermediate series. Whereas shells of unionaceans such as *Anthracosia* and *Carbonicola* are generally ovate to nasute in outline, those of myalinids like *Anthraconaia, Naiadites, Curvirimula,* and *Anthraconauta* tend to be more alate in form. But in large collections, the shell forms of the myalind genera overlap, not only each other but also variants among the aforementioned unionacean genera as well. In such circumstances, the Braidwood taxa discussed below can be neither adequately identified nor described until the range of variation within the population is determined. Eagar (e.g., 1952, fig. 5) used both pictorial and statistical diagrams to summarize the variation among polymorphic genera such as these.

Genus *Anthraconaia* Trueman and Weir, 1946
Anthraconaia ohioensis (Morningstar, 1922)
Figure 8A.3G, 8A.3H

Anthraconaia is a byssate myalinid common in Braidwood localities, especially Pit 1 and the Morris area. In eastern North America and Europe it is common in freshwater to estuarine roof shales of coal beds. Some Braidwood examples are strikingly similar to *A. ohioensis* as figured by Morningstar (1922, pl. 12, figs. 10–13), Rogers (1965, pl. 85), and Newell (1942, pl. 14, figs. 5, 6). The edentulous shells are mussellike, resembling *Myalinella meeki* but distinguished by the rounded, slightly elevated beaks that are situated distinctly behind the anterior limit of the relatively longer hinge. The valves, which are remarkably variable in shape, are comparable in size to those of *M. meeki* with maximum dimensions of usually less than 25 mm.

Genus *Anthraconauta* Pruvost, 1930
Anthraconauta sp.
Figure 8A.3I

Anthraconauta is a fresh to brackish water myalinid broadly similar to *Anthraconaia* in form and occurrence. *Anthraconauta* is distinguished by an obliquely oval outline that is less trigonal than the more common *Anthraconaia.* The three Field Museum specimens of *Anthaconauta* figured by Dickson (1974, pls. 22–24) recall both *Anthraconaia* and *Anthraconauta* in certain respects; none could be located. The specimen figured here is similar to that of the type species *Anthraconauta phillipsii* but also resembles *Curvirimula* sp. from the Upper Carboniferous of Nova Scotia as figured by Rogers (1985, pl. 1, figs. 1–4).

Superfamily Pteriacea
Family Pterineidae Miller
Genus *Leptodesma* Hall, 1883

Leptodesma ohioense (Herrick, 1887)
Figure 8A.4M

Leptodesma is a rare Essex bivalve. The various shell shapes observed in museum and amateur collections suggest that more than one species may be present. The figured specimen closely resembles *L. ohioense* (= *Pteria ohioense* [Herrick] of Morningstar, 1922, p. 212) in Hoare et al. (1979, pl. 6, fig. 7).

Leptodesma ohioense has a small to medium-sized mussel-shaped shell with a prominent posterior auricle in the form of a long pointed wing. The beaks rise slightly above the hinge line, and the body of the shell is marked by both comarginal folds and growth lines. The wing is the most distinguishing trait. Like modern pearl oysters, *Leptodesma* was an epibyssate genus that used byssal threads to attach itself to elevated surfaces such as plant stems. Winged genera like *Leptodesma* use the wing as a hydrofoil, swinging to and fro in the gentle shallow currents, the water being directed along the wing and body of the shell to the inhalant and exhalant areas of the mantle margins.

Family Posidoniidae Frech
Genus *Posidonia* Bronn, 1828
Posidonia fracta (Meek, 1875)
Figure 8A.4J

This rare Essex species is known from a single Pit 11 specimen. The shells are thin, flat, and covered with fine comarginal growth lines. The outline resembles a myalinid but fatly rounded with a short, straight oblique hinge line. The specimen compares favorably with the figure of *P. girtyi* in Morningstar (1922, pl. 12, fig. 1). Newell (1937) considered it a synonym of *P. fracta,* which he indicated was the only valid Pennsylvanian species of *Posidonia.*

Order Pectinoida Rafinesque
Superfamily Aviculopectinacea
Meek and Hayden

The pectinoids are commonly known as scallops. Identification of the Mazon Creek genera and species is based on the classic monograph of Paleozoic scallops by Newell (1937).

Family Aviculopectinidae Meek and Hayden
Genus *Aviculopecten* M'Coy, 1851

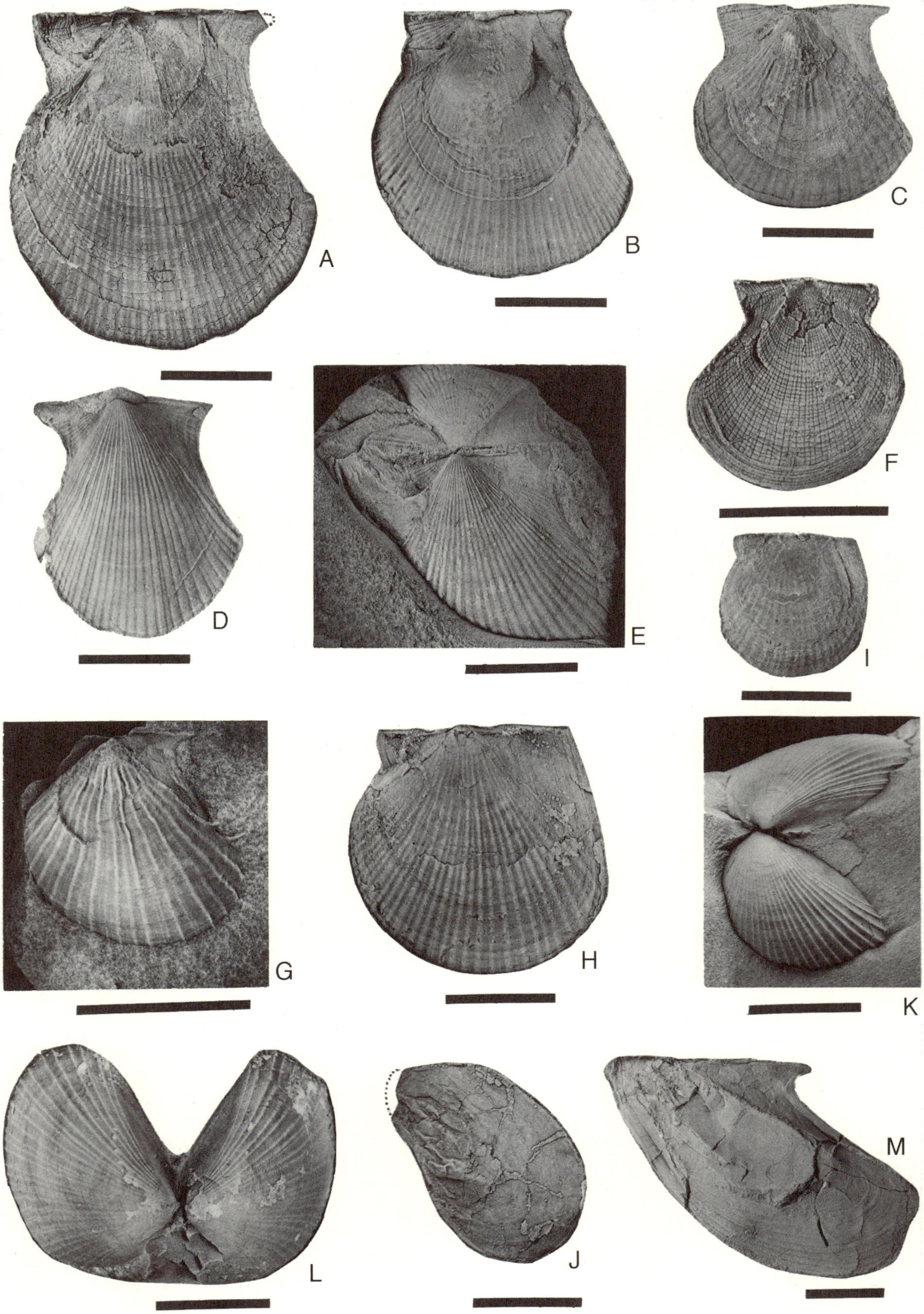
A
B
C
D
E
F
G
H
I
J
K
L
M

Aviculopecten mazonensis Worthen, 1890
Figure 8A.4A–8A.4C

The largest and most abundant of the Essex scallops is *Aviculopecton mazonensis.* Like scallops today it may have been either free living or byssally attached. Occasional specimens are encrusted by the hydroid *Drevotella proteana.* Data from Newell (1937) suggest that *A. mazonensis* had broad environmental tolerances; its common occurrence in disparate sediment types, ranging from cross-bedded oolitic limestones to pyritic carbonaceous shales suggests it was equally at home in well-oxygenated open marine environments and in marginal marine reducing muds where it occurs with solemyids.

Aviculopecten mazonensis is distinguished from other Mazon Creek scallops by well-defined posterior and anterior auricles ("ears"), strong radial ribs, and fairly large size. Newell (1937) gave a size range for this species of 17–36 mm in length and 18–35 mm in height, with a height/length ratio of 0.9–1.1. The valves are unequal, the larger left valve bearing ribs that increase in number ventrally by intercalation. The smaller right valve has ribs that increase ventrally by bifurcation. Umbos are fairly strong compared with other Essex scallops.

This species is commonly found in strip and shaft mines containing Essex concretions but is most abundant in the Will County portion of Pit 11.

Figure 8A.4. Mazon Creek bivalves (Aviculopectinacea, Limacea, and Pteriacea): Francis Creek Shale, Carbondale Formation, Pennsylvanian of Illinois. **A–C** *Aviculopecten mazonensis:* **A** left lateral view of articulated composite mold, NEIU MCP 116, Pit 11, Will County; anterior auricle is a superimposed composite of auricles of both left and right valves. **B** Left lateral view of articulated specimen, WIU MC 1105. **C** Left lateral view of articulated external mold, WIU MC 1106. **D, E** *Aviculopecten janewiedlini* new species: **D** left lateral view of holotype, NEIU MCP 187, Pit 11, Will County; **E** articulated paratype, WIU MC 52958, Pit 11, showing left and partial right valve. **F** *Euchondria pellucida,* External mold (concave) of a left valve, FMNH PE 22669, Pit 11, Will or Kankakee County. **G** *Acanthopecten* sp., a left valve with spines, NEIU MCP TVT 1764. **H, I** *Dunbarella striata:* **H** a large left valve, NEIU MCP 288, Pit 11; **I** a typical left valve, WIU MC 1108, Pit 11. **J** *Posidonia fracta,* left valve (cast), WIU MC 1109, Pit 11. **K, L** *Palaeolima retifera,* articulated specimens; the small auricles remain mostly embedded in matrix: **K** NEIU MCP 287, Pit 11; **L** NEIU MCP 115, Pit 11. **M** *Leptodesma ohioense,* left valve, NEIU MCP 113, Pit 11. All scale bars = 10 mm.

Aviculopecten janewiedlini new species
Figure 8A.4D, 8A.4E

Diagnosis. Like *Aviculopecten mazonensis,* but valves more tumid, more equilateral; valves are relatively taller but narrower, umbonal angle more acute; radial ribs are fewer, smoother, more regular, and, in the left valve, more deeply sculpted. Comarginal growth lines are weaker than in *A. mazonensis,* and the auricles are relatively larger.

Type specimens. Holotype, NEIU MCP 187; paratype, WIU MC 52958.

Etymology. Named in honor of Jane Wiedlin, singer and songwriter with the Go-Go's.

Description. Left valve: valve acline, equilateral, strongly convex; umbo low, subcentrally placed; marked by smooth even radial ribs that increase ventrally by intercalation; comarginal growth lines weak; angle of slope on either side of the umbo is between 60° and 75°; anterior and posterior auricles are prominent, the posterior one slightly larger and longer. Right valve: valve nearly flat, smaller than left valve, but similarly shaped; ornament weak, marked by smooth, even radial ribs that increase ventrally through bifurcation; posterior auricle similar to that of left valve; anterior auricle relatively larger than that of the left valve, smoothly arcuate below, with deep byssal notch. Ligament amphidetic as in *A. mazonensis.* Other features not seen.

Remarks. The 60°–70° umbonal angle is considerably narrower than the 80°–90° umbonal angle of *A. mazonensis.* The height/length ratio of *A. janewiedlini* is greater than that of *A. mazonensis,* measuring about 1.5. Ribs on the body of the shell are roughly one-third fewer to one-half of those counted in *A. mazonensis.* Along the ventral margin, the ribs in *A. janewiedlini* are evenly developed, but in *A. mazonensis* there is regular alternation of strong and weak ribs.

The swollen shells and strong relief of the regular ribs recall *Palaeolima retifera,* with which it may be confused, but the valves are more equilateral and the auricles markedly larger in *A. janewiedlini.*

Genus *Acanthopecten* Girty, 1903

Acanthopecten sp.
Figure 8A.4G

The rare "spiny scallops" of the Mazon Creek belong mostly to this genus. The figured specimen, from Pit 11 (Essex), is a small left valve 14 mm in height and 16 mm in length.

The shell of *Acanthopecten* resembles *Aviculopecten* except for the spiny ornament and slender auricles. In *Acanthopecten,* ornament consisting of weak spines or scalelike nodes marks the intersections of the growth lines with the radial costae. In addition, the costae project slightly beyond the shell margins to form short, weak spines. The spinous ornament is best developed posteroventrally, where a few of the extended costae form short hooks. The auricles are more slender than in *Aviculopecten;* in *Acathopecten* the posterior auricle narrows to a sharp point.

The morphology of this undescribed Essex species recalls the common Pennsylvanian species, *Acanthopecten carboniferus* Stevens (1858) but with some important differences. In *A. carboniferus* the spiny or scaly ornament is best developed on the left valve, whereas the right valve remains relatively smooth. But the ornament of the figured specimen (a left valve) is relatively weak and, in fact, more nearly resembles the right valve of *A. carboniferus;* the right valve of the Essex species was not seen.

Genus *Euchondria* Meek, 1874
Euchondria pellucida
(Meek and Worthen, 1860)
Figure 8A.4F

This scallop probably had a mode of life somewhat similar to *Aviculopecten.* However, its distribution is limited to pyritic black shales and carbonaceous limestones, suggesting a preference for marginal marine reducing muds. *Euchondria pellucida* is an uncommon Essex species. Its precise distribution is unknown, but most specimens are from Pit 11.

This species is smaller than *Aviculopecten mazonensis,* with lengths commonly ranging from 9 to 13 mm and heights from 8 to 12 mm. It has unequally developed left and right valves and well-defined posterior and anterior auricles. It is most easily distinguished from *A. mazonensis* by shell ornament, which on the left valve consists of widely spaced radial ribs crossed by comarginal growth lines of similar strength, producing a checkerboard appearance. The right valve is smooth or marked by comarginal growth lines. Umbos are weaker than in *A. mazonensis*

Family Pterinopectinidae Newell
Genus *Dunbarella* Newell, 1937
Dunbarella striata (Stevens, 1858)
Figure 8A.4H, 8A.4I

This small to medium-sized scallop is limited to Essex localities, with a distribution similar to but less abundant than *Aviculopecten mazonensis.* Outside the Mazon Creek, *Dunbarella* is commonly found with *Euchondria,* mostly confined to carbonaceous pyritic muds and limey muds.

The height/length ratio is about 1, with lengths and heights ranging from 11 to 21 mm. *Dunbarella* is distinguished from *Euchondria* and *Aviculopecten* by shell outline and auricle development. The umbos are weak, and the shell outline is nearly circular. Unlike other Essex scallops, only the anterior auricle is well developed; the posterior one is poorly differentiated from the body of the shell. Also, the radial ribs that mark the shell are visibly weaker in the region of the posterior auricle.

The shells are similar to those called *D. whitei* (Meek, 1872) by Newell (1937). Murphy (1967) considered *D. whitei* to be synonymous with *D. striata.*

Order Limoida Rafinesque
Superfamily Limacea Rafinesque
Family Limidae Rafinesque
Genus *Palaeolima* Hind, 1903
Palaeolima retifera (Shumard, 1858)
Figure 8A.4K, 8A.4L

Commonly called file shells, the Limidae is a mostly Mesozoic to Cenozoic marine family with relatively few representatives in the Paleozoic. Modern limids have colorful tentacles extending from the mantle edges. Although they can swim short distances by clapping the valves together, they mostly live as epifaunal nestlers, building nests constructed from shell debris held together by byssal threads.

Palaeolima retifera is a rare Essex bivalve; so far it is known only from Pit 11. The radially ribbed valves, which resemble Essex scallops, are nearly equilateral and ovate in outline, with a small auricle on each side of the beak. The valves are small, with dimensions and height/length ratios similar to those of *Euchondria* and *Dunbarella. Palaeolima retifera* is distinguished by the greater convexity of the shell, narrower, more widely spaced ribs, shorter hinge line, and weaker auricles.

Subclass Isofilibranchia Iredale
Order Modiomorphoida Newell
Superfamily Kalenteracea Marwick
Family Permophoridae van de Poel
Genus *Permophorus* Chavan, 1954
Permophorus spinulosa (Morningstar, 1922)
Figure 8A.5A

This rare species is known only from the Essex location of Pit 11. It superficially resembles modern *Cardita.* The shell is medium sized, one specimen measuring 25 mm in length and 11 mm in height. The shape is an elongated trapezoid with the low beaks placed near the anterior extremity. Distinctive features are the narrow, widely spaced ribs (about three), which radiate obliquely across the rear of the shell from the beaks. The fine comarginal growth lines abruptly change direction as they cross the ribs.

Permophorus sp.
Figure 8A.5B, 8A.5C

Other species of *Permophorus* evident in the Mazon Creek are as yet poorly known. Two rare examples from Pit 11 were donated by Mildred Scheffel-Tennant. One of these is very small, measuring only 7 mm in length, and perhaps is either juvenile or stunted. The shell is marked by a broad umbonal ridge extending posteroventrally from the beak. The other is much larger, measuring 18 mm in length and 9 mm in height. The form recalls two probable Mississippian permophorids figured by Driscoll (1965, pl. 15, fig. 3; pl. 16, fig. 2) from the Marshall Sandstone of Michigan. Driscoll called these *"Sanguinolites" herricki* and *"Parallelodon" marshallensis.*

Subclass Heteroconchia Hertwig
Order Veneroida Adams and Adams
Superfamily Crassatellacea Férussac
Family Astartidae d'Orbigny
Genus *Astartella* Hall and Whitney, 1858
Astartella concentrica (Conrad, 1842)
Figure 8A.5D

We cannot verify the presence of *Astartella* within the Essex. While it is true that various Essex collections, both private and public, contain specimens labeled as *Astartella,* the specimens seen thus far are misidentified (see *Edmondia ovata* below). Bird's (1968) statistical data for several species of *Astartella* suggest that the Essex examples are too large for this genus. Moreover, beaks of so-called astartellas from the Essex are too broad and low, and the hinge line is too long. True astartellas have a relatively short hinge line and a short shell. The umbones are relatively higher, and the comarginal lirae that mark the shell are narrower and more regular. Though not from the Mazon Creek, our figured example of *Astartella* is fairly typical. This one is from the St. David Limestone, which occurs higher in the Carbondale Formation than the Francis Creek Shale.

Order Trigonioida Dall
Superfamily Trigoniacea Lamarck
Family Schizodidae Newell and Boyd
Genus *Schizodus*
de Verneuil and Murchison, 1844

When mature, *Schizodus* was perhaps the largest of the Essex bivalves and probably an active burrower. It is relatively uncommon, evidently restricted to the Will County half of Pit 11, the Rixson shaft mine (leveled and underwater in Will County, Pit 11), and Chowder Flats.

The valves are distinguished from other Mazon Creek species by the triangular-ovate outline, high narrow beaks, and the rounded keel extending diagonally from the beaks toward the posterior, where it forms an angle in the otherwise smooth curvature of the shell margin.

Although shell forms among various genera of families Schizodidae and Eoschizodidae are often similar, dentition may differ. Because the dentition of Mazon Creek examples was not observed, the name *Schizodus* is applied here with reservation.

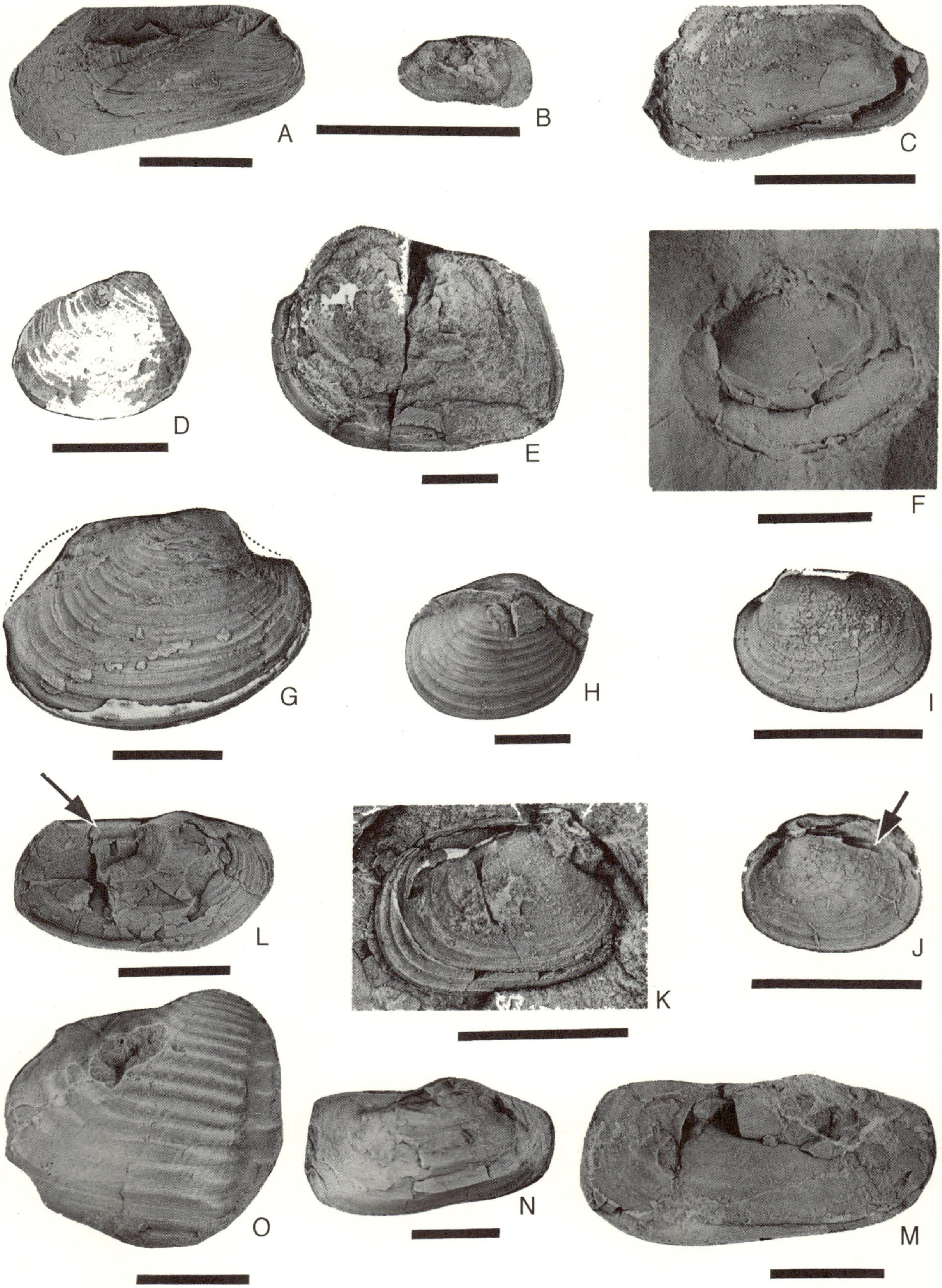
A
B
C
D
E
F
G
H
I
L
K
J
O
N
M

Schizodus cf. *S. wheeleri* (Swallow, 1863)
Figure 8A.5E

The shells here are large; one example measures 40 mm in length and 30 mm in height. Compared with other species of *Schizodus* and *Eoschizodus*, the shells of the Mazon Creek species are more equilateral with very angular posterodorsal margins, and the anterodorsal margin is higher and more emphasized than in other species we have seen. But the shells are too flattened to be certain whether the preserved shell outline resembles the original. The form compares to *S. cuneatus* Meek, but the outline of that species is too angular, the umbones are too peaked, and the small posterodorsal flange along the hinge line is lacking. The closest matches we can find are *S. wheeleri* as figured in Hoare et al. (1979, pl. 12, figs. 6–9) and *"Protoschizodus" aequilateralis* in Hind (1898, pl. 20, fig. 8). Both are like the Mazon Creek *Schizodus* in size and form but lack the high anterodorsal margin.

Figure 8A.5. Mazon Creek bivalves: (Kalenteracea, Crassatellacea, Trigoniacea, and Pholadomyacea): Francis Creek Shale, Carbondale Formation, Pennsylvanian of Illinois (except for D). **A** *Permophorus spinulosa,* right valve, FMNH PE 36734, Pit 11, Will or Kankakee County. **B, C** *Permophorus* sp.: **B** small left valve, WIU MC 1107, Pit 11; **C** right lateral view of articulated specimen showing ventral overlap of left valve, WIU MC 1110, Pit 11. **D** *Astartella concentrica,* Right lateral view of articulated specimen, WIU 9551, Carbondale Formation, St. David Limestone, Middle Grove, Illinois. **E** *Schizodus* cf. *S. wheeleri,* left lateral view of fractured specimen, NEIU MCP HTP 9534, Pit 11, Will County. **F** *Schizodus* cf. *S. affinis,* crushed interior of a right valve, WIU MC 1111. **G, H** *Edmondia ovata:* **G** Right lateral view of articulated specimen, NEIU MCP 164, Pit 11, Will County; mislabeled *Astartella* sp. **H** Right lateral view of articulated specimen, FMNH PE 36775, Mazon Creek area, Will-Grundy-Kankakee Counties; mislabeled *Astartella* sp. **I, J** *Edmondia aspinwallensis:* **I** Left lateral view of articulated specimen, FMNH PE 22318, Pit 11, Will-Kankakee Counties; **J** left lateral view of articulated specimen (partially overlapped dorsally by right valve), FMNH PE 36794, Mazon Creek area, Will-Grundy Counties; arrow shows furrowed edentulous hinge. **K** *Edmondia oblonga,* right lateral view of ?articulated specimen, FMNH PE 20390, Pit 11, Will-Kankakee Counties. **L–N** *Sedgwickia* sp.: **L** right lateral view of crushed articulated specimen showing thick edentulous hinge plate, FMNH PE 38936, Morris, Grundy County; **M** right lateral view of fractured specimen, NEIU MCP 296, Will or Kankakee County; **N** right lateral view of specimen with slightly expanded posterodorsal flange, WIU MC 1112, Pit 11. **O** *Grammysioidea hayi* new species; left lateral view of holotype, an articulated specimen showing bifurcating concentric costae, single radial sulcus, and notched posterodorsal margin; specimen is incomplete (anterior portion missing), NEIU MCP 604, Pit 11 (gift of Andrew A. Hay).
All scale bars = 10 mm.

Schizodus cf. *S. affinis* Herrick, 1887
Figure 8A.5F

This rare Essex *Schizodus* is smaller and more equidimensional than *S.* cf. *S. wheeleri.* The figured specimen is a crushed shell interior too poorly preserved for anything but tentative identification. The rounded shell outline is suggestive of *S. affinis* from the Pennsylvanian of Ohio as figured by Morningstar (1922, pl. 12, fig. 17) and Hoare et al. (1979, pl. 13, fig. 12).

Subclass Anomalodesmata Dall
Order Pholadomyoida Newell
Superfamily Pholadomyacea Gray
Family Edmondiidae King
Genus *Edmondia* de Koninck, 1841

As indicated above, the "clam-clams" of the Mazon Creek are not *Edmondia* but undescribed solemyids. Nevertheless, there are true edmondiids among the Essex, but they are uncommon and easily confused with other species. Edmondiids were shallow burrowers, perhaps with short siphons. They and related pholadomyoid clams are elsewhere common among open marine, nearshore sediments; their relative absence among the brackish environments of the Mazon Creek is not surprising. Two of the species here are smaller than normal, suggesting that their growth was stunted, perhaps by the variable salinity.

In the literature, there are many described species of *Edmondia.* Some are based on poor material; others are misidentified. Still others are either indistinguishable from one another or show ranges of shell variation that broadly overlap. North American and European edmondias show numerous similarities, but different species names are used on each continent. Although the genus itself stands as valid based on Runnegar and Newell's (1974) evaluation, the various species sorely need to be restudied. Until then, identification can be only tentative.

Edmondia ovata Meek and Worthen, 1873
Figure 8A.5G, 8A.5H

In Field Museum and amateur collections, this Essex edmondiid is more frequently seen than the others. The thin swollen shells are ovoid in outline and marked by strong comarginal rugae and growth lines of varying strengths. Specimens are often fairly large; one example measures 27 mm in length and 17 mm in height. The shell form and strongly lirate sculpture recall *Astartella,* with which it has been confused. But *Edmondia ovata* is distinguished by its more tumid valves, lower umbos, and a longer, straighter hinge line.

This relatively large species compares favorably with the figured specimen of *E. ovata* in Meek and Worthen (1873, pl. 26, fig. 13; *E. unioniformis* Phillips? earlier described by Meek and Worthen [1860, pl. 27, fig. 66] is probably also *E. ovata*). It also resembles other *E. ovata* including specimens of Girty (1915, pl. 14, fig. 13), Morningstar (1922, pl. 10, figs. 8, 9; her pl. 10, figs. 4, 5 of *E. gibbosa* are probably smaller examples of the same species), and especially, Hoare et al. (1979, pl. 15, figs. 8, 9). Although the comarginal rugae in some *E. ovata* appear to be stronger and perhaps more regular than in the Essex species, the ornament in Hoare et al.'s plate 15, figure 8, is very close if not identical to the Essex species. Similarities to an Ohio specimen of *E. meekiana* (Herrick, 1887) are evident in the illustration of Morningstar (1922, pl. 10, fig. 6), but that specimen is poorly preserved.

The Essex species also resembles certain specimens from the British Carboniferous including *E. oblonga* figured by Hind (1899, pl. 29, fig. 3, etc.) but does not closely match Runnegar and Newell's (1974, fig. 4) photograph of the same species. It also resembles others of Hind's (1899) figured specimens labeled *E. rudis* (pl. 28, fig. 12) and *Sedgwickia suborbicularis* (pl. 27, fig. 6).

Edmondia aspinwallensis (Meek, 1871)
Figure 8A.5I, 8A.5J

This is a rare Mazon Creek species; only two specimens are known. Both are from the general region of Will, Grundy, and Kankakee Counties, but the precise collecting sites are unknown. The two specimens are small, each measuring about 10 mm in length and 9 mm in height. The shells are ovoid with smoothly rounded margins and are marked by weak comarginal rugae and low umbos. The shells are inequilateral, but not strongly so. Each specimen shows the edentulous hinge bearing a diagnostic feature of *Edmondia,* namely, two weak lamellae on the hinge plate beneath the beak.

The form of the shell is in close agreement with *E. aspinwallensis* as figured by Runnegar and Newell (1974, fig. 3), but the Mazon Creek examples are much smaller, perhaps stunted. Other midwestern specimens showing similarities include *E. aspinwallensis* in Morningstar (1922, pl. 10, fig. 7) and Hoare et al. (1979, pl. 16, fig. 1) and *E. gibbosa* (M'Coy) in Hoare et al. (1979, pl. 16, figs. 3–5).

To a lesser extent this species also resembles edmondias described by Girty, including *E. gibbosa* from the Pennsylvanian Wewoka Formation of Oklahoma (Girty, 1915, pl. 14, fig. 12) and *E. crassa* from the Mississippian Moorefield Shale of Arkansas (Girty, 1911, pl. 12, figs. 8, 9).

Similar shells of the British Carboniferous include *E. rudis* (M'Coy) (Hind, 1899, pl. 28, figs. 8–11), *E. goldfussi* de Koninck (Hind, 1899, pl. 32, figs. 7, 11), and especially, *E. josepha* (de Koninck) (Hind, 1899, pl. 33, figs. 10–14, notably 11).

Edmondia oblonga M'Coy, 1855
Figure 8A.5K

This species is represented by a single specimen from the Essex assemblage at Pit 11. It is similar to *E. aspinwallensis* and may be either a biological or preservational variant of the same population. However, the more elongate and rectangular outline recalls the holotype of *E. oblonga* M'Coy shown in Runnegar and Newell (1974, fig. 4). But the Essex specimen, which measures 15 mm in length and 11 mm in height, is much smaller than *E. oblonga* and may represent yet another instance of stunting.

Family Grammysiidae Miller
Genus *Sedgwickia* M'Coy, 1844
Sedgwickia sp.
Figure 8A.5L–8A.5N

A few problematic Essex specimens are here referred to *Sedgwickia.* Runnegar (1974) based his understanding of this genus on the lectotype of *S. attenuata* figured by Hind (1899, pl. 27, fig.

9). This and other specimens of Hind (1899, pl. 27, figs. 10–13) show elongated anomalodesmatan shells with low subvertical umbos placed just behind the center and a variably expanded posterior margin that is vertically to obliquely truncate. Runnegar's (1974) plate 2, figures 10–15, show similar shells marked by variably coarse comarginal lirae and rugae.

The three Essex specimens were selected to show the variation among the one or more species present. One (Figure 8A.5N) shows a relatively narrow posterior extremity. The same specimen shows a thick (edentulous) hinge plate behind the umbos. Another (Figure 8A.5M) has an elongated form superficially like the Carboniferous pholadomyid *Wilkingia* but with more centrally placed umbos. A third variant (Figure 8A.5L) similarly recalls *Wilkingia,* but the shell is relatively shorter and the posterior margin shows a broader relative expansion.

The mode of life of *Sedgwickia* sp. was probably similar to that of *Grammysioidea* described below.

Genus *Grammysioidea* Williams and Breger, 1916

The history of usage of the many generic names of pholadomyacean clams is complex and confusing, and sorting out the various genera based on descriptions and type materials is often difficult and sometimes impossible. But papers such as Runnegar (1966), Newell's evaluation in the *Treatise* (Moore and Teichert, 1969, pp. N818–N843), Runnegar and Newell (1974), and Runnegar (1974) have done much to alleviate these problems. Among the Grammysiidae, taxonomy can be exasperating because, as Driscoll (1965) has persuasively shown, shell variation is truly remarkable and reliable qualitative characters are few. The gradational range in shell form and ornament within a given species is routinely so broad that many "species" extensively overlap. For this reason, many grammysiid species can be distinguished only through statistical testing.

Basic differences between *Grammysia* and *Grammysioidea* were reviewed by Bailey (1983, pp. 284–285). McAlester (1962) and Runnegar (1974) stressed that *Grammysia* should be applied to pholadomyaceans with strong comarginal rugae marked by a pair of radial grooves or cinctures, while *Grammysioidea* should be reserved for somewhat similar species with weaker rugae and either a single radial groove or none. But Hoare et al. (1989) proposed *Ectogrammysia* for all grammysiids lacking the radial cinctures, which, they asserted, were lost by the Carboniferous. These authors further stressed the differences between *Ectogrammysia* and *Grammysioidea* with respect to development of the lunule and comarginal rugae. However, among Devonian *Grammysioidea* populations studied by Bailey (1983), the apparent distinctions merge. To separate a single population into two separate "genera" based upon such variable criteria is obviously untenable. Furthermore, the assertion of Hoare et al. that grammysias had lost the radial groove by the Carboniferous is contradicted not only by the Essex grammysiid described below, in which the radial groove is prominent, but indeed by their own type species (*Ectogrammysia crassatis*), in which the groove is not absent as the authors claimed, but merely incipient or rudimentary. For these reasons, we reject *Ectogrammysia* and regard it as a junior synonym of *Grammysioidea.*

Grammysioidea hayi new species Figure 8A.5O

Diagnosis. Similar to *G. omaliana* (de Koninck, 1844), but with well-defined anterior radial groove. Comarginal rugae bifurcate posteriorly as they cross the groove.

Holotype. NEIU MCP 604.

Etymology. Named in honor of Andrew A. Hay, who donated the holotype.

Description. Shell medium to large, equivalve, thin, swollen, posteriorly elongated, and somewhat trapezoidal in outline with moderately elevated umbos; inequilateral, but not strongly so; dorsal margin marked by a broad lunule. Outline of posterior margin of valve is arcuate below but straightens and extends obliquely above, where it is marked by a weak sinus. Sculpture consists of widely spaced rounded comarginal rugae that markedly bifurcate behind a broad radial fold or groove extending from beaks to ventral margin. Holotype measures 30+ mm (incomplete) in length and about 25 mm in height.

Remarks. This rare Essex species is represented by a single specimen from Pit 11. It is easily distinguished from other Mazon Creek clams by the bifurcating rounded comarginal rugae and broad radial sulcus. Present limitations of material do not allow the range of variation, which was probably broad, to be investigated.

Except for the stronger radial groove, this species is basically similar in shape (including the posteriodorsal margin with its weak sinus) and ornament to *G. omaliana* (de Koninck, 1844). The presence of this highly variable European Carboniferous species in North America was noticed by Driscoll (1965), who considered it to be identical to the American species *"Sanguinolites" hannibalensis* Winchell (1865) (= *"Grammysia" hannibalensis* Walcott, 1884), from the Mississippian of the Midwest.

This species resembles two Carboniferous British species: *"Sanguinolites" omalianus* (de Koninck) (Hind, 1899, pl. 40, fig. 17) and *"S." variabilis* M'Coy (Hind, 1899, pl. 44, fig. 8). Driscoll (1965) reported the first species in the Mississippian of the midwestern United States under the name of *Grammysia omaliana.* We place the same species in *Grammysioidea* due to the lack of a double radial groove. Driscoll's (1965, text-figs. 4 and esp. 5) study is particularly significant because he documented the strong degree of variation in shell shape and ornament inherent in that species. A few of the variants possess a radial groove; most lack it. Rugae bifurcate in some but do not in others. Hoare et al. (1989) believed that Driscoll's understanding of the species was too broad, but we consider their understanding too narrow (several of the other taxa in their study are oversplit). In fact, their *Ectogrammysia crassatis* from the Imo Formation (Late Mississippian) of Arkansas is indistinguishable from *"S." (=Grammysioidea) omaliana* as figured by Hind (1899, pl. 40, fig. 8). Among Imo samples of *E. crassatis* in the Western Illinois University collections, both the distinctness of the broad radial groove and the frequency of rib bifurcation are remarkably variable (but less pronounced than in *G. hayi*). This variation tends to confirm Driscoll's original understanding of the breadth of shell variation among grammysiids.

In life, this species was likely a partially buried sedentary bivalve, attached by a byssus. *Grammysioidea,* like other pholadomyaceans, probably had short siphons. Among uncrushed Imo *G. omaliana* specimens in the Western Illinois University collections, a moderately wide, permanent siphonal gape is seen along the posterior margin; there is also a small anterior gape, probably for the byssal apparatus. These features have not been previously described.

Acknowledgments

We are indebted to John Pojeta, Jr., and Bruce Runnegar for helpful comments and criticisms of the manuscript. We are grateful to Scott Litgard and Greg Buckley of the Field Museum of Natural History and Charles Shabica, Andrew A. Hay, and Chris Ledvina of Northeastern Illinois University and the Mazon Creek Project for their help in obtaining the loan of fossil materials. We thank Mildred Scheffel-Tennant, Thomas V. Testa, and Andrew A. Hay for donations of significant fossil bivalve materials either to the collections of the Mazon Creek Project or to Western Illinois University. We thank Robert E. Johnson, resident artist and curator at the WIU Museum of Geology, for the drawings in Figure 8A.1; all photographs are by Bailey.

References

Bailey, J. B. 1983. Middle Devonian Bivalvia from the Solsville Member (Marcellus Formation), Central New York State. Bull. Amer. Mus. Nat. Hist., 174:193–325.

———. 1992. Preserved soft-part morphology and burrowing traces among Mazon Creek Bivalvia, Pennsylvanian of Illinois. Geol. Soc. Amer. Abstracts with Programs, 24(4):3.

———. In press. Mazon Creek Solemyidae (Bivalvia), Pennsylvanian of Illinois. Jour. Paleo.

Bailey, J. B., and E. C. Prosh. 1990. The oldest solemyas: occurrence of a new submodern species of the awning clam *Solemya* Lamarck in the Lower Devonian of Arctic Canada [retitled at meeting]. Geol. Soc. Amer. Abstracts with Programs, 22(5):2.

Baird, G. C., and C. W. Shabica. 1980. The Mazon Creek depositional event: examination of the Francis Creek and analogous facies in the midcontinent region, pp. 79–92. *In* R. L. Langenheim and C. J. Mann (eds.), Middle and Late Pennsylvanian

Strata on the Margin of the Illinois Basin (Guidebook). 10th Field Conf., Great Lakes Section, Soc. Econ. Paleo. Mineral., Danville, Illinois.
Baird, G. C., and S. D. Sroka. 1990. Geology and geohistory of Mazon Creek area fossil localities, Illinois, pp. C1–C70. *In* W. Hammer and D. F. Hess (eds.), Geology Field Guidebook: Current Perspectives on Illinois Basin and Mississippi Arch Geology. 24th annual meeting of the North-Central Section, Geol. Soc. Amer., 24–30 April 1990, Macomb, Illinois.
Bird, S. O. 1968. A pelecypod fauna from the Gaptank Formation (Pennsylvanian), West Texas. Bull. Amer. Paleo., 54(240):111–185.
Bretsky, P. 1968. Evolution of Paleozoic marine invertebrate communities. Science, 159:1231–1233.
Bronn, H. G. 1828. *Posidonia becheri* eine neue fossile Muschel der Ubergangs-Periode. Leonard Taschenbuch für, die gesammte Mineralogie. Zeit. Mineral., new series, 1(4):262–269.
Carter, J. G. 1990. Evolutionary significance of shell microstructure in the Palaeotaxodonta, Pteriomorphia and Isofilibranchia (Bivalvia Mollusca), pp. 135–296. *In* J. G. Carter (ed.), Skeletal Biomineralization: Patterns, Processes and Evolutionary Trends. Van Nostrand Reinhold, New York.
Cavanaugh, C. M. 1983. Symbiotic chemoautotrophic bacteria in marine invertebrates from sulphide-rich habitat. Nature, 302:58–61.
Chavan, A. 1954. Les *Pleurophorus* et genres voisins. Cahiers Gèol. Syessel, 22:200.
Conrad, T. A. 1842. Descriptions of new species of organic remains belonging to the Silurian, Devonian and Carboniferous Systems of the United States. Jour. Acad. Nat. Sci. Philadelphia, 8:183–190.
de Koninck, L. G. 1841–1844. Descriptions des animaux fossiles qui se trouvent dans le terrain carbonifére de Belgique. Liège, 650 pp.
de Verneuil, P. E. P., and R. I. Murchison. 1844. Note sur les équivalents du systéme permien en Europe, suivie d'un coup d'oeil general sur l'ensemble de ses fossiles, et d'un tableau des espéces. Bull. Soc. Géol. France, 2:475–517.
Dickson, P. A. 1974. Some studies of certain Bivalvia in the Pennsylvanian of the Mazon Creek area with emphasis on paleoecology. Unpublished M. S. thesis, University of Manchester, 111 pp.
Driscoll, E. G. 1965. Dimyarian pelecypods of the Mississippian Marshall Sandstone of Michigan. Palaeontographica Amer., 5(35):67–128.
Dunbar, C. O. 1924. Kansas Permian insects. 1, The geologic occurrence and environment of the insects. Amer. Jour. Sci., 7:171–208.
Eagar, R. M. C. 1952. Growth and variation in the non-marine lamellibranch fauna above the Sand Rock Mine of the Lancashire Millstone Grit. Q. Jour. Geol. Soc. London, 107:339–373.
Girty, G. H. 1903. The Carboniferous formations and faunas of Colorado. U.S. Geol. Surv. Prof. Paper, 16, 546 pp.
———. 1911. The fauna of the Moorefield Shale of Arkansas. U.S. Geol. Surv. Bull., 439, 148 pp.
———. 1915. Fauna of the Wewoka Formation of Oklahoma. U.S. Geol. Surv. Bull., 544, 353 pp.
Hall, J. 1883. Lamellibranchiata, plates and explanations. Nat. Hist. of New York, Paleontology, 5(1), 20 pp. (79 pls).
Hall, J., and J. D. Whitney. 1858. Report on the geological survey of the State of Iowa. Iowa Geol. Surv., 1(2):473–724.
Herrick, C. L. 1887. A sketch of the geological history of Licking County, accompanying an illustrated catalogue of Carboniferous fossils from Flint Ridge, Ohio. Denison Univ. Sci. Lab. Bull., 2:5–70, 144–148.
Hind, W. 1896–1905. Monograph of the British Carboniferous Lamellibranchiata, 1 (1896–1900), 2 (1901–1905). Palaeontographical Society, London, 708 pp.
Hoare, R. D., M. J. Heaney, and R. H. Mapes. 1989. Bivalves (Mollusca) from the Imo Formation (Mississippian, Chesterian) of North-Central Arkansas. Jour. Paleo., 63:582–603.
Hoare, R. D., M. T. Sturgeon, and E. A. Kindt. 1979. Pennsylvanian marine Bivalvia and Rostroconchia of Ohio. Ohio Geol. Surv., 67, 77 pp.
Jackson, E. N. 1982. Quantitative paleoecology: an analysis of the fossil communities of the Oak Grove Limestone Member, Liverpool Cyclothem (Pennsylvanian) at the Wolf-Covered Bridge Exposure. Unpublished undergraduate honors thesis, Knox College, Galesburg, Illinois, 56 pp.
Johnson, R. G., and E. S. Richardson, Jr. 1970. Fauna of the Francis Creek Shale in the Wilmington area, pp. 53–60. *In* W. H. Smith, N. B. Nance, M. E. Hopkins, R. G. Johnson, and C. W. Shabica (eds.), Depositional environments in parts of the Carbondale Formation—western and northern Illinois. Illinois State Geol. Surv. Guideb. Ser., 8.
Lamarck, J. B. de. 1818–1819. Histoire naturelle des animaux sans vertébrés. Paris, 5, 612 pp.; 6(1), 343 pp.
M'Coy, F. 1844. A synopsis of the characteristics of the Carboniferous limestone fossils of Ireland. University Press, Dublin, 271 pp.
———. 1851. Descriptions of some new Mountain Limestone fossils. Ann. Nat. Hist., 2nd ser., 7:167–175.
———. 1855. Descriptions of the British Palaeozoic fossils in the Geological Museum of the University of Cambridge, pp. 407–661. *In* A. Sedgwick, A Synopsis of the Classification of the British Palaeozoic Rocks. John W. Parker & Son, London.
McAlester, A. L. 1962. Upper Devonian pelecypods from the New York Chemung Stage. Peabody Mus. Nat. Hist. Bull., 16, 88 pp.
Meek, F. B. 1871. Descriptions of new species of invertebrate fossils from the Carboniferous and Devonian rocks of Ohio. Proc. Acad. Nat. Sci. Philadelphia, [1871]:57–93.
———. 1872. Report on the paleontology of eastern Nebraska with some remarks on the Carboniferous rocks of that district, pp. 85–239. *In* F. V. Hayden (ed.), Final Report of the U.S. Geol. Surv. of Nebraska and Portions of Adjacent Territories, pp. 83–239.
———. 1874. New genus *Euchondria* Meek. Amer. Jour. Sci., 3rd ser., 7:445.
———. 1875. A report on some of the invertebrate

fossils of the Waverly Group and Coal Measures of Ohio. Ohio Geol. Surv. Rep., 2(2):269–347.
Meek, F. B., and A. H. Worthen. 1860. Descriptions of new Carboniferous fossils from Illinois and other western states. Proc. Acad. Nat. Sci. Philadelphia, [1860]:447–472.
———. 1873. Fossils from the Coal Measures. *In* Descriptions of invertebrates from the Carboniferous System. Illinois State Geol. Surv., 5:560–619.
Moore, R. C., and C. Teichert, eds. 1969–1971. Treatise on Invertebrate Paleontology. Part N, Mollusca 6, Bivalvia. Geological Society of America and University of Kansas Press, Lawrence, 1224 pp.
Morningstar, H. 1922. Pottsville fauna of Ohio. Ohio Geol. Surv. Bull., 4th ser., 25:137–296.
Murphy, J. L. 1967. R. P. Stevens' pelecypod species from the Brush Creek Shale Member (Conemaugh) of Ohio. Jour. Paleo., 41:1498–1504.
Newell, N. D. 1937. Late Paleozoic pelecypods: Pectinacea. Kansas State Geol. Surv. Publ., 10, 123 pp.
———. 1942. Late Paleozoic pelecypods: Mytilacea. Kansas State Geol. Surv. Publ., 10(2), 115 pp.
Newell, N. D., and D. W. Boyd. 1978. A palaeontologist's view of bivalve phylogeny. Phil. Trans. R. Soc. London B, 284:203–215.
Pojeta, J., Jr. 1987. Class Pelecypoda, pp. 386–435. *In* R. S. Boardman, A. H. Cheetham, and A. J. Rowell (eds.), Fossil Invertebrates. Blackwell, Palo Alto, California.
———. 1988. The origin and Paleozoic diversification of solemyoid pelecypods. New Mexico Bur. Mines Min. Res. Mem., 44:201–271.
Pojeta, J., Jr., and B. Runnegar. 1985. The early evolution of diasome molluscs, pp. 295–336. *In* E. R. Trueman and M. R. Clarke (eds.), The Mollusca. Vol. 10, Evolution. Academic Press, Orlando, Florida.
Prosh, E. C. 1988. A Lower Devonian reef sequence and fauna, Disappointment Bay Formation, Canadian Arctic Islands. Unpublished Ph.D. thesis, University of Western Ontario, London, 446 pp.
Pruvost, P. 1930. La faune continentale du terrain houiller de la Belgique. Mus. R. d'Hist. Nat. Belgique Mém., 44, 282 pp.
Reid, R. G. B. 1980. Aspects of the biology of a gutless species of *Solemya* (Bivalvia: Protobranchia). Canadian Jour. Zool., 58:386–393.
Reid, R. G. B., and F. R. Bernard. 1980. Gutless bivalves. Science, 208:609–610.
Richardson, E. S., Jr., and R. G. Johnson. 1970. The Mazon Creek faunas, pp. 1222–1235. Proc. North American Paleo. Conv., 5–7 September 1969, Field Mus. Nat. Hist., vol. 1. Allen, Lawrence, Kansas.
Rogers, M. J. 1965. A revision of the species of nonmarine Bivalvia from the Upper Carboniferous of eastern North America. Jour. Paleo., 39:663–686.
———. 1985. Non-marine pelecypods from the Riverdale Group (Upper Carboniferous), Nova Scotia. Neuvième Congrès International de Stratigraphie et de Géologie du Carbonifère, Compte Rendu, 5:261–270.
Runnegar, B. 1966. Systematics and biology of some desmodont bivalves from the Australian Permian. Jour. Geol. Soc. Australia, 13:373–386.
———. 1974. Evolutionary history of the bivalve subclass Anomalodesmata. Jour. Paleo., 48:904–939.
Runnegar, B., and N. D. Newell. 1974. *Edmondia* and the Edmondiacea (Paleozoic Bivalvia). Amer. Mus. Novitates, 2533, 19 pp.
Runnegar, B., and J. Pojeta, Jr. 1992. The earliest bivalves and their Ordovician descendants. Amer. Malacol. Bull., 9(2):117–122.
Schram, F. 1979. The Mazon Creek biotas in the context of a Carboniferous faunal continuum, pp. 159–190. *In* M. H. Nitecki (ed.), Mazon Creek Fossils. Academic Press, New York.
Shumard, B. F. 1858. p. 214. *In* B. F. Shumard and G. C. Swallow, Descriptions of new fossils from the Coal Measures of Missouri and Kansas. St. Louis Acad. Sci. Trans., 1:198–227.
Sroka, S. D. 1984. Mazon Creek bivalves: taxonomy and paleoecology. Unpublished M. S. thesis, Northeastern Illinois University, Chicago, 160 pp.
Stanley, S. M. 1968. Post-Paleozoic adaptive radiation of infaunal bivalve molluscs: a consequence of mantle fusion and siphon formation. Jour. Paleo., 42:214–229.
———. 1977. Trends, rates, and patterns of evolution in the Bivalvia, pp. 209–250. *In* A. Hallam (ed.), Developments in Palaeontology and Stratigraphy. Vol. 5, Patterns of Evolution as Illustrated by the Fossil Record. Elsevier, New York.
Stevens, R. P. 1858. Descriptions of new Carboniferous fossils from the Appalachian, Illinois, and Michigan coal fields. Amer. Jour. Sci., 2nd ser., 25:258–265.
Swallow, G. C. 1863. Descriptions of some new fossils from the Carboniferous and Devonian rocks of Missouri. St. Louis Acad. Sci. Trans., 1:173–197.
Trueman, A. E., and J. Weir. 1946. A monograph of the British Carboniferous non-marine Lamellibranchia. Palaeontographical Soc. (London), 2:19–44.
Walcott, C. D. 1884. Paleontology of the Eureka District. U.S. Geol. Surv. Monog., 8, 298 pp.
Waller, T. R. 1990. The evolution of ligament systems in the Bivalvia, pp. 49–71. *In* B. Morton (ed.), The Bivalvia: Proceedings of a Memorial Symposium in Honor of Sir Charles Maurice Yonge, Edinburgh, 1986. Hong Kong University Press, Hong Kong.
Wanless, H. R. 1958. Pennsylvanian faunas of the Beardstown, Glasford, Havana, and Vermont Quadrangles. Illinois State Geol. Surv. Rep. Invest., 205, 59 pp.
Williams, H. S., and C. L. Breger. 1916. The fauna of the Chapman Sandstone of Maine including descriptions of some related species from the Moose River Sandstone. U.S. Geol. Surv. Prof. Paper, 89, 37 pp.
Winchell, A. 1865. Descriptions of new species of fossils, from the Marshall Group of Michigan, and its supposed equivalent, in other states; with notes on some fossils of the same age previously described. Proc. Acad. Nat. Sci. Philadelphia, [1865]:109–133.
Worthen, A. H. 1890. Description of fossil invertebrates: Illinois Geol. Surv., 8:69–154.

8B

Polyplacophora

Joanne Kluessendorf

Chitons, which are exclusively marine organisms, are placed in the molluscan class Polyplacophora ("bearer of many plates"). The bilaterally symmetrical body of the chiton is characterized by a dorsal shell consisting of eight aragonitic plates or valves: a head valve, a tail valve, and six nearly identical intermediate valves. These valves overlap slightly and articulate with one another by means of projections at the anterior of each valve, with the exception of the head valve. This articulation enables the animal to enroll for protection. A leathery, muscular extension of the mantle, the girdle, borders the chiton body and holds the valves in place. The girdle, which is very flexible, allows the chiton to clamp down tenaciously onto irregular, hard surfaces, so that waves and predators cannot dislodge it. The modern chiton creeps slowly along on a broad, flat, ventral foot. It scrapes algae and other organics from hard surfaces by means of a chitinous radula that can be protruded from the ventrally positioned mouth. The radula bears many rows of tiny, toothlike denticles.

Chiton fossils are rare in the geologic record, and specimens are seldom numerous at any locality. In the Essex fauna chitons, while common in a few localized "patchy" areas, are rare. Although only one genus (*Glaphurochiton*) is present, chiton individuals are found twice as often as Essex fauna gastropods and are 5 to 6 times as abundant as cephalopods (Yochelson and Richardson, 1979).

The geologic range of chitons extends from the Upper Cambrian to the Recent. The earliest chitons belong to the order Paleoloricata (Upper Cambrian–Upper Cretaceous), a group lacking good valve articulation. Chitons belonging to the order Neoloricata, including the Mazon Creek taxon and all modern forms, first appeared during the Carboniferous (Smith, 1960). Although none of the Mazon Creek specimens retains any original aragonite shell material, the external molds preserve remarkable features (Yochelson and Richardson, 1979). Girdles are preserved as a color difference within the concretion along with impressions of girdle spicules; a fossil chiton girdle is known only from one other specimen in the geologic record, from the Ordovician of Scotland (Smith, 1960). Mazon Creek chitons are the only fossil chitons known to show evidence of the radula, which, in this case, is preserved as an impression. Nearly all Mazon Creek specimens are preserved with all valves articulated, unlike most fossil chitons, which are represented only by isolated valves.

Unlike the majority of modern chitons, which are well adapted to life on rocky shores, Mazon Creek chitons probably crawled about on a muddy seafloor, feeding on organic detritus (Yochelson and Richardson, 1979). Interestingly, one modern chiton genus (*Lepidopleurus*), related to *Glaphurochiton*, lives on mud bottoms at depths of as much as 2,300 fathoms (Smith, 1960). Ferreira (1981) reported that the

Figure 8B.1. *Glaphurochiton concinnus,* FMNH MCP 1257; impression of complete specimen showing in situ radula at anterior and spicule-fringed girdle surrounding shell. Scale bar = 5 mm.

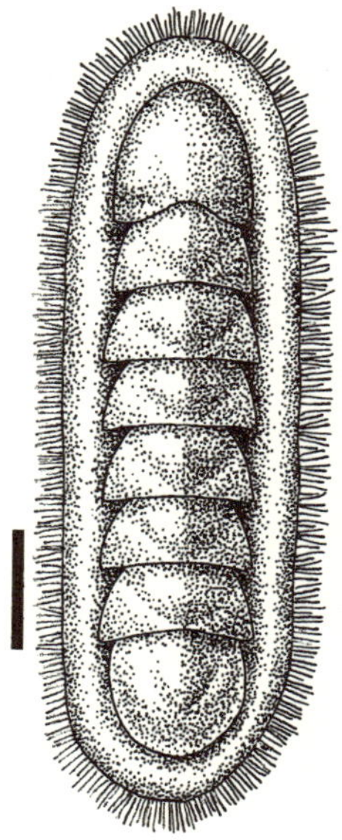

Figure 8B.2. *Glaphurochiton concinnus,* reconstruction of dorsal surface based on FMNH PE 31946. Scale bar = 5 mm.

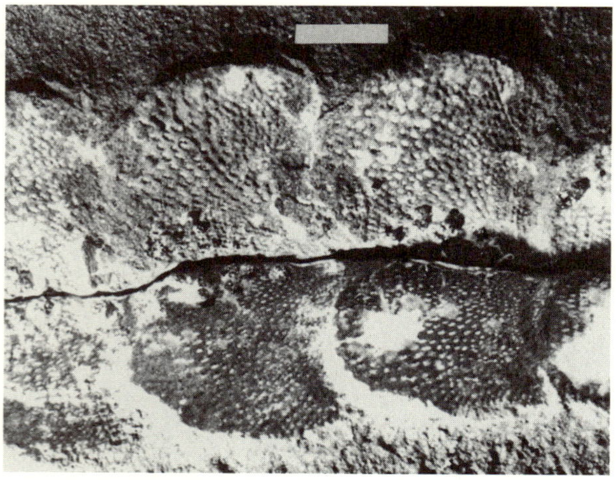

Figure 8B.3. *Glaphurochiton concinnus,* FMNH MCP 1335; enlargement of dorsal surface. Impression shows surface ornamentation of valves. Scale bar = 2 mm (estimate).

mud-dwelling *Lepidopleurus scrippsianus,* a deepwater Pacific species, lacks scales on the underside of its girdle. These ventral scales are needed by virtually all other modern chitons for protection against the hard substrates over which they crawl. The Mazon Creek chiton also appears to lack these ventral scales.

Mazon Creek chitons are often collected in proximity to other chitons. This suggests a living association of gregarious individuals that was buried in place by a sudden influx of sediment (Yochelson and Richardson, 1979). At other Pennsylvanian localities in North America, this genus typically is associated with orthoconic nautiloid cephalopods and nuculid bivalves, both of which are present although rare in the Essex fauna (Schram, 1979). Other associated organisms such as crinoids, bryozoans, sponges, and naticopsid and pleurotomariid gastropods are rare or absent from the Mazon Creek biota (Schram, 1979).

Glaphurochiton concinnus (Richardson)
Figures 8B.1–8B.5

The eight-valved shell of this lepidopleurid chiton is about five times as long as it is wide (Figures 8B.1, 8B.2). The head valve is subovoid, broadly rounded in the anterior. The tail valve, rounded in the posterior, is subovoid, and bears a blunt mucro. In small, younger specimens, the tail valve is more subcircular than ovoid. Intermediate valves are subrectangular, being wider than long, with a fairly sharp jugum. Ornamentation of all valves consists of concentric rows of fine tubercles and irregular, fine, concentric growth lines (Richardson, 1956; Figure 8B.3). The girdle surrounding the shell laterally overlaps the lower edge of each valve, extending into interplate junctions (Yochelson and Richardson, 1979; Figures 8B.1, 8B.2). Small, straight, pointed aragonite spicules covered the girdle, except at its outer edge, where a dense fringe of larger spicules occurs (Figure 8B.4).

Figure 8B.4. *Glaphurochiton concinnus,* enlargement of FMNH MCP 1249 showing dense fringe of spicules bordering the girdle. Scale bar = 1 mm (estimate).

Figure 8B.5. *Glaphurochiton concinnus,* FMNH PE 31904; impression of ventral surface showing in situ radula extending from anterior of head valve to third valve. Scale bar = 2 mm (estimate).

The ribbonlike radula, consisting of at least 100 rows of 17 denticles each, extends from the front of the head valve to the front of the third valve on the ventral surface (Figure 8B.5).

Richardson (1956) originally described this taxon as *Helminthochiton concinnus.* Hoare et al. (1972) placed *H. concinnus* into the genus *Pterochiton* and synonymized it with *Chiton carbonarius* Stevens. Yochelson and Richardson (1979) accepted the *Pterochiton* assignment but rejected the species synonymy. They believed that the majority of the Essex fauna specimens lie within the limits of *P. concinnus,* although a few individuals may represent a second, closely related species. Recently, Hoare and Mapes (1986) reassigned this taxon to the genus *Glaphurochiton,* and it is thought that *G. concinnus* was derived from *G. carbonarius.* All species of *Glaphurochiton* are considered to have been mud dwellers because they disappear rapidly from the fauna as the carbonate ratio of the sediment increases (Hoare and Mapes, 1986).

Holotype. Mr. and Mrs. John McLuckie collection, No. 3000–1.1, now at USNM.

References

Ferreira, A. J. 1981. A new species of *Lepidopleurus* Risso, 1826 (Mollusca: Polyplacophora) in the deep waters of the eastern Pacific. The Veliger 23(1):55–61.

Hoare, R. D., and R. H. Mapes. 1986. The polyplacophoran *"Chiton" carbonarius* Stevens, 1858, in North America, and related species. Jour. Paleo., 60:627–635.

Hoare, R. D., M. T. Sturgeon, and T. B. Hoare. 1972. Middle Pennsylvanian (Allegheny Group) Polyplacophoran from Ohio. Jour. Paleo., 46:675–680.

Richardson, E. S., Jr. 1956. Pennsylvanian invertebrates from the Mazon Creek area, Illinois: marine fauna. Fieldiana Geol. 12:59–76.

Schram, F. R. 1979. The Mazon Creek biotas in the context of a Carboniferous faunal continuum, pp. 159–190. *In* M. H. Nitecki (ed.), Mazon Creek Fossils. Academic Press, New York.

Smith, A. G. 1960. Amphineura, pp. I141–I176. *In* R. C. Moore (ed.), Treatise on Invertebrate Paleontology. Part I, Mollusca 1. Geological Society of America and University of Kansas, Lawrence.

Yochelson, E. L., and E. S. Richardson, Jr. 1979. Polyplacophoran molluscs of the Essex fauna (Middle Pennsylvanian, Illinois), pp. 321–332. *In* M. H. Nitecki (ed.), Mazon Creek Fossils. Academic Press, New York.

8C

Gastropoda

Joanne Kluessendorf

Class Gastropoda, comprising nearly 50,000 species of snails and slugs, is the largest molluscan class. It also is the most ecologically diverse group, exploiting marine, freshwater, and terrestrial habitats. Three gastropod subclasses are recognized: the heavy-shelled, mostly marine Prosobranchia, the detorted (straightened), and often shell-less, marine Opisthobranchia, and the air-breathing freshwater and terrestrial Pulmonata. Of these, only prosobranchs have been found in the Essex fauna.

In general, gastropods are distinguished by torsion (twisting) of the visceral mass, which contains the internal organs, producing a bilateral asymmetry. A ventral muscular foot provides locomotion and bears a distinct anterior head with sensory tentacles and eyes. Typically, when the animal is threatened it can retract its soft parts into a calcareous (mostly aragonitic), spirally coiled, unchambered, univalved shell. Once inside, the animal receives additional protection from a horny operculum attached to the foot, which can be used to seal the shell aperture.

The majority of gastropods live on the seafloor in shallow marine environments. Although most are herbivorous, scraping up algae with a radula (a structure similar to a tongue with teeth), some gastropods are carnivorous, and a few are deposit feeders.

The geologic range of gastropods extends from the Lower Cambrian to the Recent. The earliest known gastropods, which were prosobranchs, were bilaterally symmetrical with the shell coiled in a plane; asymmetrical (helically spiraled) shells first appeared in the Upper Cambrian and are the dominant shell form today (Cox, 1960). The earliest definite pulmonates and pteropod opisthobranchs, both thought to have been derived from prosobranchs, date back to the Lower Carboniferous and Lower Eocene, respectively (Cox, 1960).

Gastropods, usually a major component of Pennsylvanian marine faunas, are a very minor element in the Essex fauna. Specimens of *Strobeus* compose two-thirds of the gastropod fauna (G. C. Baird, personal communication). The remaining gastropods consist dominantly of *Euphemites, Hypselentoma, Straparollus,* and *Naticopsis* and are very rare. The general rarity and low diversity of gastropods, as well as the near absence of certain typical marine taxa, support a restricted marginal marine environment for the Essex fauna (Schindel and Yochelson, 1979). However, it is possible that gastropods are actually more abundant and diverse than is realized but are not collected with the same vigor as the more unique elements of this fauna and, therefore, are underrepresented in collections.

Most Mazon Creek snails are preserved as external molds. Only rarely was an internal mold produced by sediment infill, suggesting that the animal's soft parts may have decayed within the shell, blocking sediment infiltration. More often, voids left by later decay of soft parts were infilled by secondary calcite, pyrite, or sphalerite (Schindel and Yochelson, 1979). No impressions of soft parts have been found.

Euphemites richardsoni
Schindel and Yochelson, 1979
Figure 8C.1

This bellerophontacean snail, the most abundant of the Essex gastropods, possesses a small, bilaterally symmetrical, subglobose, bellerophontiform shell. The thin, smooth lip of the slightly flaring aperture is bisected by a deep, wide slit that extends about one-fourth of the body whorl. The selenizone, bearing simple semicircular lunulae, is slightly elevated in early whorls, but flush in later stages. The smooth, deeply impressed umbilicus becomes smaller and shallower with increasing shell size. Shell ornamentation consists of 20 to 24 fine, closely spaced, spiral lirae, which originate along the outer edge of the umbilical area.

Schindel and Yochelson (1979) reported that two distinct growth stages of *E. richardsoni,* exhibiting two distinct morphologies, are present in the Essex fauna. Two-thirds of the specimens in the Field Museum collection display a subtriangular whorl profile and are less than 7 mm in height, representing a juvenile stage. Larger, mature shells display a subglobose profile.

Bellerophontaceans range from the Upper Cambrian to the Lower Triassic, although this particular genus has a much shorter range (Devonian-Permian; Knight et al., 1960). The entire group retains a vestigial bilateral symmetry. Schindel and Yochelson (1979) concluded that *E. richardsoni* differed from other Pennsylvanian species of this widespread genus primarily in its lack of nodes at lirae inception near the umbilicus and in its mature shell shape.

Although similar in appearance to some taxa belonging to *E. carbonarius* (Cox), Schindel and Yochelson (1979) assigned the Mazon Creek taxon to a new species in order to avoid adding to the taxonomic confusion surrounding that poorly defined and misused species.

Holotype. FMNH PE 11362.

Hypselentoma aff. *H. perhumerosa*
Figure 8C.2

This pleurotomariacean gastropod is exceedingly rare in the Essex fauna; Schindel and Yochelson (1979) knew of only one specimen. Pleurotomariaceans, which range into the Recent, are thought to have been derived from the Bellerophontacea in the Upper Cambrian but have lost the bilateral symmetry of the coiled shell (Knight et al., 1960).

This Essex snail exhibits a moderately high-spired turbiniform shell consisting of four gradually expanding volutions. The flat upper whorl face slopes downward at 45° to the broadly rounded periphery. The slightly rounded outer whorl face slopes down from the periphery inward to the deeply impressed angular suture. A wide and deep peripheral sinus is reflected in the pattern of fine, closely spaced growth lines. The growth lines angle backward and downward at 60° from the suture to about halfway across the upper whorl face, where they curve backward more strongly, forming the posterior border of the sinus. The growth lines then curve gently forward across the rounded lower whorl face. The basal part of the aperture appears to be elongated and slightly flaring.

Schindel and Yochelson (1979) observed that, although similar to the type species, *H. perhumerosa,* of this Pennsylvanian genus in both shell shape and growth line pattern, the Essex taxon differs in its lack of a sharp peripheral angulation. They viewed the extreme rarity of pleurotomariacean gastropods as evidence in support of a restricted marginal marine interpretation for the Essex fauna.

Straparollus (Euomphalus) sp. indeterminate
Figure 8C.3

This cosmopolitan euomphalacean subgenus, which ranges from Silurian to Middle Permian (Knight et al., 1960), is very rare in the Essex fauna. Schindel and Yochelson (1979) knew of only two specimens, both of which were large (300 mm wide by 8 mm high) and preserved by infilling and mineralization. All features of these shells but the body whorl are obscured by matrix.

The discoidal shell bears an angular upper shoulder, the sharpness of which is emphasized by a shallow channel beneath it on the outer whorl face. The flat upper whorl face slopes gently from the suture to the upper shoulder. The nearly vertical outer whorl face exhibits a very

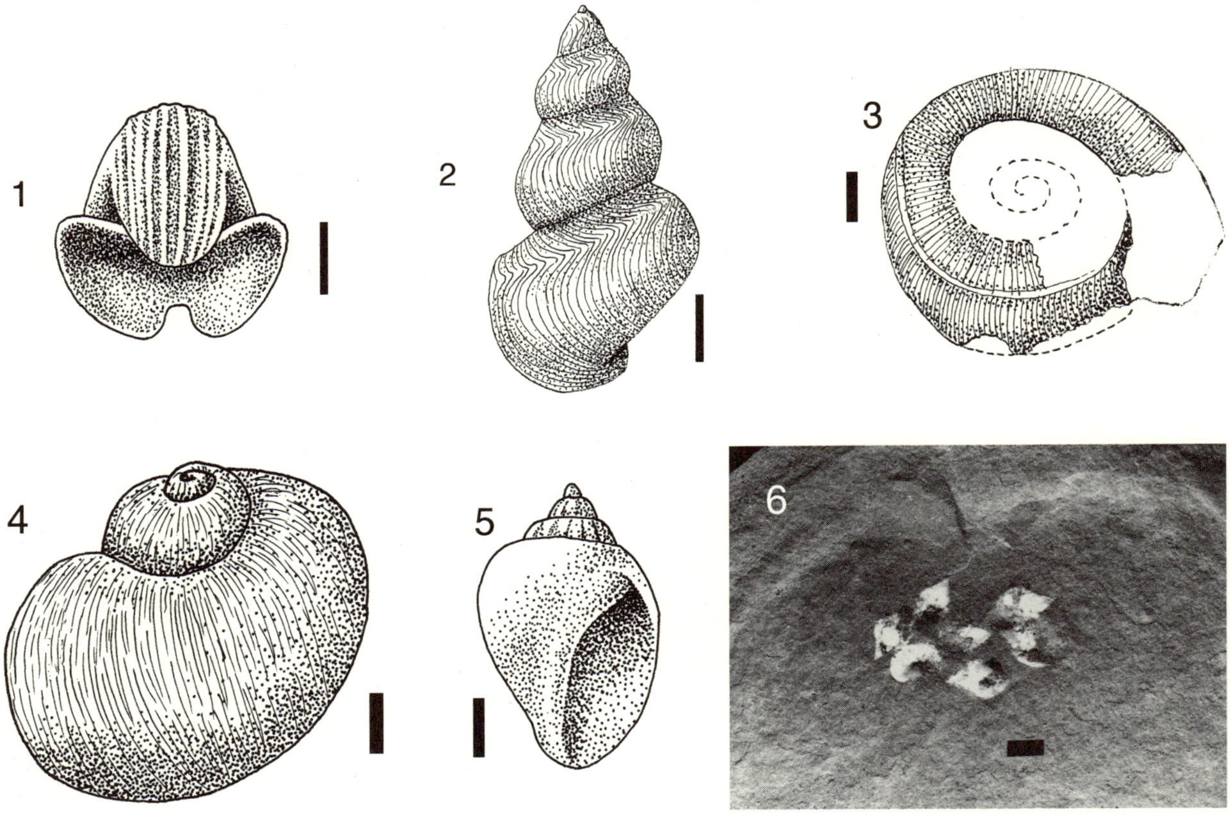

Figure 8C.1. *Euphemites richardsoni*, reconstruction based on paratype, FMNH PE 11362; apertural view. Scale bar = 2 mm.

Figure 8C.2. *Hypselentoma* aff. *H. perhumerosa*, reconstruction based on FMNH PE 9877; slightly oblique basal view. Scale bar = 5 mm.

Figure 8C.3. *Straparollus (Euomphalus)* sp. indet., reconstruction based on FMNH PE 18137; slightly oblique umbilical view. Scale bar = 5 mm.

Figure 8C.4. *Naticopsis (Naticopsis)* sp. indet., reconstruction based on FMNH PE 28976; oblique apical view. Scale bar = 5 mm.

Figure 8C.5. *Strobeus* cf. *S. primigenius*, reconstruction based on FMNH PE 36848; apertural view. Scale bar = 5 mm.

Figure 8C.6. *Strobeus* cf. *S. primigenius*, showing cluster of seven individuals on blob, D. Pearson collection, 2500. Scale bar = 5 mm (estimate).

slight outward convexity. Fine, closely spaced growth lines follow a straight course between the suture and upper shoulder on the upper whorl face. Orthoclinal growth lines on the outer whorl face curve very slightly toward the aperture.

Naticopsis (*Naticopsis*) sp. indeterminate
M'Coy
Figure 8C.4

This neritacean subgenus is known only from one small external mold in the Essex fauna (Schindel and Yochelson, 1979). The globular neritiform shell with three volutions bears a slightly protruding spire. Whorls expand rapidly; the final whorl, which composes most of the shell, coils obliquely downward. The upper whorl face is slightly flattened but becomes broadly rounded laterally. Faint prosocline growth lines suggest a simple aperture.

On the basis of its globular shell and absence of ornament besides faint growth lines, Schindel and Yochelson (1979) assigned this specimen to the typical subgenus *Naticopsis* (*Naticopsis*), a cosmopolitan taxon which ranges from Middle Devonian to Triassic. Although specific determination is impossible without preservation of the aperture and basal surface, they observed that this specimen probably is closer to *N.* (*Naticopsis*) *subovata* Worthen than to any other common Pennsylvanian species.

When first discovered, a color pattern of light and dark zig-zag stripes, fairly typical for neritaceans, was apparent on this specimen but it was removed accidentally during preparation (Schindel and Yochelson, 1979).

Strobeus cf. *S. primigenius* (Conrad)
Figures 8C.5, 8C.6

This sublitacean genus is the second most numerous gastropod taxon in the Essex fauna (Schindel and Yochelson, 1979). The rapidly expanding body whorl composes most of the shell, which possesses as many as four volutions. The suture is shallow because each whorl encroaches about halfway down the preceding whorl. The narrow spire protrudes above the larger, more rounded, later whorls. The wide oval aperture shows no basal constriction or siphonal notch. No growth lines or ornamentation are apparent. Of about two dozen specimens measured by Schindel and Yochelson (1979), the average height was 6.2 mm, the average width 4.0 mm.

Although some authors (e.g., Knight et al., 1960) include most late Paleozoic snails with this general shape in the genus *Ianthinopsis*, Schindel and Yochelson (1979) regarded the Essex taxon to be more fusiform than that genus, and they assigned it to the earlier *Strobeus*. Harper (1981) reinforced this assignment when he recommended using *Strobeus* as a replacement for *Ianthinopsis*, which he did not consider a sublitacean.

Half of the *Strobeus*-bearing concretions studied by Schindel and Yochelson (1979) contained more than one individual, usually occurring in tight clusters. They suggested that *Strobeus* may have been a gregarious scavenger feeding on decaying soft-bodied organisms such as the "blobs" with which they often are associated. At one locality, Foster (1979) observed that nearly 40 percent of the blobs, which he proposed are coelenterates, were associated with one or more of these snails, most of which occurred on the skirts of the blobs. He suggested that *Strobeus* was thin shelled and analogous to the Recent planktic snail *Janthina*, which feeds on the lower surface of living planktonic coelenterates such as siphonophores (jellyfish) and chondrophores.

References

Cox, L. R. 1960. Gastropoda, pp. I185–I1169. *In* R. C. Moore (ed.), Treatise on Invertebrate Paleontology. Vol. I, Mollusca 1. Geological Society of America and University of Kansas, Lawrence.

Foster, M. W. 1979. Soft-bodied coelenterates in the Pennsylvanian of Illinois, pp. 191–267. *In* M. H. Nitecki (ed.), Mazon Creek Fossils, Academic Press, New York.

Knight, J. B., L. R. Cox, A. M. Keen, R. L. Batten, E. L. Yochelson, and R. Robertson. 1960. Systematic descriptions, pp. I1169–I1310. *In* R. C. Moore (ed.), Treatise on Invertebrate Paleontology. Vol. I, Mollusca 1. Geological Society of America and University of Kansas, Lawrence.

Schindel, D. E., and E. L. Yochelson. 1979. Gastropoda of the Essex fauna, Francis Creek Shale (Middle Pennsylvanian), northeastern Illinois, pp. 303–319. *In* M. H. Nitecki (ed.), Mazon Creek Fossils. Academic Press, New York.

8D

Cephalopoda

Joanne Kluessendorf

All the leading conservationists insist on Rory's KLO-RID the purist yttrium chloride made for control of ammonites in swimming pools and natural waters.
No Monsters for Mary
(The Vanishing Press, 1970)

Of the five classes of molluscs, the most useful geologic index fossils are the shell-bearing cephalopods. This class developed into one of the most diverse of the fossil invertebrates. The calcareous (aragonite) shells were easily preserved, and the complex sutures became ideal "fingerprints" to help establish a geologic timeframe and evolutionary sequence of the class. The variety of shell forms adopted by these animals is illustrated in Figure 8D.1.

The class Cephalopoda is represented by approximately 650 species of squids, cuttlefish, octopods, and nautiloids in present-day oceans. Most modern cephalopods are well adapted for efficient hunting with a jet propulsion system for rapid backward movement, excellent vision, prehensile tentacles surrounding the head, and a beak and radula for tearing prey (Solem, 1974). Although the majority of the more than 10,000 species of extinct fossil cephalopods had external shells, the chambered nautilus is the only modern cephalopod so equipped. In all other present-day cephalopods, the shell is reduced and internal (squids and cuttlefish) or absent (octopods).

Cephalopod shells are subdivided into chambers by partitions called septa, and only the last chamber is occupied by the living animal. Sutures are formed where the septa meet the shell wall, and the suture pattern is used in taxonomic classification. Chambers are connected by the siphuncle, a tubelike extension of the mantle, which is used in buoyancy control. Shell shape ranges from the elongate conical shell of straight nautiloids to the planispirally coiled shell of ammonoids and modern *Nautilus.*

Cephalopods are subdivided into six subclasses: Endoceratoidea (Ordovician–Silurian), Actinoceratoidea (Ordovician–Permian), Nautiloidea (Cambrian–Recent), Ammonoidea (Devonian–Cretaceous), Coleoidea (Carboniferous–Recent; Teichert and Moore, 1964), and Bactritoidea (?Silurian, Early Devonian–Upper Triassic; Mapes, 1979). The latter four are represented in the Essex fauna (Saunders and Richardson, 1979).

Although rare, cephalopods are relatively diverse in the Essex fauna (Saunders and Richardson, 1979). The aragonitic shell material usually is lost, and specimens are preserved as internal and external molds in the nodules. Unfortunately, features such as siphuncle position and suture pattern, which are important for classification, seldom are preserved. On the other hand, preservation of chitinous elements (e.g., mandibles) and impressions or outlines of body parts is typical in the Essex specimens but rarely found elsewhere in the fossil record. Six species of fossil cephalopods have been collected from the Mazon Creek and are described below.

Subclass Bactritoidea
Bactrites sp.
Figure 8D.2

Bactritoid cephalopod with smooth orthoconic shell displaying simple straight sutures and ventral siphuncle. Saunders and Richardson (1979) assigned this taxon to the family Bactritidae because of the slow rate of shell taper it exhibits and its ventral siphuncle. The Bactritidae range from the Silurian(?), Early Devonian to Upper Triassic, and the genus *Bactrites* is common in Pennsylvanian rocks of central North America (Mapes, 1979). The bactritoids probably were derived from the nautiloids in the Devonian or earlier (Mapes, 1979), and they, in turn, may have given rise to the belemnoids and ammonoids (Erben, 1964).

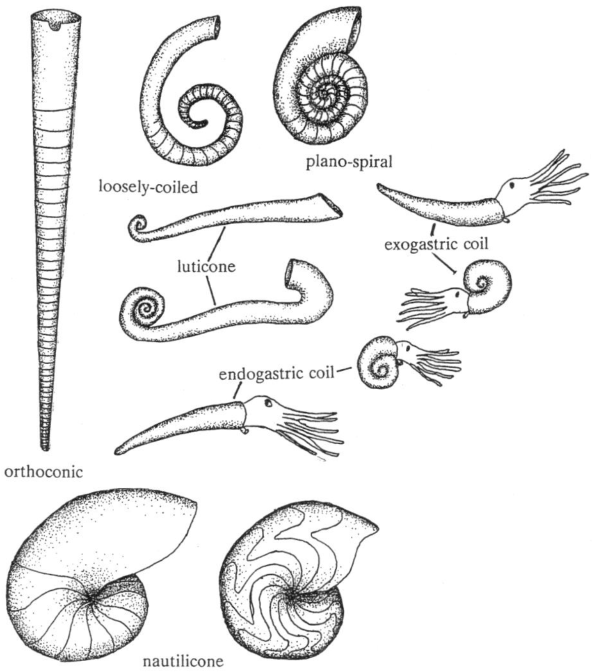

Figure 8D.1. Cartoon illustrating cephalopod shell types, modified from several sources.

Subclass Ammonoidea
Wiedeyoceras sp.
Figure 8D.3

Goniatitic ammonoid with discoidal, sublenticular shell, bearing small umbilicus and fine sinuous biconvex growth lines. Suture patterns are not preserved on known specimens, making specific identification impossible. Concentrically striated chitinous mandibles have been found in situ within the living chambers of some specimens. Calcitic jaw elements like those of *Nautilus* are not evident in the Essex *Wiedeyoceras* (Saunders and Richardson, 1979), which lends support to the proposal of Saunders et al. (1978) that ammonoid jaws may have more closely resembled those of modern coleoids. *Wiedeyoceras* is widespread in the Middle Pennsylvanian rocks of the U.S. midcontinent (Saunders and Richardson, 1979).

Schistoceratid indeterminate
Figure 8D.4

The presence of nodose umbilical shoulders and coarse reticulate sculpture led Saunders and Richardson (1979) to assign the only known and poorly preserved specimen of this taxon to the goniatitic ammonoid family Schistoceratidae. This family is known from Lower Pennsylvanian to Lower Permian rocks of the United States (Kummel, 1964).

Figured specimen. FMNH PE 21824.

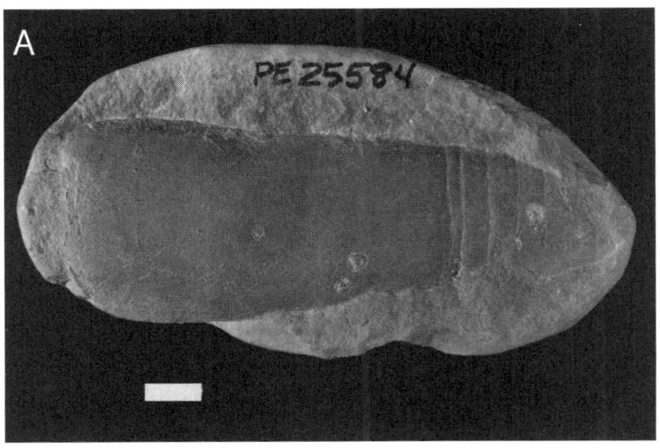

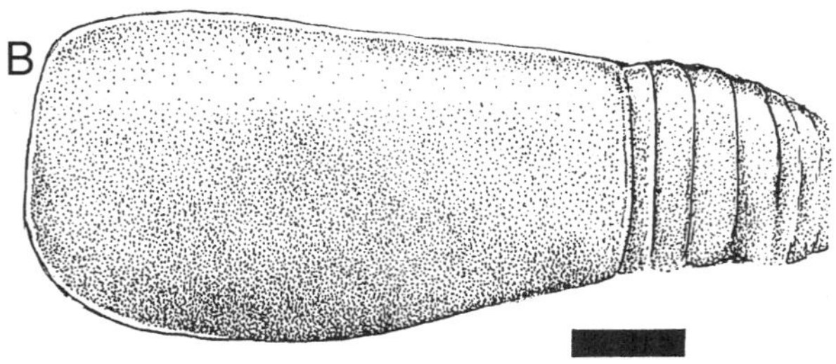

Figure 8D.2. *Bactrites.* **A** FMNH PE 25584. Scale bar = 5 mm. **B** Modified from Teichert and Moore (1964). Scale bar = 5 mm (estimate).

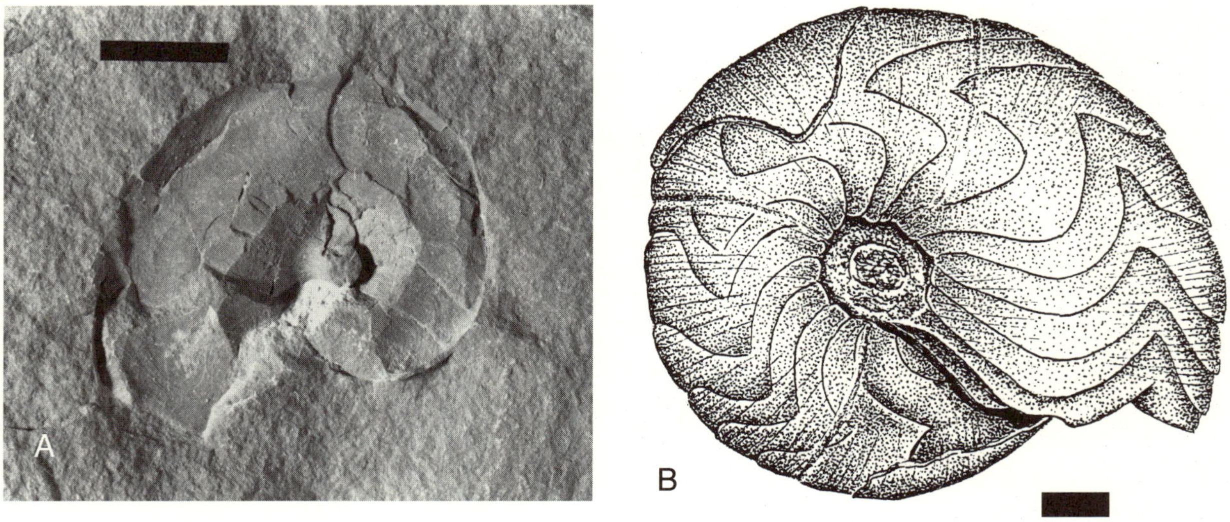

Figure 8D.3. *Wiedeyoceras.* **A** FMNH PE 10251, PE 11157. Scale bar = 5 mm. **B** Modified from Moore (1957). Scale bar = 5 mm (estimate).

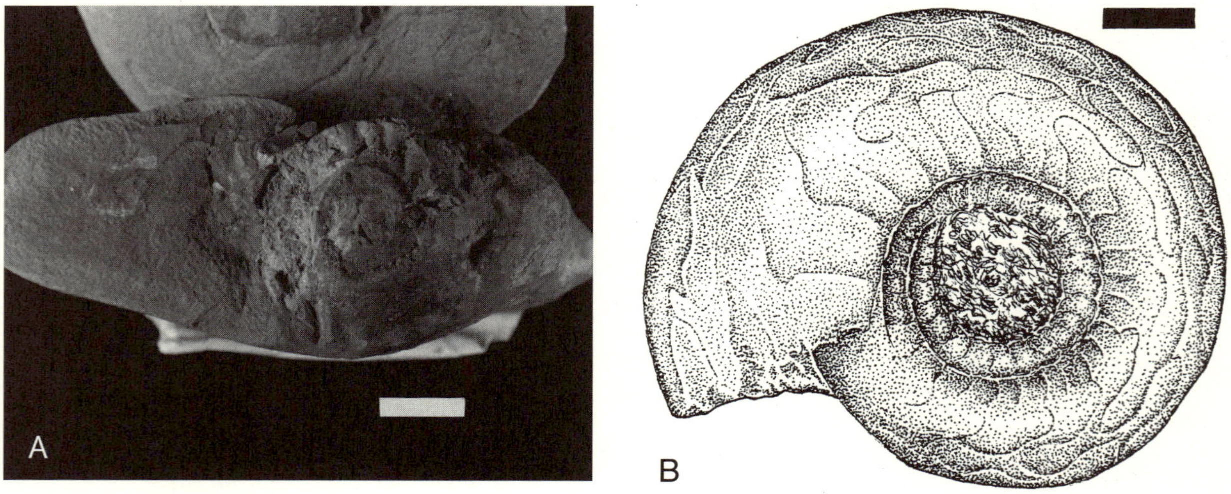

Figure 8D.4. **A** Schistoceratid indet., FMNH PE 21824. Scale bar = 5 mm (estimate). **B** *Schistoceras,* modified from Moore (1957). Scale bar = 5 mm (estimate).

Subclass Nautiloidea
Stearoceras sp.
Figure 8D.5

Coiled nautiloid with depressed quadrate whorl outline and slightly concave venter. Hyponomic sinus deep and broadly rounded with prominent ventrolateral salient. Closely spaced, fine, crenulate, longitudinal lirae on umbilical walls. The depressed quadrate outline and absence of umbilical nodes distinguish this taxon from most other coiled nautiloids (Saunders and Richardson, 1979). This genus is widely distributed in Lower Pennsylvanian–Lower Permian rocks (Kummel, 1964).

Figured specimen. FMNH PE 28953.

Titanoceras sp.
Figure 8D.6

Coiled nautiloid with subquadrate whorl cross section with flat lateral flank and subangular ventrolateral shoulder terminating in a series of axially elongate nodes (Saunders and Richard-

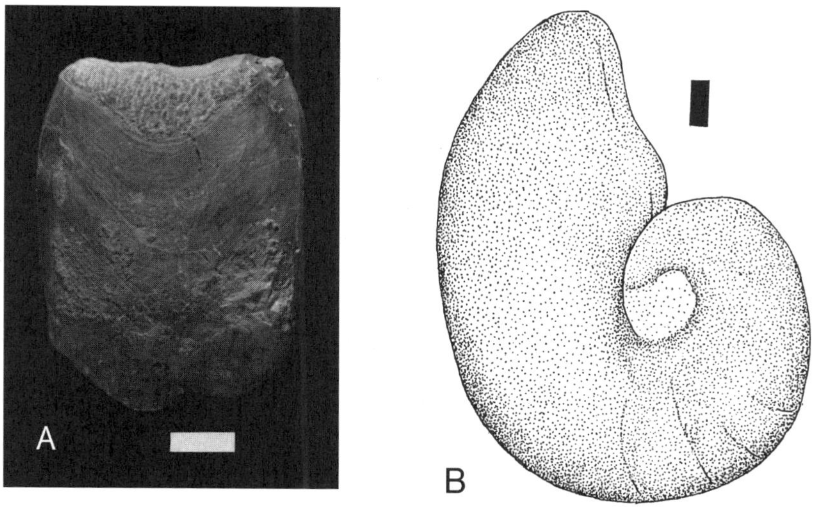

Figure 8D.5. *Stearoceras.* **A** FMNH PE 28953. Scale bar = 5 mm. **B** Modified from Moore (1957). Scale bar = 5 mm (estimate).

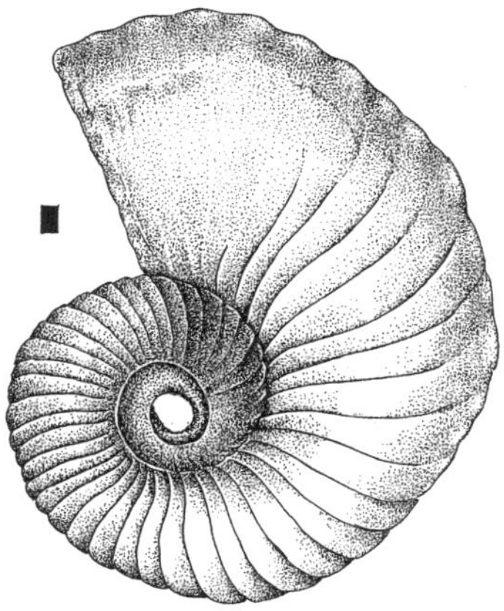

Figure 8D.6. *Titanoceras,* modified from Teichert and Moore (1964). Scale bar = 10 mm (estimate).

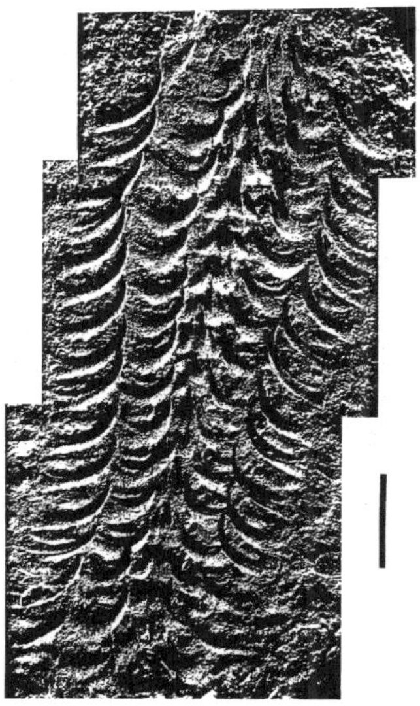

Figure 8D.7. *Paleocadmus herdinae,* holotype, FMNH PE 24538; radula, upper side impression (from Solem and Richardson, 1975). Scale bar = 1 mm.

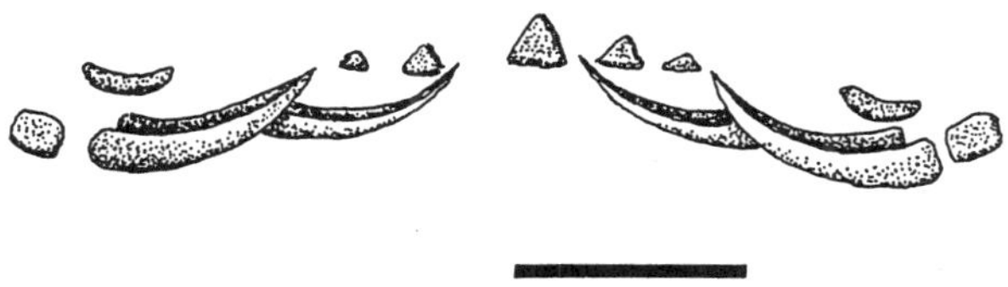

Figure 8D.8. *Paleocadmus pohli,* modified after Saunders and Richardson (1979) composite reconstruction of radular elements of holotype, NEIU MCP 512. Scale bar = 0.5 mm (estimate).

son, 1979). This genus is known from Pennsylvanian–Lower Permian rocks (Kummel, 1964).

Figured specimen. Robert Ulaszek collection.

Paleocadmus herdinae and *Paleocadmus pohli*
Figures 8D.7–8D.9, 8D.10B

Paleocadmus herdinae was described by Solem and Richardson (1975) on the basis of an isolated radular impression that consisted of 25 transverse rows of 13 elements each, which were arranged with two lateral teeth, two marginal teeth, and two support plates on either side of a central rachidian tooth. They assigned this taxon to the Nautiloidea because it is more similar to the radula of modern *Nautilus* than to any other cephalopod radula. Saunders and Richardson (1979) expanded the genus *Paleocadmus* to include nautiloid-like fossil radulae with 13 elements similar to *P. herdinae.* They named a new species (*P. pohli*) on this basis. The 13 radular elements differ between these two species as follows: *P. pohli* has a unicuspid central rachidian radular tooth, whereas that of *P. herdinae* is tricuspid; the lateral teeth of *P. pohli* are short and triangular, but those of *P.*

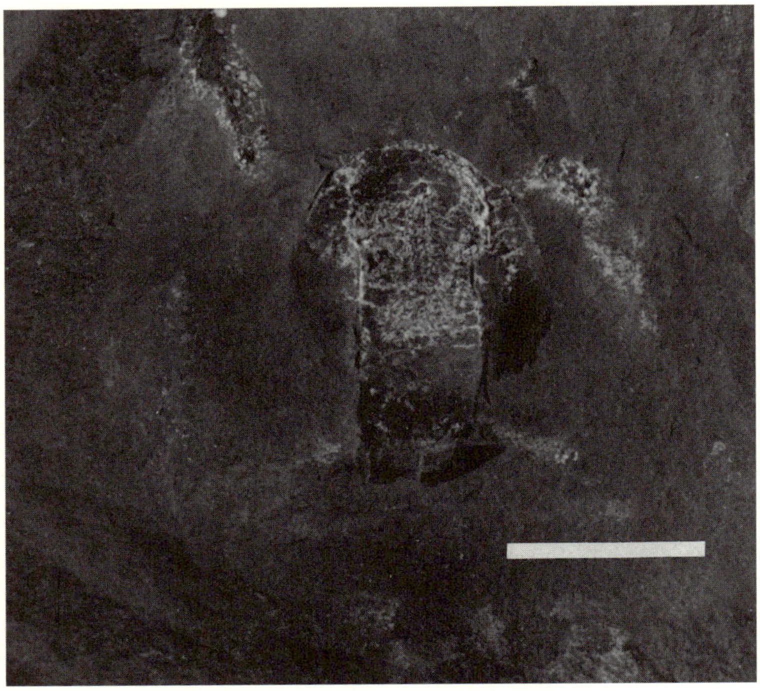

Figure 8D.9. *Paleocadmus* sp., nautiloid mandibles, FMNH PE 32522. Scale bar = 5 mm (estimate).

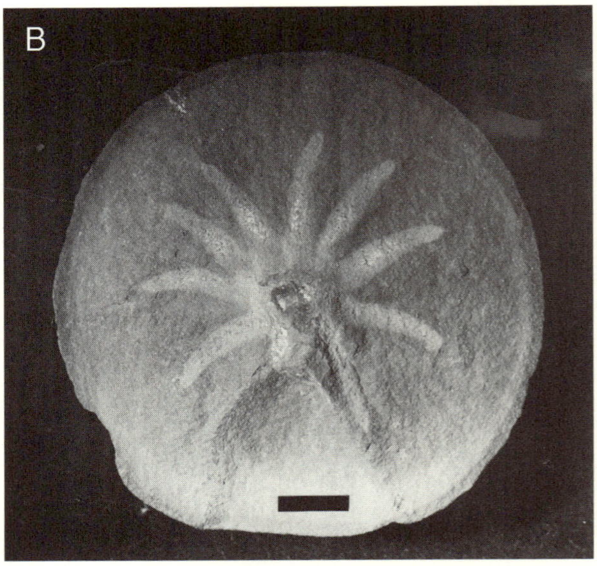

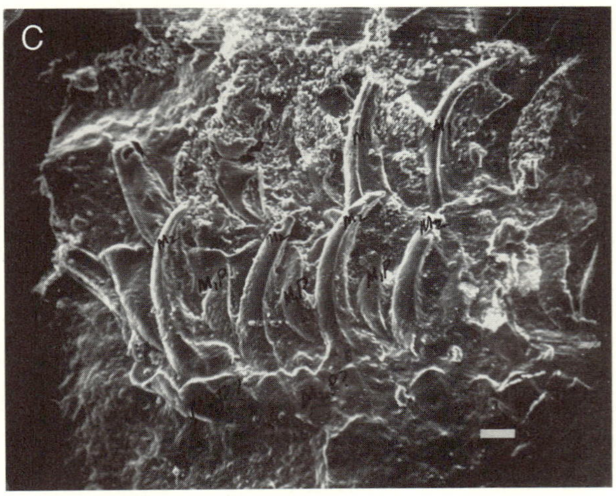

Figure 8D.10. *Jeletzkya douglassae*. **A** Shell fragment, FMNH PE 28955. Scale bar = 5 mm. **B** Holotype, Douglass collection 565. Scale bar = 5 mm. **C** SEM of "specimen no. 4," radula, FMNH MCP 2058. Scale bar = 0.1 mm.

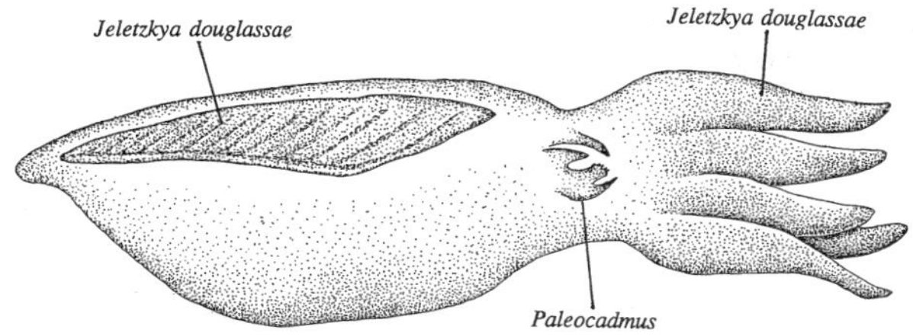

Figure 8D.11. Cartoon illustrating locations of 8D.7 and 8D.9 plus 8D.10A and 8D.10B in a hypothetical cephalopod.

herdinae are elongate and angular; and the smooth marginal teeth of *P. pohli* with their V-shaped cross section differ from those of *P. herdinae* with their deep, narrow outer groove (Saunders and Richardson, 1979).

Holotype. FMNH PE 25583.

Saunders and Richardson (1979) also described a pair of interlocking nautiloid mandibles, which they assigned to this genus. These chitinous mandibles are smooth with fine concentric growth lines; no calcitic elements like those found in Triassic–Recent nautiloids are present. The broad flanks of these mandibles are similar to those of modern *Nautilus* and ammonoids.

Figured specimen. FMNH PE 32522.

Subclass Coleoidea
Jeletzkya douglassae
Figure 8D.10, 8D.11

Ten equal-sized tentacles, each with a double row of arm hooks, encircle the head of this presumed coleoid cephalopod. The radula consists of nine elements: a triangular first lateral tooth, a similar inner marginal tooth, an elongate crescentic outer marginal tooth, and a single wide marginal plate on each side of a triangular central rachidian tooth (Saunders and Richardson, 1979). Mandibles are smooth, slightly concave and ovate, with no evidence of calcitic elements. The elongate ovate internal shell tapers posteriorly and bears longitudinal parallel lineations near the apex; however, the apical region is not well known (Saunders and Richardson, 1979). An ink sac may be present posteriorly.

Johnson and Richardson (1968) originally classified *J. douglassae* as a teuthid based on the number of arms and similarity of its internal shell to the gladius of modern squids. However, Gordon (1971) considered double rows of arm hooks to be a belemnitid-phragmoteuthid feature and suggested that the internal shell of the holotype represented a belemnitid prostracum with only the final phragmocone chamber attached.

Assignment of *J. douglassae* remains problematic. Its radular pattern is the same as modern coleoids; however, belemnoid radulae are not known (Saunders and Richardson, 1979). Single rows of arm hooks occur in modern squids, and Saunders and Richardson (1979) suggested that double rows as well as undifferentiated arms may have been primitive characteristics. Fossil ink sacs are known from both squids and belemnoids (Lehmann, 1981).

Other Taxa

Recently, Allison (1987) described a soft-bodied cephalopod from the Essex fauna that is too poorly preserved to identify. However, he interpreted the presence of a possible gut or ink sac and eye. The specimen has a streamlined body and 10 arms, as does *Jeletzkya;* however, the arms are shaped differently and bear no hooks. No internal shell is present. *Pohlsepia mazonensis* new genus and species, collected by Jim Pohl at Pit 11, bears a remarkable resemblance to modern sepiodariid (Kluessendorf, in press). Another cephalopod from Pit 11, collected by Michael Kolodzy, bears arm hooks but has no internal shell; it appears to contain a definite ink sac.

References

Allison, P. A. 1987. A new cephalopod with soft parts from the Upper Carboniferous Francis Creek Shale of Illinois, USA. Lethaia, 20:117–121.

Erben, H. K. 1964. Bactritoidea, pp. K491–K505. *In* R. C. Moore (ed.), Treatise on Invertebrate Paleontology. Part K, Mollusca 3. Geological Society of America and University of Kansas Press, Lawrence.

Gordon, Mackenzie, Jr. 1971. Primitive squid gladii from the Permian of Utah. U.S. Geol. Surv. Prof. Paper, 750-C:C34–C38.

Johnson, R. G., and E. S. Richardson, Jr. 1968. Ten-armed fossil cephalopod from the Pennsylvanian of northeastern Illinois. Science, 159:526–528.

Klussendorf, J. 1996. Oldest fossil sepioid—*Pohlsepia mazonensis* from the Carboniferous of Illinois (USA). Submitted to Palaeontology.

Kummel, Bernhard. 1964. Nautiloidea-Nautilida, pp. K383–K457. *In* R. C. Moore (ed.), Treatise on Invertebrate Paleontology. Part K, Mollusca 3. Geological Society of America and University of Kansas Press, Lawrence.
Lehmann, U. 1981. The Ammonites: Their Life and Their World. Cambridge University Press, Cambridge, 246 pp.
Mapes, R. H. 1979. Carboniferous and Permian Bactritoidea (Cephalopoda) in North America. Univ. Kansas Paleo. Contrib., 64, 75 pp.
Moore, R. C., ed. 1957. Treatise on Invertebrate Paleontology. Part L, Mollusca 4. Geological Society of America and University of Kansas Press, Lawrence.
Saunders, W. B., and E. S. Richardson, Jr. 1979. Middle Pennsylvanian (Desmoinesean) Cephalopoda of the Mazon Creek fauna, northeastern Illinois, pp. 333–359. *In* M. H. Nitecki (ed.), Mazon Creek Fossils. Academic Press, New York.
Saunders, W. B., Claude Spinosa, Curt Teichert, and R. C. Banks. 1978. The jaw apparatus of recent *Nautilus* and its paleontological implications. Palaeontology, 21:129–141.
Solem, G. A. 1974. The Shell Makers: Introducing Mollusks. Wiley, New York, 289 pp.
Solem, G. A., and E. S., Richardson, Jr. 1975. *Paleocadmus,* a nautiloid cephalopod radula from the Pennsylvanian Francis Creek Shale of Illinois. The Veliger 17:233–242.
Teichert, Curt, and R. C. Moore. 1964. Classification and stratigraphic distribution, pp. K94–K105. *In* R. C. Moore (ed.), Treatise on Invertebrate Paleontology. Part K, Mollusca 3. Geological Society of America and University of Kansas Press, Lawrence.

9

Introduction to Arthropods

Frederick R. Schram and Andrew A. Hay

The arthropods comprise about 95 percent of all named animal species. Extensive paleontological collections of this group exist from every period, Cambrian to the present day. However, the fossil record may not accurately reflect the diversity or give an accurate census of the arthropods alive at any given time due to vagaries of preservation. This vast and diverse group underwent a radiation unparalleled in the history of life, from the first arthropodlike organisms to the present day. Forms range from the almost unchanged horseshoe crab, through the "mollusclike" barnacles, to modern insects.

Although the various arthropod groups share many features, some authors believe these are not in fact features unique to arthropods. These workers maintain that arthropods are a polyphyletic group, descended from two or more different annelid or annelid-like ancestors, and form at least four distinct phyla: Uniramia, Cheliceriformes, Trilobitomorpha, and Crustacea. Other specialists believe that arthropods form a single phylum, a monophyletic group whose member species all descended from a single ancestral species. The adherents of this view consider Uniramia, Cheliceriformes, Trilobitomorpha, and Crustacea to be subphyla within the phylum Arthropoda.

Arthropod design has one basic principle, a longitudinal series of numerous, separate, and similar segments making up the body, features that arthropods share with the annelids. Specialization and diversity in arthropod form lie in the addition, fusion, or reduction of the numbers of segments in the various groups. Modification of the initially identical segments and specialization of the appendages have produced the impressive variety of designs in arthropod anatomy. The limbs or "legs" of arthropods display modifications into instruments for locomotion: walking, swimming, or flying (Kukalova-Peck, 1983), grasping, respiration, sensing, attaching, and food manipulation. Numbers of locomotory legs vary in number from the dozens, as among millipedes, to three pair as among the insects. Arthropods possess exoskeletons (external skeletons) of cuticle that are unmineralized in most groups but often contain tanned proteins. Body form varies from the wormlike millipedes, to the seemingly unsegmented body of mites, to the tiny bivalved carapace of ostracodes. It is no wonder that the disparity in anatomical design plagues the taxonomist attempting to classify arthropods and has aroused heated debates among specialists seeking to elucidate the affinities of these very diverse and interesting animals.

The aquatic arthropods are among the largest and best represented groups in the Mazon Creek fauna. Not only do they show great diversity of body types, but also the integument and the external armor of the arthropods has lead to their superb preservation. The animals themselves separate into two major groups: those with a modern affinity and those with no apparent modern analogues, known as the problematica. The former are the most common and include almost all the crustaceans and the horseshoe crabs. Some of these have distant modern affinities, like *Palaeocaris,* which is considered a syncarid, or *Belotelson,* a crustacean that is rather lobsterlike but is currently placed in an extinct group entirely by itself (Schram, 1986). The true problematica include several enigmatic forms with no easily established ties to modern species. Excellent examples of this are the euthycarcinoideans, animals that may be distant precursors to insects and millipedes (Schram and Rolfe, 1982); the thylacocephalans (Schram, 1990), whose crustacean affinities remain obscure; and the cyclids, a group that has been classified within several different subphyla. A lack of living analogues, combined with the incomplete nature of our knowledge of the anatomy of these fossils, leaves us at loose ends in attempting to understand their relationship to other arthropod forms.

The largest terrestrial arthropods ever known, fossil species of arthropleurids and dragonfly-like insects, have been found in the Francis Creek Shale of the Mazon Creek area. The earliest isopod known, a marine creature (Schram, 1970), occurs in the Mazon Creek. Today the class Isopoda is exemplified by the terrestrial sowbugs and woodlice. Insects, well represented in this biota, have a fossil record that dates at least to the Devonian. Insect wing development culminated in true flight during the Carboniferous. The first reported Paleozoic scorpion from the Western Hemisphere was found in a Mazon Creek concretion. Until the description of an unequivocally terrestrial scorpion in the Mazon Creek biota by Briggs (1987), it had long been conjectured that these scorpions were essentially aquatic, only capable of spending part of their lives on land, much like the extant horseshoe crab. Fossil spiders and spiderlike arachnids, similar in appearance to those of today, also occur in the Mazon Creek. Reportedly, spiders possessed spinnerets prior to the Carboniferous (Shear et al., 1989). An interesting question as to whether the ability to spin silk strands for prey capture had developed in the coal forests cannot be verified at present by the fossil evidence. The Mazon Creek contains other terrestrial arthropods: millipedes, centipedes, wingless insects, and onychophorans (Shear and Kukalova-Peck, 1990).

All of these animals combine to give us a fauna that appears on the one hand familiar, since the animals have some relation to our modern experience, and on the other hand slightly "out of focus," that is, sufficiently different to bring us up short and remind us that the Middle Pennsylvanian, no matter how superficially similar to modern points of reference, was indeed another place and another time.

An area of fundamental concern, aside from matters related to individual fossil species or groups of species, is how the Mazon Creek faunas fit into the continuum of Late Paleozoic ecologic communities and modern communities (Schram, 1981). These faunas are probably typical examples of community types that extended throughout the Carboniferous of the ancient continent Laurasia (North America and Europe). The basic picture derived from the study of all the occurrences of these communities is one of overall continuity and stability throughout the Carboniferous. However, it is important to note that despite the variety of the arthropods in the fauna found at Mazon Creek, the basic arthropod ecologic pattern is little changed from that found today.

References

Briggs, D. E. G. 1987. Scorpions take to the water. Nature, 326:645–646.

Johnson, R. G., and E. S. Richardson, Jr. 1966. A remarkable Pennsylvanian fauna from the Mazon Creek area, Illinois. Jour. Geol., 12:109–115.

Kukalova-Peck, J. 1983. Origin of the insect wing and wing articulation from the arthropodan leg. Canadian Jour. Zool., 61:2327–2345.

Schram, F. R. 1970. Isopod from the Pennsylvanian of Illinois. Science, 169:854–855.

———. 1981. Late Paleozoic communities. Jour. Paleo., 55:126–137.

———. 1986. Crustacea. Oxford University Press, New York.

———. 1990. On Mazon Creek Thylacocephala. Proc. San Diego Soc. Nat. Hist., 3:1–16.
Schram, F. R., and W. D. I. Rolfe. 1982. New Euthycarcinoid Arthropods from the Upper Pennsylvanian of France and Illinois. Jour. Paleo., 56:1434–1450.
Shear, W. A., et al. 1989. A Devonian spinneret: early evidence of spiders and silk use. Science, 246:479–481.
Shear, W. A., and J. Kukalova-Peck. 1990. Ecology of terrestrial arthropods. Canadian Jour. Zool., 68:1807–1834.

10

Xiphosura

Donald G. Mikulic

Xiphosurans (horseshoe crabs) are among the most distinctive and easily recognized animals in the Mazon Creek biota. Specimens are highly prized by collectors, and they generated considerable interest following F. B. Meek and A. H. Worthen's published description of *Euproops danae* in 1865. One of the specimens used in their study was collected by Joseph Even, the discoverer of the Mazon Creek fossil beds. Of all the fossils described in the eight volumes of *Geology of Illinois* (edited by A. H. Worthen, then state geologist), *Euproops danae* was chosen as the symbol to be embossed on the volume covers.

The interest in Mazon Creek xiphosurans is due in part to the general rarity of xiphosurans in the fossil record. Only the Jurassic Solnhofen Limestone of Germany has yielded xiphosuran specimens in numbers comparable to Mazon Creek. Because of this rarity, Mazon Creek specimens are of special importance in the study of the evolution, functional morphology, and behavior of this unique group of arthropods.

Xiphosurans belong to the arthropod superclass Chelicerata, which comprises two classes: the Arachnida (spiders, mites, scorpions) and Merostomata, which includes the Order Xiphosurida and the extinct Order Eurypterida. The Merostomata, which range from the Cambrian to the Recent, are aquatic in general and are much less diverse than the predominantly terrestrial and highly varied arachnids.

The limited morphologic change exhibited by xiphosurans throughout their long history and the existence of well-known living representatives present a good basis for interpreting the biology of fossil forms. The following information on modern xiphosurans is derived primarily from Størmer (1955) and Barnes (1974).

The xiphosuran exoskeleton is divided into two parts. The anterior highly vaulted semicircular prosoma, which is a fused cephalothorax, contains most of the internal organs, the eyes, the chelicerae, and the walking legs. The posterior opisthosome, or abdomen, which contains the gill plates, articulates with a long terminal telson. *Limulus,* the well-known extant horseshoe crab, has five pairs of walking legs and a single pair of anterior chelicerae.

Limulus is a benthic organism that spends most of its life crawling across the shallow seafloor or burrowing partially into soft sand and mud in search of food. Young and, to a lesser extent, older individuals can swim by rapidly flexing their appendages while oriented upside down. The telson is used primarily as a lever to allow the organism to right itself if overturned, but it may aid in burrowing; it is not a defensive weapon.

Horseshoe crabs are scavengers or low-level benthic predators. Food particles are picked up by a pincer, or chela, on the walking leg, passed to the spiny gnathobase where they are ground, and then passed forward to the mouth. Xiphosurans, like all arthropods, must molt their mineralized exoskeletons periodically in order to grow. The marginal suture along the front of the prosoma splits, and the animal crawls forward and out of its old shell. The animal then enlarges and forms a new shell. Each individual may produce tens of molts throughout its lifetime, all of which may be fossilized intact. Therefore, it is a strong possibility that fossils of even complete xiphosures may represent molts and not dead individuals. Exceptions to this, however, possibly include some enrolled specimens and certainly the spectacular Solnhofen specimens preserved at the end of their trails (Müller, 1979), which are obviously dead individuals.

Limulus and two other closely related genera are the only living members of the Merostomata. Recent xiphosurans live in shallow marine, nearshore, and estuarine environments, and many fossil forms probably lived in similar habitats. Modern horseshoe crabs are able to crawl a short distance out of water, and Fisher (1979) suggested that the Mazon Creek taxon *Euproops danae* may have lived in moist leaf litter on the floor of Pennsylvanian swamps and forests based on distribution, associations, and morphology. Anderson (1994), however, effectively argued against Fisher's theory, concluding that Fisher's evidence does not demonstrate subaerial activity. He points out that *Euproops* would be unable to breathe or eat on land. Fisher (1975, 1977) also speculated that the size and placement of spines and the enrollment mode enabled a swimming individual to drop to the seafloor in a controlled manner in order to avoid detection by predators.

Limulus reaches a length of 60 cm, which is much larger than any xiphosuran from Mazon Creek. The small size of Mazon Creek specimens may be due to the paucity of large concretions or the absence of large adult individuals.

Hundreds, if not several thousands, of xiphosurans have been collected from the Mazon Creek area during the last 140 years. *Euproops danae,* the most common xiphosuran taxon, is better preserved and more abundant in the Braidwood biota (Fisher, 1979), whereas *Palaeolimulus* sp. is more typical of the Essex biota. The overall abundance of xiphosurans in these two biotas is 27 percent and less than 1 percent, respectively (Richardson and Johnson, 1971), but the molting process of these animals suggest that their abundance probably is exaggerated in comparison with nonarthropod taxa.

Mazon Creek horseshoe crabs display a number of important preservational features. Complete individuals are rare, as the telson frequently projected beyond the concretion and is often missing. The ophthalmic spines, which project off the eye ridge, seldom are visible. Specimens commonly are preserved as external molds, and most are compressed slightly, particularly in the area of the high convex prosoma. Fisher (1975) noted that both dorsal and ventral features may appear on the same surface because of this compression. The appendages, which probably are present on most specimens, are rarely observed because the plane of weakness through the concretion is across the shell's convex dorsal surface. Although articulated specimens are typical, disarticulated parts also occur; Fisher (1979) described coprolitic concretions containing several disarticulated xiphosures, other taxa, and organic matter. However, the large size of some of these concretions and the low density of the scattered fossil debris suggest that not all are coprolites, or at least they are not coprolites of animals that fed selectively on xiphosurans. Local fortuitous accumulations of xiphosure molts and transportation of these molts by tides and storms into the terrestrial environment are observed along present-day coasts; these factors may account for the "coprolites" and more terrestrial occurrences of the Mazon Creek taxa.

The following xiphosurans have been identified in Mazon Creek collections. The descriptions are based on papers published by Meek and Worthen (1865, 1866), Meek (1867a, 1867b), Packard (1886), Raymond (1944, 1945), Langford (1963), and Darrah (1969).

Euproops danae (Meek and Worthen, 1865)
Figures 10.1, 10.2

This horseshoe crab, a member of the family Euproopidae, was the first Mazon Creek xiphosuran discovered, and it is the most common. This taxon is restricted in the Braidwood biota, where nearly complete individuals are typical (Fisher, 1979). Disarticulated and possibly bro-

Figure 10.1. *Euproops danae* (Meek and Worthen, 1865).
A FMNH 5769UC;
B FMNH MCP 489, Pit 1.
Both scale bars = 10 mm.

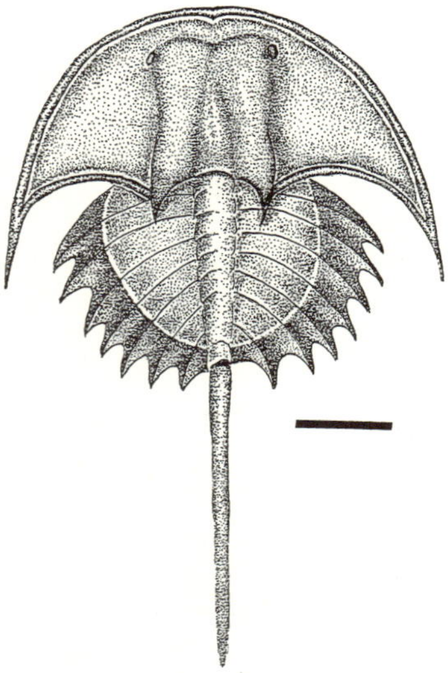

Figure 10.2. *Euproops danae,* modified from Packard (1886). Scale bar = 10 mm.

ken parts are found more rarely in the Essex biota, where coprolitic concretions contain parts of several individuals.

Euproops danae originally was described under the genus *Bellinurus* (Meek and Worthen, 1865), then *Prestwichia* (Meek, 1867a), and finally as the genotype of *Euproops* (Meek, 1867b).

The prosoma (cephalothorax) is very short and wide compared to other Mazon Creek horseshoe crabs; its width is 2.5 to 3 times the length. It has a broad, rounded appearance, tapering backward into long, flat genal spines (i.e., spines on the posteriolateral corner of the cephalothorax) that are as long as the entire prosoma. Small compound eyes are located near the anterior end of the ophthalmic ridges. These ridges converge backward slightly, extending into a pair of ophthalmic spines, which project backward over the opisthosoma. The ophthalmic spines usually are observed only on the external molds. The cardiopthalmic area, located between the ophthalmic ridges, is well marked. The central cardiac lobe is bordered by distinct axial furrows and tapers rapidly forward, with the anterior one-third forming a narrow ridge. Paired ocelli are present at the anterior end of the cardiac lobe.

The opisthosoma (abdomen) is broad, rounded, with six distinct fused segments. The central axis, composed of six rings, is well defined by dorsal furrows. The first and third rings possess rounded tubercles; the last ring forms an enlarged rounded boss that bears a short spine. The pleural (lateral) lobes of the abdomen are marked by narrow ridges that extend beyond the border into long, projecting marginal spines. The long, narrow telson seldom is preserved completely. Its length is the same as the combined lengths of the prosoma and opisthosoma. The appendages of *Euproops* are similar to those of modern horseshoe crabs.

Small (juvenile) specimens differ from larger specimens in that the cardiac lobe exhibits a more constant taper, the prosoma is only 2 times as wide as long, and the abdomen is proportionately wider and has weakly developed spines. Raymond (1944) observed that the

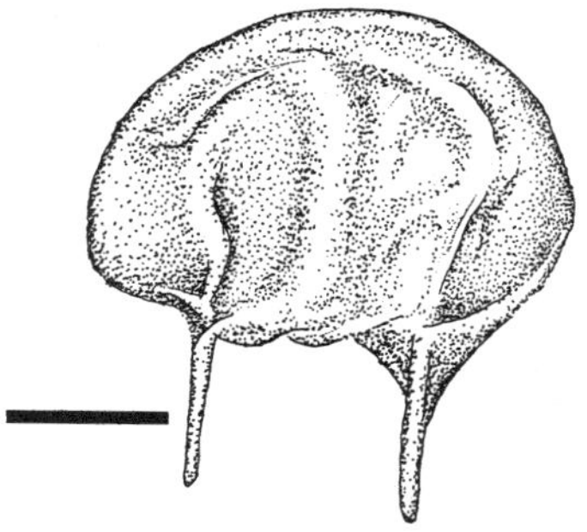

Figure 10.3. *Pringlia bispinosa* Raymond, modified from Raymond (1944).
Scale bar = 10 mm.

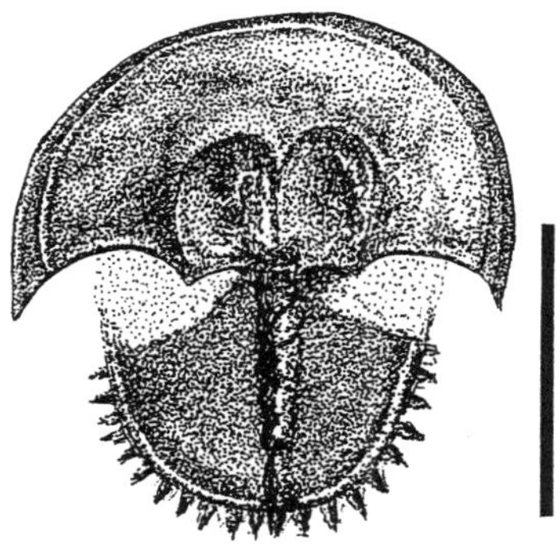

Figure 10.4. *Streptocyclus langfordi* Raymond, modified from Langford (1963).
Scale bar = 10 mm.

young could enroll. Specimens, without telson, range in size from 16 to 44 mm in length.

Raymond (1944) described two other species of *Euproops* from the Mazon Creek area. *Euproops thompsoni* differs from *E. danae* in possessing a longer and narrower prosoma, which is only about twice as wide as long. The prosoma of *E. laevicula* is similar to that of *E. thompsoni,* but the opisthosoma has only weak features. The validity of these two species is questionable, and restudy of all material is needed. Preservational differences and population variation may account for the features used by Raymond to erect these and other Mazon Creek xiphosuran species. Both Fisher (1975) and Anderson (1994) believe that only one species (*E. danae*) occurs at Mazon Creek.

Euproops occurs widely in rocks of Pennsylvanian (Upper Carboniferous) age and may also occur in Permian rocks. The genus has been found in Europe and North America, including Illinois, Indiana, Oklahoma, and Pennsylvania.

Pringlia bispinosa Raymond
Figure 10.3

This taxon is a rare member of the family Euproopidae found in the Mazon Creek biota. It differs from *Euproops* in lacking genal spines and possessing a triangular opisthosoma with weak axial furrow, weak pleural ridges, and possibly no marginal spines.

Streptocyclus langfordi Raymond
Figure 10.4

This rare species differs from *Euproops* in that the posterior portions of the ophthalmic ridges are outwardly convex and, like *Pringlia,* in possessing weakly defined abdominal features.

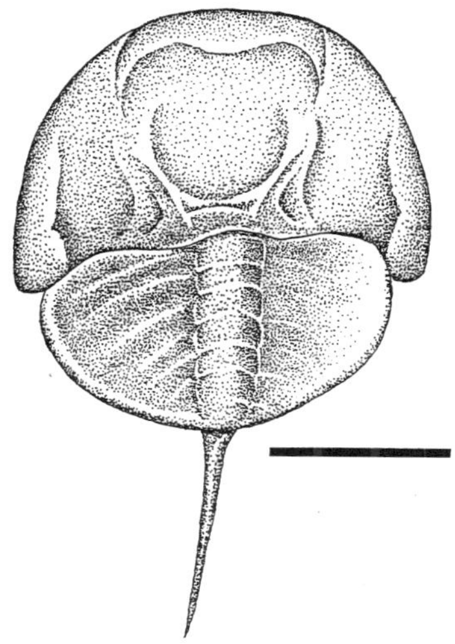

Figure 10.5. *Liomesaspis laevis* Raymond, modified from Raymond (1944).
Scale bar = 5 mm.

Liomesaspis laevis Raymond
Figure 10.5

A member of the family Liomesaspidae, this species was erected by Raymond on specimens in the Yale University and Harvard University collections. Raymond implied that this taxon was common in Mazon Creek collections; however, Langford (1963) disputed this. The prosomal cardiac lobe is bulbous, and the prosoma, lacking genal spines or angles, is rounded. The axial lobe on the opisthosoma is narrow, expanding into a high boss. The lateral margins are smooth, lacking marginal spines.

Figure 10.6. *Palaeolimulus* sp., FMNH MCP 765. Scale bar = 5 mm.

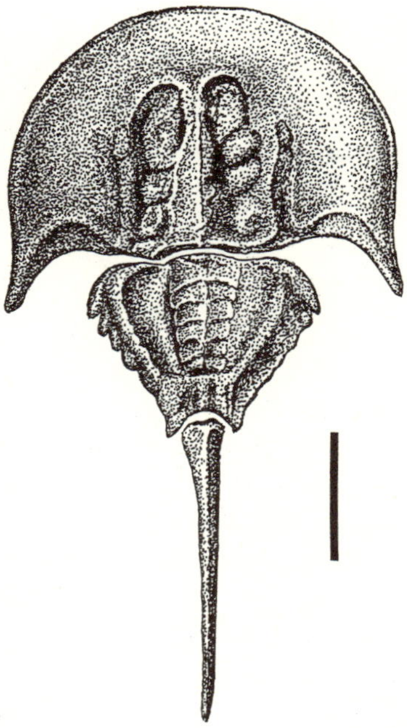

Figure 10.7. *Palaeolimulus,* modified from Shimer and Shrock (1944). Scale bar = 5 mm.

Palaeolimulus sp.
Figures 10.6, 10.7

This undescribed taxon is a member of the family Paleolimulidae within the superfamily Limulacea, which includes the recent horseshoe crabs. It is characterized by eyes that are set in a more medial position compared to other Mazon Creek xiphosurans. The ophthalmic ridges run subparallel posterior from the eyes. The abdomen has an annulated axis with a movable segment in front of the telson and movable marginal spines.

References

Anderson, L. I. 1994. Xiphosurans from the Westphalian D of the Radstock Basin, Somerset Coalfield, the South Wales Coalfield and Mazon Creek, Illinois. Proc. Geol. Assoc., 105:265–275.

Barnes, R. D. 1974. Invertebrate Zoology, 3rd ed. Saunders, Philadelphia. 870 pp.

Darrah, W. C. 1969. A critical review of the Upper Pennsylvanian flora of the eastern United States with notes on the Mazon Creek flora of Illinois. Published by the author, 220 pp.

Fisher, D.C. 1975. Evolution and functional morphology of the Xiphosurida. Ph.D. thesis, Harvard University, Cambridge, Massachusetts, 403 pp.

———. 1977. Functional significance of spines in the Pennsylvanian horseshoe crab *Euproops danae.* Paleobiology, 3:175–195.

———. 1979. Evidence for the subaerial activity of *Euproops danae* (Merostomata, Xiphosurida), pp. 379–448. *In* M. H. Nitecki (ed.), Mazon Creek Fossils. Academic Press, New York.

Langford, George. 1963. The Wilmington coal fauna and additions to the Wilmington coal flora from a Pennsylvanian deposit in Will County, Illinois. ESCONI, Downers Grove, Illinois, 200 pp.

Meek, F. B. 1867a. Note on *Bellinurus danae,* from the Illinois Coal Measures. Amer. Jour. Sci., 2nd ser., 43:257–258.

———. 1867b. Note on a new genus of fossil Crustacea. Amer. Jour. Sci., 2nd ser., 43:394–395.

Meek, F. B. and A. H. Worthen. 1865. Notice of some new types of organic remains, from the Coal Measures of Illinois. Proc. Acad. Nat. Sci. Philadelphia, 17:41–53.

———. 1866. Invertebrate fossils of the Coal Measures. Illinois State Geol. Surv., 2:310–411.

Müller, K. H. 1979. Fossilization (taphonomy), pp. A2–A78. *In* R. Robison and C. Teichert (eds.), The Treatise on Invertebrate Paleontology. Part A, Introduction. Geological Society of America and the University of Kansas, Lawrence.

Packard, A. S., Jr. 1886. On the Carboniferous xiphosurous fauna of North America. Nat. Acad. Sci. Mem., 3, pt. 2:145–157.

Raymond, P. E. 1944. Late Paleozoic xiphosurans. Bull. Mus. Comp. Zool., 94(10):475–508.

———. 1945. Xiphosura in the Langford collection. Illinois State Mus. Sci. Paper, 3(3):4–8.

Richardson, E. S., Jr., and R. G. Johnson. 1971. The Mazon Creek faunas, pp. 1222–1235. Proc. North American Paleo. Conv., 5–7 September 1969, Field Mus. Nat. Hist., vol. 1. Allen, Lawrence, Kansas.

Shimer, H., and Shrock, R. 1944. Index Fossils of North America. Wiley, London, 708 pp.

Størmer, L. 1955. Merostomata, pp. P4–P41. *In* R. C. Moore (ed.), Treatise on Invertebrate Paleontology. Part P, Arthropoda 2. Geological Society of America and University of Kansas, Lawrence.

11

Arachnida

Bret S. Beall
Including a Section on the Ricinuleida
by Paul A. Selden

The Mazon Creek region has yielded over 250 arachnid specimens, more than from any other Paleozoic arachnid locality. Only the Cenozoic amber deposits of the Baltic and the Dominican Republic have produced more arachnid fossils. The Mazon Creek arachnids occur with a great variety of other terrestrial animal and plant taxa. This co-occurrence offers an almost unique opportunity to investigate the role of arachnids in a major Pennsylvanian terrestrial biota. Although this sort of investigation is possible at other localities from which arachnids have been collected, the abundance of the Mazon Creek arachnids in association with other taxa necessarily reduces the influence of sampling bias on estimates of relative taxic abundance of the original biota.

The Mazon Creek arachnids are invaluable for studying the evolution (including phylogeny and functional morphology) of the class primarily because of their exquisite preservation. They are usually preserved as relatively uncompressed external molds in siderite concretions, although the entire suite of arachnid fossils actually exhibits a gradient of compression and distortion. The fine texture of the siderite concretions and the minimal distortion of the specimens together preserve a variety of morphological features that either are not visible on specimens from other localities or are possible to interpret only ambiguously. Consequently, the morphology of most of the taxa of Mazon Creek arachnids can be described more accurately. Additionally, the unusual abundance of material permits the comparison of multiple specimens with similar morphology, further reducing the probability that distortion due to taphonomic processes will be interpreted as original morphology. Arachnid fossils are often incomplete, but the presence of multiple specimens aids composite reconstruction at several taxonomic levels. However, such reconstructions involve the risk of underrepresenting actual variation of a particular taxon and so must be made with caution. Certain aspects of the arachnids from Mazon Creek and other localities suggest that many of the extant orders that are also present as fossils at these localities are quite conservative morphologically in spite of the more than 300-million-year histories of their clades. The excellent preservation of the Mazon Creek arachnids reduces influence of distortion-induced measurement error of morphological stasis or bradytelic rates of evolution. The Mazon Creek specimens are capable of providing substantial evidence bearing on the evolutionary history of arachnids, their relationship to other arthropods, and their interactive roles in Late Paleozoic terrestrial biotas.

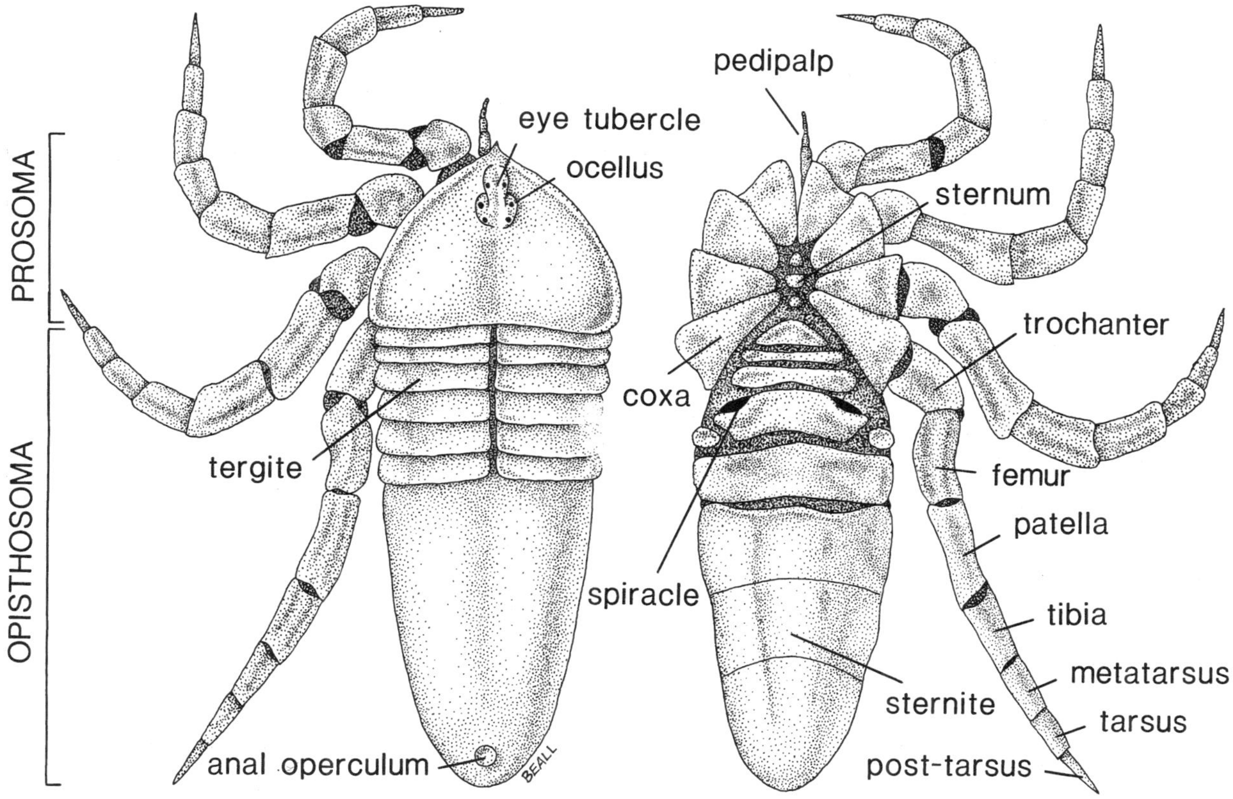

Figure 11.1. General arachnid morphology: *Phalangiotarbus lacoei,* FMNH PE 32197.

Morphology

The general morphology of an arachnid is illustrated in Figure 11.1. The arachnid body consists of distinct anterior and posterior regions called the prosoma (head and thorax) and opisthosoma (abdomen), respectively. The prosoma is composed of the unsegmented acron and the first six somites (segments) and is usually covered dorsally by a single tagma (peltidium or carapace). This tagma may bear up to eight single eyes, or ocelli, some or all of which may be arranged on a tubercle. Each of the six prosomal segments bears a pair of appendages; the chelicerae are the anteriormost pair, followed by the pedipalps and four pairs of legs. In some, the chelicerae have pincerlike claws (chelae). Each appendage is composed of segments (podomeres) that articulate at one or more points (condyles). The segment of the leg that attaches it to the body is the coxa. The segment at the end farthest from the body is the tarsus.

The abdomen (opisthosoma) consists of 12 segments (somites) and a telson at the end. Primitively, each segment has a dorsal tergite and a ventral sternite that are modified in various ways in each order.

Respiration occurs cutaneously, or by means of book lungs, sieve tracheae, or tube tracheae. The openings to these respiratory structures are called spiracles.

Systematic Paleontology

Class Arachnida is in the phylum Chelicerata. Petrunkevitch (1913, 1945, 1949, 1955) recognized 10 orders of the class Arachnida (subphylum Chelicerata), and numerous lower taxa among the Mazon Creek fossils. Subsequently, only Kjellesvig-Waering (1986, in preparation) and Selden (1983, 1984) have reviewed parts of Petrunkevitch's classification. Much work is still necessary to evaluate the validity of many of the genera and species of Mazon Creek arachnids.

A number of problems plague the lower level taxonomy of Paleozoic arachnids from Mazon

Creek and other localities. Both Petrunkevitch and Kjellesvig-Waering developed classification systems in which the limits of the taxa are relatively arbitrary, making their taxonomies quite subjective. Kjellesvig-Waering (1986, in preparation) has emphasized Petrunkevitch's apparent failure to consider taphonomic influences on arachnid morphology; because of this failure, many of Petrunkevitch's taxa are artificial, being distortions of morphology. Finally, because most of the genera and species of extant arachnids have been diagnosed using characters that are rarely preserved in fossils, comparing these extant taxa with extinct taxa of the same rank is extremely difficult. Because of these and other problems at lower taxonomic levels, this chapter focuses on the orders of arachnids known from the Mazon Creek region.

Figure 11.2. *Bicarinitarbus pieckorum,* FMNH PE 107. Scale bar = 5 mm (estimate).

Order Phalangiotarbida

A phalangiotarbid was one of the first arachnids to be described from the Mazon Creek region (Scudder, 1868). At least 150 specimens have been collected, making the Phalangiotarbida the most abundant (61 percent) of Mazon Creek arachnids that can be identified to order.

Morphology. Prosoma semicircular (primitive) to rhomboidal (derived), with pointed anterior projection; eye tubercle trifoliate with three pairs of ocelli along its flanks. Ornamentation consists of the fine pustules (interpreted as setal bases). Ventrally, the basal segment (coxa) of appendages III to VI meet at a sternum composed of up to four sclerites, some of which are medially divided.

Chelicerae unknown. Pedipalps incompletely preserved, but are rather short, antenniform appendages with at least four segments; coxae unknown. Third pair of prosomal appendages raptorial (used for grabbing, based on kinematic analyses of limb structure; Beall, 1984, 1985); corresponding coxae abut medially, with medial margins sometimes slightly spinose. Last three pairs pedal, with leg length increasing posteriorly.

Dorsally, abdomen with six anteroposteriorly narrow, slightly overlapping, medially divided tergites, followed by a large posterior tagma that may bear relics of fusion of tergites; all surfaces finely pustulated. Posterior tagma bears a round anal operculum near posterior margin (Kjellesvig-Waering, 1978, in preparation). Ventral surface with nine sternites connected to tergites by pleated, unsclerotized integument. First three sternites fit between fourth coxae. Fourth sternite as wide as abdomen (opisthosoma), divided into three transverse sections. Fourth through ninth sternites each bearing a pair of transversely arranged, shallow depressions that represent the external expression of entapodemes for dorsoventral musculature.

One or two pairs of spiracles present. One pair of relatively large spiracles always situated along the anterior margin of the median sclerite of the fourth sternite. Another pair of smaller openings sometimes present at lateral edges of anterior margin of second sternite.

Taxonomy. Kjellesvig-Waering (in preparation) assigned all phalangiotarbids to a single family, Phalangiotarbidae, with characteristics of the order. He recognized the following seven genera among the Mazon Creek phalangiotarbids.

Genus *Phalangiotarbus*
Figure 11.1

Posterior margin of prosoma straight to slightly concave anteriorly.

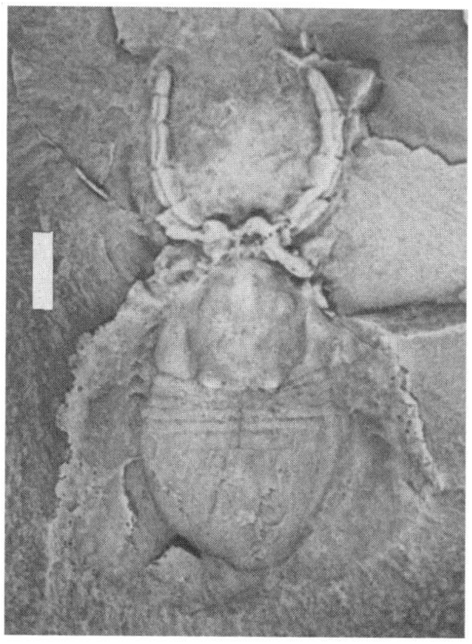

Figure 11.3. *Paratarbus carbonarius,* ISM 14864. Scale bar = 5 mm (estimate).

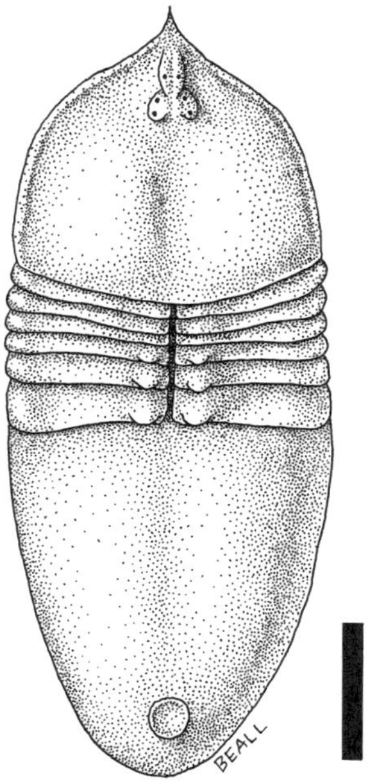

Figure 11.4. *Mesotarbus extraneus,* dorsal reconstruction, FMNH PE 32189. Scale bar = 5 mm.

Genus *Bicarinitarbus*
Figure 11.2

Posterior margin of prosoma straight; opisthosomal median line flanked by pair of "keels" that extend anteroposteriorly.

Genus *Paratarbus*
Figure 11.3

Posterior margin of prosoma slightly convex; two protuberances occurring at posterior edge of prosoma.

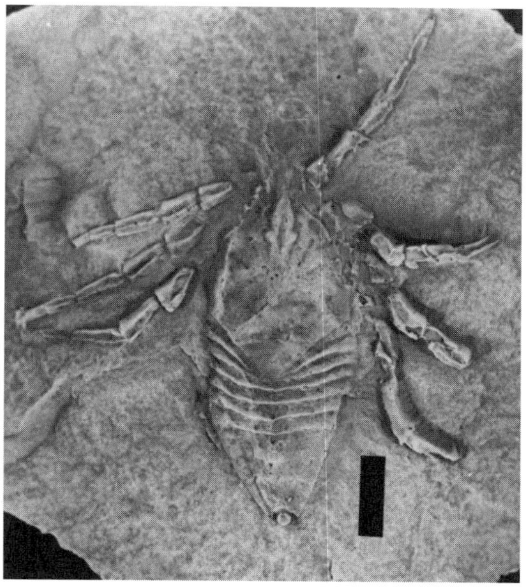

Figure 11.5. *Architarbus rotundatus,* YPM 185. Scale bar = 5 mm.

Genus *Mesotarbus*
Figure 11.4

Posterior margin of prosoma straight to slightly convex posteriorly; six anterior tergites each with a pair of small protuberances flanking the median sulcus.

Genus *Architarbus*
Figure 11.5

Posterior margin of prosoma strongly convex posteriorly; anterior two tergites similarly curved.

Genus *Mazonitarbus*
Figure 11.6

Posterior margin of prosoma more strongly convex posteriorly that in *Architarbus;* anterior four tergites similarly curved. *Mazonitarbus* probably is a juvenile instar of *Architarbus.*

Genus *Hadrachne*
Figure 11.7

Posterior margin of prosoma slightly convex, with only the first abdominal tergite similarly curved. Known from only a single specimen (largest known phalangiotarbid), this genus may represent an extremely old instar of *Architarbus.*

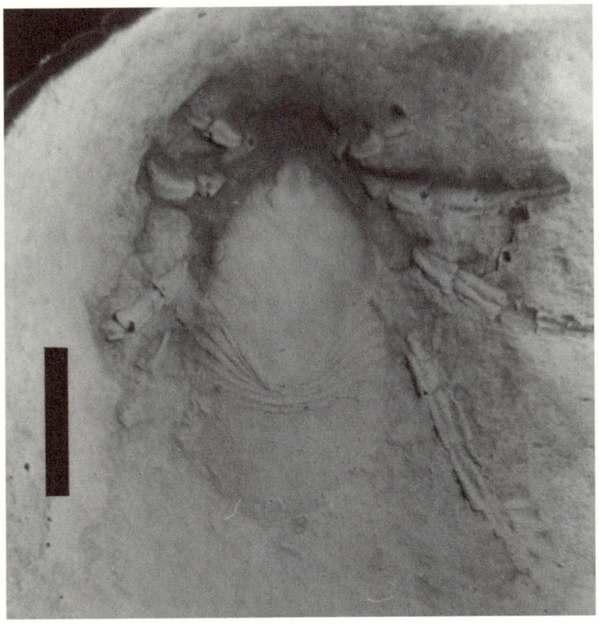

Figure 11.6. *Mazonitarbus minor,* FMNH PE 32155. Scale bar = 5 mm (estimate).

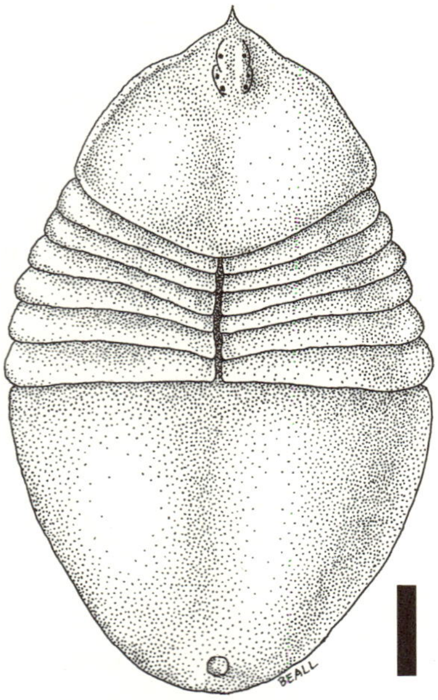

Figure 11.7. *Hadrachne horribilis,* dorsal reconstruction, FMNH OC 9232. Scale bar = 5 mm.

Order Anthracomartida

Originally the trigonotarbids and anthracomartids were placed in the same order, Anthracomarti (Karsch, 1882). Later, Petrunkevitch (1949) placed these arachnids in separate orders. Shear and Selden (1986) have questioned whether these orders are truly distinct, but the

Figure 11.8. Unnamed arthracomartid, FMNH PE 32156. Scale bar = 5 mm (estimate).

trigonotarbids and anthracomartids are treated separately here pending additional analysis.

At least 14 Mazon Creek anthracomartid specimens are known, representing 5.4 percent of the arachnids. Most are too poorly preserved to be described at the generic or specific level but they do provide important data regarding morphological diversity of the order (Figure 11.8).

Morphology. Prosoma subtrapezoidal to subtriangular; convex dorsally. Narrow median ridge sometimes bisects anterior third of prosoma. Pustules form a reticulate pattern. Eyes unknown.

Chelicerae unknown in Mazon Creek forms. Pedipalps antenniform with five segments (podomeres) excluding coxae (Petrunkevitch, 1955). Remaining appendages pedal, with six segments (podomeres) excluding coxae.

First and ninth abdominal tergites undivided; remaining seven tergites divided into five sclerites arranged in five fields. Opisthosomal perimeter smoothly curved or slightly scalloped. Tergites finely pustulose. Ventrally, opisthosoma with 10 undivided, smooth sternites; anal operculum in center of tenth sternite.

Petrunkevitch (1949, 1955) used a series of British anthracomartids to infer aspects of the modes of reproduction and respiration in these arachnids. He interpreted topographic variation of the second, third, and fourth sternites as indicating sexual dimorphism. However, Petrunkevitch does not defend his interpretations of female or male, nor does he argue that the features were not taphonomically induced. Similar structures are absent in North American specimens.

Petrunkevitch (1913, 1949, 1955) also illustrates "lungs" and "pulmonary apodemes" on the second, third, and fourth sternites. Actually, the "lungs" are lung covers, which do not necessarily indicate book lungs (Levi, 1969). Also, Petrunkevitch believed the fossils displayed the internal (rather than external) ventral opisthosomal surface; if he drew the "pulmonary apodemes" accurately, they resemble spiracles.

Taxonomy. No Mazon Creek anthracomartids have been described. As stated above, most of the specimens are too poorly preserved to allow adequate comparison with existing taxa or to justify describing them as new taxa.

Figure 11.9. *Aphantomartus pustulatus,* UMMP 7218. Scale bar = 5 mm (estimate).

Order Trigonotarbida

The trigonotarbids, the last of the three extinct orders from Mazon Creek, constitute 5.8 percent of the arachnids. Selden and Romano (1983) suggest that a taxonomic review may be necessary.

Morphology. Prosoma semicircular to subtriangular, with posterior margin straight; ovoid eye tubercle with two ocelli may be present in center of prosoma. Depression behind eye tubercle may be homologous with foveal depression (external expression of entapodemes for the origination of muscles inserting on the "pumping" stomach) in certain extant orders. Ornamentation often pustulose.

Chelicerae known only in Devonian forms, in which they are clawlike. Pedipalps antenniform, with five segments excluding coxae; coxae abut medially. Remaining appendages pedal with six podomeres, excluding coxae; sternum with independent sclerites or single large tagma.

Opisthosoma with eight tergites and eight to nine sternites; anal operculum terminal or in center of last sternite. Sternites smooth, increasing in curvature (convex anteriorly) toward posterior. Tergites divided into three fields; dorsal surface may be smooth to heavily tubercular. Posterolateral corners of sixth and seventh tergites may be elongated into spines.

Taxonomy. Trigonotarbid fossils are usually less distorted than specimens of other orders. Although the three genera recognized by Petrunkevitch (1955) are defined in part using primitive characters, each is also distinguished by certain derived characters, and so the taxa can be retained with redescription.

Genus *Aphantomartus* (= *Trigonomartus*) Selden and Romano, 1983 Figure 11.9

Prosoma tuberculate. Sternum wide with sclerites fused into single tagma. Opisthosoma pustulose with many tubercles. Short spines pro-

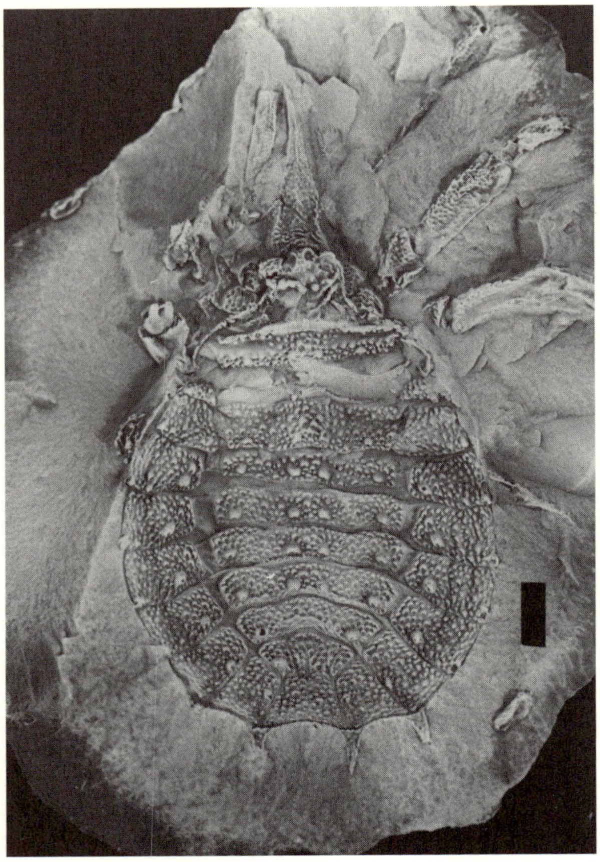

Figure 11.10. *Pleophrynus ensifer,* ISM 14873.
Scale bar = 5 mm.

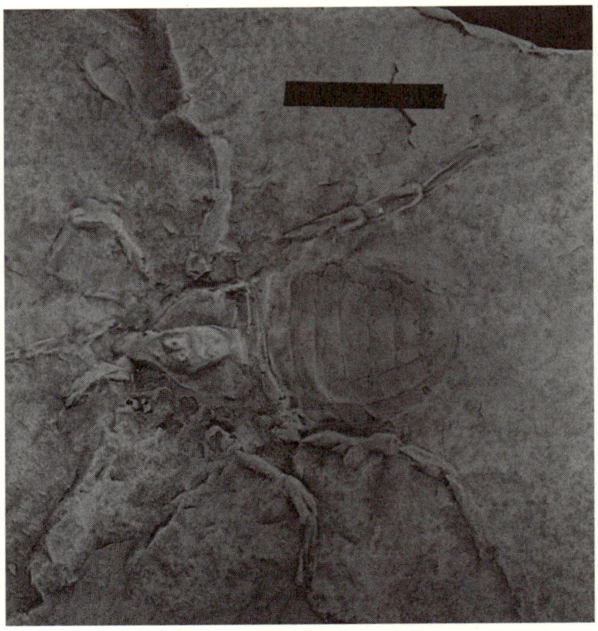

Figure 11.11. *Lissomartus schucherti,* YPM 169.
Scale bar = 5 mm (estimate).

ject from posterolateral corners of the sixth and seventh tergites, a synapomorphy with several other trigonotarbids (cover).

Genus *Pleophrynus*
Figure 11.10

Prosoma pustulose with sharp anterior process; sternum narrow. Opisthosoma finely pustulose with four pairs of larger tubercles along posterior margins of each tergite. Spines project from posterolateral corners of sixth and seventh tergites.

Genus *Lissomartus*
Figure 11.11

Prosoma with rounded anterior process. Sternum narrow, with three sclerites. Opisthosoma smooth. Fourth sternite with anterior lobe, and corresponding curvature in second and third sternites.

Order Uropygida

The uropygids (whip scorpions) represent less than 3 percent of the Mazon Creek arachnids. The whip scorpions are among the most morphologically conservative of all arachnid orders.

Morphology. Prosoma roughly hexagonal, somewhat elongate anteroposteriorly. One pair of medial ocelli and two groups of three closely arranged anterolateral ocelli present. Foveal depression on posterior half of prosoma.

Ventrally, robust coxae flanking sternal sclerites. Chelicerae chelate, bearing two segments. Massive, subchelate pedipalps with five podomeres. Massive, subchelate pedipalps with five podomeres, excluding coxae; coxae abut medially. Third appendages tactilosensory, with slender, elongate podomeres; tarsi subdivided. Remaining appendages pedal, with six podomeres (tarsi subdivided).

Opisthosoma with 12 somites, posterior three reduced to narrow pygidium. Telson "whiplike." Tergites and sternites well sclerotized.

Book lungs present in both second and third opisthosomal somites; spiracles at posterolateral margins of sternites. Anus terminal, immediately below telson.

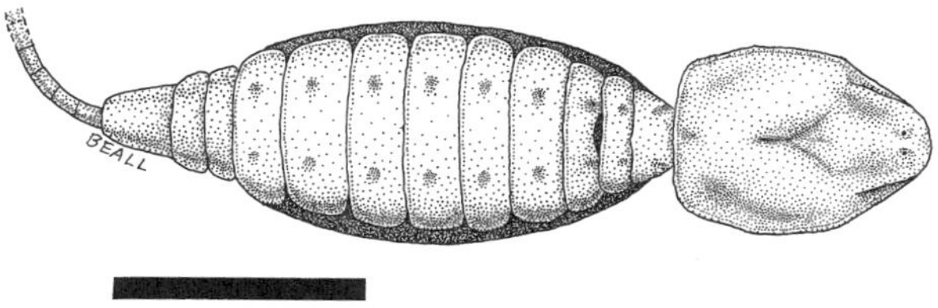

Figure 11.12. *Geralinura carbonaria,* dorsal reconstruction, USNM 37985. Scale bar = 5 mm.

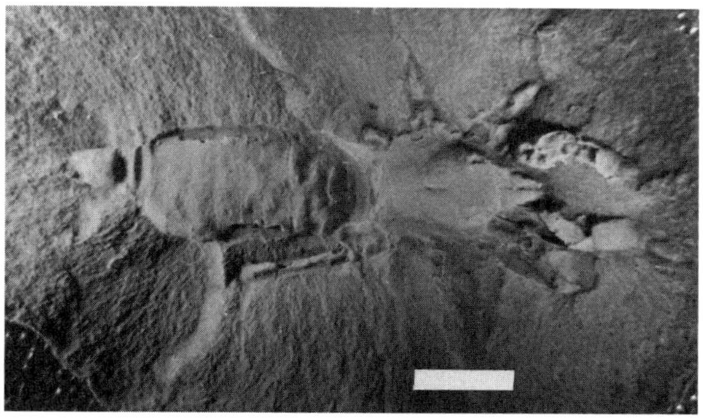

Figure 11.13. *Prothelyphonus* sp., YPM unnumbered specimen. Scale bar = 5 mm (estimate).

Taxonomy. Petrunkevitch (1913, 1949, 1955) recognized two genera of Mazon Creek uropygids. Generic separation may be inappropriate, but additional work is necessary before the genera can be synonymized.

Genus *Geralinura*
Figure 11.12

Prosoma ovate, without eyes; sternum consists of single large sclerite between the first through third pedal coxae.

Genus *Prothelyphonus*
Figure 11.13

Prosoma approximately hexagonal, bearing four pairs of ocelli as described for order.

Order Amblypygida

Like the closely related uropygids, the amblypygids are a rather morphologically conservative order. They constitute only 2.3 percent of the Mazon Creek arachnids.

Morphology. Prosoma slightly heart shaped, with median eye tubercle bearing two ocelli, and two lateral clusters of three ocelli each. Foveal depression on posterior half of prosoma. Ventrally, coxae radiate from four sternal sclerites. Chelicerae with two podomeres, modified into fangs. Pedipalps raptorial, with six podomeres. Third pair of appendages tactile (as in uropygids). Remaining appendages pedal, with six podomeres (tarsi subdivided).

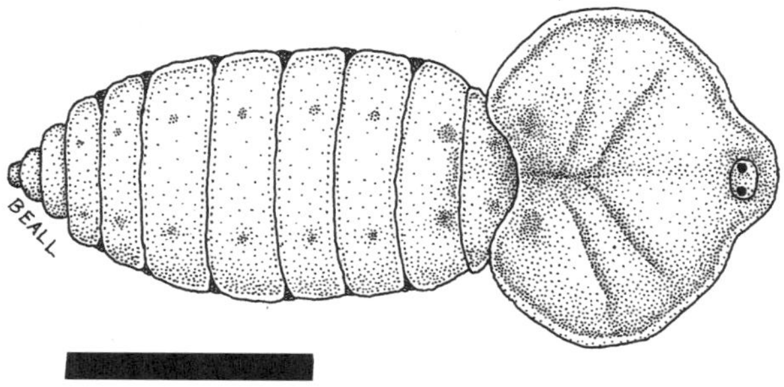

Figure 11.14. *Graeophonus carbonarius,* dorsal reconstruction, USNM 37969. Scale bar = 5 mm.

Opisthosoma with 12 well sclerotized tergites and sternites. First somite forms pedicle; twelfth somite is a reduced pygidium with terminal anus. Book lungs on third opisthosomal somite.

Taxonomy. Petrunkevitch (1955) recognized three genera of Mazon Creek amblypygids. The main differences between the genera (reported number of eyes) is probably due to taphonomic influence.

Genus *Graeophonus*
Figure 11.14

Prosoma with median tubercle bearing single pair of ocelli; groove present in posterior half of carapace.

Genus *Protophrynus*
Figure 11.15

Anterior projection of carapace truncated, with single pair of ocelli; three pairs of lateral ocelli.

Genus *Thelyphrynus*
Figure 11.16

Carapace kidney shaped, without anterior process and without ocelli.

Order Araneida

The araneids (true spiders) represent another minor (2.3 percent) component of the Mazon Creek arachnids. Although poorly preserved, these specimens provide important data regard-

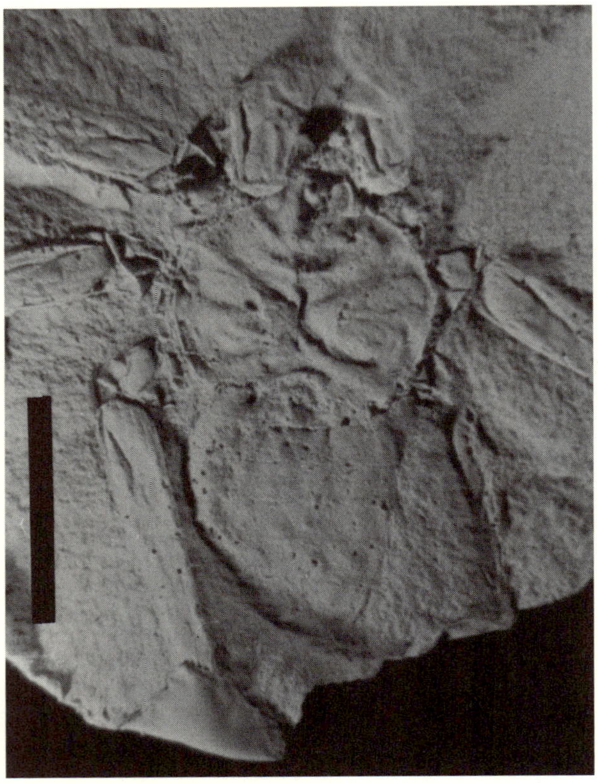

Figure 11.15. *Protophrynus carbonarius,* UMMP 7222. Scale bar = 5 mm (estimate).

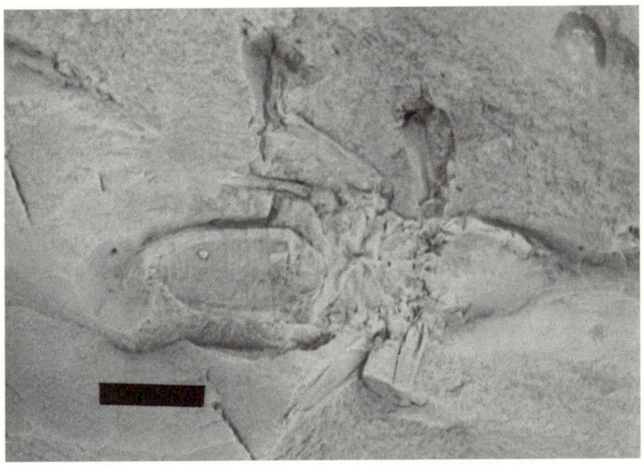

Figure 11.16. *Thelyphrynus elongatus,* UMMP 7220. Scale bar = 5 mm (estimate).

ing morphological diversity of the order during the Middle Pennsylvanian.

Morphology. Prosoma ovoid to slightly hexagonal or rectangular, with up to eight ocelli variably arranged on or around an anteromedian tubercle. Foveal depression present. Ventrally, coxae radiate from large sternum.

Chelicerae with two podomeres; terminal podomeres modified as fangs. Pedipalps pediform, with six podomeres. Remaining appendages pedal or raptorial, with seven podomeres.

Opisthosoma with 12 lightly sclerotized tergites, and a variable, smaller number of sternites. First opisthosomal somite forms pedicle. Book lungs or tracheae on second and third opisthosomal somites. Fourth and fifth somites each with up to two pairs of spinnerets, the structures connected to the silk glands and used for spinning webs among other things (a unique trait linking the araneids). Anus terminal.

Taxonomy. All Mazon Creek araneids have segmented opisthosomata, a character state found only in the suborder Mesothelae (= Liphistina) among extant araneids. However, none of these specimens exhibit the shared derived characters described by Plotnick and Gertsch (1976) as necessary to place them within either of the extant suborders. Because the Mazon Creek araneids are so poorly preserved, the absence of shared derived character states in the fossil specimens may be due either to a loss of morphology caused by taphonomic processes or to an original absence of the characters in the specimens. Because neither of these two interpretations can be rejected, the Mazon Creek araneids may be assigned either to the Mesothelae using primitive character states (segmented opisthosoma; medial, rather than terminal, spinnerets) or to Araneida incertae sedis. At the present time I favor the latter option, pending additional work to identify characters in the Mazon Creek specimens that are obligatorily correlated with the shared derived characters of the extant suborders.

The named Mazon Creek araneids are both placed in the genus *Arthrolycosa* (Figure 11.17). Petrunkevitch (1955) stated that this genus possessed two pairs of eyes on a median tubercle, and a carapace with convex sides.

Order Ricinuleida

Ricinuleida is the only extant arachnid order to be described first from a fossil (Cooke and Shadab, 1973). Mazon Creek ricinuleids have provided much information about the extinct

Figure 11.17. *Arthrolycosa danielsi,* UMMP 7219. Scale bar = 5 mm (estimate).

forms, even though they represent only 3.5 percent of the arachnid specimens.

Morphology. Movable hood (cucullus) hinged to front of carapace and covering chelicerae is diagnostic of the order. Cuticle commonly tuberculate; pair of lateral eyes usually present on fossils at lateral margins of carapace (vestigial in living ricinuleids). Foveal and other depressions on carapace.

Chelicerae chelate, of two podomeres. Pedipalps chelate, of six podomeres with few or no tubercles. Remaining limbs pedal; tarsi (poorly known in fossils) subdivided; third walking leg modified for sperm transfer in males.

Pedicle bearing sexual organs links prosoma and opisthosoma but is commonly hidden by locking of these tagmata. Ventral pouches on anterior opisthosoma receive coxae of walking leg 4 during locking. Transverse tergites in living forms and fossil Poliocheridae (Figures 11.18, 11.20); other fossils without transverse tergites but with longitudinal line or sulcus on dorsal abdomen (Figures 11.19, 11.20). Posterior three somites form a telescopic pygidium.

Taxonomy. A recent revision by Selden

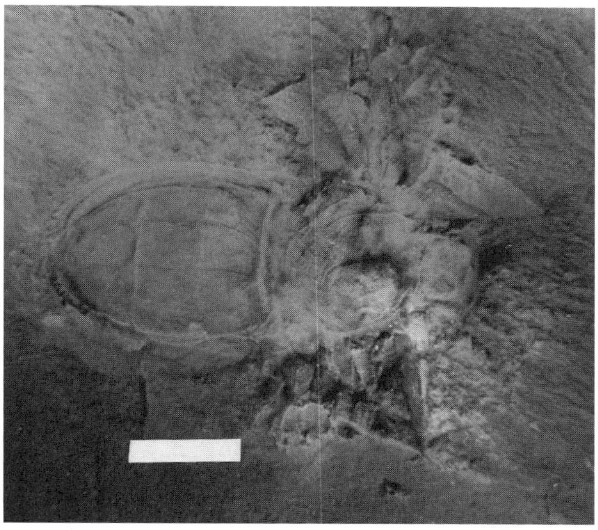

Figure 11.18. *Poliochera glabra,* UMMP 7223. Scale bar = 5 mm.

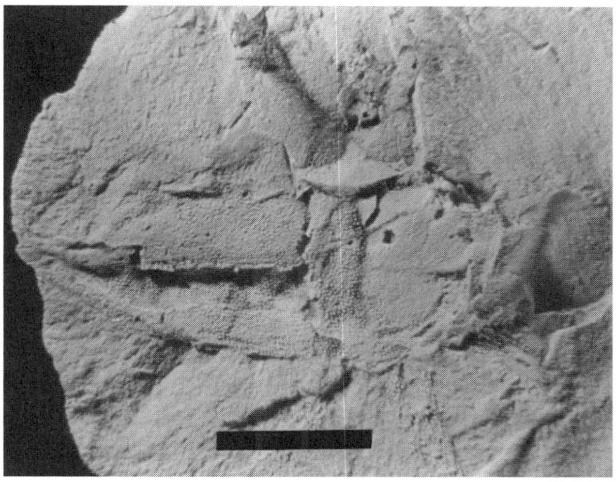

Figure 11.19. *Curculioides mcluckiei,* FMNH PE 780. Latex mould of dorsal surface. Scale bar = 5 mm.

(1992) erected two new suborders based on presence or absence of functional eyes and relative sizes of coxae of walking legs 3 and 4. All living ricinuleids are in Neoricinulei, all fossils in Palaeoricinulei. Three of the four genera of Palaeoricinulei occur at Mazon Creek. For species identification see Selden (1992).

Genus *Curculioides*
Figures 11.19, 11.20b, 11.20f

Dorsal abdomen divided by a median line. *Curculioides mcluckiei, C. scaber,* and *C. gigas* occur at Mazon Creek.

Genus *Amarixys*
Figure 11.20d, 11.20e

Dorsal abdomen divided by a median sulcus. *Amarixys sulcata, A. stellaris,* and *A. gracilis* occur at Mazon Creek.

Genus *Poliochera*
Figures 11.18, 11.20a, 11.20c

Dorsal abdomen divided by transverse tergites. *Poliochera punctulata, P. gibbsi,* and *P. glabra* occur at Mazon Creek.

Order Opilionida

Petrunkevitch (1913, 1949, 1955) described two Mazon Creek arachnids as opilionids, or daddy longlegs, *Nemastomoides longipes* (Figure 11.21) and *N. depressus*. The former is an opilionid, but the latter specimen is a phalangiotarbid.

Petrunkevitch (1913) erected the order Kustarachnida to include three Mazon Creek arachnids. However, all of the diagnostic characteristics listed by Petrunkevitch as distinguishing the Kustarachnida either are absent in the fossils or are consistent with reinterpreting the specimens as opilionids (Beall, 1986). When the "kustarachnids" are included with other opilionid fossils, the order represents almost 4 percent of the Mazon Creek arachnids.

Morphology. Prosoma subtriangular to semicircular, with a median eye tubercle bearing two ocelli. Coxae radiate from a central region that may bear a sternum.

Chelicerae chelate, with three podomeres. Pedipalps pediform or raptorial, with six podomeres. Remaining appendages pedal, with seven rather elongate podomeres.

Opisthosoma broadly attached to prosoma; somites difficult to identify externally. A pair of tracheal tubes open ventrally on second opisthosomal somite. Terminal anus operculate.

Taxonomy. The genus *Kustarachne* should be regarded as a nomen nudum. Petrunkevitch (1955) described the family Nemastomoididae as having an oval sternum; the fourth pedal coxae abut, separating the sternum from the first opisthosomal sternite.

Genus *Nemastomoides*
Figure 11.21

Carapace wider than long, with evenly rounded front and sides; eyes on tubercle considerably anterior to center of carapace (Petrunkevitch, 1955).

Order Solpugida

The Solpugida (wind scorpions) are represented in the Mazon Creek biota by a single extremely poorly preserved specimen (YPM 155). Except for rare specimens preserved in Cenozoic amber, no other solpugid fossils are known.

Morphology. Prosoma with large anterior tagma (propeltidium) and three separate posterior sclerites. Propeltidium with median pair of ocelli, and two pairs of lateral ocelli. Pedipalpal coxae large; rostrum projects anteriorly from between the pedipalpal coxae. Remaining coxae arranged in two almost parallel rows; sternum absent.

Chelicerae massive, chelate with two podomeres. Pedipalps tactile and thickly pediform with six podomeres. Third pair of appendages also tactile, antenniform with seven podomeres. Remaining appendages pedal, with eight podomeres. Racket organs (malleoli) situated along coxae and first two podomeres of sixth appendages.

Opisthosoma with 11 somites, each with separate tergites and sternites. Third and fourth sternites with paired tracheal openings, and fifth sternite with a single opening (another pair of openings between third and fourth coxae). Last opisthosomal somite with slitlike anus.

Taxonomy. The holotype and only known specimen of the Mazon Creek solpugid, *Protosolpuga carbonaria,* is severely compressed, and much of the morphology is obscured by irregular sections of carbonized organic material. Although the specimen displays characters that are diagnostic of the order (e.g., massive chelicerae, large pediform pedipalps, rounded posterior margin of anterior prosomal tagma), characters diagnostic of lower taxonomic levels cannot be observed in the specimen (Figure 11.22).

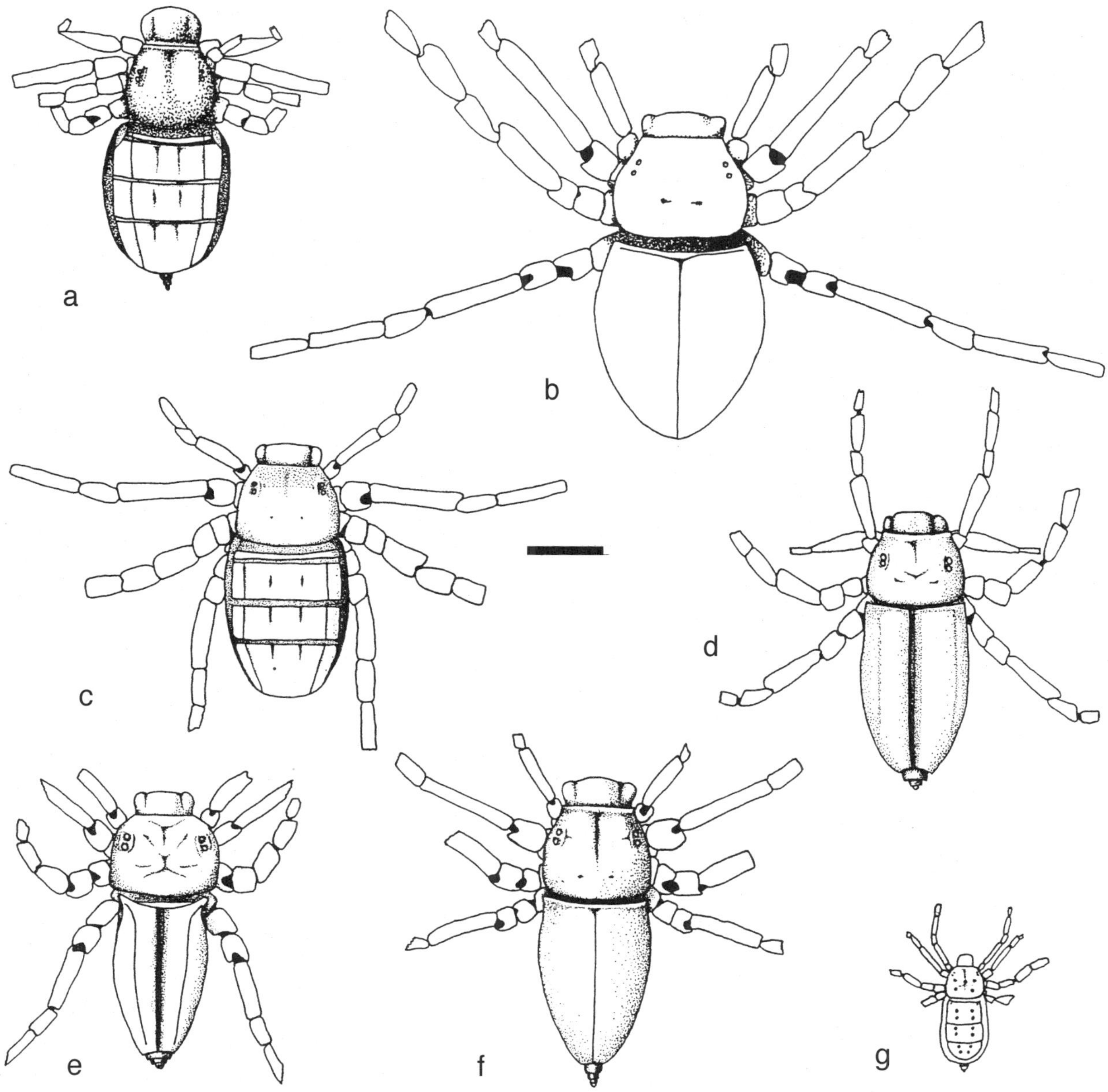

Figure 11.20. Reconstruction of a range of fossil Ricinulei: **a** *Poliochera glabra,* based on UMMP 7223; **b** *Curculioides gigas,* based on FMNH PE 784; **c** *Poliochera punctulata,* based on USNM 37971 and FMNH PE 32206; **d** *Amarixys gracilis,* based on ISM 14862; **e** *Amarixys stellaris,* based on FMNH PE 32184; **f** *Curculioides mcluckiei,* based on USNM 440519; **g** *Terpsicroton alticeps,* based on In 31238, collection of BM (NH). Scale bar = 5 mm. Reproduced by permission of the Royal Society of Edinburgh from Trans. R. Soc. Edinburgh Earth Sci. 83 (1992): 595–634.

Order Scorpionida

The phylogenetic position of scorpions has been the source of some controversy. Traditionally, workers have referred scorpions to the Arachnida (e.g., Petrunkevitch, 1913, et seq.), but Bergstrom (1980) argued for their merostome (specifically, eurypterid) affinities. Although the latter interpretation deserves serious consideration, I will follow tradition by including the scorpions in this chapter.

Almost 14 percent of the Mazon Creek arachnid fossils that can be assigned to an order are scorpions; only the phalangiotarbids are more abundant. Kjellesvig-Waering (1986) proposed a new classification of the Scorpionida that acknowledged the extreme morphological diversity of the Carboniferous scorpions; his restudy

Figure 11.21. *Nemastomoides longipes,* YPM 171. Scale bar = 5 mm (estimate).

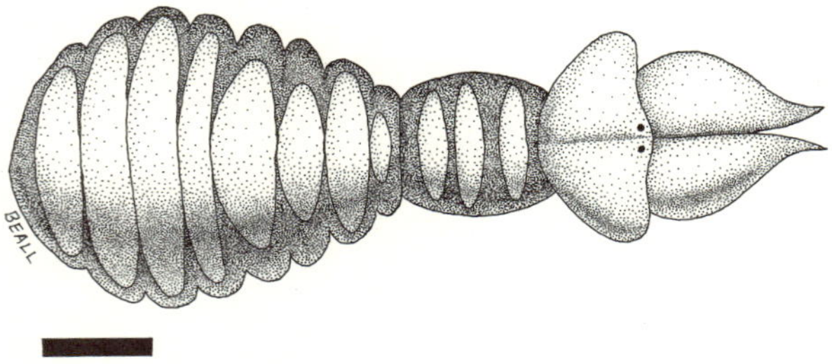

Figure 11.22. *Protosolpuga carbonaria,* preliminary dorsal reconstruction, YPM 155. Scale bar = 5 mm (estimate).

is an excellent summary of the morphology of all fossil scorpions. Although certain aspects of his classification must be questioned, I will use his higher taxa here, using quotation marks to indicate where morphological interpretations may need reconsideration.

Morphology. Prosoma pentagonal or trapezoidal, with anteromedian tubercle bearing two ocelli; single pair of compound eyes or three pairs of ocelli present laterally. Ornamentation pustulose.

Opisthosoma divided into broad mesosoma (seven somites) and narrow metasoma (five somites), with telson modified into sting. Mesosomal tergites and sternites separated by lateral pleural membrane. Kjellesvig-Waering (1986) interpreted the ventral mesosomal sclerites of all but one Mazon Creek scorpion (*Palaeopisthocanthus schucherti;* Figure 11.26) as "abdominal plates"; the homology of the ventral sclerites is a controversial topic that requires more space than is available here.

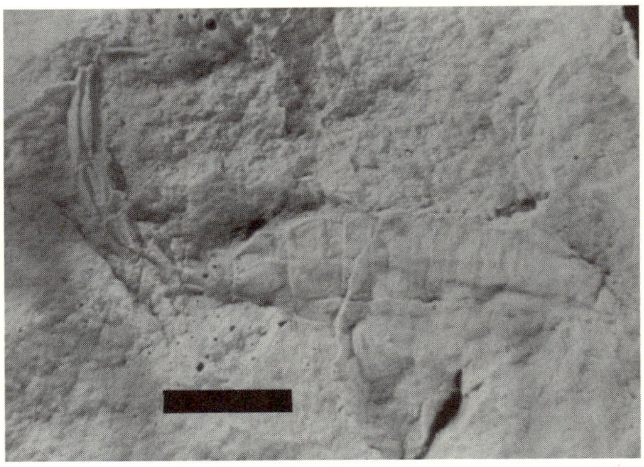

Figure 11.23. Holostern branchioscorpion (*Eoctenus miniatus*), YPM 131. Scale bar = 5 mm.

Figure 11.24. Meristostern branchioscorpion (*Palaeobuthus distinctus*), YPM 133. Scale bar = 5 mm.

Taxonomy. Research addressing the controversies regarding scorpion respiration and the homology of the ventral mesosomal sclerites may modify the following taxa to which Kjellesvig-Waering has assigned all of the Mazon Creek scorpions.

Suborder Branchioscorpionina

Presumed aquatic scorpions respiring by gills. "Gill openings" at intersection of "abdominal plates" and doublure, or on doublure.

Infraorder Holosternina

Branchioscorpions with five rectangular, undivided ventral "abdominal plates" enclosing "gill chamber." Figure 11.23 illustrates *Eoctenus miniatus,* a holostern branchioscorpion.

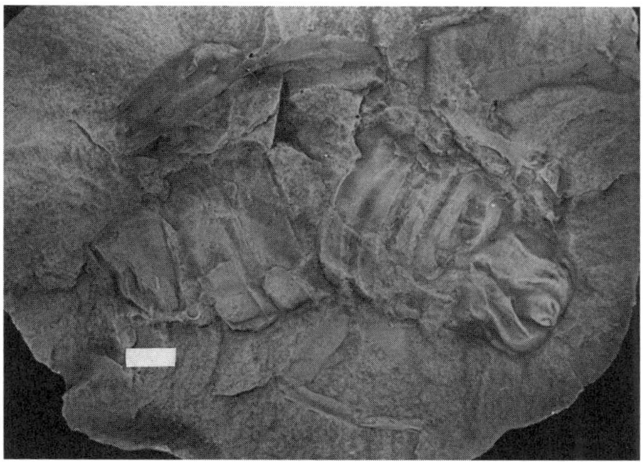

Figure 11.25. Lobostern branchioscorpion (*Kronoscorpio danielsi*), UMMP 7216. Scale bar = 5 mm.

Figure 11.26. Orthostern neoscorpion (*Palaeopisthocanthus schucherti*), YPM 140. Scale bar = 5 mm.

Infraorder Meristosternina

Branchioscorpions with "abdominal plates" with median suture and enclosing "gill chamber." Figure 11.24 illustrates *Palaeobuthus distinctus,* a meritostern brachioscorpion.

Infraorder Lobosternina

Branchioscorpions with five bilobate "abdominal plates" enclosing "gill chamber." Figure 11.25 illustrates *Kronoscorpio danielsi,* a lobostern branchioscorpion.

Suborder Neoscorpionina

Presumed terrestrial scorpions with sternites, four of which bear a pair of spiracles leading to book lungs

Infraorder Orthosternina

Neoscorpions with rectangular sternites bear spiracles; doublures absent. Figure 11.26 illustrates *Palaeopisthocanthus schucherti,* an orthostern neoscorpion.

Acknowledgments

I wish to thank the editors for inviting me to write this chapter. D. C. Fisher and H. Van Iten commented on early drafts of the manuscript; their comments have improved the text greatly, but all oversights remain my responsibility. The additions by P. Selden subsequent to the completion of the original manuscript are highly appreciated. The following institutions permitted me to examine specimens deposited in their collections: FMNH, USNM, ISM, YPM, NEIU MCP, and UMMP. Many individuals provided information on specimens either directly or indirectly: J. and J. Anderson, C. Berry, P. Caponera, D. Chlipala, J. Dedina, K. Dedina, A. and J. Hay, K. Holm, J. Konecny, S. LeMay, W. Lietz, H. and T. Piecko, J. Pohl, K. and S. Ramsdell, W. Rieger, G. Siwik, M. Smith, T. Testa, and M. Young. I am indebted to these individuals and to the faculty and staff of the above institutions for their gracious and continued assistance.

References

Beall, B. 1984. Functional and autecological analysis of phalangiotarbid arachnids. Geol. Soc. Amer. Abstract with Program, 16:457.

———. 1985. Functional and ecological analyses of phalangiotarbid arachnids (Carboniferous). Unpublished M.S. thesis, University of Michigan, Ann Arbor, 75 pp.

———. 1986. Reinterpretation of the Kustarachnida. Amer. Arachnol., 34:4.

Bergstrom, J. 1980. Arthropod phylogeny and the fossil record, pp. 3–56. *In* A. P. Gupta (ed.), Arthropod Phylogeny. Van Nostrand Rheinhold, New York.

Cooke, J., and M. Shadab. 1973. New and little known ricinuleids of the genus *Cryptocellus* (Arachnida, Ricinulei). Amer. Mus. Novitates, 2530:1–25.

Karsch, F. 1882. Ueber ein neues Spinnenthier aus der schlesischen Steinkohle und die Arachniden der Steinkohlenformation überhaupt. Zeit. Deutsch Geol. Gesell., 34:556–561.

Kjellesvig-Waering, E. 1978. The Phalangiotarbida (Arachnida) of Mazon Creek. Geol. Soc. Amer. Abstract with Program., 10:258–259.
———. 1986. A monograph of the fossil Scorpionida. Palaeontographica Amer., 55:1–287.
———. In preparation. Phalangiotarbid arachnids from the Mazon Creek region of Illinois (Carbondale Formation, Pennsylvanian). Manuscript.
Levi, H. 1969. Adaptations of respiratory systems of spiders. Evolution, 21:571–583.
Petrunkevitch, A. 1913. A monograph of the terrestrial Palaeozoic Arachnida of North America. Trans. Connecticut. Acad. Arts Sci., 18:1–137.
———. 1945. Palaeozoic Arachnida of Illinois: an inquiry into their evolutionary trends. Illinois State Mus. Sci. Papers, 3(2):1–72.
———. 1949. A study of Palaeozoic Arachnida. Trans. Connecticut Acad. Arts Sci., 37:60–315.
———. 1955. Arachnida, pp. P42–P162. *In* R. C. Moore (ed.), Treatise on Invertebrate Paleontology. Part P, Arthropoda 2. Geological Society of America and University of Kansas, Lawrence.
Plotnick, N., and W. Gertsch. 1976. The suborders of spiders: a cladistic analysis (Arachnida, Araneae). Amer. Mus. Novitates, 2605:1–15.
Scudder, S. 1868. Descriptions of fossil insects found on Mazon Creek, and near Morris, Grundy Co., Ill. Illinois State Geol. Surv. Rep., 3:556–561.
Selden, P. A. 1983. The fossil ricinuleids (abstract). 9th Int. Cong. Arachnol., 1983, Panama.
———. 1984. A comparison of the functional morphology of living and fossil ricinuleids (abstract). Symp. "Fossil Arthropods as Living Animals," p. 6.
———. 1992. Revision of the fossil ricinuleids. Trans. R. Soc. Edinburgh Earth Sci., 84:595–634.
Selden, P. A., and M. Romano. 1983. First Paleozoic arachnid from Iberia: *Aphantomartus areolatus* Pocock (basal Stephanian; prov. Leon, N. W. Spain), with remarks on aphantomartid taxonomy. Biol. Geol. Mineral., 94(2):106–112.
Shear, W., and P. Selden. 1986. Phylogenetic relationships of the Trigonotarbida, an extinct order of arachnids. Act. 10th Int. Cong. Arachnol., 1986, Jaca/Espana, 1:393–397.

12

Crustacea

Frederick R. Schram,
W. D. Ian Rolfe, and Andrew A. Hay

The subphylum Crustacea includes arthropods that have a head composed of five segments: two sets of antennae (sometimes the second pair has a scale or scaphocerite), a pair of mandibles, and two sets of maxillae (see Figure 12.1). Occasionally, additional segments from the trunk or thorax can be incorporated into the head, whereupon one or more sets of maxillipedes may be associated with the head. Conversely, the second pair of maxillae occasionally may become almost leglike in form, thus giving the secondary appearance of the head's having only four segments. Eyes may or may not be present and may be either stalked or sessile (flat against the head). Advanced forms of crustaceans (most types are found in Mazon Creek concretions) subdivide the trunk behind the head into an anterior thorax and a posterior abdomen.

Crustaceans may or may not have a posterior outgrowth of the head called the carapace, or head shield. If it is present, the size and nature of the carapace may differ widely among crustaceans. In the shrimp and their relatives, it may cover the thorax in whole or in part and have an anteriorly directed projection called the rostrum. In the brine shrimp, barnacles, and their allies, the carapace typically encompasses some or all of the abdomen as well. The carapace is a distinctive crustacean structure (though similar but nonhomologous structures in other arthropods are sometimes also loosely referred to as carapaces), but it appears to have evolved several times independently in the group.

Thoracic limbs can exhibit several forms. The most primitive arrangement is that of two-branched, or biramous, paddles. The outer branch is called the exopod and the inner branch is the endopod. These can become altered in advanced groups. Sometimes the exopod is lost, to produce a uniramous limb, common in walking (reptant) forms. The distal ends of the anteriormost endopods can be modified for grasping, sometimes as a chelate claw, sometimes as a subchelate jackknife. Additional branches may be developed at the base of the thoracic limbs. These additions in the malacostracans are typically small and used for respiration, while in the phyllopodans these branches are quite large and used to assist the exopods and endopods in food procurement and swimming.

The abdominal limbs are typically biramous, are reduced in size over that noted on the thorax, and, especially in highly advanced forms, may not even be developed on all the segments. The posterior or tail end of the abdomen can be specialized into a tailfan. In this case, the terminal segment is developed as a telson, sometimes bearing one or more lobes or furca, and this is flanked by a special set of abdominal limbs called uropods.

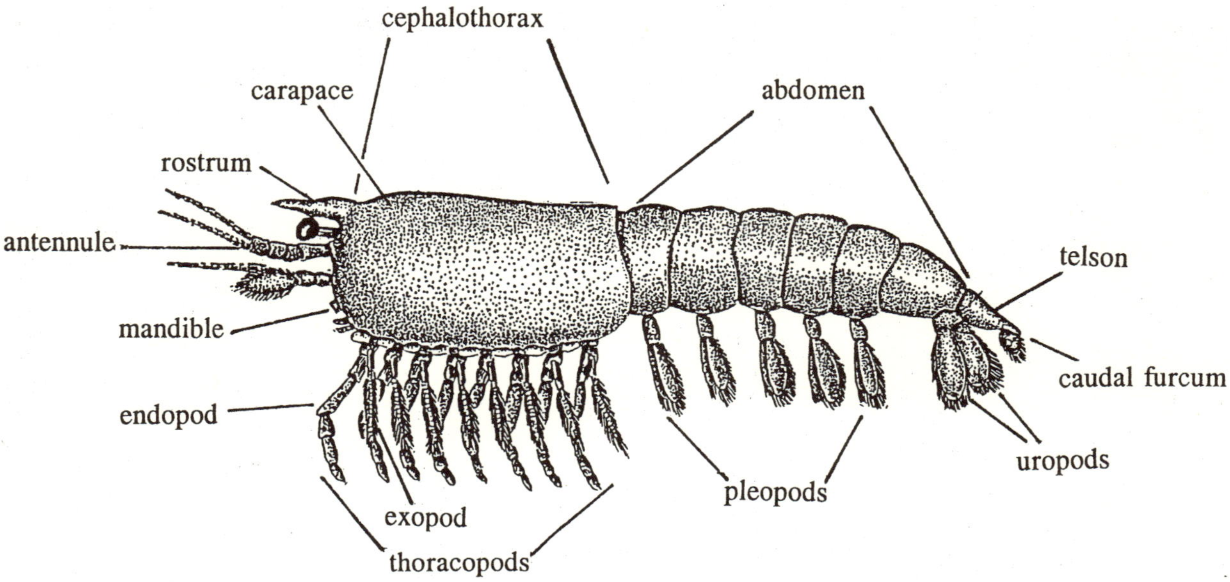

Figure 12.1. Descriptive terms used for Crustacea.

Figure 12.2. *Cryptocaris hootchi* Schram (1974), holotype, FMNH PE 32503, Pit 11, Will-Grundy-Kankakee Counties. Scale bar = 5 mm.

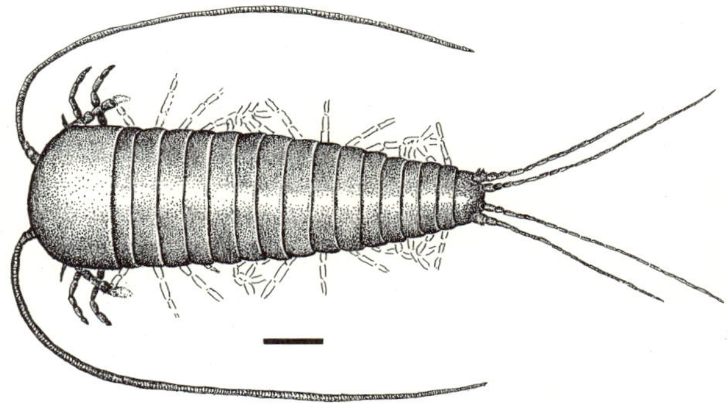

Figure 12.3. *Cryptocaris hootchi,* modified from Schram (1991). Scale bar = 5 mm.

Class Remipedia

Body divided into a head and trunk (the latter without any differentiation of a thorax and abdomen). Posterior mouthparts as subchelae.

Cryptocaris hootchi Schram, 1974
(see Emerson and Schram, 1991)
Figures 12.2, 12.3

Head shield narrower anteriorly than posteriorly; antennules very long and well developed, antennae somewhat smaller; trunk without any subdivisions into thorax and abdomen, composed of about 14 equally developed segments; caudal rami extremely long.

This was the holotype specimen of what is now called *Eucryptocaris asherorum,* a tanaidacean, but it appears to have nothing to do with the description based on other specimens of that tanaidacean. Rather, the holotype (which must continue to bear the original name) has more in common with a strange crustacean from the uppermost Mississippian of Texas, *Tesnusocaris goldichi,* which in turn is closely related to living, nectiopodan, remipede crustaceans (Schram, 1986). Clearly preserved limbs on *C. hootchi* have never been observed and, if found, would go a long way to clarifying the relationships of this species to other crustaceans.

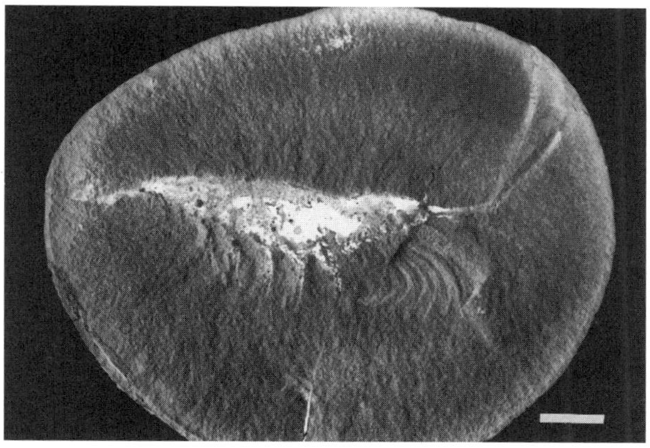

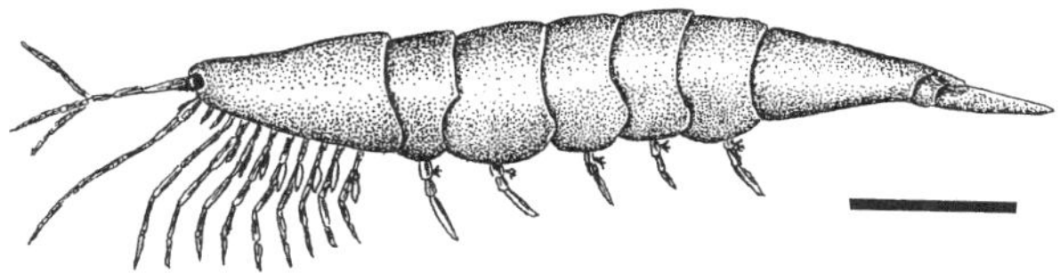

Figure 12.4 (*Left*). *Kallidecthes richardsoni* Schram (1969), holotype, FMNH PE 9771, Pit 11, Will-Grundy-Kankakee Counties. Scale bar = 10 mm.

Figure 12.5 (*Above*). *Kallidecthes richardsoni,* modified from Schram (1969). Scale bar = 10 mm.

Class Malacostraca

Trunk subdivided into an anterior eight-segmented thorax and a posterior six- or rarely seven-segmented abdomen plus a telson. Usually with a well-developed tailfan; may have a carapace, but this does not cover anything other than the head and thorax.

Subclass Hoplocarida

Antennules triramous, abdomen much longer than thorax, abdominal limbs bearing gills, tailfan with bladelike uropods and robust telson.

Order Aeschronectida

Carapace completely covers the thorax, maxilla and thoracic limbs as delicate walking limbs, thoracic limbs simple and not specialized as claws or subchelae.

Kallidecthes richardsoni Schram, 1969
Figures 12.4, 12.5

Shrimplike body; antennule with three flagella; carapace subtriangular in outline with apex toward antennules, distinct optic notch and compound eye tightly affixed in it; thoracic legs longer anteriorly than posteriorly; abdomen long (about two-thirds the length of the body), abdominal limbs with tufts of branched gills at their base; tailfan with bladelike uropods and broad, spatulate telson.

This species is the second most numerous crustacean in the Essex biota. It represents an entirely extinct radiation of crustaceans whose closest living relatives today are the mantis shrimp. Originally, these beasts were thought to have been filter feeders; it is suspected now that they may have been carnivores or scavengers. Some of the somewhat larger specimens of this species, seemingly characterized by a slightly longer and deeper abdomen and a more rectangular carapace, may in fact be a separate species.

Order Stomatopoda

Carapace not covering posterior segments of thorax, anterior limbs as subchelae.

Tyrannophontes theridion Schram, 1969
Figures 12.6, 12.7

Mantis shrimp body; carapace very short, completely exposing the three or four most posterior thoracic segments; antennules with three flagella, antenna with large oval scale attached to the base; anterior thoracic limbs subchelate (i.e., claws turned back on themselves), posterior thoracic limbs stiltlike; abdomen more than half the length of the body, abdominal limbs with tufts of branched gills at their base; tailfan with bladelike uropods and a robust subtriangular telson.

The tyrannophontids were ancient precursors of the modern-day mantis shrimp. These animals are today, and undoubtedly were then, active rapacious carnivores, one of the few such known among the crustaceans. More primitive versions, known as palaeostomatopods, occur in Late Devonian and the middle and lower parts of the Carboniferous. During the Mesozoic, the tyrannophontids gave way to the modern mantis shrimp.

Subclass Eumalacostraca

Antennules uni- or biramous, gills if present usually on the thorax, abdomen about equal to or shorter than the thorax, tailfan often broad with delicate uropods.

Order Syncarida

Lacks a carapace, thoracic limbs biramous.

Acanthotelson stimpsoni
Meek and Worthen, 1865
Figures 12.8, 12.9

First thoracic limb markedly reduced; outer ramus of thoracic limb as a single flaplike segment; second and third thoracic limbs robust and spiny; abdominal limbs with two flaplike branches; telson and uropods styliform, with telson equal to or slightly longer than the uropods.

The cuticle of these animals was apparently very well sclerotized. As a result, these fossils are often well preserved with distinct segmental rings in evidence along the length of the body. This species has been the subject of much taxonomic confusion. In the older literature, it was often referred to by the incorrect generic name *Eileticus.* It is quite abundant in the Braidwood assemblage, but relatively rare and poorly preserved in the Essex assemblage.

Palaeocaris typus Meek and Worthen, 1865
Figures 12.10, 12.11

Mandibles massive, head-to-thorax ratio 1:4, antennal scale longer than the two peduncular segments of the antennal flagellum or whiplike

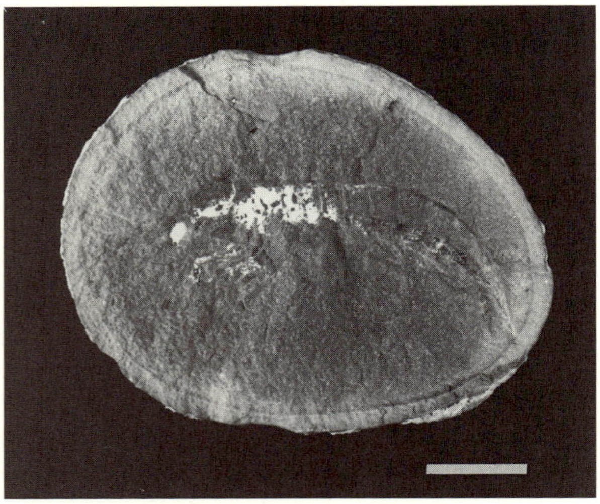

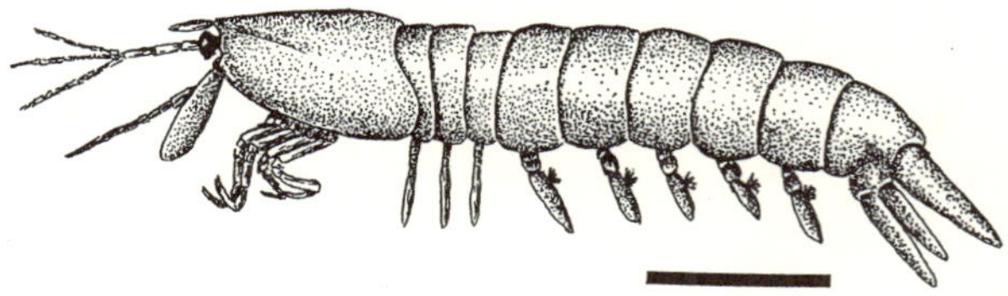

Figure 12.6 (*Left*). *Tyrannophontes theridion* Schram (1969), holotype, FMNH PE 12098 (coll. Calvin George), Pit 11, Grundy-Will-Kankakee Counties. Scale bar = 10 mm.

Figure 12.7 (*Above*). *Tyrannophontes theridion,* modified from Schram (1969). Scale bar = 10 mm.

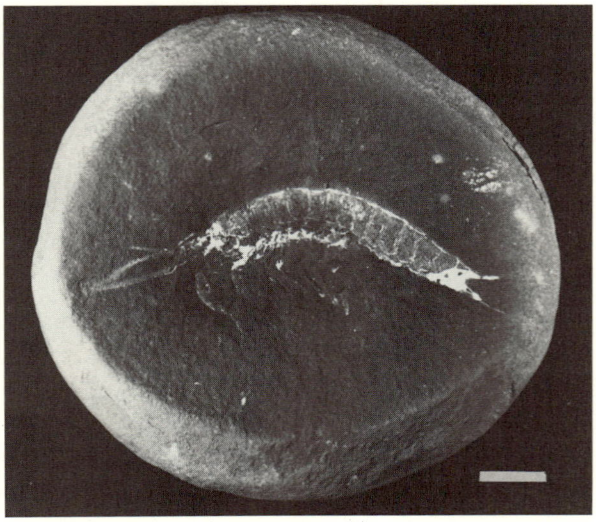

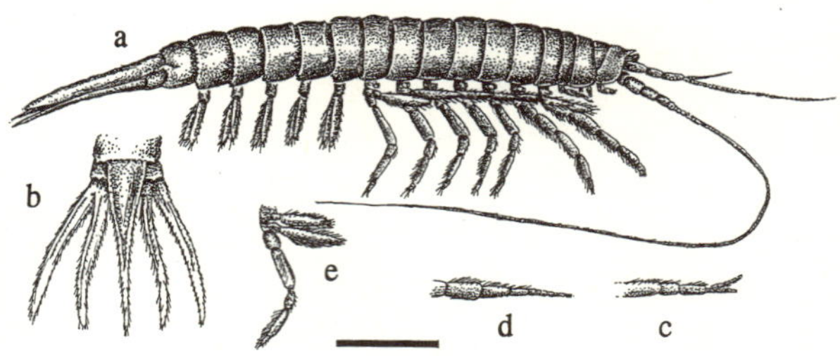

Figure 12.8 (*Left*). *Acanthotelson stimpsoni* Meek and Worthen (1865), lectotype, UI X346, Mazon Creek area, Will County. Scale bar = 5 mm.

Figure 12.9 (*Above*). *Acanthotelson stimpsoni,* modified from Schram (1984): **a** side view; **b** tailfan; **c** antennule; **d** antenna; **e** posterior thoracopod. Scale bar = 5 mm.

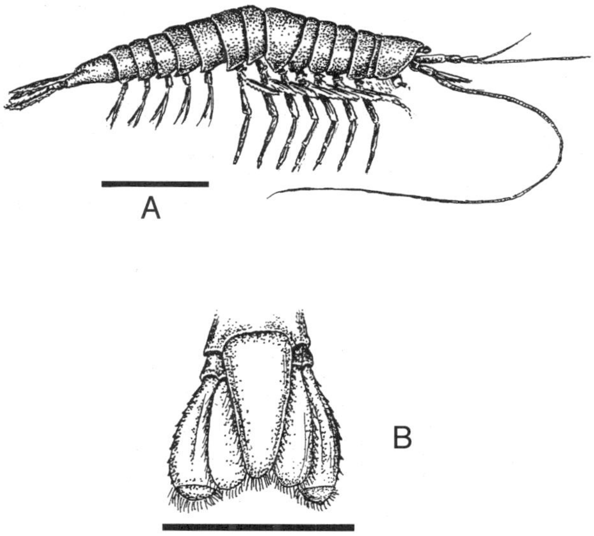

Figure 12.10 (*Above*). *Palaeocaris typus* Meek and Worthen (1865), holotype, UI X338, Mazon Creek area, Will County; see also Brooks (1962) and Schram (1984). Scale bar = 5 mm.

Figure 12.11 (*Above right*). *Palaeocaris typus,* modified from Schram (1984): **a** side view, **b** tailfan. Scale bar = 5 mm.

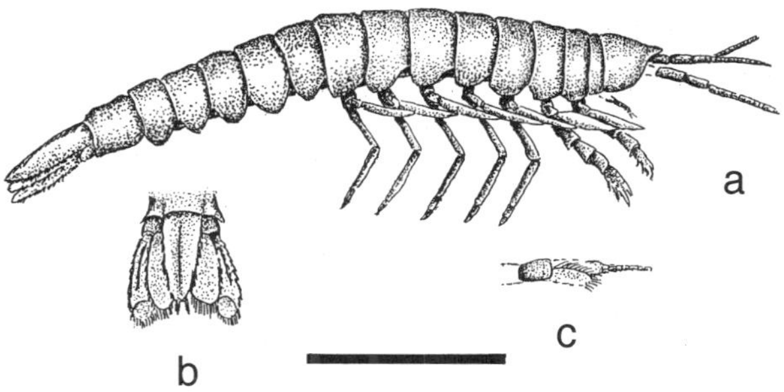

Figure 12.12 (*Left*). *Palaeosyncaris micra* Schram (1984), tail of FMNH PE 12168, Pit 11, Will-Grundy-Kankakee, Counties. Scale bar = 1 mm.

Figure 12.13 (*Above*). *Palaeosyncaris micra,* modified from Schram (1984): **a** side view; **b** tailfan; **c** antenna. Scale bar = 5 mm.

branch; first thoracic segment greatly reduced, sixth thoracic segment longer dorsally than any other segment, exopod branches of the thoracic limbs flaplike; abdominal limbs are multisegmented; uropods very setose (hairy), with a two-segmented outer branch or exopod, exopod longer than the endopod, which in turn is longer than the telson, outer margin of exopod with slightly developed and widely placed spines; the telson is ovoid but wider at its base than near its tip.

This is the second most common crustacean from the classic Braidwood localities, although rare and poorly preserved in the Essex fauna.

The reconstruction of Brooks (1962) had the abdominal limbs as flaplike appendages; however, these are now known to be rather delicate and multisegmented.

Palaeosyncaris micra Schram, 1984
Figures 12.12, 12.13

Head small, body size relatively small (in comparison with other Mazon Creek syncarids); first thoracic segment reduced, first thoracic limb very reduced, second and third slightly robust and with terminal spines, thoracic limbs with flaplike outer branches; all body segments smooth with little or no decoration, pleural

Figure 12.14. *Belotelson magister* Packard (1986), holotype, USNM 3844 (401 a,b) (coll. J. C. Carr), Mazon Creek, Will County. One or two specimens came from the subsurface mines near Braidwood. Subsequent to its description, thousands of specimens were collected from Pit 11; also see Brooks (1962) and Schram (1974a). Scale bar = 10 mm.

Figure 12.15. *Belotelson magister,* modified from Schram (1974): **a** side view; **b** bottom view; **c** tailfan. Scale bar = 10 mm.

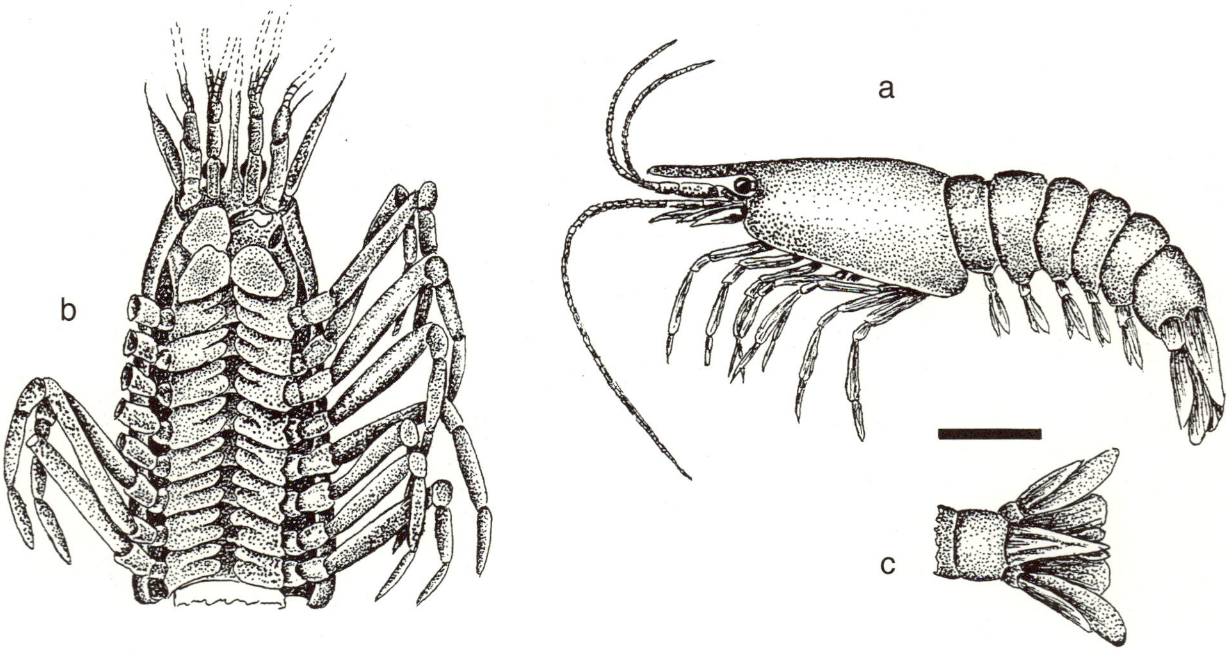

plates on side of abdomen have smooth margins; telson is oval with spinose margins that increase in size from the base toward the tip; uropods with spines on the lateral margin.

These small and generally incompletely preserved fossils have, in the past, been lumped in the sorting process with the small and poorly preserved *Palaeocaris typus.* Though quite rare, subtle differences in the tailfan (a more rounded character to the uropods and telson, and a markedly rounded form of the two segments' boundary in the uropodal exopod) and the spiny and robust anterior thoracic limbs can be used to distinguish this species from *P. typus.*

Order Belotelsonidea

Well-developed carapace and tailfan, all eight thoracic limbs as well-developed, uniramous walking legs.

Belotelson magister Packard, 1886
Figures 12.14, 12.15

Robust crustacean; rostrum long and robust; antennules and antennae large and well developed; carapace subrectangular in lateral outline, originally very well sclerotized and undecorated except for a small pore on the posterior midline; thoracic limbs large and composed of only a single branch or endopod; abdomen well

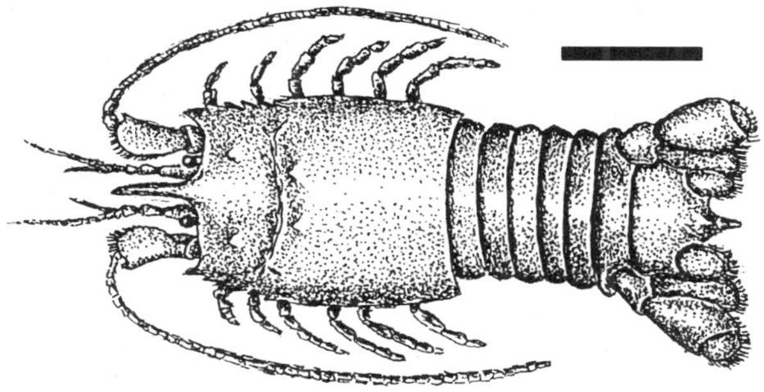

Figure 12.16 (*Left*). *Anthracaris gracilis* Meek and Worthen (1865), holotype, ISM 3067 (Illinois Geol. Surv.), Will County. See also Brooks (1962) and Schram (1974a). Scale bar = 5 mm.

Figure 12.17 (*Above*). *Anthracaris gracilis,* modified from Schram (1981). Scale bar = 5 mm.

developed, with pleural lobes set off and somewhat pointed along their ventral margins; pleopods biramous and flaplike; tailfan well developed, uropods broadly rounded delicate flaps (except along the reinforced lateral margins of the exopods), telson large and subtriangular with terminal pair of moderately developed, delicate, lobate caudal furca.

This species was known originally from only three specimens in the Braidwood fauna but is by far the most common type of crustacean in the Essex fauna. The species is apparently distinguished by a great variety of types of preservation including juvenile and adult forms as well as a great variety of cast-off exoskeletons.

This lobsterlike form represents an extinct order of crustaceans that flourished during the Coal Age of Europe and North America. A second yet undescribed species, possibly related to *B. magister,* is common in Mazon Creek collections and has a slightly smaller size with an extremely long and rather delicate rostrum.

Order Pygocephalomorpha

Well-developed carapace and tailfan; first and second thoracic limbs often (but not always) modified as maxillipedes, posterior six to seven thoracic limbs biramous; females with thoracic brood plates, males often with cones on ventral surface of eighth thoracic segment.

Anthracaris gracilis Meek and Worthen, 1865
Figures 12.16, 12.17

Carapace with well-developed branchial region covering the gills, anterolateral spines large and followed by a row of eight or more marginal serrations, hepatic and gastric spines present, arms of the cervical groove meeting in the midline; tailfan with well-developed flaplike uropods, exopodal ramus (outer branch), with two segments and the joint between the two (diaresis) diagonally directed; telson truncated at the point of attachment of the caudal furca (tail lobes), which are in turn posteriorly directed, accessory projection of telson directed over the lobate caudal spine.

This pygocephalmorph mysidacean is a characteristic component of the Braidwood biota. It is rarely found in Essex localities. Like its close relative, *Mamayocaris jaskoskii,* this creature was a bottom-dwelling scavenger.

Mamayocaris jaskoskii Schram, 1974a
Figures 12.18, 12.19

Carapace with well-developed branchial region covering the gills, anterolateral spines short and followed by a row of five marginal serrations, gastric spines present cervical groove posteriorly directed; tailfan with well-developed flaplike uropods, exopodal ramus with two segments and the joint between the two (diaresis) s-shaped; telson subtriangular in outline, bear-

Figure 12.18. *Mamayocaris jaskoskii* Schram (1974a), NEIU MCP 610 (gift of W. Lietz), Pit 11, Will-Grundy-Kankakee Counties. Scale bar = 5 mm.

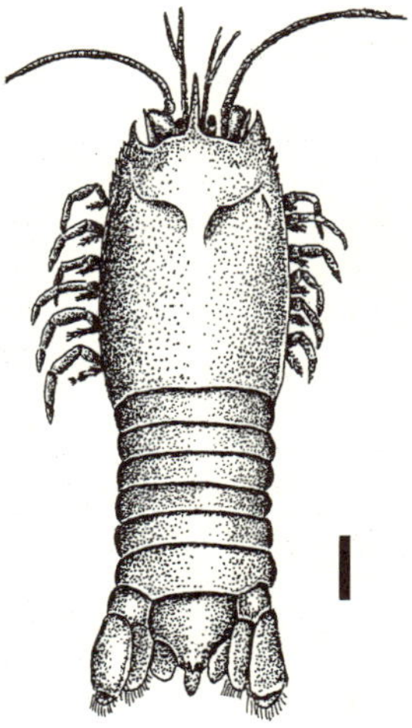

Figure 12.19. *Mamayocaris jaskoskii,* modified from Schram (1974). Scale bar = 5 mm.

ing small lobate caudal furca and an accessory projection of the telson over the lobate caudal spine.

These crayfishlike mysidaceans are rarely encountered and occur only in the Essex fauna. However, because they were bottom-dwelling scavengers with a cuticle that was originally well sclerotized, they are typically well preserved.

Order Lophogastrida
Peachocaris strongi (Brooks, 1962)
Figures 12.20, 12.21

Shrimplike body; rostrum moderate in size, slightly keeled, cervical groove slightly developed; eight thoracic limbs all equally developed, with segmented and setose exopods; abdominal segments equally developed, with pleural lobes all similar to each other and rounded; tailfan with broadly rounded uropods and subtriangular telson that appears to bear a set of terminal caudal furca.

This species is often confused with *Anthracophausia ingelsorum,* especially when poorly preserved (see comments on *A. ingelsorum*). This species was once placed among the "eocarids" (a catchall order) but has since been compared to, and placed among, the mysidacean types (Schram, 1986).

Incertae sedis (possible mysidacean)
Anthracophausia ingelsorum Schram, 1976
Figures 12.22, 12.23

The moderate-sized shrimplike body is characterized by a distinct hooklike rostrum; carapace is subtrapezoidal in outline; abdominal pleura with rounded lobate form.

Generally poor preservation makes this species difficult to separate from the similar *Peachocaris strongi.* There is no way to distinguish these forms except in specimens in which the rostrum is clearly preserved. *Anthracophausia ingelsorum* is a rare species, while *P. strongi* appears in moderate numbers.

Essoidia epiceron Schram, 1974a
Figures 12.24, 12.25

Small shrimplike form; rostrum moderately developed, prominent optic notch, carapace subrectangular in outline without any decoration; thoracic legs long and delicate in form; abdomen with first segment greatly inflated, deflecting the distal parts of the abdomen downward; uropods bladelike; telson subrectangular,

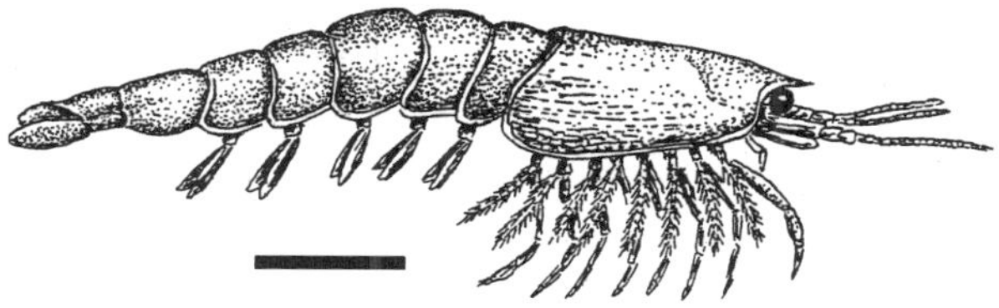

Figure 12.20 (*Left*). *Peachocaris strongi* (Brooks, 1962), holotype, YPM 18821 (coll. S. S. Strong), Mazon Creek area, Grundy County. Originally placed in the genus *Anthracophausia* by Brooks (1976; not *Peachella* Schram, 1974). Scale bar = 5 mm.

Figure 12.21 (*Above*). *Peachocaris strongi,* modified from Schram (1974). Scale bar = 5 mm.

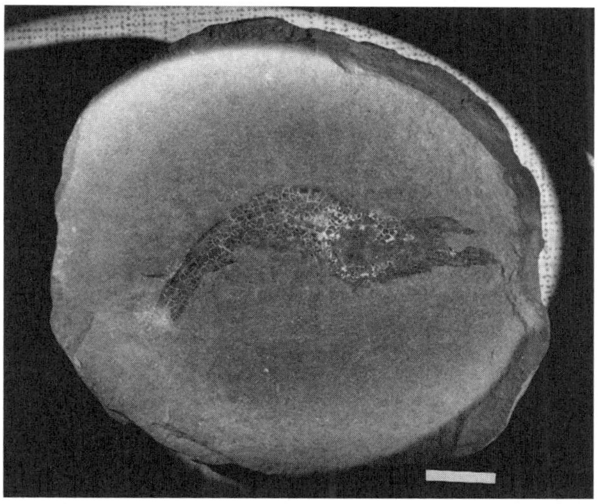

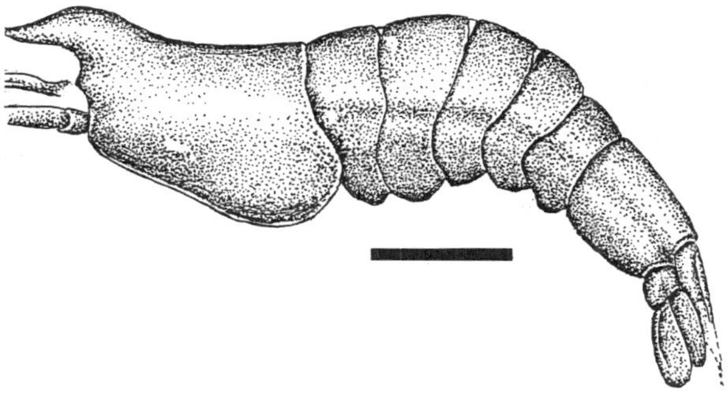

Figure 12.22 (*Left*). *Anthracophausia ingelsorum* Schram (1976), holotype, FMNH P 32085 (coll. Corliss, Elva, and Eric Ingels), holotype from Sunspot Mine, near Astoria, Fulton County. The holotype and best specimen comes from the Francis Creek Shale in western Illinois. More poorly preserved material has come from the Pit 11 Essex locality. Scale bar = 5 mm.

Figure 12.23 (*Above*). *Anthracophausia ingelsorum,* modified from Schram (1976). Scale bar = 5 mm.

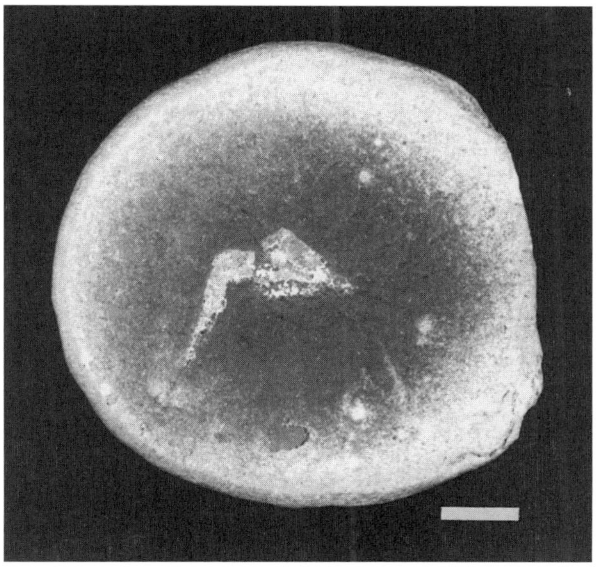

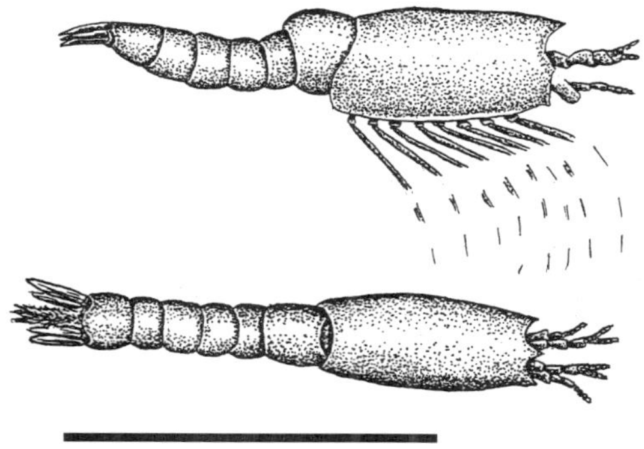

Figure 12.24 (*Left*). *Essoidia epiceron* Schram (1974a), holotype, FMNH PE 12211, Pit 11, Will-Grundy-Kankakee Counties. Scale bar = 5 mm.

Figure 12.25 (*Above*). *Essoidia epiceron,* modified from Schram (1974). Scale bar = 5 mm.

Figure 12.26. *Eucryptocaris asherorum* Schram (1989), holotype, FMNH PE 51405; see also Schram et al. (1986). Scale bar = 5 mm.

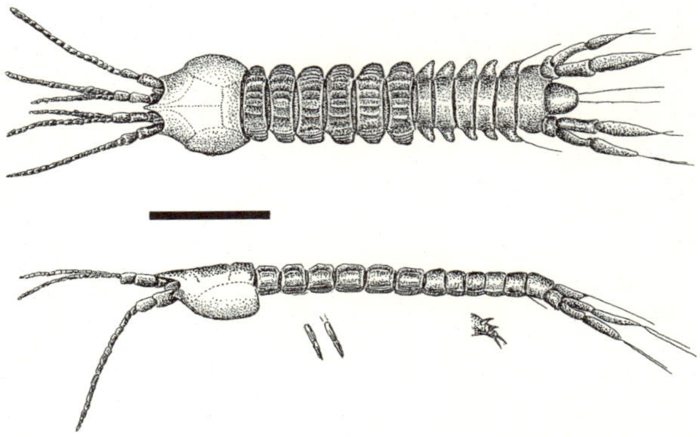

Figure 12.27. *Eucryptocaris asherorum,* reconstruction. Scale bar = 5 mm.

lateral margins serrate, the terminus is pointed and flanked by two pairs of spinelike caudal furca.

This animal was placed among the "eocarids" but has many structural similarities to mysidan mysidaceans and may in fact be one. The animal is relatively rare among the Essex fauna crustaceans, though this fact may not reflect its original abundance since it was probably a pelagic form that was rarely preserved in the Essex biota.

Order Hemicaridea

Carapace covering only one or two anteriormost thoracic segments; abdomen often shorter than the thorax.

Suborder Tanaidacea

Abdomen very reduced, tailfan poorly developed.

Eucryptocaris asherorum Schram, 1989
(see also Schram, 1974b)
Figures 12.26, 12.27

Carapace inflated laterally, covering the first two thoracic segments, rostrum prominent, prominent anterolateral corners, optic notch well developed; six posterior thoracic segments flattened, with paired longitudinal ridges on either side of the midline, pleural lobes or body segments well developed to extend laterally and each bearing three longitudinal ridges; abdominal segments undecorated, with well-developed pleural lobes; uropods (limbs of last abdominal segment) with two equally developed, two-segmented, flaplike rami (branches); telson or terminal segment moderately developed and rounded in form.

Order Edriophthalma

Lacks a carapace, eyes sessile, thoracic limbs uniramous, anterior thoracic limb as maxillipede.

Suborder Isopoda

Thoracic limbs all very similar; uropods poorly developed.

Hesslerella shermani Schram, 1970
Figures 12.28, 12.29

Head short, antennules and antennae with long single flagella, body without a carapace covering, eyes sessile on sides of head; thoracic segments with anterior and posterior ridges, thoracic limbs equally developed and composed of one endopodal ramus except for the first limb, which is reduced to a large maxillipede; four anterior abdominal segments with rounded pleural lobes and bladelike pleopods (abdominal limbs), fifth segment large and elongate, sixth segment fused with telson and bearing a pair of biramous uropods.

Though not frequently encountered in collections of Mazon Creek fossils, the quality of preservation of this species is usually excellent. The rounded head capsule with its protruding

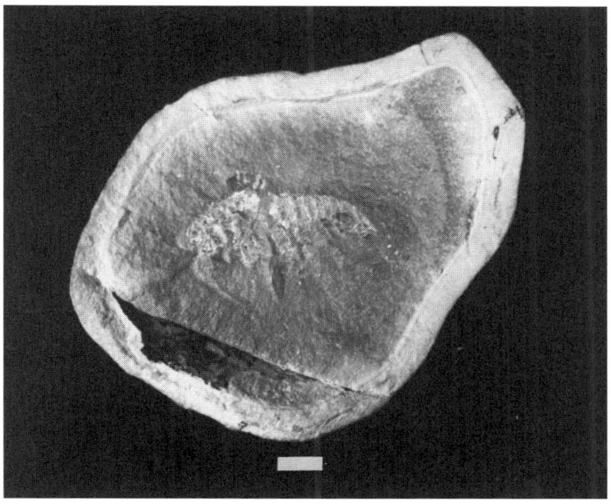

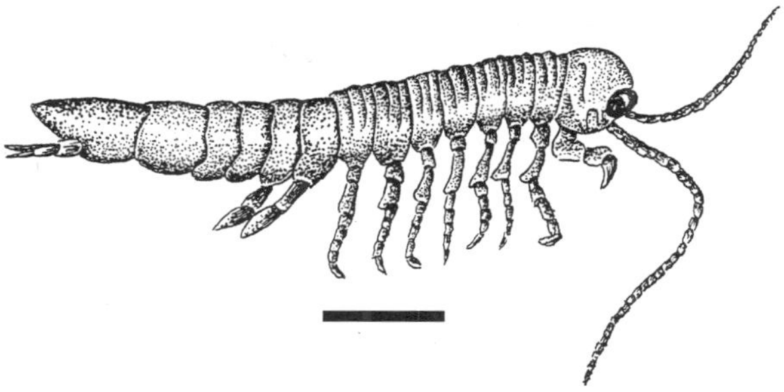

Figure 12.28 (*Left*). *Hesslerella shermani* Schram (1970), holotype, FMNH PE 16527 (coll. Levi Sherman), Pit 11, Will-Grundy-Kankakee Counties. Scale bar = 5 mm.

Figure 12.29 (*Above*). *Hesslerella shermani,* modified from Schram (1970). Scale bar = 5 mm.

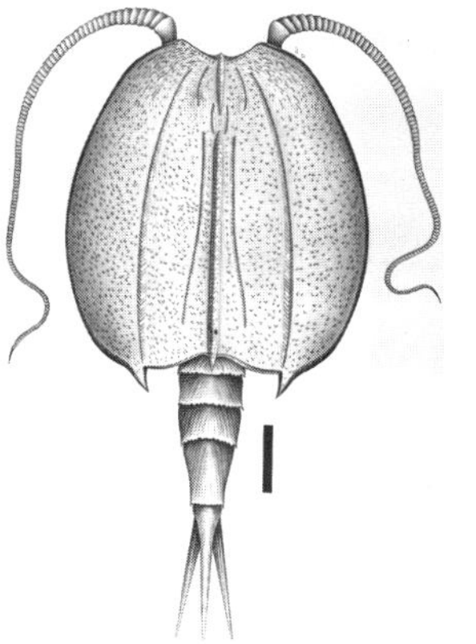

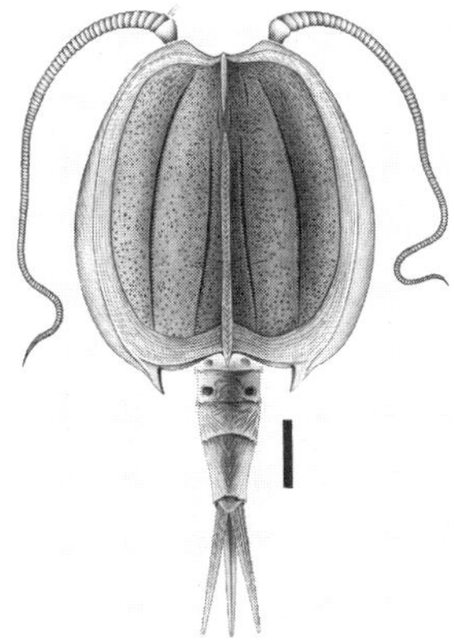

Figure 12.30 (*Above*). *Dithyrocaris* n. sp., top view with valves folded out, FMNH PE 29363. Scale bar = 5 mm.

Figure 12.31 (*Above right*). *Dithyrocaris* n. sp. Rolfe and Richardson MS, new reconstruction, top view. Scale bar = 5 mm.

Figure 12.32 (*Above far right*). *Dithyrocaris* n. sp. Rolfe and Richardson MS, new reconstruction, bottom view. Scale bar = 5 mm.

eyes on the side of the head is easily recognized. This is the earliest known isopod crustacean. It is the member of a subgroup of "pill bugs" that was widely spread in Paleozoic seas and is today restricted to habitats in Australia and Tasmania.

Class Phyllopoda

Carapace, if present, generally covering most of body; thoracic limbs thin and leaflike; abdomen often lacking limbs in whole or in part.

Subclass Phyllocarida

Thorax with eight segments, abdomen with seven segments plus a telson.

Dithyrocaris new species
Rolfe and Richardson MS
Figures 12.30–12.32

Carapace shieldlike with narrow median plate and rostral plate; posterior margin sinuous; mes-

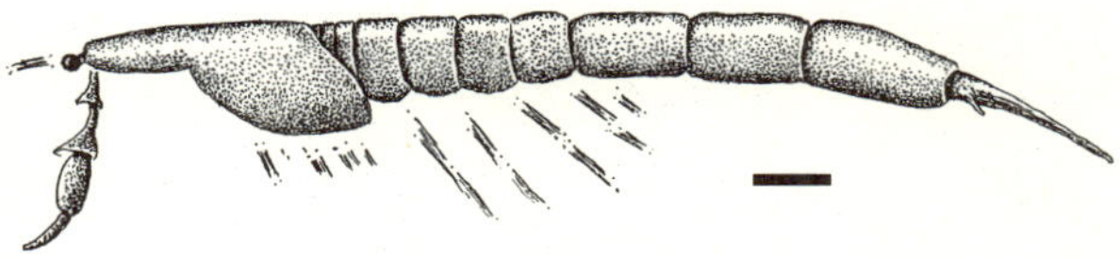

Figure 12.33 (*Left*). *Kellibrooksia macrogaster* Schram (1973), holotype, FMNH PE 12201, Pit 11, Will-Grundy-Kankakee Counties. Scale bar = 5 mm.

Figure 12.34 (*Above*). *Kellibrooksia macrogaster,* modified from Schram (1973). Scale bar = 5 mm.

olateral carina gently arcuate, terminating in a large posteroventral spine; dorsolateral carinae flanking median plate diverge posteriorly in adults; doublure very broad; surface with cusplike ornamentation; last abdominal segment somewhat longer than those anterior, abdominal segments with a chevronlike ornament on surface; telson and caudal rami subequal and marked by chevronlike ornament seen on abdominal segments. Carapace lengths vary from 5 to 40 mm with a mean of 20 mm. *Dithyrocaris* from other localities are generally fully marine. This is a form characteristic of the Essex fauna.

A rare fossil, Rolfe only saw 34 specimens in his survey of collections in 1981 and felt that there were probably no more than 50 in all collections. In life this animal was probably convexly arched; and fossils are often preserved with the marginal doublure of the carapace displaced laterally. The creature was most likely benthic. No limbs have been preserved on the specimens at hand, which might suggest that they were poorly sclerotized in life if not rather thin and delicate. Individual carapaces of this animal can be preserved showing either the outer or inner surface.

Kellibrooksia macrogaster Schram, 1973
Figures 12.33, 12.34

Carapace short without hinged rostrum; compound eyes stalked; antenna (antennule?) with robust articles, possibly subchelate; thoracic limbs very short and apparently delicate; abdomen very long with segments progressively longer distally in the sequence; telson as a long spike with tiny caudal rami flanking the base. No clear preservation of thoracic and abdominal legs has ever been seen, suggesting that these were possibly thin, leaflike structures.

The form of these strange animals is similar to that of the stomatopod hoplocaridans, and suggest these too may have been bottom dwellers that buried themselves lying in wait for prey.

Kellibrooksia is a phyllocarid named after Harold Kelly Brooks, who published the first well-illustrated modern account of Mazon Creek crustaceans.

Subclass Branchiopoda
Order Conchostraca

Carapace resembling a clam shell completely enclosing the body; trunk undifferentiated into thorax or abdomen.

Leaia and *Pemphilimnadiopsis* are conchostracans, or "clam shrimp," tiny, nonmarine crustaceans that live suspended by a ligament from the dorsal portion of their bivalved carapace. They have been collected in both Braidwood and Essex localities. Considered to be freshwater to brackish water organisms (Petzold and Lane, 1988), their presence in Essex localities is probably due to their having been transported into the site of deposition.

Leaia tricarinata Meek and Worthen, 1868
Figures 12.35, 12.36

Bivalved, laterally compressed carapace that is equivalved, inequilateral, and subovate to subquadrate (squarish) in form. Valves decorated with 12 to 14 coarse lamillae or growth lines.

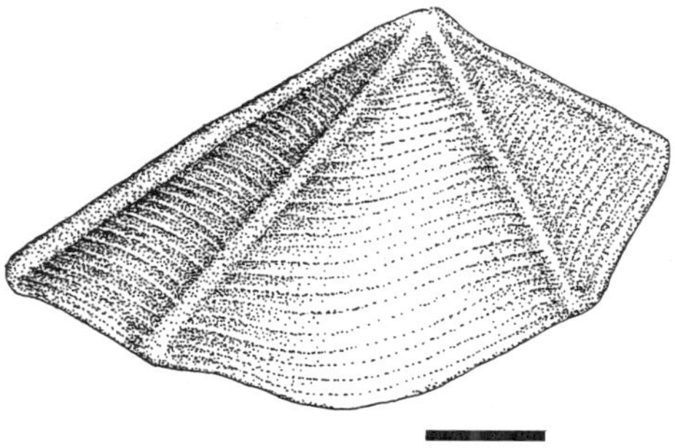

Figure 12.35. *Leaia tricarinata* Meek and Worthen (1868). Type species is NYSM 4444–1. Scale bar = 5 mm.

Figure 12.36. *Leaia tricarinata,* modified from Petzold and Lane (1988). Scale bar = 1 mm.

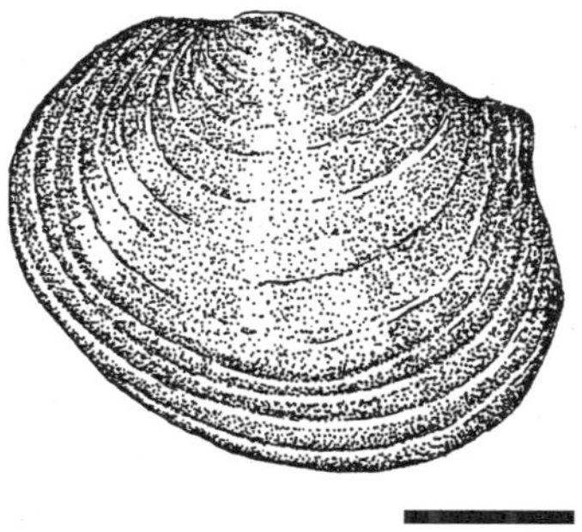

Figure 12.37. *Pemphilimnadiopsis ortoni* Tasch (1961), NEIU MCP 609 (donated by W. Lietz), Pit 11, Will-Grundy-Kankakee Counties. Mistakenly called *Cyzicus* by many collectors. Scale bar = 1 mm.

Figure 12.38. *Pemphilimnadiopsis ortoni,* modified from Tasch (1969). Scale bar = 1 mm.

Carapace marked by three, smooth, diagonal, carinate ridges; ridges extend from near beak to lower margin. Alpha angle of central ridge is approximately 45°. Anterior ridge curved, dorsal ones defining a shieldlike area. Size of adult valves, about 9 mm long by 6 mm high.

According to D. Petzold (personal communication to A.A.H., 1989), this species is the only leaid noted from the Pennsylvanian strata of the Illinois Basin, and it is uncommon in Mazon Creek collections, possibly due to collector bias (small size?) and poor preservation.

Pemphilimnadiopsis ortoni Tasch, 1961
Figures 12.37, 12.38

Bivalved carapace subquadrate with straight hinge line half the shell length, hinge confined to the left valve. Valves thin and often wrinkled due to crushing during preservation; surface has 10 to 12 prominent concentric ridges between which are seen numerous branching striae. Key characteristics for identification are a node on beak area and posterior recurvature of growth lines. Unfortunately, most specimens observed are poorly preserved and do not show these features. Overall shell size about 5 mm high by 6 mm wide.

Conchostracan specimens from the Illinois Basin previously referred to as *Estheria* or *Cyzicus* are all probably *P. ortoni* (D. Petzold, personal communication to A.A.H., 1989). The genus *Estheria,* or the term "estheriids," have appeared in the geologic literature of the Illinois Basin for more than 100 years (e.g., Meek and Worthen, 1868). *Estheria* is now a junior synonym for *Cyzicus,* a form identified on the basis of micro-ornamentation between growth lines. As no ornamentation has been noted for Illinois Basin specimens, Petzold and Lane

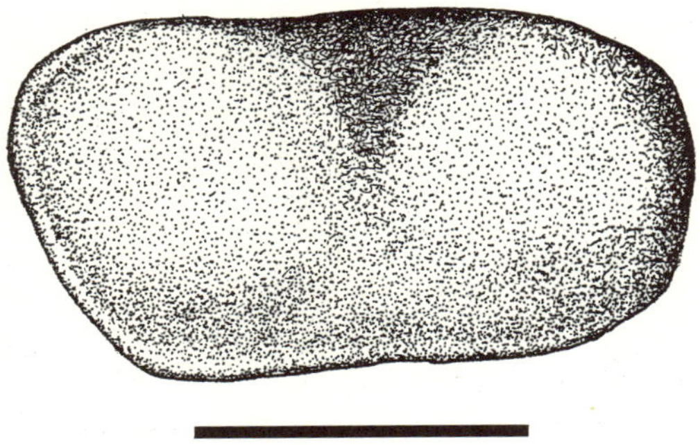

Figure 12.39. *Geisina (Hastifaba) robusta* Cooper (1946), modified from Langford (1963). Scale bar = 1 mm.

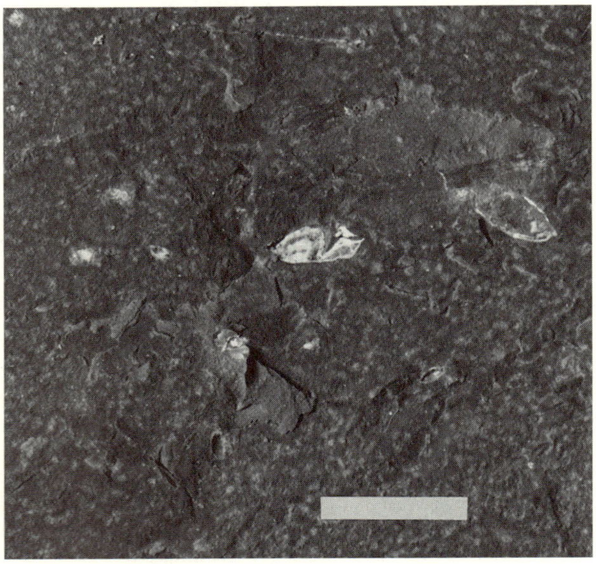

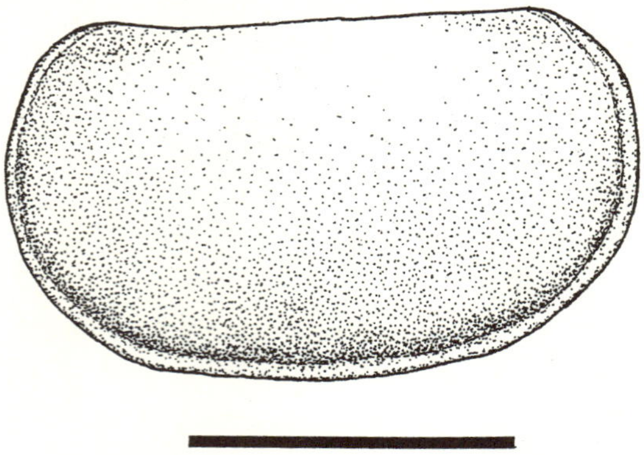

Figure 12.40. *Paraparchites mazonensis* Sohn (1977), NEIU MCP 611, Pit 11, Will-Grundy-Kankakee Counties. Scale bar = 5 mm.

Figure 12.41. *Paraparchites mazonensis,* modified from Sohn (1977). Scale bar = 1 mm.

(1988) conclude that the Illinois Basin specimens are not *Cyzicus.*

Class Maxillopoda

Antennules generally uniramous; thorax generally of no more than six segments; abdomen lacking limbs.

Subclass Ostracoda

Body greatly reduced and carapace very small.

Ostracodes are tiny crustaceans possessing an all-enveloping carapace that completely encloses the body. The carapace is practically the only source of knowledge about fossil ostracodes; their record extends from the Cambrian(?) to the present. Interestingly, although Mazon Creek concretions are found that may contain hundreds of individuals (see Sohn, 1977), they are almost never mentioned by species name ("ostracodes" suffices) in the literature of that biota. Is this perhaps a bias against their inconspicuous size?

Geisina (Hastifaba) robusta Cooper, 1946
Figure 12.39

Small, 1 mm long, straight-backed ostracode with sulcus (groove or depression) on dorsal surface of subelliptical valves. Right valve overlaps left valve. May have small spine on posterior dorsal surface. Langford (1963) stated that this ostracode, which is abundant in the Illinois Pennsylvanian, may be found in the Wilmington (Mazon Creek) area.

Paraparchites mazonensis Sohn, 1977
Figures 12.40, 12.41

Like *Geisina* in outline except that it lacks a sulcus. This species is a robust, subovate, smooth, unequivalved, 2.4 mm long, thick-shelled ostracode. The larger valve overlaps the smaller. The straight hinge line is depressed in a channel. Sohn (1977) suggested that the presence of this genus in the Braidwood fauna denoted a brackish water environment. To date this is the only ostracode species described in a formal paper as being present in the Mazon Creek biota.

Subclass Cirripedia

Body reduced and enclosed in a multiplated shell.

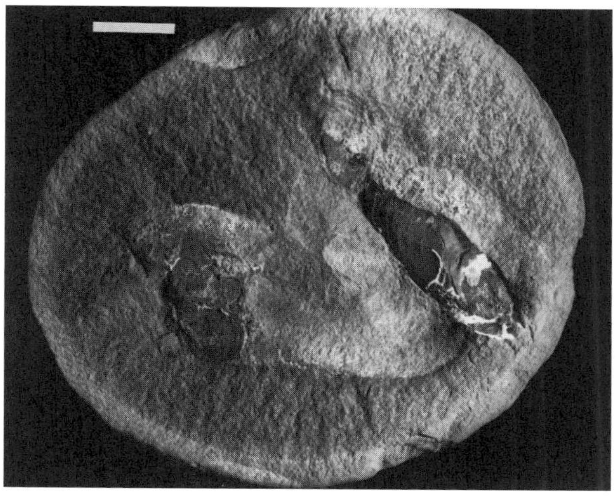

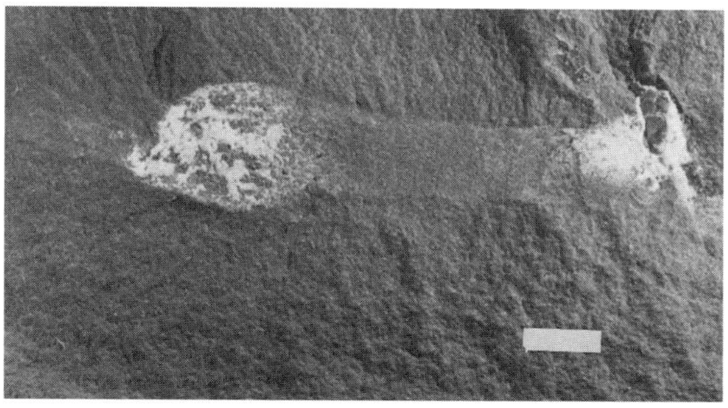

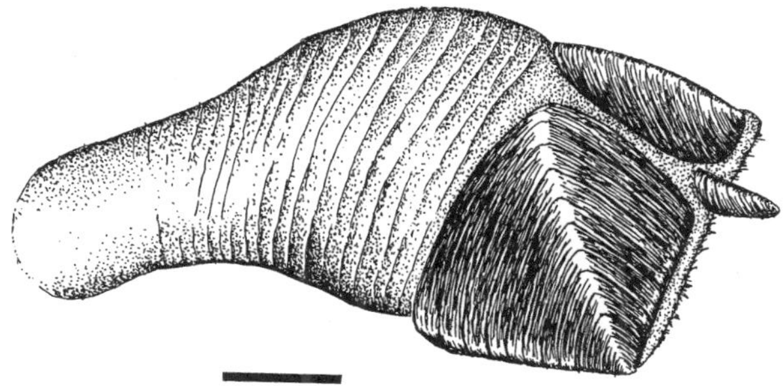

Figure 12.42 (*Top left*). *Illilepas damrowi* (Schram, 1975) on Myalina, NEIU MCP 192, Pit 11, Will-Grundy-Kankakee Counties. Originally assigned to the genus *Praelepus.* Scale bar = 2 mm.

Figure 12.43 (*Bottom left*). *Illilepas damrowi* (Schram, 1975), paratype, NEIU MCP 328, Pit 11, Will-Grundy-Kankakee Counties. Scale bar = 2 mm.

Figure 12.44 (*Above*). *Illilepas damrowi,* modified from Schram (1975). Scale bar = 2 mm.

Illilepas damrowi (Schram, 1975)
Figures 12.42–12.44

Capitulum well developed, scutum large and subrectangular, tergum moderate in size and oval, together covering almost the entire capitulum, capitular margin marked by a row of fine spines and a large platelike spine between the scutum and tergum (originally thought to be a reduced tergum); peduncle well developed and without any armature.

This barnacle species was originally placed in the family Praelepadidae and related to *Praelepas jaworskii* of the Soviet Carboniferous. Subsequently, the form of the fossil was reevaluated and found to be closer to that seen in the living Iblidae, stalked barnacles of problematic affinities.

Subclass Halicyna
Cyclus americanus Packard, 1883
Figures 12.45, 12.46

Body flattened, circular in outline and about 1 cm in diameter, anterior end developed as a

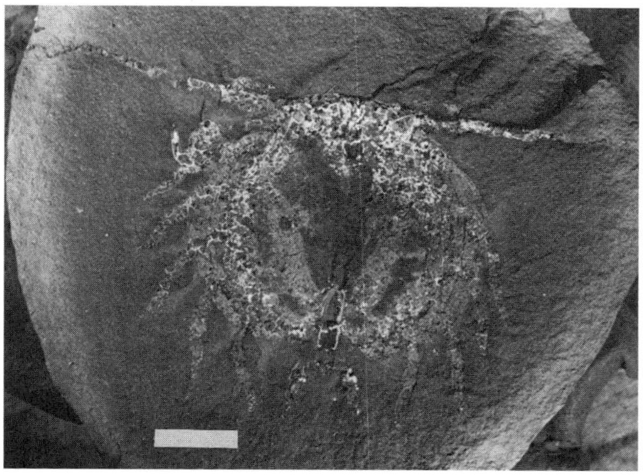

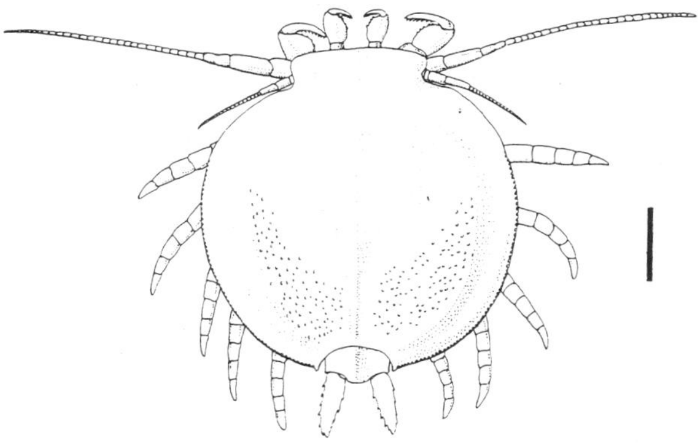

Figure 12.45. *Cyclus americanus* Packard (1883), NEIU MCP 507 (gift from Hay collection), Pit 11, Will-Grundy-Kankakee Counties. Scale bar = 5 mm.

Figure 12.46. *Cyclus americanus,* reconstruction. Scale bar = 5 mm.

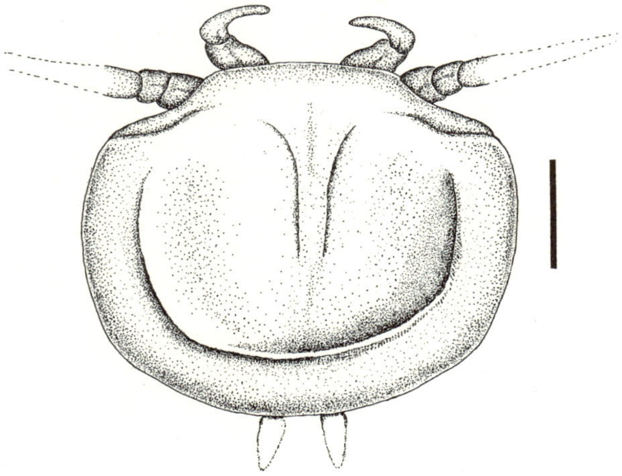

Figure 12.47. *Cyclus obesus* n. sp., reconstruction. Scale bar = 5 mm.

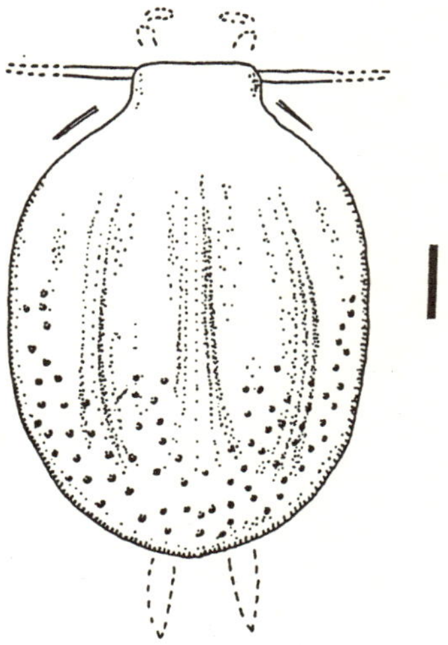

Figure 12.49. Cycloid *Apionicon apioides* n. gen. n. sp., reconstruction. Scale bar = 5 mm.

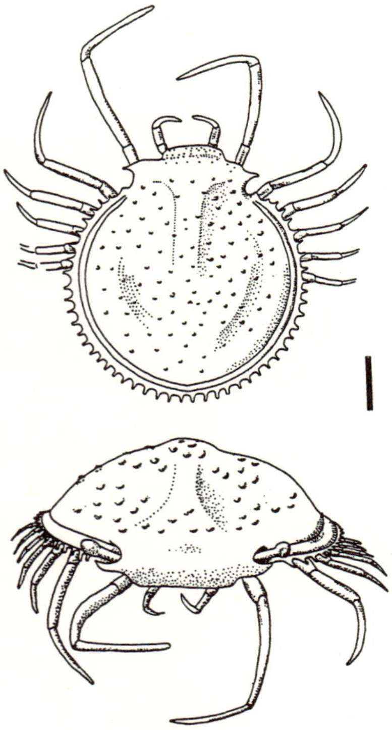

Figure 12.48. Cycloid *Halicyne max* n. sp., reconstruction. Scale bar = 5 mm.

projecting rostral plate that bears antennae; thoracic limbs robust and spinose; abdomen as two segments, with a pair of large caudal rami.

The taxonomic affinities of *Cyclus* have been uncertain since its first being described by Packard. Alternately considered crustaceans, agnostid trilobites, or chelicerates, it now appears, after cladistic analysis, that cycloids are specialized maxillopodans and a possible sister group to the Copepoda. Although sometimes thought possibly to be parasitic crustaceans that lived off various forms of fish, they are often associated with plant fragments, indicating they may have filled a niche similar to modern crabs.

The Mazon Creek biota contains four cycloid taxa: *Cyclus americanus* (Figures 12.45, 12.46) is among the most common fossil arthropods found here; a rarer form, somewhat larger and wider, *Cyclus obesus* n. sp. exists (see Figure 12.47); a form with a distinct oval, caplike, and decorated carapace, apparently is the genus *Halicyne* (see Figure 12.48); the fourth form is *Apionicon apioides* n. gen. n. sp. (see Figure 12.49). A paper describing the cycloids is currently in press (Schram et al.).

References

Brooks, H. K. 1962. Paleozoic Eumalacostraca of North America. Bull. Amer. Paleo., 44:163–338.

Cooper, C. L. 1946. Pennsylvanian ostracodes of Illinois. Illinois State Geol. Surv. Bull., 70.

Emerson, M. J., and F. R. Schram. 1991. Remipedia. Part 2, Paleontology. Proc. San Diego Nat. Hist. Soc., 7:1–52.

Langford, G. 1963. Wilmington Coal Fauna and Additions to the Wilmington Coal Flora. ESCONI, Downers Grove, Illinois, 280 pp.

Meek, F. B., and A. H. Worthen. 1865. Notice of some new types of organic remains, from the Coal Measures of Illinois. Proc. Acad. Nat. Sci. Philadelphia, 17:41–53.

———. 1868. Paleontology. Illinois State Geol. Surv., 3, pt. 2:291–572.
Packard, A. S. 1883. A monograph on the phyllopod Crustacea of North America with remarks on the order Phyllocarida. 12th Ann. Rep. U.S. Geol. Geog. Surv. Terr., 1[1878]:295–592.
———. 1886. On the Syncarida, a hitherto undescribed synthetic group of extinct malacostracous Crustacea. Mem. Nat. Acad. Sci., 3:123–128.
Petzold, D., and N. G. Lane. 1988. Stratigraphic distribution and paleoecology of Pennsylvanian conchostracans (Crustacea: Branchiopoda) on the east side of the Illinois Basin. Jour. Paleo., 62:799–808.
Rolfe, W. D. I. 1981. Phyllocarids and the origin of the Malacostraca. Geobios., 14:17–25.
Schram, F. R. 1969. Some Middle Pennsylvanian Hoplocarida. Fieldiana Geol., 12(14):235–289.
———. 1970. Isopod from the Pennsylvanian of Illinois. Science, 169:854–855.
———. 1973. On some phyllocarids and the origin of Hoplocarida. Fieldiana Geol., 26(2):77–94.
———. 1974a. Mazon Creek caridoid Crustacea. Fieldiana Geol., 30(2):9–65.
———. 1974b. Paleozoic Peracarida of North America. Fieldiana Geol., 33(6):95–124.
———. 1975. A Pennsylvanian lepadomorph barnacle from the Mazon Creek area, Illinois. Jour. Paleo., 49:928–930.
———. 1976. Some notes on Pennsylvanian crustaceans in the Illinois Basin. Fieldiana Geol., 35(3):21–28.
———. 1981. Late Paleozoic communities. Jour. Paleo. 55:126–137.
———. 1983. Remipedia and crustacean phylogeny. Crust. Issues, 1:23–28.
———. 1984. Fossil Syncarida. Trans. San Diego Soc. Nat. Hist., 20:189–246.
———. 1986. Crustacea. Oxford University Press, New York.
———. 1989. Designation of a new name and type for the Mazon Creek tanaidacean. Jour. Paleo., 63:536.
Schram, F. R., R. Vonk, and C. H. J. Hof. In press. Mazon Creek Cycloidea. Jour. Paleo.
Sohn, I. G. 1977. *Paraparchites mazonensis* (Ostracoda) from Middle Pennsylvanian of Illinois. Fieldiana Geol. 37(2):43–59.
Tasch, P. 1961. Pemphilimnadiopsidae, a new family of fossil conchostracans. Jour. Paleo., 35:117–120.

13

Myriapods and Arthropleurids

Joseph T. Hannibal

Millipedes, known also as "centipedes" or "multipedes," are insects belonging to the earthworm genus, hairy, with numerous feet, forming curves as they crawl, and contracting themselves when touched: the Greeks give to this insect the name of "oniscos," others, again, that of "tylos" ["iulos"].[1] Boiled with leek-juice in a pomegranate rind, it is highly efficacious, they say, for pains in the ears.
The Natural History of Pliny.

Myriapods (many-legged animals) are secretive arthropods typically associated with the forest floor. They possess a head with a single pair of antennae, many basically similar trunk segments, and nine or more pairs of legs. Their body segments may be divided into distinct dorsal (top), lateral (side), and ventral (bottom) parts known as tergites, pleurites, and sternites. The tergites, pleurites, and sternites may be partially or completely fused.

There are four extant classes of myriapods: the Diplopoda (millipedes or "thousand-leggers"), Chilopoda (centipedes or "hundred-leggers"), Symphyla (symphylids), and Pauropoda (pauropods), the first three of which have also been found as fossils. The evolutionary relationships between these classes is at present uncertain. There is at least one similar, extinct class of arthropods, the Arthropleuridea (arthropleurids).

Classification of fossil myriapods and allied forms at the species level is based on external characters such as the number of segments, ornamentation, and arrangement of spines, but that of extant forms is typically based on characters of the male genitalia. The characteristics of the main groups include:

Diplopoda (millipedes). Medium-sized, primarily herbivorous myriapods, up to about 300 mm (12 in.) in length. The trunk segments of these slow-moving arthropods are composed of two partially coalesced segments, one anterior and the other posterior. Millipedes thus have two pairs of legs per trunk segment (Silurian–Recent).

Chilopoda (centipedes). Medium-sized myriapods, up to 380 mm (15 in.) in length. These are fast-moving, carnivorous forms with generally one pair of legs per trunk segment (Silurian–Recent).

Symphyla. Very small myriapods, 2–8 mm (0.1–0.3 in.) in length, with 15 to 22 trunk segments and 12 pairs of legs (Oligocene–Recent).

Pauropoda. Minute myriapods, usually less than 2 mm (0.1 in.) long, with 9 to 11 trunk segments, which each bear a pair of legs, and a tail segment (Recent; the lack of fossil forms is probably due to their small size).

Arthropleuridea. Minute to gigantic myriapod-like arthropods, up to about 2 m (6.6

1. This account, like other Greco-Roman references, appears to confuse true myriapods, which are many-legged arthropods, with caterpillars, which are insects, and other animals. Such confusion continues to some extent among the general populace today. Greek names such as *iulos* and *oniscos* are now parts of the names of various myriapods.

ft.) in length, with about 30 body segments and two pairs of legs per body segment, at least on some segments. Probably herbivorous. These forms share several major characteristics (multiple body segments, many pairs of legs, etc.) with millipedes and centipedes, but their exact relationship with them is problematical (Silurian–Pennsylvanian).

Fossil Myriapods

A multilegged marine animal that at least superficially resembles a myriapod has been reported from the Cambrian (Robinson, 1990), but this form may instead be a polychaete annelid (W. Shear, personal communication, 1992). Possible, but not likely, millipede burrows have been reported from the Ordovician (Retallack and Feakes, 1987). The first body fossils of definite millipedes are Late Silurian in age (Almond, 1985), as are possible centipedes (Jeram et al., 1990). Arthropleurids are first found in Upper Silurian rocks. Early occurrences of myriapods in the fossil record are reviewed by Gray and Shear (1992) and the ecology of Paleozoic myriapods is reviewed by Shear and Kukalova-Peck (1990).

The first diverse assemblages of myriapods and arthropleurids to appear in the fossil record are those found in the Carboniferous rocks of North America and Europe. A large number of species have been described from these strata, especially from the Mazon Creek deposits and the "gas coal" of Nýřany, Czech Republic. More species of fossil myriapods, including both millipedes and centipedes, have been described from the Mazon Creek deposits than from any other locality in North America. Arthropleurids are also known from these deposits.

Initial descriptions and interpretations of Mazon Creek myriapods were made by famed paleontologists Fielding Bradford Meek and Amos Henry Worthen and the esteemed entomologist and paleontologist Samuel H. Scudder. Their pioneering work (including Meek and Worthen, 1868; Scudder, 1882, 1890) was of great importance. Until recently, most of our knowledge of Carboniferous myriapods was based on the work of Scudder and of Anton Fritsch, who studied myriapods found in the "gas coal" as well as some forms found at Mazon Creek (Fritsch, 1907). The early work of Scudder and others, however, is often lacking in detail and accuracy; the back ends of some forms were even described as the front ends! More than 20 species at one time referred to as myriapods have been described from the Mazon Creek deposits, but not all of these species are valid. Also, there has been debate over whether some of these were in fact myriapods. Recently, *Ilyodes* Scudder, 1890, and *Palaeocampa* Meek and Worthen, 1868, once thought to be myriapods, were found to belong to other groups. *Ilyodes* is now known to be an onychophoran and *Palaeocampa* a polychaete worm (Rolfe et al., 1982). See Chapter 7A for a discussion of *Palaeocampa,* and Chapter 15C for *Ilyodes.*

The classification and systematics of species found at Mazon Creek, as well as fossil myriapods in general, are badly in need of revision. The latest comprehensive review of fossil myriapods in general is that of Hoffman (1969), although Silurian and Devonian forms have been reviewed by Almond (1985).

The millipedes, centipedes, and arthropleurids found at Mazon Creek are interpreted as fully terrestrial forms. Despite claims to the contrary, no Pennsylvanian myriapod is known to have been amphibious. The Mazon Creek forms are typically found in the Braidwood (freshwater) fauna, but some forms are also found in the Essex (marine) fauna. The presence of myriapods in a marine environment can be explained by the deltaic mode of deposition of the deposits (Baird et al., 1985) and possibly also by the presence of myriapods within coprolitic concretions (Fisher, 1979, p. 436, fig. 4f) that may have been produced by amphibious or aquatic predators moving from freshwater to saltwater habitats. Predators of myriapods who would have been capable of producing large coprolites include amphibians, reptiles, and fish. (Fish would have eaten myriapods that had fallen into the water.)

Mazon Creek myriapods are sometimes found preserved in shale matrix but are most often found preserved in concretions. These myriapods are preserved in several modes within concretions: some with their bodies outstretched, others loosely coiled (Figure 13.4).

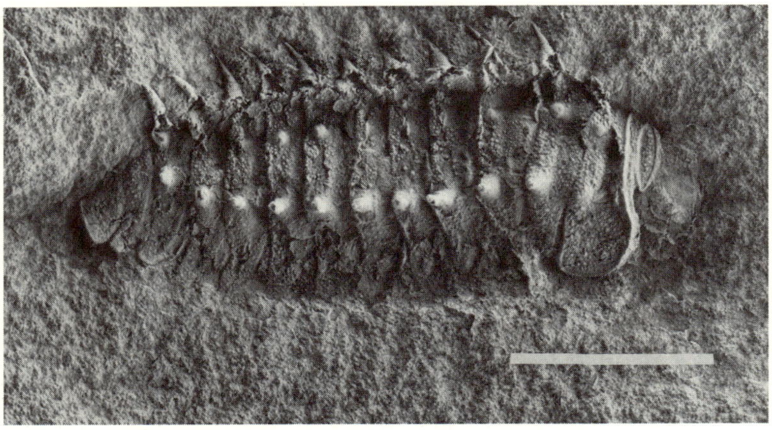

Figure 13.1. *Amynilyspes wortheni* Scudder, 1882, FMNH PE 12808; view of top and right side. The head is to the right, just beyond the small first tergite (collum). Latex cast of natural mold, from Hannibal (1984). Scale bar = 10 mm.

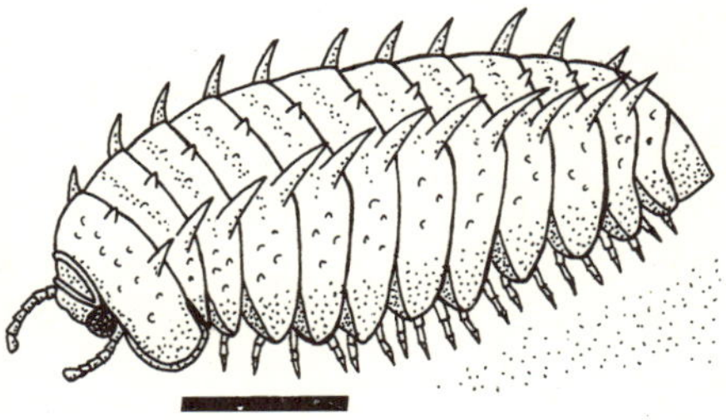

Figure 13.2. *Amynilyspes wortheni* Scudder, 1882, reconstruction based on Mazon Creek specimens, adapted from Hannibal (1984). Scale bar = 10 mm.

Coiling is a common defensive position but is also used to prevent desiccation. Complete specimens, especially of small and medium-sized forms, are sometimes found, but more often only partial specimens, especially of large forms, are found. Often body parts, such as heads and portions of the trunk, are detached. While this may be the result of natural decay before fossilization, it may also be the result of predation. Some modern predators of millipedes leave headless carcasses and disarticulated corpses behind. Some partial fossils may also represent molted cuticles.

Myriapods are typically found preserved alone but are also found preserved with plants, for instance *Pecopteris,* and occasionally with other animals, including tube worms and horseshoe crabs (Dawson, 1882; Fisher, 1979, fig. 4f).

Because many described genera and species are imperfectly known, only the well-established taxa of Mazon Creek myriapods are discussed below.

Class Diplopoda

Several types of millipedes occur at Mazon Creek. These include euphoberiids, large, long, many-segmented forms with prominent spines; the oniscomorphs, pill millipedes, so-called because of their ability to coil into a ball; and xyloiulids, more typical, garden variety millipedes, elongate, cylindrical, and many segmented. Millipedes, in part due to well-mineralized exoskeletons, are the most common myriapods found in the Mazon Creek deposits.

Most of these forms, like modern millipedes, were probably primarily detritivorous or saprophagous, ingesting decaying matter. Millipedes today are important producers of soil humus (Blower, 1985, p. 33), and this was probably the case during the Carboniferous. Burke (1979, p. 19) suggested that some forms, such as *Acantherpestes,* may have been carnivorous. There is no good evidence for such carnivory, however. While modernlike millipedes of presumably cryptic habitats are present in the Mazon Creek fauna, many Carboniferous millipedes (including euphoberiids and pill millipedes) had large eyes and spines, indicating that they lived in more open, surficial habitats (Kraus, 1974).

Superorder Oniscomorpha

Oniscomorphs (pill millipedes) have 12 to 14 body segments, a very small first tergite (collum), and a very large second tergite ("shield"). They have the ability to coil into a ball so tightly that, when enrolled, they are even watertight. Pleurites are found on the ventral side of the animal, on the outside of the legs.

Amynilyspes wortheni Scudder
Figures 13.1, 13.2

Amynilyspes wortheni is a medium-sized, spiny pill millipede with 14 tergites (one more than any modern pill millipede) and large eyes.

Figure 13.3. Nonspinose pill millipede, FMNH PE 29386; top view of specimen preserved in a concretion along with a front view of *Pecopteris* cf. *P. arborescens*. The exposed part of this millipede is about 13 mm (0.5 in.) long (from Hannibal, 1984). Scale bar = 10 mm.

Figure 13.4. Euphoberiid millipede, FMNH PE 32243; part and counterpart. This millipede is coiled into a loose ball. The concretion is about 30 mm wide (from Hannibal, 1981). Scale bar = 10 mm. Reprinted courtesy Cleveland Museum of Natural History.

Figure 13.5. Close-up of eye of euphoberiid. The individual parts making up the aggregate eye can be seen. Scale bar = 1 mm. Photograph Bruce Frumker, Cleveland Museum of Natural History. Photograph copyright Cleveland Museum of Natural History.

This animal was armored by large spines along the edge of its back and by a row of small spines down the middle of its back.

Most modern oniscomorphs are associated with the forest floor. Based on the presence of stout spines on this species, however, it may have lived in more open habitats, possibly including arboreal habitats (Hannibal and Feldmann, 1981, p. 744). *Amynilyspes* is the most common oniscomorph found in the Mazon Creek fauna. The type specimen of *A. wortheni* was presumably found along Mazon Creek; specimens have since been found in Pit 11 and Pits 1 and 6—thus, in both the Essex and Braidwood faunas.

Nonspinose oniscomorph
Figure 13.3

This oniscomorph, with smooth segments, originally was described (Hannibal and Feldmann, 1981) as having 13(?) tergites. Like most Carboniferous oniscomorphs, however, this form may have had 14 tergites (Hannibal, 1984). Specimens are found partly coiled. This form has been found in pits in Will and Grundy Counties, in the Braidwood fauna.

Modern forms of nonspinose pill millipedes, most of which have smooth tergites, are chiefly cryptic, adapted to burrowing and digging, but are also known to occasionally climb trees. A more cryptic, burrowing habit may explain why nonspinose oniscomorphs are less common than *Amynilyspes*. Forms living in more open habitats may be more easily washed into areas that are conducive to preservation.

Subclass Helminthomorpha

These are slender, elongate forms, with many segments. Helminthomorphs are what one thinks of as "typical" millipedes. There is

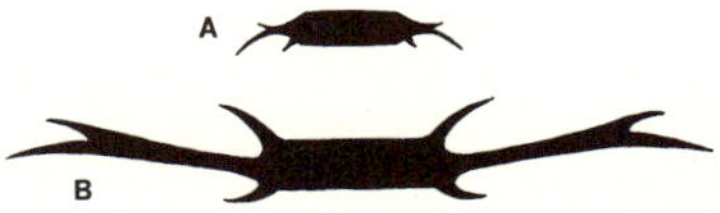

Figure 13.6. Diagram showing top view of the metazonites (posterior portions of the pleurotergites) of two of the euphoberiid genera reported from Mazon Creek: A *Euphoberia;* B *Myriacantherpestes.*

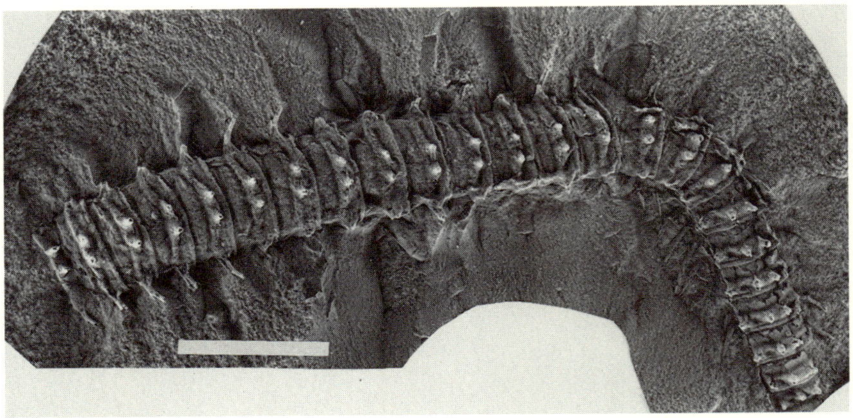

Figure 13.7. *Euphoberia carri* Scudder, 1882, USNM 38031a, in dorsal view. Specimen, including spines, is about 60 mm (about 2.3 in.) long. Note the large modified legs located on the ninth preserved body segment. Photograph is of a latex cast of a natural mold. Scale bar = 10 mm. Photograph Bruce Frumker, Cleveland Museum of Natural History. Photograph copyright Cleveland Museum of Natural History.

much variability in the configuration of the trunk segments of forms considered to be in this very large group. Some forms have clearly demarcated tergites, pleurites, and sternites; in other forms these segments are coalesced.

Order Euphoberiida
Figures 13.4–13.12

The euphoberiids (millipedes in the order Euphoberiida) are long, spinose millipedes with large heads, large aggregate eyes, and long legs. They have more than 20 and fewer than 60 body segments, ranging in length to about 300 mm (12 in.), although most forms are shorter. Each body segment is composed of two fused pleurotergal elements (coalesced coverings of the top and sides), of which the anterior is termed the prozonite and the posterior the metazonite, and has a pair of wide sternites (bottom plates). Most pleurotergites have a pair of lateral spines as well as additional, smaller spines and spinelets. The sternites have a pair of subrounded medial pores (ventral sac pits) as well as two slitlike spiracles (breathing pores), one located to the outside of each of the leg bases (coxae). Some (presumably male) specimens of euphoberiids have a pair of large modified legs (clasping appendages?) located along the midbody (Hannibal, 1995).

Euphoberiids are among the most common millipedes found at Mazon Creek, attesting to their prominence in the forests near or on the Mazonian Delta. These millipedes were probably adapted to cursorial activity in more open habitats (see Rolfe, 1985).

A number of species of euphoberiids were described from the Mazon Creek deposits by Scudder and Meek and Worthen, but not all are valid. More recent work on euphoberiids by Burke (1973, 1979) corrected a number of errors in interpretation made by these earlier workers, proving that these forms were true millipedes.

Euphoberiids typically occur in the Braidwood fauna. The relationship of this group to other helminthomorphs is unresolved.

Genus *Euphoberia*
Figures 13.6A, 13.7

The genus *Euphoberia* includes small to medium-sized euphoberiids, typically 80–150 mm (3–6 in.) in length. Metazonites of these millipedes have relatively short lateral spines, lacking the small anterior spines in front of these spines that are present in the genus *Myriacantherpestes.* Also, there are a pair of spines near the middle of the back (paramedian spines), each with an anterior spinelet. A large number of species of *Euphoberia,* not all of them valid, have been described. Meek and Worthen (1868, p. 26) described the origin of this generic name, which at first included the genera now known as *Euphoberia* and *Acantherpestes,* as follows: "For this uncouth looking creature, we would propose the generic name *Euphoberia* ["spiny creeper"], in allusion to the formidable appearance a living example, more than a

Figure 13.8. *Acantherpestes major* (Meek and Worthen), holotype, UI X504, mostly in top view. Specimen is about 60 mm (about 2.3 in.) long. Most of the spines of this specimen have been broken off; several legs can be seen extending from the bottom of the specimen (from Burke, 1973, fig. 3). Scale bar = 10 mm. Reprinted courtesy Cleveland Museum of Natural History.

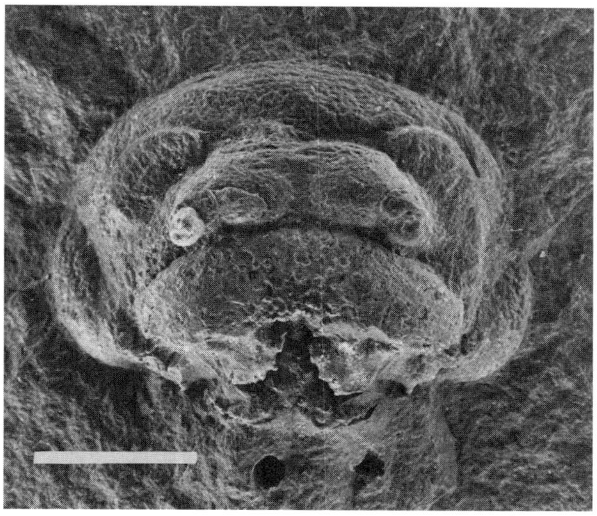

Figure 13.9. *Myriacantherpestes*?, Wallis Triebull collection. The eyes, antennal bases, and mandibles can be seen. Scale bar = 5 mm. Photograph Bruce Frumker, Cleveland Museum of Natural History. Photograph copyright Cleveland Museum of Natural History.

foot in length, must have presented, when alive and moving about, with its back bristling with forked spines, and its 150 legs in motion."

Genus *Acantherpestes*
Figure 13.8

This large euphoberiid ranges in size to about 300 mm (12 in.). It bears lateral spines, lacking small anterior spines, and has simple paramedian spines. The holotype of *Acantherpestes major* (Meek and Worthen) has been restudied by Burke (1973) and is illustrated here as Figure 13.8. Many of the well-known specimens of *A. major* were removed from the genus *Acantherpestes* by the actions of Burke (1979) when he erected the new genus *Myriacantherpestes.*

Burke (1973, p. 22) hypothesized that "protected from most predators by sheer size, *Acantherpestes* was probably able to move about freely" and that it may have "ventured into open areas of the lowlands bordering the Carboniferous swamps." He also suggested that this form may have been carnivorous.

Genus *Myriacantherpestes*
Figures 13.6B, 13.9–13.12

This medium-sized to large euphoberiid has very long lateral spines, anterior spines, and simple paramedian spines or nodes. It can be distinguished from *Euphoberia* and *Acantherpestes* by the presence of anterior spines (Burke, 1979, p. 16). One of the largest species of this genus is *Myriacantherpestes hystricosus* (Scudder), which may have reached 300 mm (12 in.) in length.

Burke (1979) found this form to be adapted to a "bulldozer" function, akin to that of some modern millipedes that push through the leaf litter of forests. This is unlikely, however, as the large spines of the animal would have made this practically impossible (W. Shear, personal communication, 1992). It probably lived in more open habitats.

Order Spirobolida
Family Xyloiulidae

These are Paleozoic millipedes with cylindrical body segments composed of mostly coalesced pleurites, tergites, and sternites. The pleurotergites are covered, or mostly marked, with a series of horizontal or oblique ridges and grooves (Hoffman, 1969, p. R587). Similarly shaped modern millipedes are sometimes known as "snake millipedes."

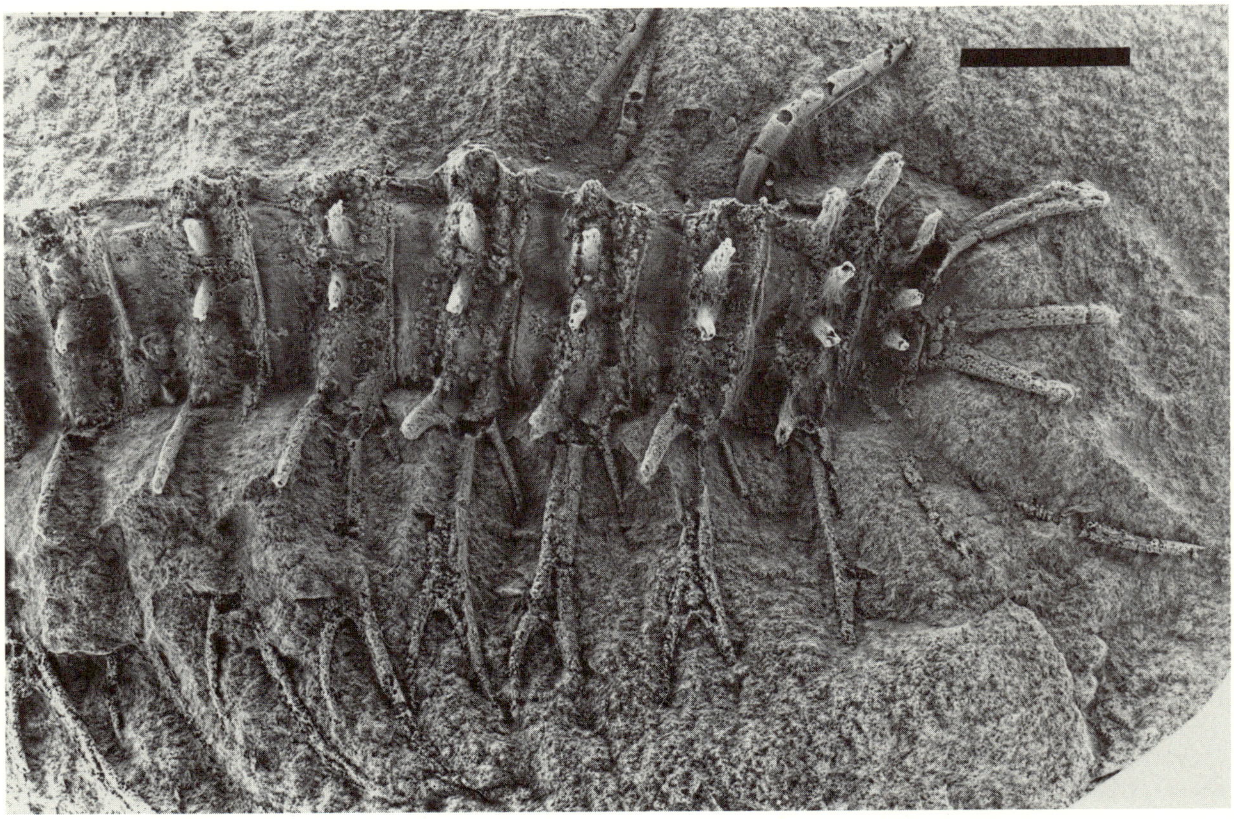

Figure 13.10. *Myriacantherpestes hystricosus* (Scudder), CMNH 3725 (coll. Wilbert A. Rath); top view of posterior of the millipede. This form has especially long, elegant lateral spines. The smaller spines, just to the anterior and posterior of the large lateral spines, that characterize this species can be seen along the left side of the animal. Three legs can be seen along the right side. Spines included, this millipede was about 65 mm (2.5 in.) wide in life. Photograph of a latex cast of a natural mold (from Hannibal, 1986). Scale bar = 10 mm. Reprinted permission *Earth Science* magazine, published by the American Geological Institute.

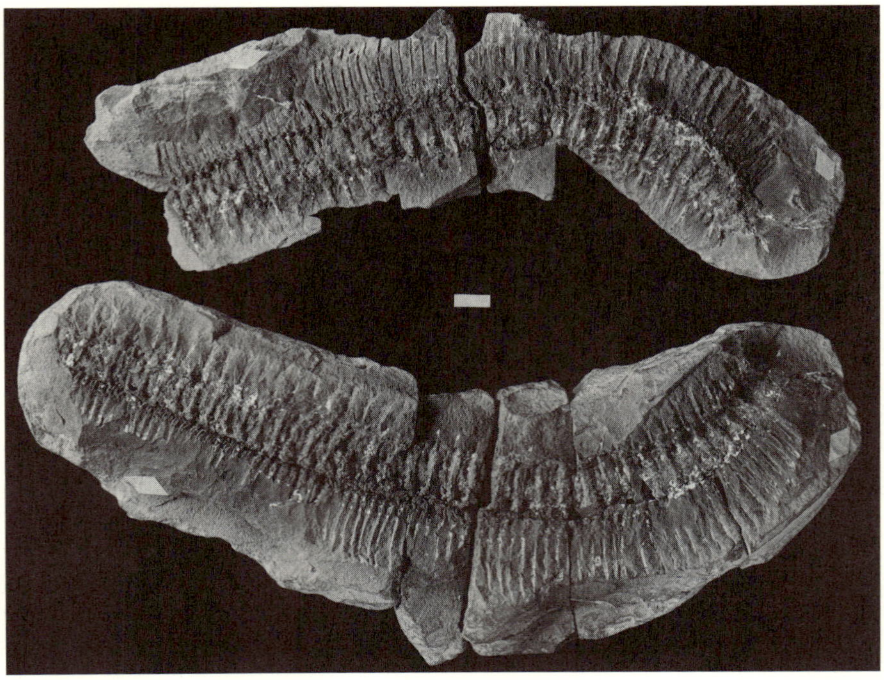

Figure 13.11. *Myriacantherpestes hystricosus* (Scudder), USNM 38038. This large specimen, about 200 mm long, is enclosed in a 220 mm (about 9 in.) long concretion. It was originally classified as *Acantherpestes major* by Scudder (1882), who called it "the most perfect large specimen that has been discovered." It has been one of the most commonly illustrated specimens of euphoberiid millipede, often referred to as "giant spined myriapods." From Hannibal, 1986. Scale bar = 10 mm. Reprinted permission *Earth Science* magazine, published by the American Geological Institute.

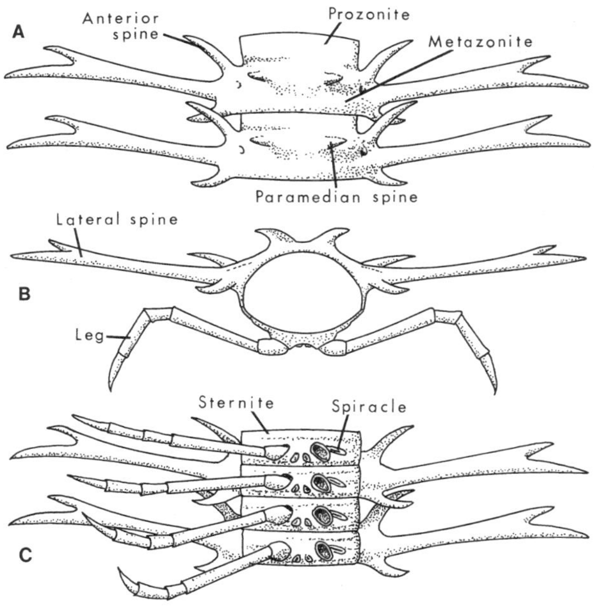

Figure 13.12. Drawing showing a typical midbody segment of a generalized *Myriacantherpestes* in top (*A*), cross-sectional (*B*), and bottom (*C*) views. *A* and *C* adapted from a drawing by Brant Gebhart, Cleveland Museum of Natural History; *B* adapted from Burke (1979, fig. 1). The interpretation of the morphology shown in these drawings is based on the work of Burke (1973, 1979).

Genus *Xyloiulus*

This form is elongate and has more than 50 segments. Horizontal ridges and grooves mark the pleurotergites.

Like most modern long, elongate millipedes, this millipede was probably cryptic, burrowing into the leaf litter and upper layers of the soil. Xyloiulids are typically found in the Braidwood fauna. A photograph of a loosely coiled specimen can be found in Baird et al. (1985, fig. 5.3).

Class Chilopoda

Centipedes are quick-moving, carnivorous myriapods with from 15 to 181 pairs of legs, one pair per segment. Their bodies are flattened from top to bottom, and they have large pincer-like claws through which they inject poison into their prey (small insects, insect larvae, and other small invertebrates). Centipedes live in a wide variety of habitats, typically in the leaf litter of the forest and in the upper layers of the soil, but some forms are even known to live in the intertidal zone along the seashore (Lewis, 1981, pp. 384–385). They have relatively soft

Figure 13.13. *Mazoscolopendra.* Most of the legs and body segments of this centipede can be seen in this view.
Scale bar = 10 mm (estimate).

and unmineralized exoskeletons and are not nearly as common as fossils as are millipedes. At least three genera of centipedes, however, have been found at Mazon Creek.

Order Scolopendromorpha

Scolopendromorph centipedes are small to large centipedes with 21 to 27 trunk segments, 21 or 23 pairs of long legs, and antennae with up to 30 segments. The largest modern centipedes belong to this group. Modern scolopendrids live in a variety of habitats including the leaf litter and upper layers of the soil (Hoffman, 1982, p. 683). They are known to eat a wide variety of prey, including worms, nymphal cockroaches, flying insects, mites, spiders, toads, snakes, and mice (Lewis, 1981, pp. 172–174).

Mazoscolopendra richardsoni Mundel
Figure 13.13

This centipede, up to 53 mm (2 in.) long has 21 leg-bearing trunk segments, long legs, and antennae with at least 14 segments (Mundel, 1979, pp. 363–366). Its name, loosely translated, is "Richardson's Mazonian centipede." The second most common centipede at Mazon Creek, *M. richardsoni* has been reported from Pit 11, an Essex fauna locality.

Palenarthrus impressus Scudder

Palenarthrus impressus is characterized by flattened rectangular tergites. It was originally described by Scudder (1890) based on a 46 mm (almost 2 in.) long specimen with about 23

Figure 13.14. *Latzelia primordialis* Scudder, CMNH 8672. Most of the tergites (rectangular plates covering the back) and many of the long legs of this centipede can be seen in this view. This specimen is almost 30 mm (about 1.2 in.) long. Scale bar = 10 mm. Photograph copyright Cleveland Museum of Natural History.

preserved segments and long legs. Mundel (1981) reaffirmed its identity as a scolopendromorph, noting that Scudder had originally described the rear of the animal as the front! It is found among the Braidwood fauna.

Order Scutigeromorpha

Scutigeromorph centipedes are swift-moving predators, with very long antennae and 15 pairs of very long legs. They dwell along the surface of the forest and are excellent climbers. They have spiracles (breathing pores) located along the back edge of the top of the tergites. The common house centipede of North America, *Scutigera coleoptrata* (often observed while stalking prey on ceilings or scurrying across floors), is a member of this group.

Scutigeromorphs are known to consume a variety of prey, including house flies, moths, spiders, worms, cockroaches, and sow bugs (Lewis, 1981, pp. 183–186).

Latzelia primordialis Scudder
Figure 13.14

This species had seven smooth, gently arched tergites; a moderately sized head; and very long legs. It has been redescribed by Mundel (1979, pp. 366–373), who has shown this form to be quite similar in general aspect to, but larger and more robust than, *Scutigera coleoptrata.* Like its modern relatives it was a swift-moving predator. Its generic name is in honor of Robert Latzel, a nineteenth-century Viennese myriapod worker. The original specimen was presumably collected along Mazon Creek. The collection at the Field Museum of Natural History examined by Mundel (1979) contained specimens from Pit 11, from the Essex fauna. This is the most common centipede found at Mazon Creek (Mundel, 1981).

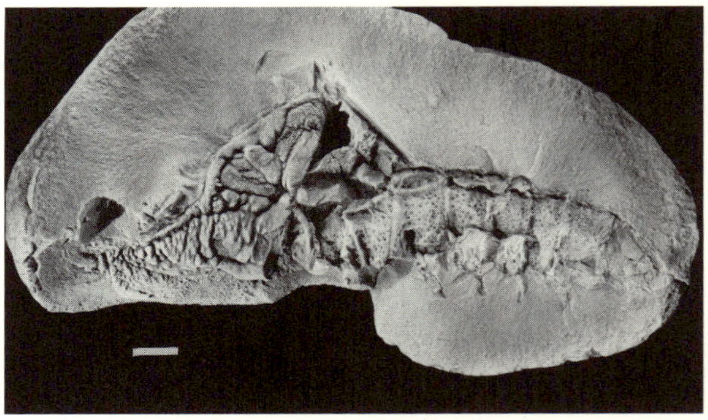

Figure 13.15. *Arthropleura cristata* Richardson, USNM 439582 (coll. Mrs. John McLuckie). Left leg of a medium-sized specimen. Scale bar = 10 mm.

Class Arthropleuridea
Order Arthropleurida
Figures 13.15–13.17

These extinct, many-legged, animals are among the very largest known terrestrial arthropods. They ranged in size up to 2 m and were possibly even longer. Arthropleurids had about 30 spinose, tuberculate tergites divided into three lobes by longitudinal grooves (a superficial resemblance to trilobites that resulted in their once being classified in the Trilobitomorpha). Each trunk segment bore two pairs of legs, at

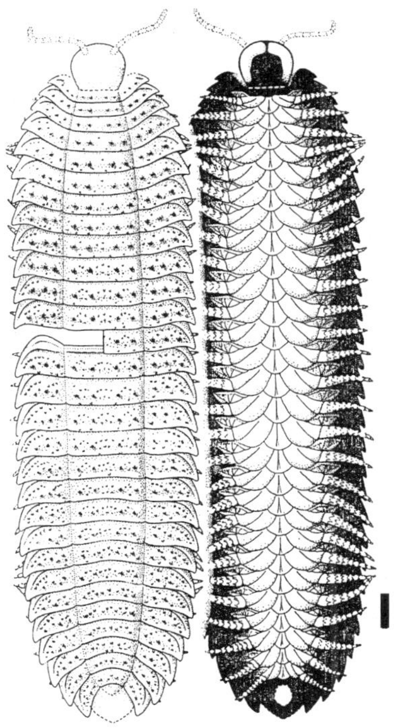

Figure 13.16. *Arthropleura armata* Jordan and Meyer; reconstruction, top (*left*) and bottom (*right*) views. Half of one tergite removed to show anterior border of underlying tergite; no limbs are shown on first trunk somite. Interrupted lines indicate restored regions of body. Study of complete specimens by Hahn et al. (1986) shows that the animal tapered more toward the front and the rear than shown here. It also appears likely that (Almond *in* Briggs et al., 1984) *Arthropleura* possessed more than one pair of limbs per segment, on average. Some individuals grew to more than 1 m in length. Illustration by J. K. Ingham from W. D. I. Rolfe and J. K. Ingham (1967, fig. 2). Scale bar = 100 mm (estimate). Reprinted from the Scottish Journal of Geology; courtesy Scottish Academic Press (Journals) Limited.

Figure 13.17. Reconstruction of 2 m long *Arthropleura* moving through a Carboniferous forest. Drawing by Annemarie Burzynski from Briggs, Plint, and Pickerill (1984, text-fig. 5). Reproduced courtesy Palaeontological Society.

least on some segments (Almond *in* Briggs et al., 1984; Briggs and Almond, 1994). Large rosette plates (convex, irregularly oblong plates divided into several lobes, bearing a resemblance to flower petals) are found near the base of the legs.

Fossils of whole or partial skeletons of arthropleurids, as well as large trackways made by them, are known from a number of localities in North America and Europe (Briggs et al., 1984). Only partial specimens, consisting of portions of legs, rosette plates, and portions of tergites, have been described from Mazon Creek.

Arthropleura cristata Richardson
Figure 13.15

Richardson (1956) was the first to accurately describe specimens of *Arthropleura* from Mazon Creek. Later (1959), he named a new species, *A. cristata,* designating a specimen consisting of the dorsal surface of a single paratergal fold (lateral part of a tergite) as the holotype. This very large species is characterized by a row of large spines near the back border of the paratergal fold accompanied by a row of smaller spines even nearer to the border, and other characters. This rare form is found within the Braidwood fauna. Although there have been a number of good papers (e.g., Rolfe, 1969; Hahn et al., 1986) on arthropleurids, there has been little subsequent work on the classification of arthropleurids from Mazon Creek since Richardson's work. Most specimens from Mazon Creek consist of isolated limb fragments and rosette plates.

The flattened body form of *Arthropleura* and the layers of lycopod tree debris sometimes found fossilized in its guts suggest a life in the surface layers of litter along the forest floor, burrowing within the rotting trunks of lycopod trees (Rolfe, 1980, 1985). However, trace fossils made by these large beasts indicate that they could roam about sandy areas as well (Rolfe, 1985; Briggs et al., 1984).

Acknowledgments

Information on *Arthropleura* was kindly provided by W. D. Ian Rolfe, Royal Museum of Scotland. This chapter was improved by the helpful comments of William A. Shear, Hampden-Sydney College, and Rodney M. Feldmann, Kent State University.

References

Almond, J. E. 1985. The Silurian-Devonian fossil record of the Myriapoda. Phil. Trans. R. Soc. London B, 309:227–237.

Baird, G. C., C. W. Shabica, J. L. Anderson, and E. S. Richardson, Jr. 1985. Biota of a Pennsylvanian muddy coast: habitats within the Mazonian Delta Complex, northeast Illinois. Jour. Paleo., 59:253–281.

Blower, J. G. 1985. Millipedes. Published for the Linnean Society of London and the Estuarine and Brackish-Water Sciences Association by E. J. Brill. 242 p. (Synopses of the British Fauna series, edited by D. M. Kermack and R. S. K. Barnes)

Briggs, D. E. G., and J. E. Almond. 1994. The arthropleurids from the Stephanian (Late Carboniferous) of Montceau-les-Mines (Massif Central, France), pp. 127–135. *In* C. Poplin and D. Heyler (eds.), Quand le Massif Central était sous l'équateur: un écosystème Carbonifère à Montceau-les-Mines. Mémoires de la Section des Sciences 12. Éditions du Comité des Travaux Historiques et Scientifiques, Paris.

Briggs, D. E. G., A. G. Plint, and R. K. Pickerill. 1984. *Arthropleura* trails from the Westphalian of eastern Canada. Palaeontology, 27:843–855.

Burke, J. J. 1973. Notes on the morphology of *Acantherpestes* (Myriapoda, Archypolypoda (*sic*)) with the description of a new species from the Pennsylvanian of West Virginia. Kirtlandia, 17:1–24.

———. 1979. A new millipede genus, *Myriacantherpestes* (Diplopoda, Archipolypoda) and a new species, *Myriacantherpestes bradebirksi*, from the English Coal Measures. Kirtlandia, 30:1–24.

Dawson, J. W. 1882. Note on *Spirorbis* contained in an ironstone nodule from Mazon Creek, with millipede. Proc. Boston Soc. Nat. Hist., 21, pt. 2:157–158.

Fisher, D.C. 1979. Evidence for subaerial activity of *Euproops danae* (Merostomata, Xiphosurida), pp. 379–447. *In* M. H. Nitecki (ed.), Mazon Creek Fossils. Academic Press, New York.

Fritsch, A. 1907. Miscellanea Palaeontologica. Vol. 1, Palaeozoica. Prague, 23 pp.

Gray, J., and W. Shear. 1992. Early life on land. Amer. Sci., 80:444–456.

Hahn, G., R. Hahn, and C. Brauckmann. 1986. Zur Kenntnis von *Arthropleura* (Myriapoda; Ober-Karbon). Geol. Palaeo., 20:125–137.

Hannibal, J. 1981. The spinose archipolypods: giant millipedes of the Coal Age. The Explorer, 23(2):15–17.

———. 1984. Pill millipedes from the Coal Age. Field Mus. Nat. Hist. Bull., 55(8):12–16.

———. 1986. The enduring myriapods. Earth Sci., 39(1):21–23.

———. 1995. Modified legs (clasping appendages?) of Carboniferous euphoberiid millipeds (Diplopoda: Euphoberiida). Jour. Paleo., 69:932–938.

Hannibal, J. T., and R. M. Feldmann. 1981. Systematics and functional morphology of oniscomorph millipedes (Arthropoda: Diplopoda) from the Carboniferous of North America. Jour. Paleo., 55:730–746.

Hoffman, R. L. 1969. Myriapoda, exclusive of Insecta, pp. R572–R606. *In* R. C. Moore (ed.), Treatise on Invertebrate Paleontology. Part R, Arthropoda 4. Geological Society of America and University of Kansas, Lawrence.

———. 1982. Chilopoda, pp. 681–688. *In* S. P. Parker (ed.), Synopsis and Classification of Living Organisms, vol. 2. McGraw-Hill, New York.

Jeram, A. J., P. A. Selden, and D. Edwards. 1990. Land animals in the Silurian; arachnids and myriapods from Shropshire, England. Science, 250:658–661.

Kraus, O. 1974. On the morphology of Palaeozoic diplopods: Symp. Zool. Soc. London, 32:13–22.

Lewis, J. G. E. 1981. The Biology of Centipedes. Cambridge University Press, Cambridge, 476 pp.

Meek, F. B., and A. H. Worthen. 1868. Preliminary notice of a scorpion, a *Eurypterus*?, and other fossils from the Coal Measures of Illinois. Amer. Jour. Sci., 2nd ser., 45:19–28.

Mundel, P. 1979. The centipedes (Chilopoda) of the Mazon Creek, pp. 361–378. *In* M. H. Nitecki (ed.), Mazon Creek Fossils. Academic Press, New York.

———. 1981. New and little known fossil myriapods from the Mazon Creek area of Illinois (Carboniferous, Westphalian D), p. 40. 5th Int. Cong. Myriapodol., Abstracts of Papers.

Retallack, G. J., and C. R. Feakes. 1987. Trace fossil evidence for Late Ordovician animals on land. Science, 235:61–63.

Richardson, E. S. Jr. 1956. Pennsylvanian invertebrates of the Mazon Creek area, Illinois. Trilobitomorpha: Arthropleurida. Fieldiana Geol., 12(4):69–76.

———. 1959. Pennsylvanian invertebrates of the Mazon Creek area, Illinois. Trilobitomorpha: Arthropleurida, II. Fieldiana Geol., 12(5):77–82.

Robison, R. A. 1990. Earliest-known uniramous arthropod. Nature, 343:163–164.

Rolfe, W. D. I. 1969. Arthropleurida, pp. R607–R620. *In* R. C. Moore (ed.), Treatise on Invertebrate Paleontology, Part R, Arthropoda 4. Geological Society of America and University of Kansas, Lawrence.

———. 1980. Early invertebrate terrestrial faunas, pp. 117–157. *In* A. L. Panchen (ed.), The Terrestrial Environment and the Origin of Land Vertebrates. Academic Press, London.

———. 1985. Aspects of the Carboniferous terrestrial arthropod community, pp. 303–316. *In* J. T. Dutro, Jr., and H. W. Pfefferkorn (eds.), 9th Cong. Int. Strat. Géol. Carbonifère, Compte Rendu. Vol. 5, Paleo., Paleoecol., Paleogeog. Southern Illinois

University Press, Carbondale and Edwardsville.
Rolfe, W. D. I., and J. K. Ingham. 1967. Limb structure, affinity and diet of the Carboniferous "centipede" *Arthropleura.* Scottish Jour. Geol., 3(1):118–124.
Rolfe, W. D. I., F. R. Schram, G. Pacaud, D. Sotty, and S. Secretan. 1982. A remarkable Stephanian biota from Montceau-les-Mines, France. Jour. Paleo., 56(2):426–428.
Scudder, S. H. 1882. Archipolypoda, a subordinal type of spined myriapods from the Carboniferous formation. Mem. Boston Soc. Nat. Hist., 3(5):143–182.
———. 1890. New Carboniferous Myriapoda from Illinois. Mem. Boston Soc. Nat. Hist., 4(9):417–442.
Shear, W. A., and J. Kukalova-Peck. 1990. The ecology of Paleozoic terrestrial arthropods: the fossil evidence. Canadian Jour. Zool., 68:1807–1834.

14A

Insecta

Frank M. Carpenter

There are so many bugs running around underfoot that there is hardly any sense at all in learning their names.
S. Mortis Rigueur (Steven M. Richardson), *The Gullible Hiker's Nature Atlas* (The Perennial Press, 1972)

Insects are hexapodous (six legged) arthropods, with a body consisting of three major divisions, the head, thorax, and abdomen. Each of these parts consists of several segments: six fused segments in the head, three distinct segments in the thorax, and eleven more or less distinct segments in the abdomen (Figure 14A.12). Insects have one pair of antennae and several mouthparts, which are exposed externally (ectognathous). The respiratory system consists of paired, valvelike structures or spiracles along the sides of several thoracic and abdominal segments and leading to a complex series of tubes (tracheae) that carry air throughout the body of the insect.

There is much diversity of structural detail within the class Insecta. The primitive mouthparts were apparently mandibulate and used for chewing, but in several orders, such as the true bugs (Hemiptera, "half wing") and the flies (Diptera, "two wings"), they have been modified to form a haustellate or sucking beak. The most generalized insects, such as mayflies and related species, have a pair of long, segmented processes, the cerci, articulated to the eleventh abdominal segment (Figure 14A.12). In most insects now living the cerci are relatively short or even vestigial. The earliest insects, as far as is known, were terrestrial, but many species, such as mayflies, mosquitoes, some bugs, and beetles, have become adapted to living in water, at least during immature stages, and possess tracheal gills.

The most remarkable structures developed by the insects are their wings. The first insects, possibly as far back as the Devonian, are presumed to have been primitively wingless, like the modern silverfish (Thysanura, subclass Apterygota). The evolutionary development of the wings was certainly the most significant event in the history of the insects. Over 99 percent of the species of insects now living (subclass Pterygota) have wings or have secondarily lost them again. Functional wings are now typically present on the second and third thoracic segments—the pair of forewings on the mesothorax and the pair of hind wings on the metathorax. In some species, as in the true flies (order Diptera), the hind wings are virtually absent. In others, the forewings have become thickened and serve as a cover for the abdomen and the hind wings when the insect is not flying, as in the grasshoppers and beetles. The wings of insects provide many structural features that are useful in classification, especially for the fossils since the wings are often the only parts preserved. The wings consist of a membrane supported by a series of cuticular thickenings, termed veins. The basic pattern of the veins is observable in most insects, and their homologies satisfactorily determined (Figure 14A.2). The vein along the anterior margin is termed the costa (C); the vein just below that, the subcosta (SC). This is followed by the radius (R) and the radius sector (RS), then the media (M) and the cubitus (CU). There are also one or more anal veins, in the most posterior or anal part of the wing, although their homologies are not always determinable. In addition, cross veins are usually present, joining the main veins to their branches; and often the venation becomes complicated by the partial fusion or anastomosis of certain branches.

The class Insecta is usually divided into two subclasses: the Apterygota, to include the primitively wingless species, and the Pterygota, for the winged or secondarily wingless species. The subclass Pterygota, in turn, is divided into two infraclasses. One of these, the Palaeoptera, includes species that are unable to flex their wings back over their abdomen in resting position, so that the wings are always extended laterally or upward, as in the dragonflies and mayflies. The infraclass Neoptera includes those species having an articular mechanism that allows the wings to be flexed over the abdomen at rest. About 98 percent of the existing insects belong to the Neoptera. Two divisions of the Neoptera are usually recognized. The division Exopterygota includes the species whose immature stages or nymphs resemble the adults and live in the same environment and whose wings develop externally in the nymphal stages (incomplete metamorphosis). These are the grasshoppers, cockroaches, and their relatives. The division Endopterygota includes those species whose immature stages or larvae are very different from the adults and live in different environments and whose wings develop internally in the larval stages, becoming external in a quiescent or pupal stage (complete metamorphosis). About 88 percent of living species of insects are in this category. Representatives of all of these categories have been found in Upper Carboniferous strata with the exception of the Endopterygota, which is not known until the Permian.

Insects are only rarely preserved as fossils in most deposits and are especially scarce in Upper Carboniferous beds, in which the oldest insects have been found. A few Upper Carboniferous deposits in Europe and Asia have produced a substantial number of specimens, but the Francis Creek Shales have produced most of the evidence on which our knowledge of the Upper Carboniferous insects of North America is based. Nearly 150 species, belonging to 30 families and 102 genera, have been described and named from the concretions. Fully half of the genera, however, are based on small fragments that cannot be placed in families until better fossils have been found. In the following account only the better known genera are mentioned. Up to the present time no species of the subclass Apterygota have been described and named, but they are present in the concretions and are being treated in an article now in preparation.

The Pterygota described from Mazon Creek belong to the following orders, all of which are extinct except for the Blattaria (cockroaches):

Subclass Pterygota (winged)

A. Infraclass Palaeoptera (wings extended)
   1. Order Palaeodictyoptera
   2. Order Megasecoptera
   3. Order Diaphanopterodea
   4. Order Protodonata
B. Infraclass Neoptera, division Exopterygota (wings flexed)
   1. Order Protorthoptera
   2. Order Blattaria

Infraclass Palaeoptera
Order Palaeodictyoptera

These were palaeopterous insects of moderate to large size. All main veins were present in the wing, and anastomosis of M with RS or M with CU was exceptional; cross veins were numerous and commonly formed a dense network (archedictyon). The fore and hind wings were similar, though the hind wing was usually broader basally. The antennae were slender and multisegmented; the mouthparts formed a sucking beak, with five stylets. The prothorax typically had a pair of small, winglike lobes (paranota); the legs were adapted for running and had five tarsal segments. An external ovipositor was well developed in the female and the cerci were long and multisegmented. The immature stages (nymphs) were terrestrial, with mouth-

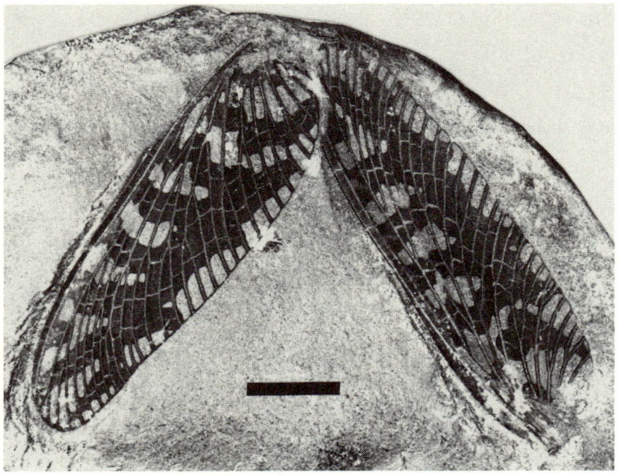

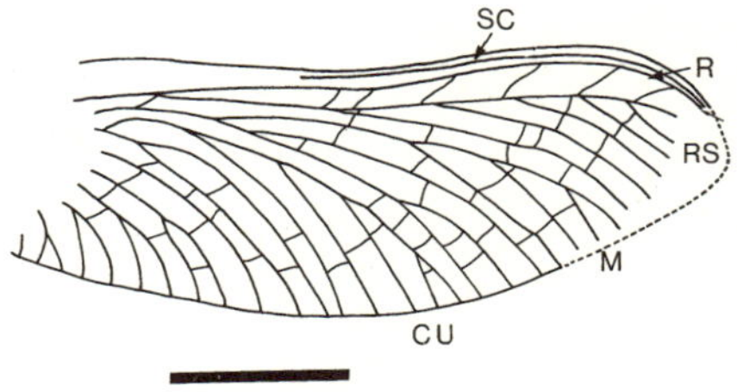

Figure 14A.1 (*Left*). *Homaloneura dabasinskasi,* holotype, Dabasinskas collection, near Braidwood, Will County; fore and hind wings. Scale bar = 10 mm.

Figure 14A.2 (*Above*). *Spilaptera americana,* drawing of holotype, H413—FMNH PE 24851, Pit 6. Scale bar = 10 mm. Abbreviations: CU, cubitus; M, media; R, radius; RS, radial sector; SC, subcosta.

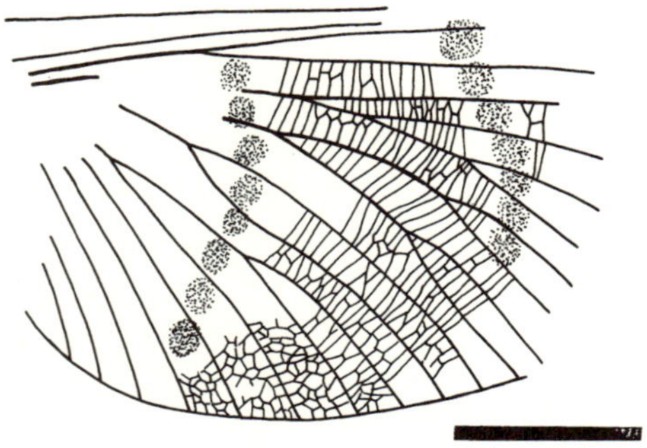

Figure 14A.3. *Neofouquea suzannae,* holotype, Dabasinskas collection, near Braidwood; hind wing. Scale bar = 10 mm.

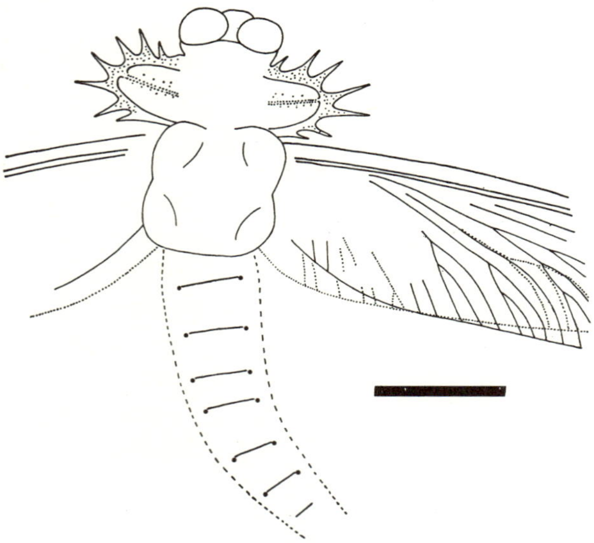

Figure 14A.4. *Notorachis wolfforum,* drawing of holotype, FMNH PE 21699, Pit 11. Scale bar = 10 mm.

parts like those of the adults. The developing wings were attached to the thorax only at the articular area, and there was apparently at least one subimaginal stage. The adults and nymphs probably fed on the juices or soft tissues of plants, such as pteridosperms, cordaites, or lycopods. The order is known from the Upper Carboniferous and Permian.

This was one of the major orders of the Carboniferous and, to a lesser extent, of the Permian. Twenty families, including about 70 genera, are usually recognized in the order. An additional 40 genera have been tentatively assigned to the order, but since they are poorly known, their relationships are uncertain. Six families have been found in the Mazon Creek nodules.

The family Spilapteridae is represented by two genera. *Homaloneura dabasinskasi* Carpenter (1964) is a typical member of the family, but with unusual wing markings (Figure 14A.1). *Spilaptera americana* Carpenter and Richardson (1971) is a smaller species, apparently lacking all markings (Figure 14A.2). Both genera occur in the Upper Carboniferous shales of Commentry, France.

The family Fouqueidae is represented by a single species, *Neofouquea suzannae* Carpenter and Richardson (1971), which has most of the cross veins reticulate, and several rows of dark spots across the wings (Figure 14A.3). This is typical of the family.

Two genera appear to belong to the family Lycocericidae. *Notorachis wolfforum* Carpenter and Richardson (1971) has a strikingly modified

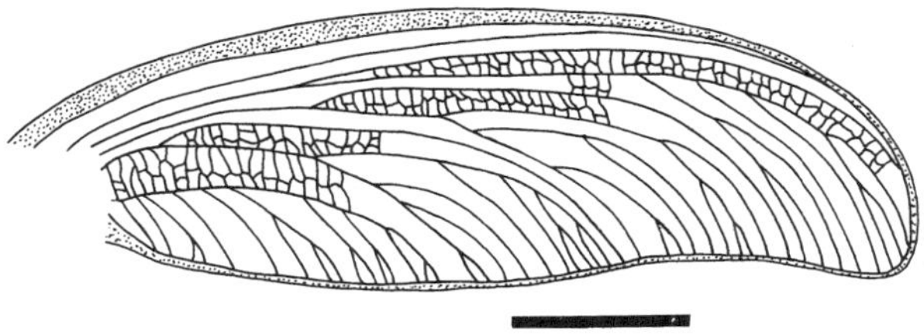

Figure 14A.5 (*Left*). *Lycodemas adolescens,* holotype, H413—FMNH PE 24852, Pit 11. Scale bar = 5 mm.

Figure 14A.6 (*Above*). *Lycodemas adolescens,* Pit 11; drawing of forewing of older nymph. Scale bar = 5 mm.

prothorax, the pronotum being heavily sclerotized and bearing long spines (Figure 14A.4). However, since its venation is not completely known, its family position is uncertain. The second species, *Lycodemas adolescens* Carpenter and Richardson (1971), is represented by a nymph, the first immature form that could be definitely associated with the Palaeodictyoptera (Figures 14A.5, 14A.6). The wings are of course small and divergent from the body, with a clear venation, and are enclosed in a sheath or wing pad. The venation is very close to that of the Commentry genus *Lycocerus.*

The family Homoiopteridae includes the largest of the Palaeodictyoptera, with the hind wings broader than those of other members of the order. Five genera, each based on a single species, have been found in the Mazon Creek beds. *Thesoneura americana* Carpenter (1944), with a wing length of 101.6 mm (4 in.), is the smallest of the Mazon Creek species (Figures 14A.7, 14A.8). *Mazonopterum wolfforum* Kukalova-Peck and Richardson (1983) is the largest, with a wing length of 177.8 mm (7 in.; Figure 14A.9). *Larryia osterbergi* Kukalova-Peck and Richardson (1983), with a cavity of the front margin of the forewing, is apparently a nymphal homoiopterid (Figure 14A.10).

The family Eubleptidae, known only from Mazon Creek beds, includes a single species at present, *Eubleptus danielsi* Handlirsch (1906). This is one of the smallest members of the order Palaeodictyoptera, with a wing length of only 12.7 mm (0.5 in.) (Figures 14A.11, 14A.12). It is also one of the best known members of the Mazon Creek insect fauna, being represented by 17 specimens, of which all but two are from Pit 11 Carpenter (1983). The antennae and cerci are very long and slender. The fore and hind wings have a distinct line of four maculations along the anterior margins.

The family Synonopteridae is known only from the Mazon Creek fauna. It was originally placed by Handlirsch in the Palaeodictyoptera, which, in the present state of our knowledge, still seems the best place for it, but its affinities have been questioned by other workers. *Lithoneura mirifica* Carpenter (1944), a large species with unusually broad wings, is the best known representative (Figures 14A.13, 14A.14).

Order Megasecoptera

These palaeopterous insects were of small to large size. In many respects they were similar to the Palaeodictyoptera, but the wings were very slender basally and often subpetiolate. The two pairs of wings were much alike. Veins SC and R were very close together and very near

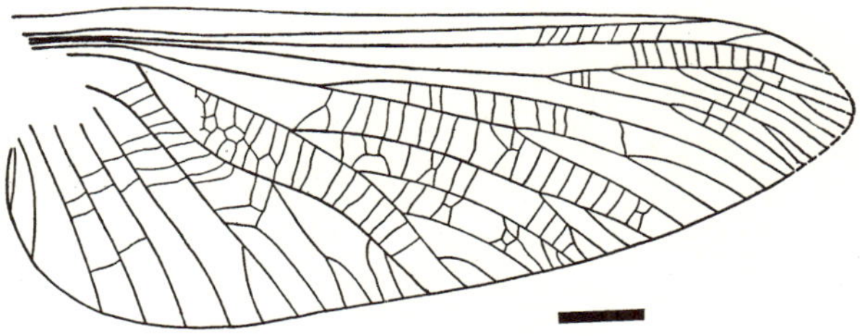

Figure 14A.7 (*Left*). *Thesoneura americana,* holotype, ISM 14876, near Braidwood; hind wing. Scale bar = 10 mm.

Figure 14A.8 (*Above*). *Thesoneura americana,* drawing of holotype. Scale bar = 10 mm.

Figure 14A.9. *Mazonopterum wolfforum,* holotype, FMNH PE 15138, Worthington Township, Will County. Scale bar = 10 mm.

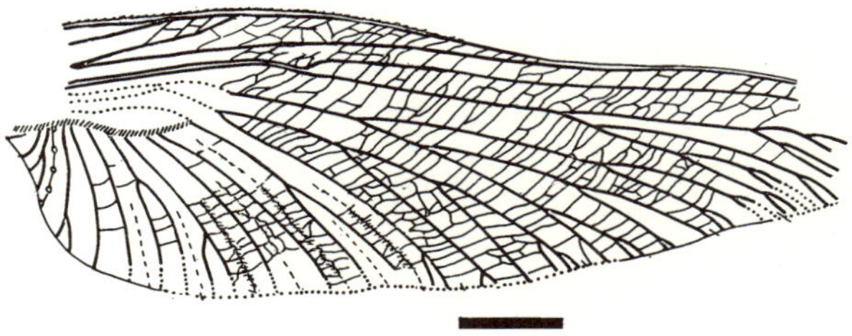

Figure 14A.10. *Larryia osterbergi,* holotype, Osterberger collection P9B 674; drawing of holotype, a nymphal hind wing (Kukalova-Peck and Richardson, 1983). Scale bar = 10 mm.

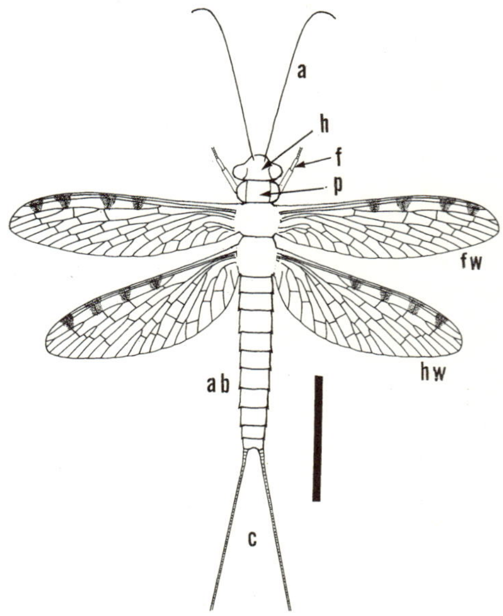

Figure 14A.11. *Eubleptus danielsi,* FMNH PE 40223, (Paul Harris specimen), Pit 11. Scale bar = 10 mm.

Figure 14A.12. *Eubleptus danielsi,* reconstruction by Carpenter (1983). Scale bar = 10 mm. Abbreviations: a, antenna; ab, abdomen; c, cerci; f, foreleg; fw, forewing; h, head; hw, hind wing; p, prothorax.

the wing margin. Pronotal lobes like those of the Palaeodictyoptera were apparently absent, although in some families the pronotum was broad and spinose. The mouthparts and cerci were like those of the Palaeodictyoptera, but the legs were apparently relatively short. The nymphs had divergent wing pads like those of the Palaeodictyoptera. The order is known from the Upper Carboniferous and Permian, with its maximum development in the Permian.

Twenty-one families, with 32 genera, are known in this order, two of which are represented in the Mazon Creek fauna. *Mischoptera douglassi* Carpenter and Richardson (1968), of the family Mischopteridae, is based on a nymph with a striking resemblance to the

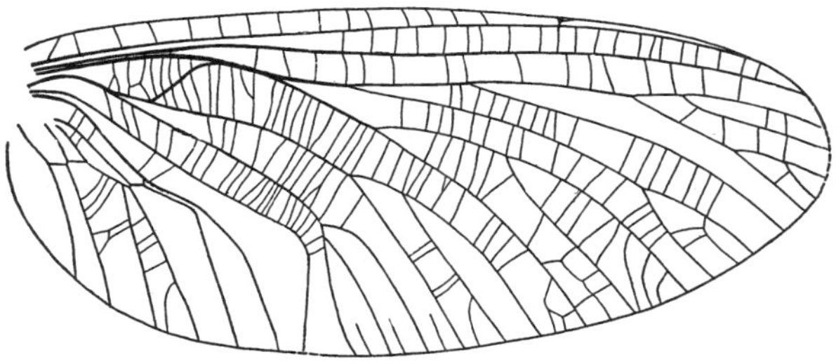

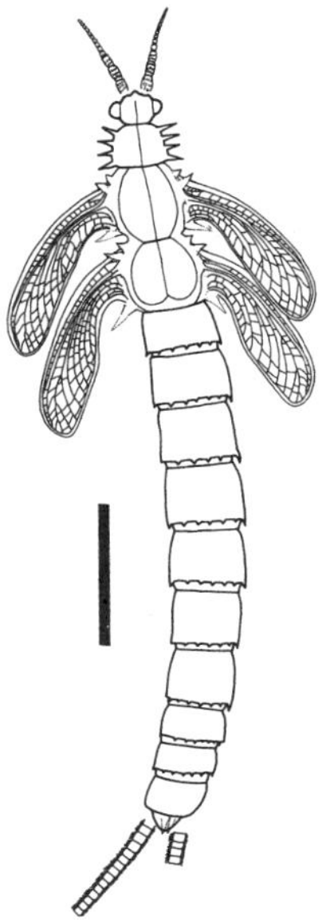

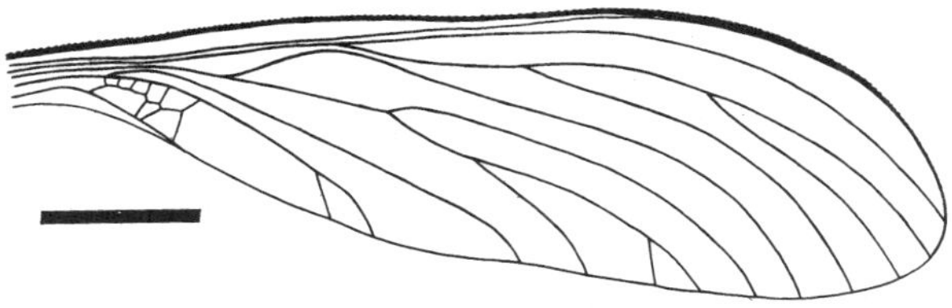

Figure 14A.13. *Lithoneura mirifica,* holotype, ISM 14880, near Braidwood; hind wing. Scale bar = 10 mm.

Figure 14A.14. *Lithoneura mirifica,* drawing of holotype. Scale bar = 10 mm.

Figure 14A.15. *Mischoptera douglassi,* drawing of holotype, Douglass collection 39, Pit 6. Scale bar = 10 mm.

Figure 14A.16. *Eubrodia dabasinskasi,* drawing of holotype, MCP 336, Pit 11. Scale bar = 10 mm.

adults of *Mischoptera* found in the Commentry Shale in France, including the spinose prothorax and the characteristic wing venation (Figure 14A.15). This was the first nymph to be found that could be definitely associated with Megasecoptera. The only other species of the order found at Mazon Creek is *Eubrodia dabasinskasi* Carpenter (1967) of the family Brodiidae, previously known only from the Upper Carboniferous of England (Figure 14A.16). The venation is typical of that of the Megasecoptera, but instead of a few cross veins the wing has a very fine network or archedictyon, which is unusual in this order.

Order Diaphanopterodea

These are small to medium-sized, palaeopterous insects, closely related to the Megasecoptera. Their outstanding, distinctive trait is their ability to fold their wings back over the abdomen at rest, much like the Neoptera. However, the articular mechanism by which this is achieved is very different from that of the Neoptera (Kukalova-Peck, 1974). Hence, these are apparently palaeopterous insects, related to the Megasecoptera, that have developed a method of flexing the wing independently of that in the Neoptera. The wing venation is essentially like that of the Megasecoptera, although there are very few cross veins and no archedictyon. The cerci are very long. The order is a small one, with 10 families and 17 genera. It is known from the Upper Carboniferous and Permian, with the greatest development in the Permian.

Only one species of this order, *Prochoroptera calopteryx* Handlirsch (1911), of the family Prochoropteridae, has been found in the Mazon Creek beds (Figure 14A.17). Four specimens have been studied, and all have the wings flexed over the abdomen (Carpenter, 1978). There is only one other species of the order known from the Upper Carboniferous of North America: *Euchoroptera longipennis* Carpenter, from the Stanton Formation, Kansas.

Order Protodonata

These are large to very large palaeopterous insects, closely related to the dragonflies (Odonata) but lacking several specializations on the wings, such as a nodus and pterostigma. Some of these were the largest insects known, with a wing length of 355.6 mm (14 in.). Immature stages have not been found but are assumed to have been aquatic like those of the existing Odonata. The order was apparently a small one; it is known from the Upper Carboniferous and

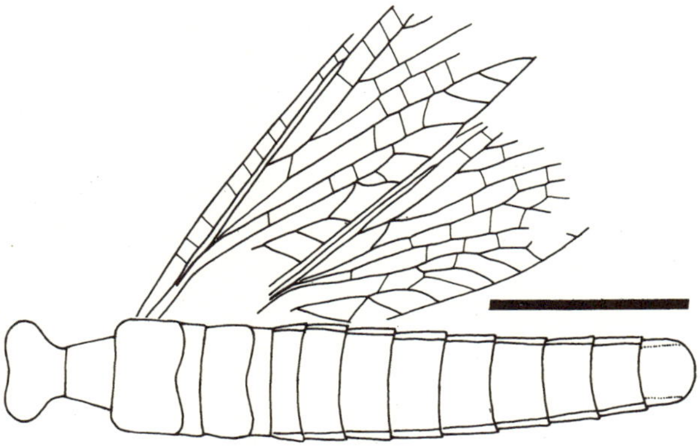

Figure 14A.17. *Prochoroptera calopteryx,* reconstruction by Carpenter and Richardson (1978); near Braidwood and Pit 11. Scale bar = 10 mm.

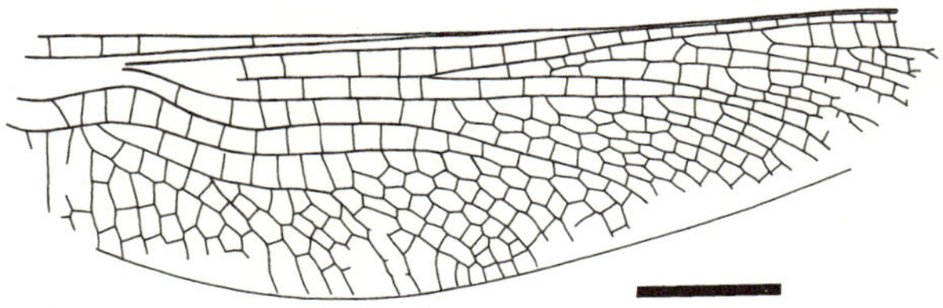

Figure 14A.18. *Oligotypus makowskii,* drawing of holotype, Makowski collection, IWJM-I-C, Pit 11. Scale bar = 10 mm.

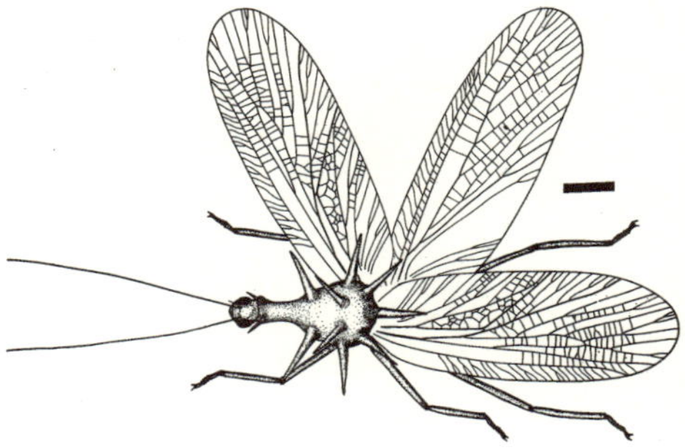

Figure 14A.19. *Gerarus danielsi,* holotype, USNM 35574, near Coal City. Scale bar = 10 mm.

Figure 14A.20. *Gerarus danielsi,* reconstruction by Burnham (1983). Scale bar = 10 mm.

Permian and includes only two families and eight genera.

Two species have been found in the Mazon Creek nodules. The better known of these is *Oligotypus makowskii* Carpenter and Richardson (1971) of the family Paralogidae, a relatively small protodonate with a wing length of about 101 mm (4 in.; Figure 14A.18).

Infraclass Neoptera
Order Protorthoptera

These were neopterous insects with mandibulate mouthparts. They were obviously related to the members of the existing order Orthoptera but lacked the jumping hind legs characteristic of that order. In some species both fore and hind wings were membranous, but in many species the forewings were thickened or coriaceous, and the hind wings had an enlarged anal area. The wing venation is usually well preserved in the fossils, and it obviously was very diversified throughout the order. The body structure, however, is not so well known because the folding of the wings over the thorax and abdomen tends to conceal those parts in the fossils. Those species that do have the body preserved show much diversity of structure.

This order is known only from the Upper Carboniferous and Permian, but it was the largest of all the extinct orders of insects, in terms of numbers of species and genera. Over 50 families have been described; these include 125 genera. Another 85 genera have been named, but they have not been assigned to families because of poor preservation. Twelve families are known from Mazon Creek, represented by 25 genera. An equal number of genera are in the incertae sedis category.

The family Geraridae is one of the best known of the Protorthoptera in Mazon Creek. The prothorax was very long and exceptionally broad posteriorly, with long, stout spines. The

Figure 14A.21. *Nacekomia rossae,* holotype, FMNH PE 791, "strip mine area." Scale bar = 10 mm.

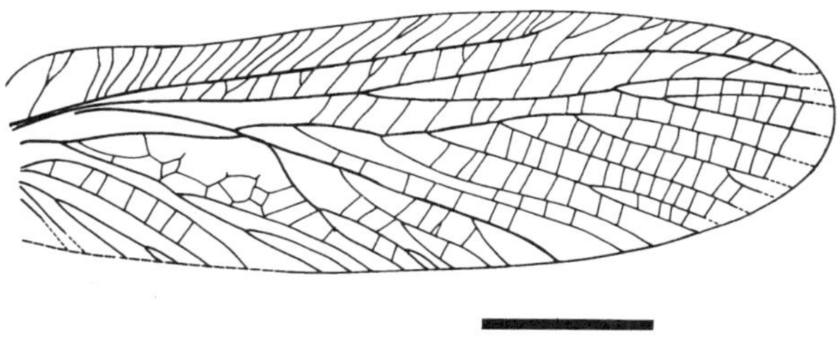

Figure 14A.22. *Nacekomia rossae,* drawing of holotype by Burnham (1983). Scale bar = 10 mm.

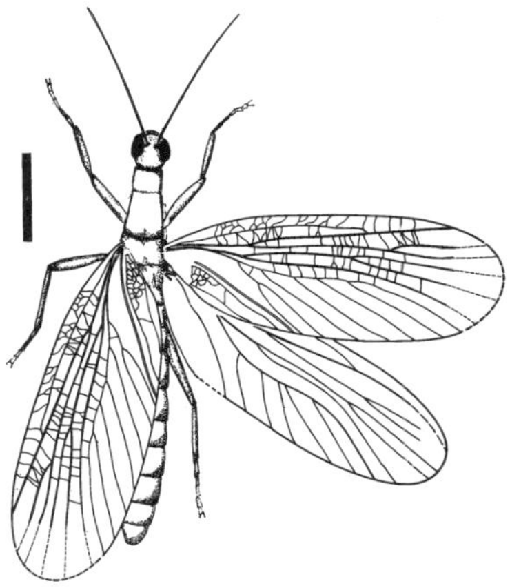

Figure 14A.23. *Protecticus ambulans,* reconstruction by Burnham (1986); near Braidwood. Scale bar = 10 mm.

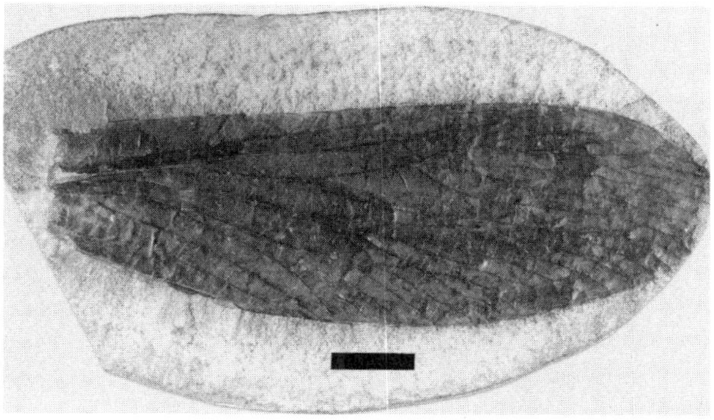

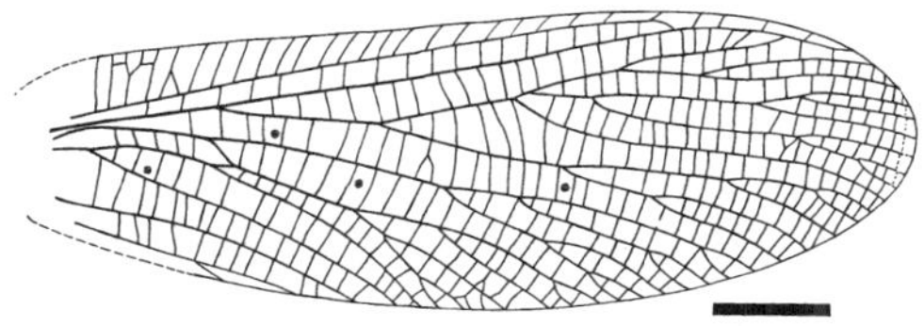

Figure 14A.24. *Heterologus langfordorum,* holotype, ISM 14879, near Braidwood; forewing. Scale bar = 10 mm.

Figure 14A.25. *Heterologus langfordorum,* drawing of holotype. Scale bar = 10 mm.

subcostal vein, SC, ended on the front margin of the wing. Vein M was the most extensive vein in the forewings, but RS was the largest vein in the hind wing. *Gerarus danielsi* Handlirsch is probably the most frequently collected insect in the Mazon Creek nodules (Figures 14A.19, 14A.20). *Nacekomia rossae* Richardson (1956) was a similar species but had the front margin of the forewing distinctly concave basally (Figures 14A.21, 14A.22).

The family Spanioderidae includes several species from Mazon Creek. They resemble the Geraridae in some respects, but vein SC of the forewing always ends on the R vein. *Protecticus ambulans* (Handlirsch, 1906) is the type of the family and one of the best known species (Figure 14A.23).

Heterologus langfordorum Carpenter (1944) of the family Cacurgidae is the largest of the Protorthoptera from Mazon Creek (Figures 14A.24, 14A.25). The forewing is elongate oval; SC is very straight, and RS is more extensive than M. Cross veins were numerous but did not form a network.

Eucaenus ovalis Scudder (1885) of the family Eucaenidae is another well-known species at Mazon Creek (Figure 14A.26). The oval forewing has numerous slanted veinlets on the costal area of the wing. The legs were short, with slightly swollen femora, and the cerci were very short, almost vestigial. One pair of the appendages of the mouthparts, the maxillae, were exceptionally long, nearly the length of the antennae. Specimens of this insect have been found at Pit 11, as well as at the strip mine near Coal City, and on the bed of Mazon Creek, near Coal City.

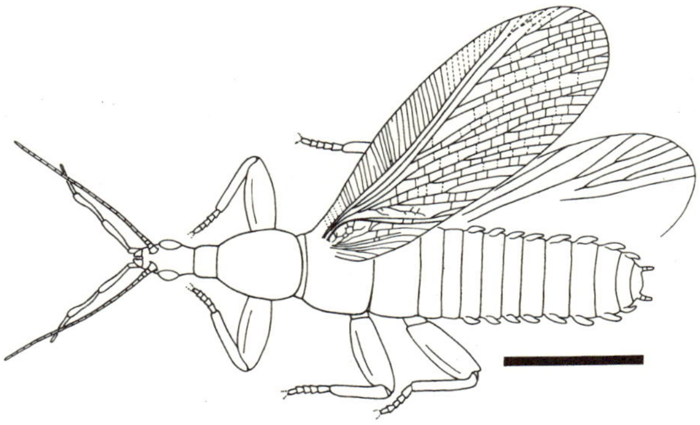

Figure 14A.26. *Eucaenus ovalis,* reconstruction by Carpenter and Richardson (1976); near Coal City and Pit 11. Scale bar = 10 mm.

Figure 14A.29. Cockroach, order Blattaria, family Archimylacridae; Pit 11. Scale bar = 10 mm (estimate).

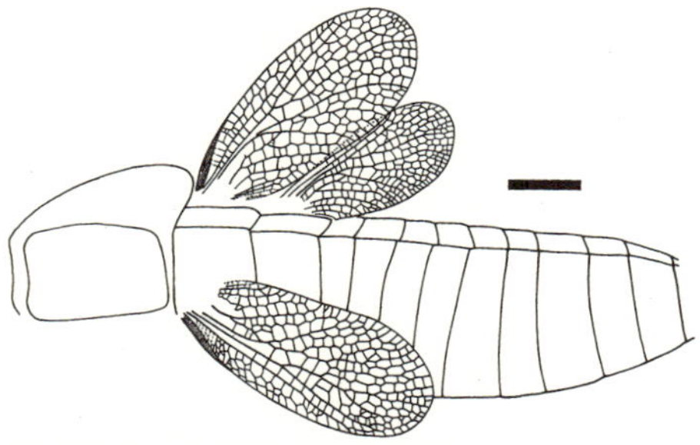

Figure 14A.27. *Herdina mirificus,* drawing of holotype, H412—FMNH PE 24853. Scale bar = 2 mm.

Figure 14A.28. *Herdina mirificus,* wings of holotype. Scale bar = 2 mm.

Herdina mirificus Carpenter and Richardson (1971), of the family Herdinidae, is the most unusual of the Protorthoptera found at Mazon Creek (Figures 14A.27, 14A.28). Both fore and hind wings were very short, not sufficient for flight. The veins, however, were very strong and formed a series of cells over the wings. Several specimens of this insect have been found at Pit 11. At first the specimens were thought to be nymphal forms, but since the wings were heavily sclerotized and the veins strongly formed, it is more likely that this insect was subhapterous, as are many species in the order Orthoptera.

Order Blattaria

Specimens of these insects, the cockroaches, were the most numerous of all of the orders in the Upper Carboniferous, although they are not so abundant at Mazon Creek. Most specimens consist of isolated wings and only rarely is a complete specimen preserved. In such cases, however, the wings are folded back over the abdomen, obscuring the venation and the body structures. Since the venation is very variable within species, the systematics of the fossil roaches is in a confused state. The accompanying photograph is of a complete specimen of the extinct family Archimylacridae, collected at Pit 11 (Figure 14A.29).

References

Burnham, Laurie. 1983. Studies on Upper Carboniferous insects. 1, The Geraridae (Order Protorthoptera). Psyche, 90:1–57.

———. 1986. Revisional studies on Upper Carboniferous insects in the order Protorthoptera. Ph.D. thesis, Cornell University, Ithaca, New York.

Carpenter, F. M. 1944. Carboniferous insects from the vicinity of Mazon Creek, Illinois. Illinois State Mus. Sci. Papers, 3:1–20.

———. 1964. Studies on North American Carbonifer-

ous insects. 3, A spilapterid from the vicinity of Mazon Creek, Illinois (Palaeodictyoptera). Psyche, 71:117–124.
———. 1967. Studies on North American Carboniferous insects. 5, Palaeodictyoptera and Megasecoptera from Illinois and Tennessee, with a discussion of the order Sypharaopteroidea. Psyche, 74:58–64.
———. 1976. Structure and relationships of the Upper Carboniferous insect *Eucaenus ovalis* (Protorthoptera: Eucaenidae). Psyche, 83(3–4):223–242.
———. 1983. Studies on North American Carboniferous insects. 7, The structure and relationships of *Eubleptus danielsi* (Palaeodictyoptera). Psyche, 90(1–2):81–95.
Carpenter, F. M., and E. S. Richardson, Jr. 1968. Megasecopterous nymphs in Pennsylvanian concretions from Illinois. Psyche, 75:295–309.
———. 1971. Additional insects in Pennsylvanian concretions from Illinois. Psyche, 78(4):267–295.
———. 1978. Structure and relationships of the Upper Carboniferous insect *Prochoroptera calopteryx* (Diaphanopterodea, Prochoropteridae). Psyche, 85:219–228.
Handlirsch, Anton. 1906. Revision of American Paleozoic insects. Proc. U.S. Natl. Mus., 29:661–820.
———. 1911. New Paleozoic insects from the vicinity of Mazon Creek, Illinois. Amer. Jour. Sci., 4th ser., 31:297–326, 353–377.
Kukalova-Peck, Jarmila. 1974. Pteralia of the Paleozoic insect orders Palaeodictyoptera, Megasecoptera, and Diaphanopterodea. Psyche, 81:416–430.
Kukalova-Peck, Jarmila, and E. S. Richardson, Jr. 1983. New Homoiopteridae (Insecta: Palaeodictyoptera) with wing articulation from Upper Carboniferous strata of Mazon Creek, Illinois. Canadian Jour. Zool., 61:1670–1687.
Richardson, E. S., Jr. 1956. Pennsylvanian invertebrates of the Mazon Creek area, Illinois. Insects. Fieldiana Geol., 12(2):13–56.
Scudder, S. H. 1885. Palaeodictyoptera: or the affinities and classification of Paleozoic Hexapoda. Mem. Boston Soc. Nat. Hist., 3:319–351.

14B

Mazon Creek Insect Fossils:

The Origin of Insect Wings and Clues about the Origin of Insect Metamorphosis

Jarmila Kukalova-Peck

Among the many significant advances that Mazon Creek fossils have made possible in paleontology are those relating to insects and insect evolution. There is a wealth of well-preserved morphological details found in the wing pads and wing articulations of primitive Carboniferous insect nymphs and adults. These details have offered significant new clues to the original nature of protowings and the beginning of insect flight and metamorphosis.

Origin of Insect Wings

The problem of insect wings is simple to state: Where did they come from and how and why did they develop? It has nevertheless, been the subject of intense debate for a long time. Many contradictory hypotheses have been proposed, but very little evidence has been offered to support them. To develop the reasoning, I have to give background information. Most winged insects have two pairs of wings, and these are articulated above two pairs of fully functional legs (but in flies only one pair of wings is used for flying). Contrary to this, all other animals that acquired powered (flapping) flight (birds, bats, and extinct flying reptiles) had to sacrifice their legs in exchange for wings. Typically, their wings used an already existing, intricate leg articulation and musculature. In birds, the wings are formed by the forearms, which are covered with feathers, while the fingers are reduced to small stubs; in bats, the wings are mostly long, thin, widely spread fingers interspaced with membrane; and in flying reptiles, the wings are mostly enormously elongated fourth fingers that serve as a mast for a membranous sail. Of all living creatures, only insects managed to acquire wings *without* losing their legs. This makes the identity of insect wings very puzzling indeed. Ever since the study of insects (entomology) became a scientific discipline, entomologists have struggled to solve the problem of where insect wings came from.

An insect wing is a very special appendage. The flight agility and endurance of insects is legendary. Some massive beetles and bumblebees have proportionally very small wings and should be unable to fly according to aerodynamic principles. Yet, they are accomplished fliers. Dragonflies and mosquitoes can fly in all directions, even backward and sometimes upside down. Winged insects make up some 75 percent of all animals, and this astonishing evolutionary success is mostly due to the presence of wings. Little wonder that the enigma of the origin of insect wings has been pursued by many over the past 150 years.

In the last century it was widely believed that insect wings evolved from the gills of aquatic insect ancestors (Oken, 1911). This hypothesis has wings developing from a series of already articulated and twistable "winglets," of which only two pairs on the thorax became enlarged. It was revived and developed by Wigglesworth (1973, 1976) and Kukalova-Peck (1978, 1983, 1987). Some people, however, pre-

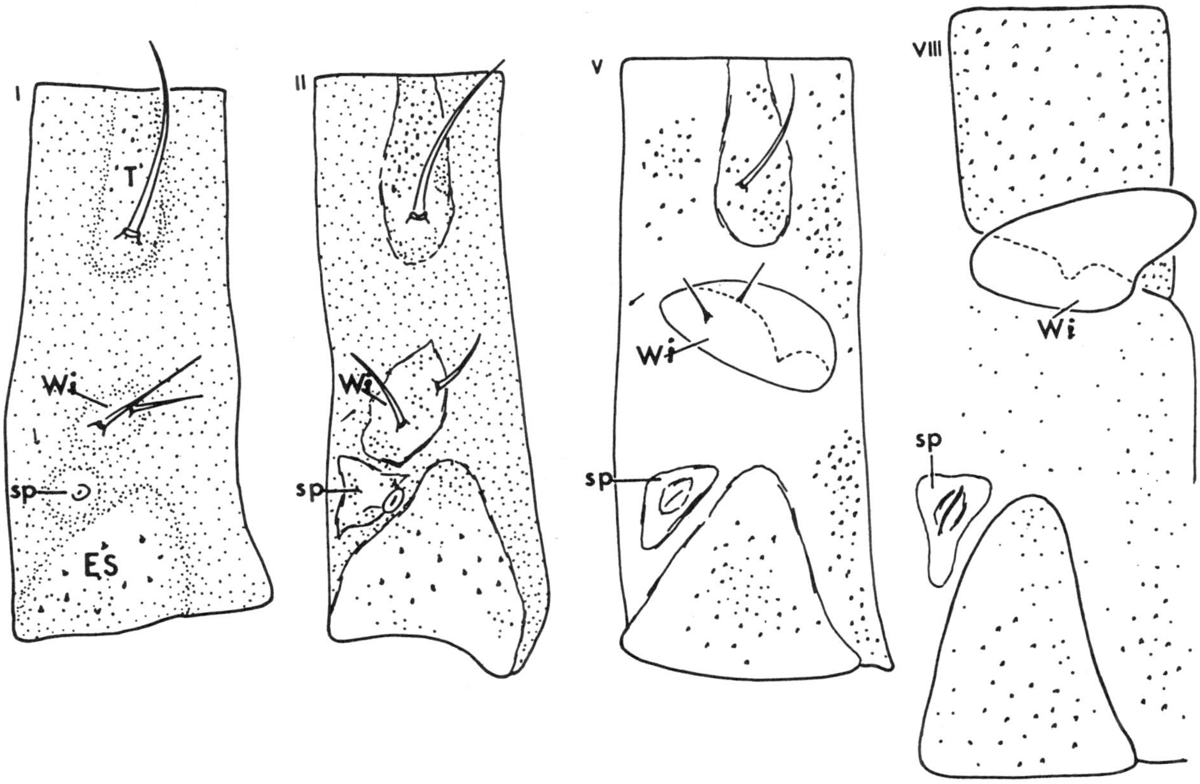

Figure 14B.0 Development of wing (Wi) in modern dragonflies. In the first instar (I) wing is just a small thickening of membrane above the spiracle (sp); wing bud becomes larger in the second instar (II); it evaginates and shifts upward in the fifth instar (V); wing joins the dorsal skeleton (T) and changes into a solid pad that looks like, but is not, a sidelobe on the dorsal plate in the eighth instar (VIII). Living *Libellula quadrimaculata,* adapted from Bocharova-Messner (1959).

ferred a rival hypothesis, formulated in 1875 by Müller and developed by Forbes (1943), Hinton (1963), Flower (1964), and Rasnitsyn (1981). This hypothesis proposed that the wings originated from a series of rigid, lobelike expansions (paranotal lobes) of the dorsal skeleton, which only later became articulated and musculated. Both evolutionary models had been about equally favored until 1916. In this year G. C. Crampton, one of the best known entomologists of his time, severely criticized the "gill" hypothesis and argued in favor of the dorsal skeleton as the source of wings. Since then, the idea that insect wings are expansions of the rigid dorsal skeleton dominated the field; this idea became known as the "paranotal theory." The evolutionary scenario went like this: The ancestors of winged insects lived in tall vegetation. They frequently fell or jumped off the trees. First, the rigid skeletal plates on the insect dorsum expanded to the sides and formed solid lobes; these lobes were useful for stable parachuting and later gliding from trees to the ground. Second, the lobes became separated from the dorsal plate and hinged; they could then be tilted. Third, the lobes became fully articulated and enlarged, and improved gliding and powered flapping flight followed. In this form, the paranotal theory was generally accepted by specialists for the following 65 years.

However, the paranotal theory had several weak points that contradicted some well-known biological facts. For instance, if the wings are expansions of the dorsal skeleton, why would the winged insects move their wings by using the leg muscles, the same as those that move locomotory appendages (= leg branches, exites) on arthropodan legs. Insect wings are completely encircled by the leg muscles and do not receive any muscles from the dorsal skeleton. Also, rigid skeletal plates have never been known to produce appendages involved in locomotion. Why should they do so in insects? And, as shown in experiments with flying models by Wootton and Ellington (1992), side lobes can be used in controlled gliding only if they are appropriately inclined—that is, they have to be articulated *before* they can be twisted. This puts the horse behind the cart in the scenario of the paranotal theory.

However, one potentially important argument seemingly gave good support to the paranotal theory: all nymphs of living insects, such as baby cockroaches, have wing pads completely rigid, immobile, and firmly fused to the dorsal skeleton. It is a well-known fact that animal embryos and immatures often "repeat" a few ancestral evolutionary stages that have been lost in the adults (i.e., young human embryos bear distinct tails). The rigid wing pads of modern nymphs thus seem to be "repeating" the presence of rigid protowings in the ancestors of winged insects, as in the scenario offered by the paranotal theory.

Since the 1960s, accumulated paleontological evidence shows that the "repetition" of rigid ancestral protowings in modern nymphs must be a misinterpretation. Many Carboniferous and Permian nymphs have been discovered that bore fully articulated, agile little wings, as if they were miniature adults. Some Paleozoic nymphs were probably even capable of flying. Clearly, Paleozoic nymphs are showing a real and early stage of wing evolution, while contemporary nymphs must have developed their rigid wing pads after the Paleozoic as a secondary condition, for protection and better hiding ability.

However strong the fossil evidence was, some entomologists were not convinced by the paleontological data because many questions remained unanswered. For instance, if the wings evolved from mobile, articulated protowings, what were they used for when they were still small? In experiments with flying models, small wings are useless for powered flapping flight. Flying following jumping from level ground requires high speeds, far beyond the jumping speed of locusts. If insect wings did not originate from rigid skeletal lobes, where did they come from?

Clearly, to replace the paranotal theory, a new, detailed, comprehensive, and credible evolutionary hypothesis was needed that would put together and integrate all paleontological and other evidence that has accumulated during the past 50 years. At this turning point, some uniquely preserved insect fossils from Mazon Creek nodules provided crucial clues to what kind of appendages might be the source of insect wings. I also found many scattered data in genetics, organ transplants, neurophysiology, developmental morphology, and comparative morphology that supported and strengthened the paleontological evidence and likewise showed that insect wings originated from movable appendages coming from within components of the leg structure. The paranotal theory was shown to be erroneous from many different viewpoints. A new hypothesis integrating all the multifaceted evidence offered by Kukalova-Peck (1983, 1987, 1991) has been gaining ground and replacing the paranotal theory in modern textbooks. The main aspects of the hypothesis are discussed below.

The wings of primitive Mazon Creek Palaeodictyoptera insects are consistently encircled by a horseshoelike band of densely crowded small sclerites. These sclerites bear the attachments of leg musculature and serve as the wing articulation (Figures 14B.1, 14B.2). It is known that the skeleton of arthropods frequently becomes fissured around places to which the muscles are attached. In arthropods, the first one, two, or three leg segments, inserted by leg muscles, often change into a plate and enter the body wall to strengthen the sides of the body. Insects have *two* flattened leg segments engaged in strengthening the body wall: the first plate is above the wing, and the second one is placed immediately under the wing (Figure 14B.3). I think that the first leg segment became fissured into the wing articulation around the attachments of muscles executing wing movements, and the second leg segment now serves as the body wall support on which the wing rests and pivots.

The next question is, What is the origin or identity of the wing itself? Numerous Mazon Creek insects have small, conical appendages articulated between the leg segments. In some insects there are up to five such appendages on the outer side of the leg (Figures 14B.4–14B.6). Similar appendages, usually called exites or ex-

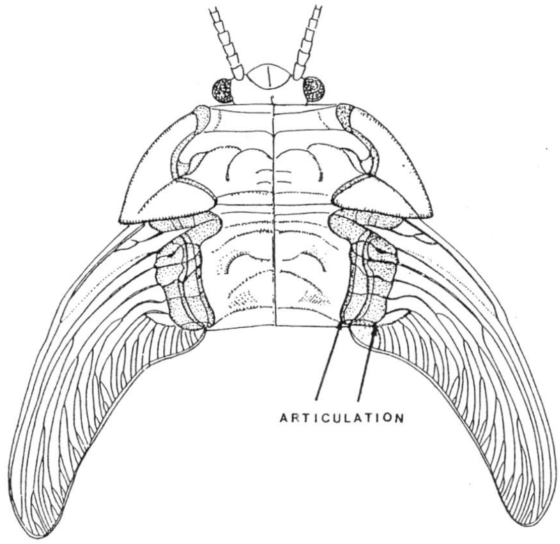

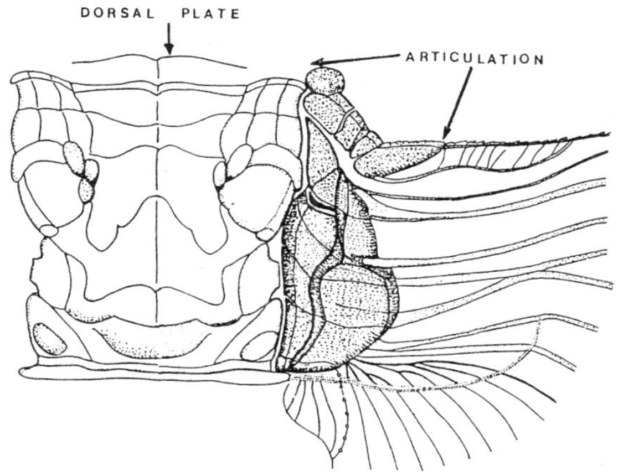

Figure 14B.1 (*Left*). The primitive wing articulation of a flying nymph consisted of densely crowded articular sclerites (*stippled*) arranged in a horseshoelike band around the wings. Paleodictyoptera nymph, dorsal view; *Adolarryia bairdi,* FMNH PE 30272.

Figure 14B.2 (*Above*). The same densely crowded band of sclerites (*stippled*) as it occurs in adult Paleodictyoptera. Dorsal plate, articulation, and right wing base of *Mazonopterum wolfforum,* K. Dagget collection, Chicago.

opodites, are known in other arthropods, such as crustaceans, chelicerates, and trilobites. They are moved by leg muscles arising from two neighboring segments and are used mainly for swimming and breathing. Swimming exites mobilized by leg muscles are especially important in the crustaceans. They are often transformed into powerful oars that operate above the true legs, which may be used for walking. The similarity between crustacean swimming "legs" (= exites) and insect wings is quite remarkable: the wings are articulated between two leg segments, and so are the exites; the wings are twisted and mobilized by leg muscles from two neighboring segments, and so are the exites; the wings are used for powered flapping flight in the air, the exites for powered flapping "flight" (swimming) underwater; the leg shafts in insects are used for walking as they are in crustaceans. The main difference is that the wings are articulated very high on the leg, while crustacean swimming "legs" (= exites) are articulated somewhat lower, usually (but not always) between the third and fourth leg segment. These strong similarities indicate that the insect wings very probably originated from leg exites. The most likely candidates are the exites between the first and second leg segments. The common ancestor of insects and myriapods (millipede, centipede, etc.) is the sister group of Crustacea and, like Crustacea, it must have originated before the Cambrian. At

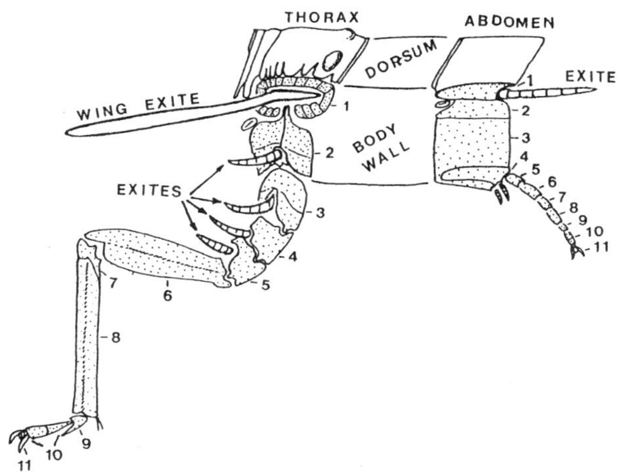

Figure 14B.3. Basic plan of insect legs (*stippled*) as seen in Mazon Creek fossils. Two leg segments in the thorax and four in the abdomen became incorporated into the body wall for strength. Thoracic legs bear five articulated appendages, the exites, and abdominal legs bear only one. The first thoracic exite is transformed into the wing and is surrounded by a horseshoelike band of articulate sclerites, which originated by fissuring of leg segment 1.

that time all arthropods and other animals lived in water. These yet undiscovered aquatic ancestors swam around by using swimming "legs" (exites), like crustaceans. The exites that gave rise to wings developed a sophisticated articulation and musculature over millions of years of adaptation to efficient rowing. Thus,

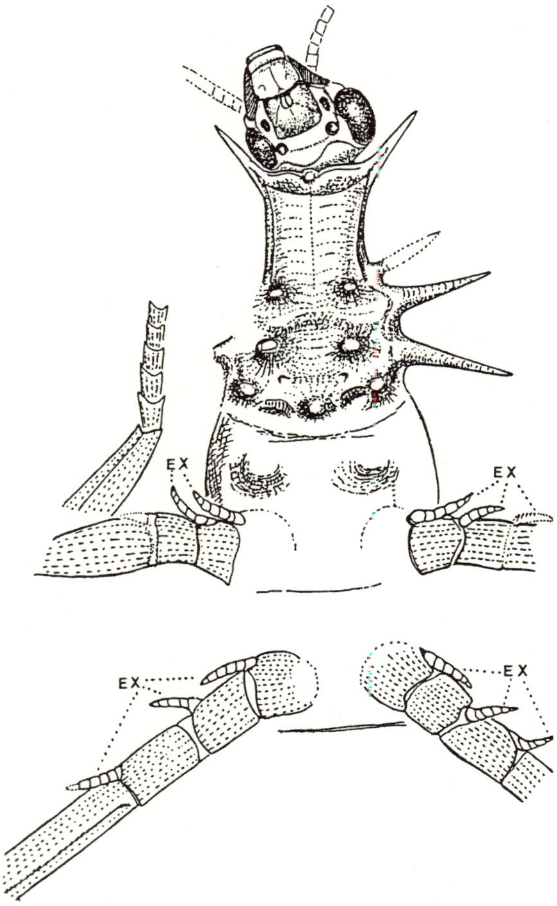

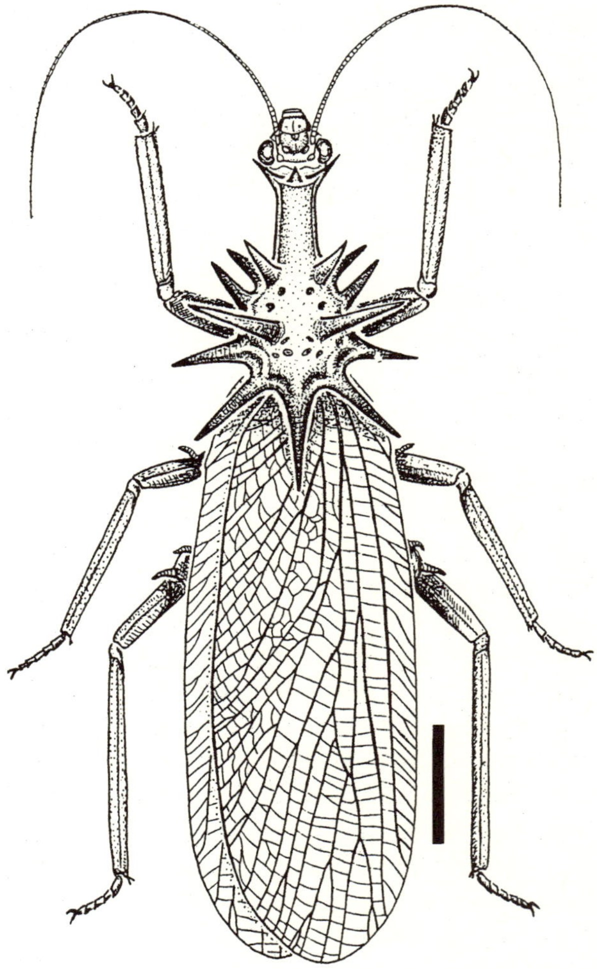

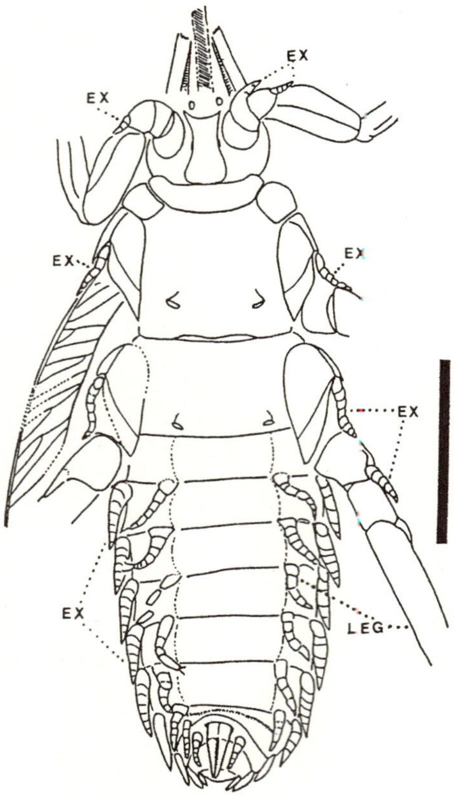

Figure 14B.4 (*Above left*). Mazon Creek fossil with superbly preserved leg exites (EX). The presence of exites is strong proof that insect legs closely resemble the legs of other arthropods. *Gerarus danielsi* (Paraneoptera), FMNH PE 40126 (coll. F. A. Wolff 491).

Figure 14B.5 (*Above*). *Gerarus danielsi,* complete insect. Note the presence of a highly domed bulge on the front of the head, which together with the wing form, venation, and type of genitalia characterizes a distant relative of today's bark lice. The prothorax, adorned with 16 spines, offered protection against predators. Original reconstruction based on FMNH PE 40126, PE 32029, PE 5276, PE 31973, and PE 38729 (genitalia). Scale bar = 10 mm.

Figure 14B.6 (*Left*). Fossil insect showing thorax with legs and exites (EX), and abdomen with reduced legs (LEG) and abdominal exites (EX). Cf. Figure 14B.3. Order unknown, FMNH PE 30370. Scale bar = 10 mm.

the wings of insects, as in other animal groups, very probably evolved from a preexisting, old, articulated, and fully mobile appendage, with its own sophisticated original musculature.

The above model based on paleontology can be verified by genetics, embryology, and neurophysiology. 1) Insect wings are part of the original leg structure; this can be genetically tested. 2) There was originally a pair of protowings on each of the body segments, arranged in an uninterrupted series, which might show up in certain genetic experiments. 3) The protowings were movable; there could be remnants of their original nerve connections also in abdominal segments. 4) The wing articulation is derived from the first leg segment; it should therefore show an association with the leg and the wing in experiments.

The search for supporting data showed that in the past two decades diverse biological disciplines have obtained much experimental and morphological data that strongly support the new evolutionary model. Developmental geneticists discovered that the wing disc in insect embryos, which will form the wing, separates from the leg disc, not from the dorsal skeleton, and that the wing sac in insect larvae is directly connected to the leg sac. Neurophysiologists proved that the neural impulses for flight in locusts are fired not only from thoracic neurons but also from abdominal neurons, where the wings are missing. The implication is that the wings (protowings) of insect ancestors also occurred on the abdomen but became reduced, leaving behind only their neural system. In another timely experiment, geneticists developed a method of shifting legs and leg-derived appendages between head, thorax, and abdomen (e.g., a head antenna can be replaced by a leg because insect antennae originated from legs). In one such treatment, they were able to shift rudiments of wings to the head and to the abdomen. This provides a convincing, definitive proof that wings behave as leg appendages; if they were part of the dorsal skeleton, the shift would not be possible.

Morphologists also found that the movable, abdominal gills of living mayfly nymphs are articulated at the same high level and are operated by the same (coxal) leg muscles as those that mobilize the wings. This indicated that these particular mayfly winglets are the continuation of the wing series on the abdomen. They are used by primitive mayfly nymphs like the original exites, for swimming and breathing.

Developmental morphologists found that the wings of newly hatched dragonfly nymphs start developing quite far from the dorsal skeleton. These little fingerlike wings then travel upward with each successive molt. After the nymphs molt eight times, the wing-fingers reach the dorsal plate and turn into rigid lobes. Only then do they fuse with the dorsal skeleton and start giving the impression that they are lateral outgrowths of the dorsum. Thus, very young contemporary dragonfly nymphs in fact "repeat" past evolution in the same way as the Paleozoic nymphs. Older dragonfly nymphs are now changed to protect their growing wings inside rigid, injury-resistant pads that become fused to the dorsal skeleton. But this relatively recent adaption does not give any support to the paranotal theory.

Experimental morphologists discovered that the wing buds of a caterpillar, if transplanted, develop not only into a wing but also into a complete wing articulation. This gives the ultimate proof that the wing and the entire wing articulation belong together and that none of the wing's articular sclerites have separated from the dorsal skeleton. Many developmental studies now show that insect wings and legs come from a common primordial structure in early insect embryos. This shows that the wings and the legs are closely related in origin. Additional support comes from evidence in developmental and molecular genetics. Averof and Cohen (1997) studied the genes in insect wings and compared them with those in the thoracic limbs of a primitive crustacean, Artemia brine shrimp. The limbs of Artemia, like those of Paleozoic insects, bear short branches called exites. The researchers isolated two genes expressed in the wings in insects, called pdm (nubbin) and apterous, and looked for their expression in Artemia. They found that the equivalents of both pdm and apterous genes occur in the first, dorsal branch of crustacean limbs, which has a respiratory and osmoregulatory function. This was precisely the type of structure that is expected to have given rise to insect wings. This strongly supports the idea that insect wings have evolved from gill-like swimming appendages homologous to the branches (exites) of crustacean limbs. These movable appendages on the limbs of ancestral arthropods are very old, as old as the arthropodan leg. Of course, there have been many genetic developmental changes in legs and exites since their common origin. But the important outcome is that the appendage that gave rise to insect wings existed already in the most primitive and oldest, probably Precambrian, arthropods.

The discoveries listed above are important but were made by various scientists without their seeing how the results propose a new hypothesis for the origin of insect wings (for details, see Kukalova-Peck, 1983, 1987, 1991, 1992). The hypothesis that integrates this scientific evidence was formulated by me over several years, but the main inspiration for it was the fossil insects in the Mazon Creek deposits. The hypothesis is rapidly gaining acceptance among entomologists. However, various aspects should be studied further, updated, and continuously improved, as discussed below.

One of the most difficult problems in the question of the origin of insect flight is explaining the conditions under which the winglets changed into fully adapted (and much larger) wings. If the early protowing acted as an oar, even a short oar would be quite efficient in propelling an animal underwater. Compared to this, a wing must be quite large to be effective in powered flight in the air. Diverse aspects of flight mechanics and aerodynamics in experimental models of ancestral insects have been carefully studied by Wootton and Ellington (1992). Their verdict is clear: Powered flight could not have started without large, aerodynamic wings with specialized wing venation, twistable, articulated wing base, and sophisticated leg and body musculature (dorsal longitudinal, dorsoventral, and oblique dorsal muscles). Such a massive build-up of all-structural adaptation is quite impossible without a long, intense preadaptation. In this long period, the protowings must have been "kept busy" and involved in vigorous locomotion; otherwise, they would probably have vanished the way unused organs often do in animals. If the experimental flight model is true, then the ancestors of winged insects would have remained fully aquatic until the Late Devonian and used their winglets for rowing. However, this may be hard to accept. Insects (and other six-legged animals) have legs with two gripping claws, and they breathe with tracheae, which are generally considered to be the typical adaptation for climbing on vegetation and walking on dry land. This strongly indicates that first Silurian or Ordovician insects could have spent at least some time out of water, on floating vegetation mats, or climbing on low plants rooted in swamps. But, why would they have kept their winglets and the ability to twist them if they were totally useless?

Marden and Kramer (1994) observed modern, short-winged stoneflies with their legs and abdomen rested on the water surface, as they skimmed away from predators by flapping vigorously their short wings. This opened a new, attractive possibility. Ancestral (?Ordovician) insects lived in isolated patches of ancient vegetation and in floating algal mats. Young juveniles lived in water; older juveniles and adults lived on the water surface and skimmed away from predators by flapping vigorously their protowings. Escape and survival were the incentives for the wings to enlarge and to improve flapping efficiency until they could be used in powered, flapping flight.

Whatever the outcome of experiments with flight models may be, the initial stage of the evolution of flight was also the time when the winged insects diversified. Various wing characters strongly suggest that all major evolutionary lineages of winged insects must have already formed at the stage of advanced protowing, *before* the wings were fully adapted for flapping flight (Kukalova-Peck, 1991). Diverse groups of winged insects developed for their own kind of powered flight their own special version of wing venation and wing articulation. They also developed for flight a highly varied mixture of leg, longitudinal, dorsoventral, and oblique dorsal musculature. To sum up all present discoveries, all winged insects share the same original protowing, wing venation, and articulation, but these features later became transformed, profoundly and diversely, as different groups became fully adapted to powered flight. Therefore, the main lineages of winged insects (Paleoptera: paleodictyopteroids, ephemeroids, odonatoids; Neoptera: plecopteroids, orthopteroids, blattoids, hemipteroids, endopterygotes) differ more profoundly in wing venation and articulation than they do in other body characters.

Between 1965 and 1987 I studied, with the kind assistance of custodian Eugene S. Richardson, all fossil insects from the Mazon Creek area deposited in the Field Museum of Natural History, as well as about 60 additional specimens that were lent to me by private collectors. Since that time Charles W. Shabica of Northeastern Illinois University has generously helped me to study all insect specimens available in the Mazon Creek Project collection, and he also encouraged private collectors to send their specimens to me for examination. Some 400 so far undescribed specimens are in my files and are gradually being published. Mazon Creek fauna have yielded some 50 nymphs of Paleodictyopteroidea, 2 nymphs of Protodonata, 3 nymphs of Ephemeroptera, and some 30 nymphs of various Neoptera; most of the nymphs are preserved well enough to bring supporting morphological evidence to the hypothesis that insect wings originated from the arthropodan swimming "legs" (exites).

The specimens that contributed the most to the new hypothesis are the following: *Mazonopterum wolfforum* (Paleodictyoptera), owned by J. Daggett, for the best preserved bandlike articulation in adults (Figure 14B.2); *Adolarryia bairdi* (Paleodictyoptera), found by G. C. Baird, for the bandlike articulation in nymphs (Figure 14B.1); *Gerarus danielsi* (hemipteroids) found by the F. A. Wolffs, for the best preserved series of leg exites (Figures 14B.4, 14B.5); and a hemipteroid insect, found by G. C. Baird, for exites serial to those of the wings, also preserved on the abdomen (Figure 14.B.6)

Mazon Creek Clues to the Origin of Insect Metamorphosis

All immature living insects have immobile wings, which are kept either in hardened cuticular cases on their backs, as in young cockroaches and grasshoppers, or in minute sacs inside the body, as in larval bees, beetles, and butterflies. In order to transform these nonfunctional wing buds into full-sized, functionally flapping wings, the insects must metamorphose. The most sophisticated kind of metamorphosis involves larval and pupal stages, as in the caterpillar, which spins a cocoon, forms a pupa, and emerges as an adult butterfly. This is called complete metamorphosis. Other, more primitive insects, such as mayflies, dragonflies, cockroaches, and grasshoppers, also experience metamorphosis. The nymphs of these insects have their wings well protected in hard wing cases that are fused to their backs. These wings become slightly larger with each molting until the last nymphal stage, which is called the metamorphic instar. This last nymph looks like the others but is in fact very special because it has the whole of the adult wings delicately folded inside the small wing cases, like a parachute in a bag. During the last molt, the emerging adult pulls the wings out and pumps the wing veins full of blood; the wing then expands (like an air mattress), and the insect flies away. This is called incomplete metamorphosis.

The most striking discovery about metamorphosis, as learned from Paleozoic winged insects, was that some primitive groups did not metamorphose. But there was no need because the nymphal wings were not hidden inside rigid cases on the back. They looked and acted like miniature adult wings. Contemporary wingless insects such as silverfish and bristletails also do not metamorphose, but only molt and grow with each molt until they reach adulthood.

Immatures of Paleozoic insects are rarely preserved as fossils, but the Mazon Creek area has delivered the world's largest collection of fossil nymphs with bodies and even the first known true insect larva from the Carboniferous. A fine collection of nymphs of Neoptera, which are insects capable of flexing their wings backward over their abdomens, was assembled by the most successful collectors, especially Tom Testa, the Ramsdells, the Wolffs, Walter Lietz, and Gene Siwik. The *Herdina* group, very distantly related to bark lice, is introduced in Figure 14B.7, and the same series is reconstructed in Figure 14B.8. The specimens were found by

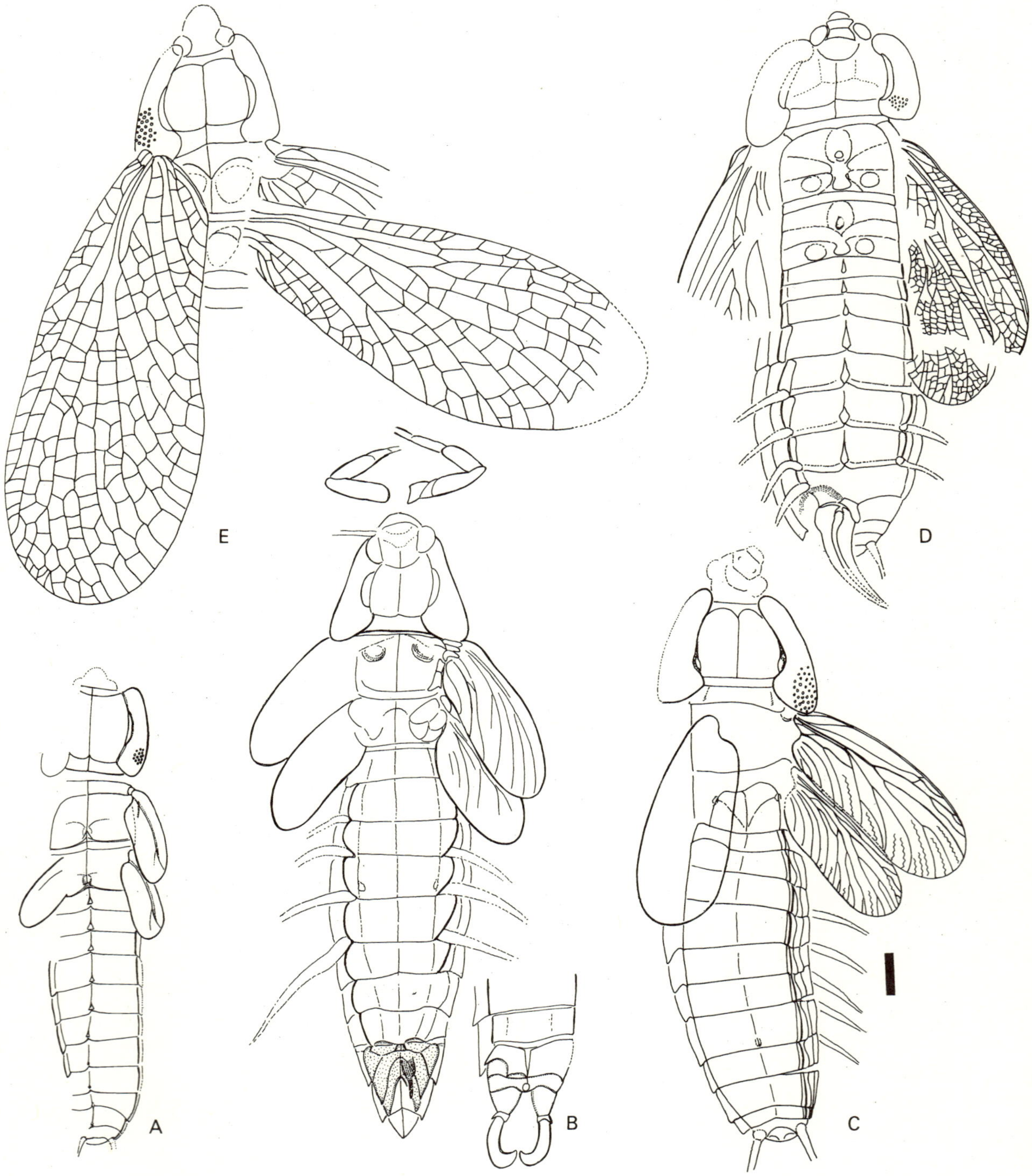

Figure 14B.7. Neoptera without metamorphosis. This is the finest known series of Paleozoic nymphs documenting articulated and independently movable small wings. The youngest nymph (**A**) had only incipient venation, which became fully formed in **C**; **E** was the adult. No metamorphic stage was necessary to activate the wings. *Herdina* group (Paraneoptera): **A** Walter H. Lietz collection, L 1699A; **B** Frank Greene collection, 650; **C** *Herdina mirificus,* H412A—FMNH PE 248853; **D** FMNH PE 21809; **E** NEIU unnumbered specimen. Scale bar = 2 mm.

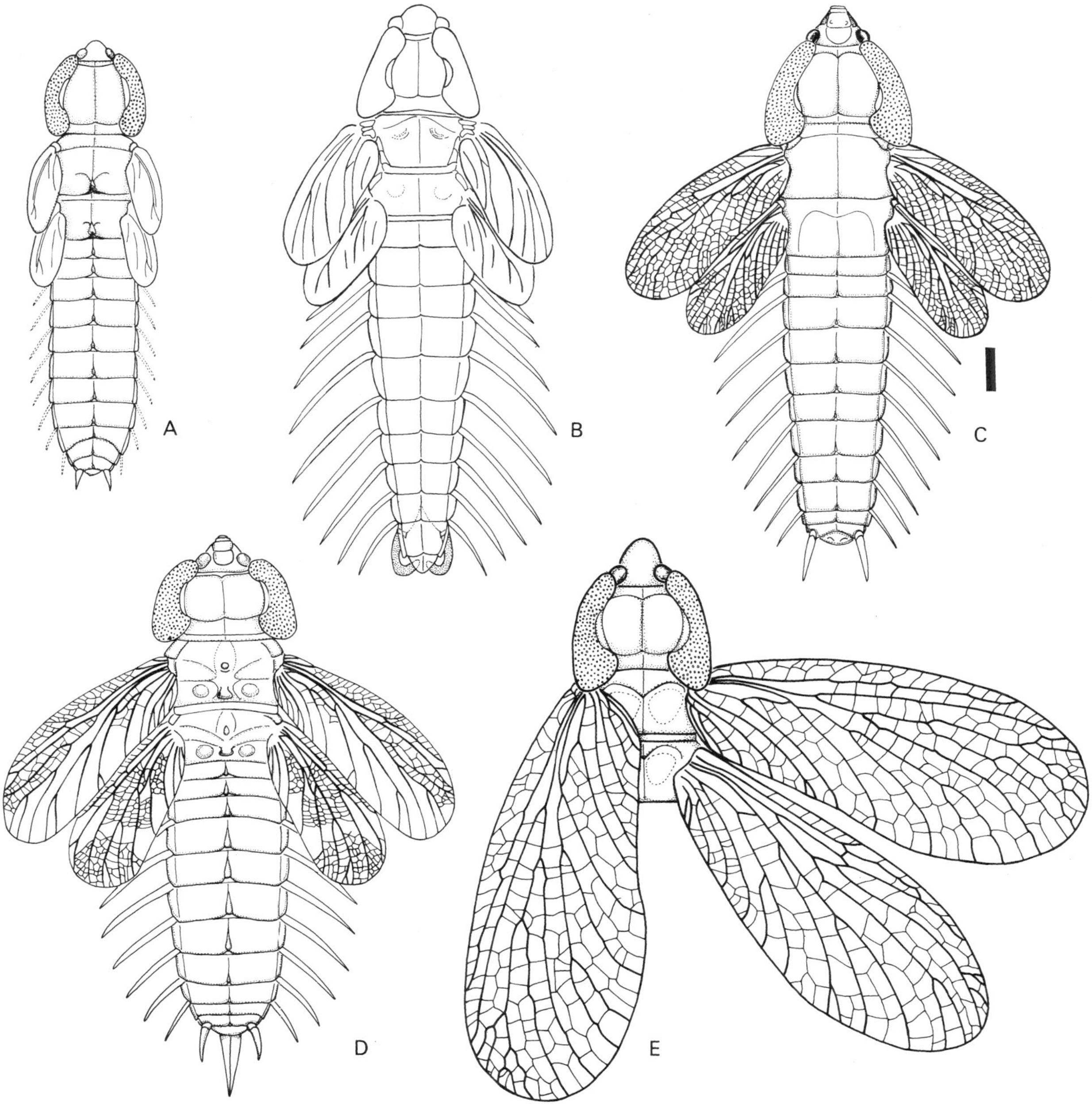

Figure 14B.8. The same series as in Figure 14B.7, reconstructed. Note the presence of abdominal exites. The *Herdina* group is probably very distantly related to bark lice, as shown by the shape of the wings and the vein pattern. Scale bar = 2 mm.

Walter Lietz, Frank Greene, Jerry Herdina, and Karlene and Stephen Ramsdell. Note that the fossil nymphs have their wings oriented at different angles to the body, showing that they were movable and functional and sometimes available for use in short, limited flights. The nymphs may have also hopped and simultaneously used the wing blades as some contemporaneous short-winged scorpionflies do. The *Herdina* group did not need metamorphosis because wings present in the immatures were articulated and fully functional.

The development of wings without metamorphosis in those groups of Paleozoic Paleoptera, insects that could not flex their wings backward over their abdomens, is unique and especially interesting. These insects include the orders Paleodictyoptera, Megasecoptera, and

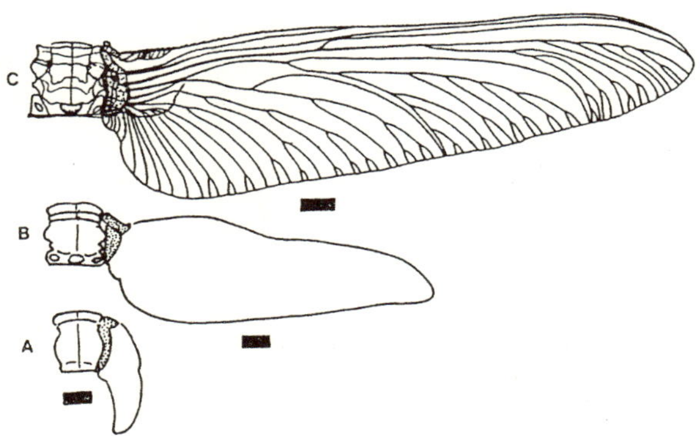

Figure 14B.9. Paleoptera without metamorphosis. A developmental series of wings of Paleodictyoptera, showing **A** strongly curved wings in a young nymph; **B** slightly curved wings in a subadult; and **C** straight wings in an adult. Wing articulation (*stippled*) was present, and the wings were movable. **A** *Adolarryia bairdi,* FMNH PE 30272; **B** *Larryia osterbergi,* Larry Osterberger collection, P9B 674; **C** *Mazonopterum wolfforum,* K. Dagget collection. All scale bars = 10 mm.

Protodonata, which became extinct, and mayflies and dragonflies, which are still living. Their nymphs would potentially have had a problem because the awkwardly protruding wings may have prevented them from hiding in thick vegetation. But there was a solution to this predicament of 300 million years ago. The young nymphs had wings strongly curved backward into a position that brought them closer to the body, allowing the nymphs greater agility in hiding. As the nymphs grew, the curve straightened gradually with each molting. The subadults still had slightly curved wings, but the adults had wings that were perfectly straight. A series showing progressive wing development in Paleodictyoptera, based fully on Mazon Creek material, is shown in Figure 14B.9. The nymphal wing of *Adolarryia bairdi* (A) was collected by G. C. Baird, the subadult *Larryia osterbergi* (B) by Larry Osterberger, and the adult *Mazonopterum wolfforum* (C) by J. Daggett.

Another very interesting developmental series, unique in that it contains young nymphs, has been documented in Megasecoptera (Figure 14B.10). The nymphs are reconstructed from Mazon Creek material. The youngest nymph, *Mischoptera* sp. (A) was found by Helen Piecko, the older nymph, *Mischoptera douglassi* (B), by Lincoln Douglass.

Another fortunate discovery for the understanding of the evolution of insects was that of the first Carboniferous larva by JoLynn Berthe-Traub. This oldest known larva resembles the larvae of living scorpion flies in its overall appearance and is covered with hair. It also shows some features of primitive Paleozoic nymphs, such as a segmented head, relatively large eyes, thoracic legs and abdominal leglets with double claws, and a pair of short annulated cerci at the body end (Figure 14B.11). This specimen documents that insects with pupae, by far the most plentiful and successful group in the animal kingdom, were already present in Carboniferous communities.

Why was the early occurrence of pupating insects, before the discovery of the Berthe-Traub specimen, so uncertain? The reason is that adults of these insects do not differ much from those of insects without pupae. In the present fauna, we know by experience which groups of insects pass through the pupal stage, but in the unknown, fossil fauna we have no such clues. Some scientists suspected that pupating insects must have been around by the Carboniferous, but others vigorously denied this possibility. There was no way to resolve the argument (which has great implications for the evolution of plants and terrestrial communities) until their occurrence was verified with the finding of the Berthe-Traub larva.

The presence or absence of metamorphosis in Carboniferous insects depends on the state of the wings in the immatures. If the little wings are movably articulated as in *Herdina* nymphs (Figures 14B.7, 14B.8), then development was smooth and gradual, the wings grew a little with each molting, and metamorphosis was not necessary. If the wings lost their articulation or became rigidly fused with the dorsal skeleton as in Carboniferous baby cockroaches (Figure 14B.12), metamorphosis must have taken place to restore to the adults the large size, articulation, and mobility of flying wings. Note that some Carboniferous cockroaches still show remnants of the vanishing articulation, especially the Mary Helmus-Smith specimen (Figure 14B.12B); also note that in modern cockroaches the boundary between the wing pads and dorsal skeleton is completely smoothed over. The Berthe-Traub larva also must have undergone metamorphosis because larval wings

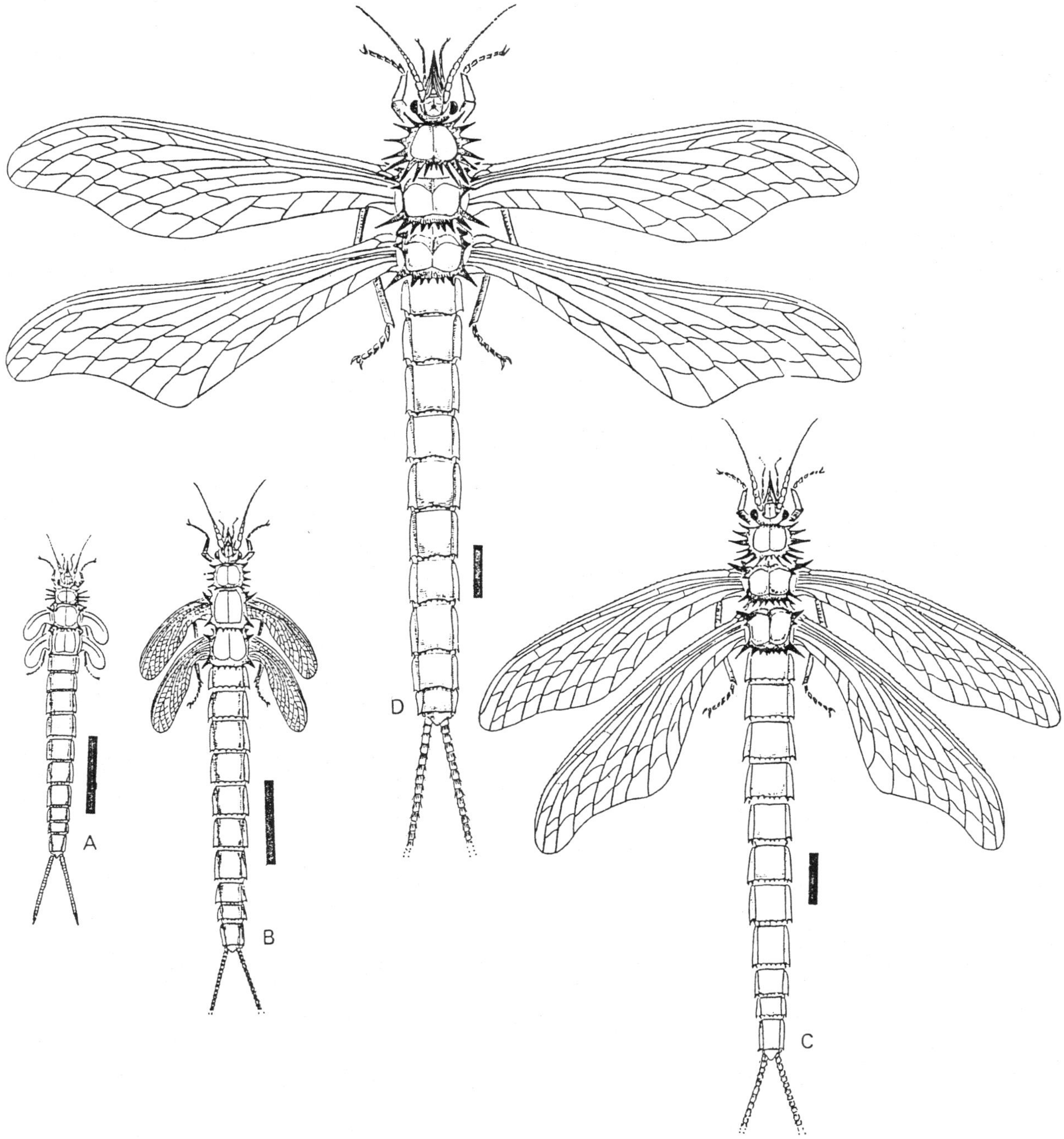

Figure 14B.10. Paleoptera without metamorphosis, original reconstruction. A developmental series of Megasecoptera. The youngest nymph (**A**) has not yet developed wing venation. Narrow wing bases show very well how the wings became straightened gradually with each molt. **A** *Mischoptera* sp., HTP 12327, Carleton University, Ottawa; **B** *Mischoptera douglassi,* David Douglass collection, 36; **C** *Mischoptera nigra* subadult, based on typical subadults from the Lower Permian of Moravia; **D** *Mischoptera nigra* from the Upper Carboniferous of France. All scale bars = 10 mm.

are not visible and must be hidden in minute sacs inside the body.

As shown above, Carboniferous insects had a highly varied development. Some lacked metamorphosis: the primitive Paleoptera (Paleodictyoptera, Megasecoptera, Permothemistida, Diaphanopterodea, Ephemeroptera [mayflies], and Protodonata [ancestral dragonflies]) and the primitive Neoptera (i.e., ancestral true bugs known as "Protorthoptera" and ancestral Plec-

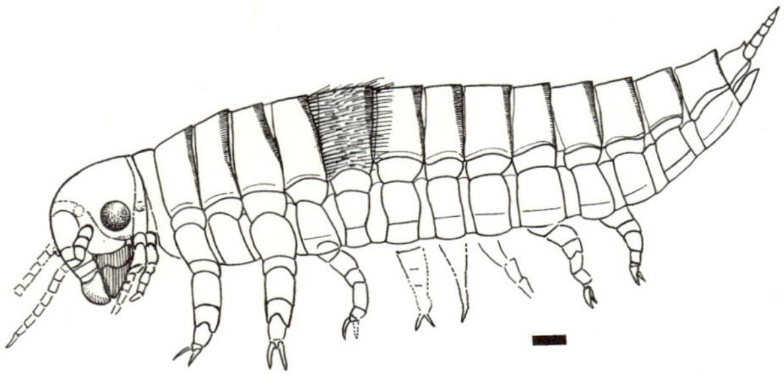

Figure 14B.11. Neoptera with metamorphosis. This oldest known insect larva (*Srokalarva berthei*) documents that pupating insects already existed in the Upper Carboniferous. It resembles in body form the larvae of modern scorpion flies, but the segmented head, abdominal legs, and cerci are very primitive. The wings possibly developed in minute sacs under the skin. NEIU MCP 322 (coll. JoLynn Berthe-Traub and Steven D. Sroka).

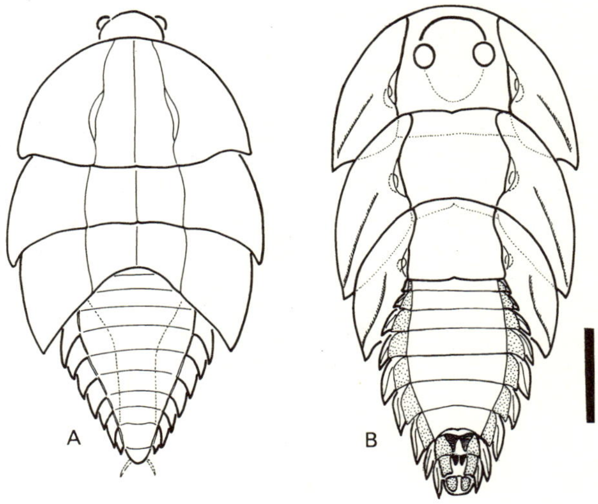

Figure 14B.12. Neoptera with metamorphosis. Nymphs of cockroaches had their wings enclosed in hardened cases and fused with the dorsal skeleton, resembling the thoracic shield side lobe of crustaceans, as do the nymphal wings of all present-day insects. A metamorphic instar was needed to return the wings to a working condition. **A** Blattoidea, nymph, FMNH PE 218827; dorsal view. **B** Nymph, Mary Helmus-Smith collection; ventral view. Scale bar = 10 mm (estimate).

optera [stoneflies]). Other Neoptera such as Blattodea (cockroaches) went through incomplete metamorphosis, and still other Neoptera had wormlike larvae that required a complete metamorphosis with a pupal stage.

In contrast to this, all of today's winged insects undergo metamorphosis. What was the biological need that pushed all insect groups, so varied at the end of the Paleozoic, in the same direction, toward developing metamorphosis? The answer almost certainly lies in the survival of their tender and vulnerable immatures. Originally, Paleozoic insects hatched from their eggs with very small wings, and for much of their development the nymphs could not fly away from predators because their wings were too short. Young and middle-aged nymphs had to hide among vegetation and in crevasses or they would be eaten. To be able to hide their wings had to be held close to the body and also had to be protected against mechanical damage. One solution was bending the wings backward and enclosing them in small hardened cases firmly attached to the dorsal skeleton. Developing them inside the body in little sacs was a still better solution because a larva could burrow and stay permanently hidden from predators while feeding. During many millions of years of natural selection, those winged insects without a good hiding strategy probably lost so many immatures that they became extinct. But those that did develop a successful strategy had to restore the large size and mobility of their wings through metamorphosis. The evidence in the fossil record helped to solve this old question: namely, did winged insects always undergo metamorphosis, and if not, did metamorphosis originate once or several times? The answer is that the capability to metamorphose evolved in living insects independently and several times. However, it was probably instigated by the same phenomenon, the need to protect the vulnerable young from predators.

Thus, the contribution of Mazon Creek fossils to the understanding of insect development and metamorphosis is outstanding. For the first time a developmental series of completely preserved nymphs and adults has been documented in Neoptera without metamorphosis (Figures 14B.7, 14B.8). The morphology of many nymphs, including very young stages, has become known in the Paleoptera (Figure 14B.10). And the oldest true larva of pupating insects was discovered (Figure 14B.11).

References

Averof, M., and Cohen, S. M. 1997. Evolutionary origin of insect wings from ancestral gills. Nature, 385:627–630.

Bocharova-Messner, O. M. 1959. Development of the wing in the early post-embryonic stage in the ontogeny of dragonflies (order Odonata). Trudy Inst. Morf. Zhiv. Im. Severtsova, 27:187–200.

Flower, J. W. 1964. On the origin of flight in insects. Jour. Insect Physiol., 10:81–88.

Forbes, W. T. M. 1943. The origin of wings and venational types in insects. Amer. Midland Naturalist, 29:381–405.

Hinton, H. E. 1963. The origin of flight in insects. Proc. R. Entomol. Soc. London C, 28:24–25.

Kukalova-Peck, J. 1978. Origin and evolution of insect wings and their relation to metamorphosis, as documented by the fossil record. Jour. Morphol., 156:53–126.

———. 1983. Origin of the insect wing and wing articulation from the arthropodan leg. Canadian Jour. Zool., 61:1618–1669.

———. 1987. New Carboniferous Diplura, Monura, and Thysanura, the hexapod ground plan, and the role of thoracic lobes in the origin of wings (Insecta). Canadian Jour. Zool., 65:2327–2345.

———. 1991. Fossil history and the evolution of hexapod structures. *In* T. D. Naumann and CSIRO (eds.), The Insects of Australia, 2nd ed. Melbourne University Press, Melbourne.

———. 1992. The "Uniramia" do not exist: the ground plan of the Pterygota as revealed by Permian Diaphanopterodea from Russia (Insecta:Paleodictyopteroidea). Canadian Jour. Zool., 70:236–255.

Marden, J. H., and Kramer, M. G. 1994. Surface-skimming stoneflies: a possible intermediate stage in insect flight evolution. Science, 226:427–430.

Müller, F. 1875. Beiträge zur Kenntniss der Termiten. Jenaische Zeit. Naturwiss., 9:241–165.

Oken, L. 1811. Lehrbuch der Naturphilosophie. Jena.

Rasnitsyn, A. P. 1981. A modified paranotal theory of insect wing origin. Jour. Morphol., 168:331–338.

Shear, W. A., and Kukalova-Peck, J. 1990. The ecology of Paleozoic terrestrial arthropods: the fossil evidence. Canadian Jour. Zool., 68:1807–1834.

Wigglesworth, V. B. The origin of flight in insects. Proc. R. Entomol. Soc. London C, 28:23–24.

———. 1973. Evolution of insects wings and flight. Nature, 246:127–129.

———. 1976. The evolution of insect flight, pp. 255–269. *In* R. C. Rainey (ed.), Insect Flight (Symp. R. Entomol. Soc., 7). Oxford: Blackwell.

Wootton, R. J., and Ellington, C. P. 1992. Biomechanics and the origin of insect flight, pp. 99–112. *In* J. M. V. Rayner and R. J. Wootton (eds.), Biomechanics and Evolution. Cambridge University Press, Cambridge.

15A

Eurypterida

Roy E. Plotnick

Eurypterids were a group of ancient arthropods closely related to xiphosurans (horseshoe crabs) and arachnids (spiders, scorpions, etc.). They first appear in the Late Ordovician, reach their peak diversity in the Late Silurian, and then gradually decline. Never common in the fossil record, they are especially rare in Upper Paleozoic rocks and apparently became extinct by the Middle Permian.

Ordovician–Devonian age eurypterids are most commonly found in sediments deposited in nearshore, shallow-water marine environments, although some deepwater and nonmarine forms are known. Carboniferous and Permian forms, however, are found exclusively in freshwater rocks. This is the case for Mazon Creek eurypterids, which are components of the nonmarine Braidwood fauna.

According to Richardson and Johnson (1971), less than 3 percent of the Braidwood arthropods are eurypterids. Nevertheless, as a result of years of extensive collecting, Mazon Creek has yielded one of the largest and best studied collections of eurypterids in the world. Bearing a strong resemblance to the closely related modern scorpions, eurypterids are popularly known as sea scorpions. Several Silurian and Devonian members of the group reached lengths of at least 2 m (Clark and Ruedemann, 1912) and were probably the largest arthropods of all time. Most eurypterids, including those of Mazon Creek, were quite a bit smaller.

The eurypterid body is distinctly divided into an anterior head region, or prosoma, and a posterior abdomen, or opisthosoma (Figure 15A.1). The dorsal surface of the prosoma is covered by the prosomal shield, or carapace, on which are located a pair of large lateral compound eyes and a medial pair of small simple eyes, or ocelli (note that the carapace is *not* the same as the similarly named feature in crustaceans). Under the carapace are six pairs of appendages. The first pair of appendages, called the chelicerae, are small pincers. This is in contrast to the antennal appendages of crustaceans or insects, which occupy the same relative position in the body. The presence of chelicerae is a key feature relating eurypterids to xiphosurans and arachnids. The next four or five pairs of appendages are usually considered to be walking legs. The sixth pair of prosomal appendages can form wide, flat swimming paddles that resemble those of modern-day blue crabs. A large medial plate, the metastoma, covers the bases of the appendages.

The opisthosoma, like that of scorpions, is made up of 12 segments plus the terminal tail spine or telson. In many eurypterids there is a noticeable difference in width between the first seven segments, the preabdomen, and the last five segments, or postabdomen. An alternative division of the eurypterid opisthosoma is between the first six segments, termed the mesosoma, and the last six segments, or metasoma. The mesosoma possesses the gills and reproductive organs of the eurypterids. This fact was

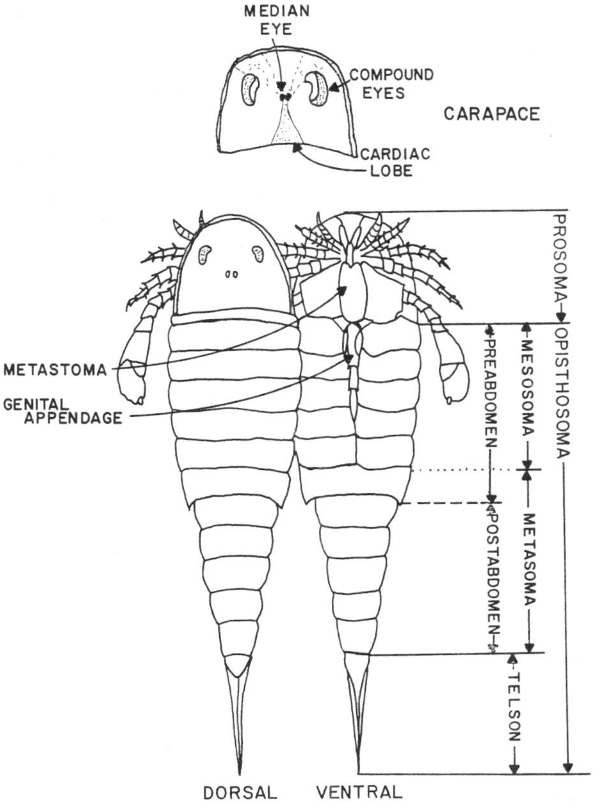

Figure 15A.1. Generalized eurypterid, based primarily on *Parahughmilleria* (Størmer, 1973).

Figure 15A.2. *Adelophthalmus mazonensis* (Meek and Worthen, 1868), FMNH PF 32378. Scale bar = 10 mm.

partly determined from the work of Wills (1964), who studied *Adelophthalmus mazonensis* from Mazon Creek.

Three taxa of eurypterids have been described from Mazon Creek and were extensively reviewed by Erik N. Kjellesvig-Waering (1948, 1963). Of these, only *A. mazonensis* is well known; it is represented by numerous specimens. There are poorly preserved fragments possibly belonging to the genus *Mycterops*. The third described taxon, *Mazonipterus cyclophthalmus*, was believed by E. S. Richardson (personal communication) to probably be a part of a tree fern of the genus *Caulopteris*. I concur with this identification.

Adelophthalmus mazonensis
(Meek and Worthen)
Figure 15A.2

Adelophthalmus is the most common and widespread Late Paleozoic eurypterid. Its stratigraphic range is from the Upper Devonian to the Lower Permian. This genus has been found throughout the Northern Hemisphere, including Great Britain, France, Spain, Portugal, Belgium, and China. In the United States, specimens are known from Illinois, Pennsylvania, Nebraska, Oklahoma, Kansas, and New Mexico.

Adelophthalmus was exclusively nonmarine and is usually found associated with freshwater and terrestrial plants and animals. It was probably an active, agile swimmer. The food preference is unknown, though modern relatives (horseshoe crabs, arachnids) are predators and scavengers.

The closest well-known relative of *Adelophthalmus* is probably the Silurian and Devonian genus *Parahughmilleria* (Størmer, 1973).

The prosoma of *A. mazonensis* is subrectangular, with relatively small, kidney-shaped compound eyes and strongly rounded anterolateral margins. The sixth pair of appendages are formed into wide and flat swimming paddles. A marked division exists between the preabdomen and postabdomen. The posterior lateral corners of the preabdominal segments are extended to form short spines. Most of the animal's surface is covered with a distinctive scale-like ornamentation.

Wills (1964) exhaustively detailed the anatomy of *Adelophthalmus.* Growth and variation in this genus were examined by Van Oyen (1956) and by Kues and Kietzke (1981).

Synonymies (for *Adelophthalmus*). *Lepidoderma, Anthraconectes, Eurypterus* (some species).

Holotype. Eurypterus (Anthraconectes) mazonensis, UI X345 (part), AMNH 8538 (counterpart).

References

Clarke, J. M., and R. Ruedemann. 1912. The Eurypterida of New York. New York State Mus. Mem., 14, 628 pp.

Kjellesvig-Waering, E. N. 1948. The Mazon Creek eurypterid: a revision of the genus *Lepidoderma.* Illinois State Mus. Sci. Paper, 3:1–46.

———. 1963. Pennsylvanian invertebrates of the Mazon Creek area, Illinois: Eurypterida. Fieldiana Geol., 12:85–106.

Kues, B. S., and K. K. Kietzke. 1981. A large assemblage of a new eurypterid from the Red Tanks Member, Madera Formation (Late Pennsylvanian–Early Permian) of New Mexico. Jour. Paleo., 55:706–729.

Meek, F. B., and A. H. Worthen. 1868. Preliminary noticed of a scorpion, a *Eurypterus*? and other fossils from the Coal Measures of Illinois. Amer. Jour. Sci., 2nd ser., 46:19–28.

Richardson, E. S., Jr., and R. G. Johnson. 1971. The Mazon Creek faunas, pp. 1222–1235. Proc. North American Paleo. Conv., 5–7 September 1969, vol. 1. Allen, Lawrence, Kansas.

Størmer, L. 1973. Arthropods from the Lower Devonian (Lower Emsian) of Alken an der Mosel, Germany. Part 3, Eurypterids, Hughmilleriidae. Senckenbergiana Lethaea, 54(2/4):119–209.

Van Oyen, F. H. 1956. Contribution à la connaissance de genre *Adelophthalmus* Jordan et Meyer, 1854. Meded. Geol. Stict., ser. C-IV-3, 7:1–98.

Wills, L. J. 1964. The ventral anatomy of the Upper Carboniferous eurypterid *Anthraconectes* Meek and Worthen. Palaeontology, 7:474–507.

15B

Euthycarcinoids and Thylacocephalans

Frederick R. Schram and W. D. Ian Rolfe

The euthycarcinoids are a problematic group of arthropods currently assigned to the uniramians as marine cousins of the myriapods and insects. The group is distinguished by the large numbers of diplo- and triplosomites in the preabdomen. The head is apparently primitive, composed of an anterior procephalon bearing antennae and lobed eye and a wide posterior gnathocephalon. The two cephalic units are noteworthy in that they are articulated in relation to each other. The exact phyletic affinity of the group is yet to be adequately evaluated. The Euthycarcinoidea appear to have a geologic range from the Silurian to the Middle Triassic.

Class Euthycarcinoidea
Pieckoxerxes pieckoae
(Schram and Rolfe, 1982)
Figures 15B.1, 15B.2

Preabdomen with at least 14 tergites preceded by a short monosomite behind the cephalon, up to 28 sternites, postabdomen probably of about six segments and a short styliform telson; preabdomen and gnathocephalon with well-developed, paratergal lobes rounded anteriorly and acutely developed posteriorly.

This species was originally thought to be a juvenile *Kottixerxes gloriosus.* Study of Montceau-les-Mines fauna in the Upper Carboniferous of France revealed a similar form that allowed these euthycarcinoids with long preabdomens to be recognized as a separate family. They appear to be more primitive than the euthycarcinids, which have shorter preabdomens. Originally named *Sottyxerxes pieckoae* by Schram and Rolfe (1982).

Smithixerxes juliarum Schram and Rolfe, 1982
Figures 15B.3, 15B.4

Preabdomen with four multisomite tergites, each with longitudinal crests; 10 or 11 ventral sternites, first postabdominal segment short,

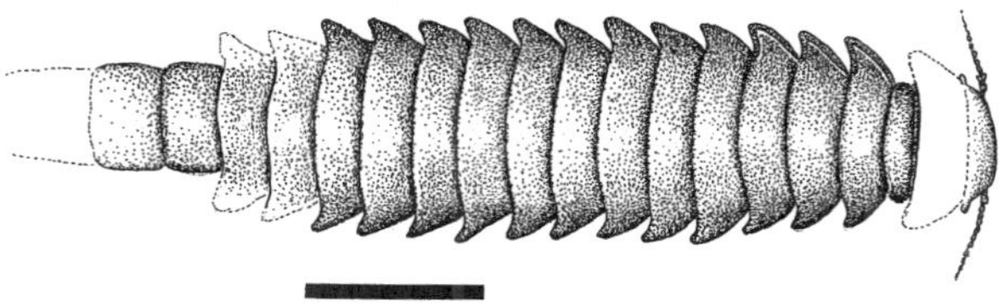

Figure 15B.1 (*Left*). *Pieckoxerxes pieckoae* (Schram and Rolfe, 1982), FMNH PE 31238. Scale bar = 5 mm.

Figure 15B.2 (*Above*). *Pieckoxerxes pieckoae* (Schram and Rolfe, 1982), modified from Schram and Rolfe (1982). Scale bar = 5 mm.

distal segments long and narrow and each with posteriorly directed paratergal prolongations; telson approximately equal to postabdomen in length.

The main difference between this species and *Kottixerxes* is the fewer number of preabdominal tergites, which would seem to imply that this species has larger numbers of triplosomites rather than diplosomites that have been noted in other Euthycarcinidae. Of all the Mazon Creek euthycarcinoids, *Smithixerxes* is more common than *Pieckoxerxes pieckoae* but less common than *Kottixerxes gloriosus.*

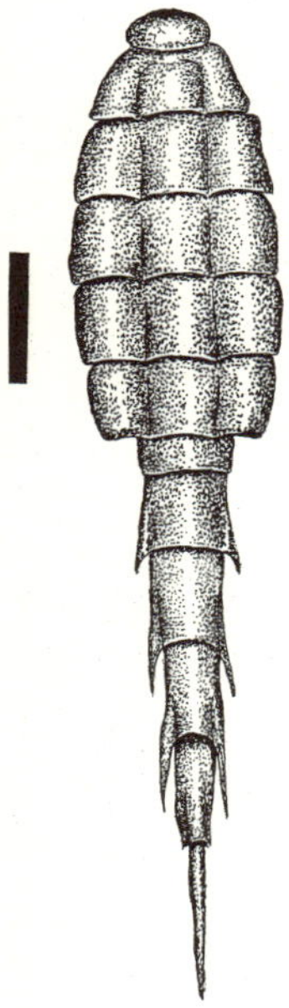

Figure 15B.3. *Smithixerxes juliarum* Schram and Rolfe, 1982, holotype, FMNH PE 23420. Scale bar = 5 mm.

Figure 15B.4. *Smithixerxes juliarum* Schram and Rolfe, 1982, modified from Schram and Rolfe (1982). Scale bar = 5 mm.

Kottixerxes gloriosus Schram, 1971
Figures 15B.5, 15B.6

Preabdomen with 11 ventral sternites, five tergites; tergite surfaces smooth and undecorated, posterior margins slightly sinuous, all tergal plates of approximately the same width; length of telson less than length of postabdomen; uniramous preabdominal limbs flagellate and equipped with large, scalelike marginal setae.

Class Thylacocephala

The Thylacocephalans (flea shrimp) are another strange group in the Arthropoda (subphylum uncertain), with a range that extends from the Cambrian to the Cretaceous. Long known from individual shells or carapaces, only recently have fossils been found that supply information on other aspects of the body. The best preserved material is from the Jurassic of France and includes a strange thylacocephalan that appears to have been a rather formidable predator. "Flea shrimp" are common in the Mazon Creek Essex fauna, though they may not in fact be shrimp or crustaceans.

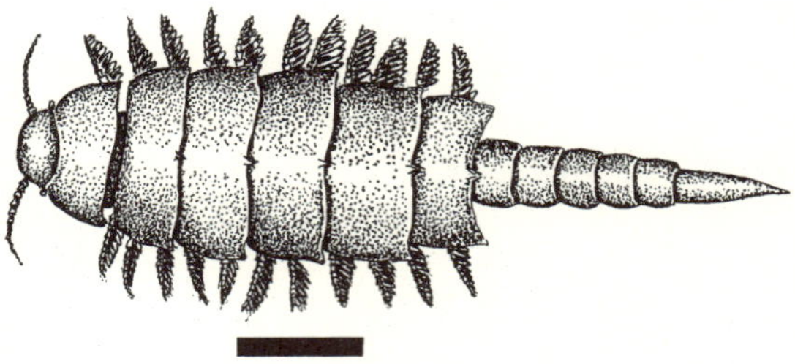

Figure 15B.5 (*Left*). *Kottixerxes gloriosus* Schram, 1971, Tully collection. Scale bar = 5 mm.

Figure 15B.6 (*Above*). *Kottixerxes gloriosus* Schram, 1971, modified from Schram and Rolfe (1982). Scale bar = 5 mm.

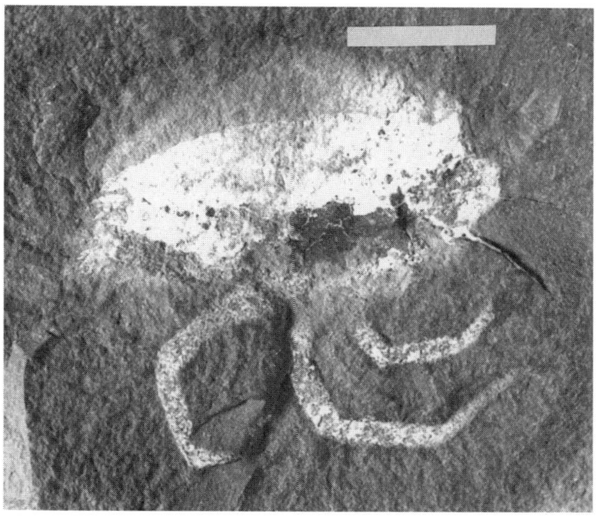

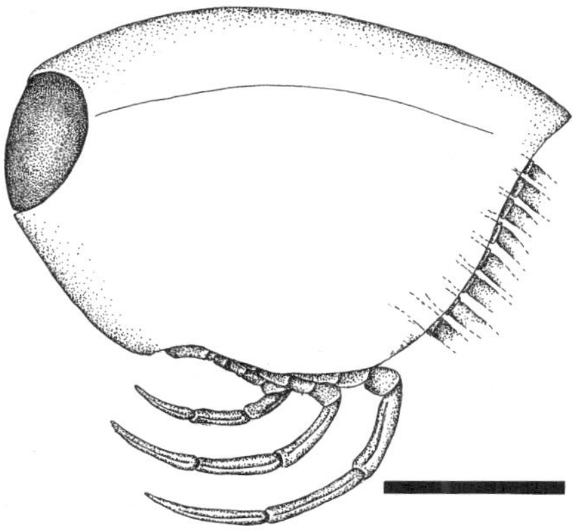

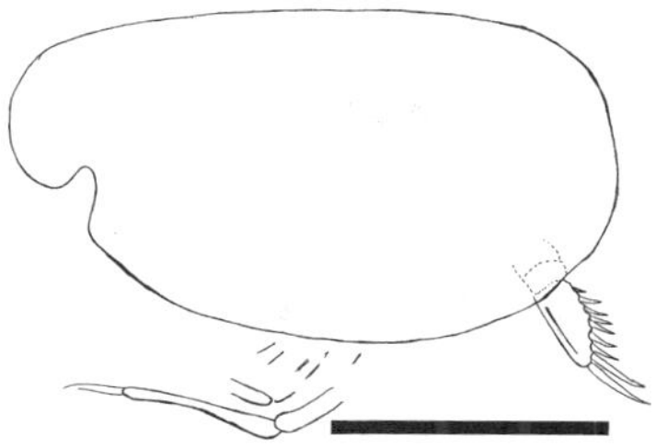

Figure 15B.9. *Concavicaris remipes* Schram, 1990, reconstruction. Scale bar = 5 mm.

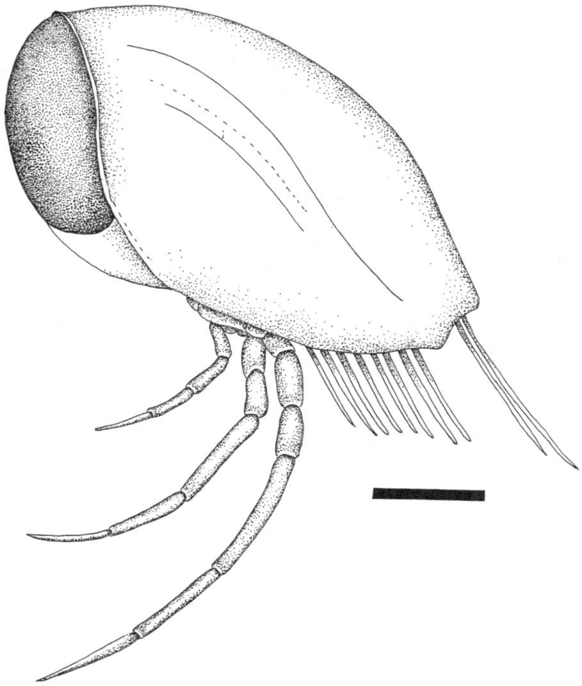

Figure 15B.7. *Concavicaris georgeorum* Schram, 1990, NEIU MCP 559. Scale bar = 5 mm.

Figure 15B.8. *Concavicaris georgeorum* Schram, 1990, modified from Schram (1990). Scale bar = 5 mm.

Figure 15B.10. *Convexicaris mazonensis* Schram, 1990, NEIU MCP 594. Scale bar = 10 mm.

Figure 15B.11. *Convexicaris mazonensis* Schram, 1990, modified from Schram (1990). Scale bar = 5 mm.

Concavicaris georgeorum Schram, 1990
Figures 15B.7, 15B.8

Body almost completely enclosed within a carapace that is notched anteriorly to accommodate large, lobate, compound eyes, ventral edge of carapace with slight encavation, carapace oval in outline and about as long as it is deep; anterior limbs located posterior to the eyes (mouthparts) as three sets of subchelate raptorial appendages; posterior part of body apparently undifferentiated, each segment bearing a set of rather delicate limbs.

Concavicaris remipes Schram, 1990
Figure 15B.9

A small concavicarid with an oval carapace, a large, deep, ventrally deflected rostrum, and a small optic notch. The body terminates in a large spinose paddle, the function of which is unknown. The paddle is similar to structures in ostracods, cladocerans, and conchostracans; used for locomotion. Parts of a raptorial limb are seen in some specimens.

Convexicaris mazonensis Schram, 1990
Figures 15B.10, 15B.11

Relatively large carapace with sinuous outline. A pair of ridges run parallel to the sinuous axis of the carapace. The compound eyes are large compared to the size of the body and protrude anteriorly from the carapace on a cephalic sac. The raptorial limbs are large.

References

Schram, Frederick R. 1971. A Strange Arthropod from the Mazon Creek of Illinois and the Trans Permo-Triassic Merostomoidea (Trilobitoidea). Fieldiana Geol., 20(6):85–102.

———. 1990, On Mazon Creek Thylacocephala. Proc. San Diego Soc. Nat. Hist., 3:1–16.

Schram, Frederick R., and W. D. Ian Rolfe. 1982. New euthycarcinoid arthropods from the Upper Pennsylvanian of France and Illinois. Jour. Paleo., 56:1434–1450.

15C

Onychophora

Andrew A. Hay and Samuel Kruty

Onychophora is a small taxon consisting of slightly more than 70 species contained within two families: Peripatidae and Peripatopsidae. The worm or caterpillar-like body is covered by a very thin layer of flexible, permeable chitin. The exterior surface of the body is ornamented with numerous tubercles and ring markings or bands (Figure 15C.5). Locomotion is achieved by the paired, conical, stubby walking legs. The onychophoran blend of arthropod and annelid characteristics has led some authors to consider the Onychophora to be a "missing link" between the Annelida and the Arthropoda. This hypothesis is put in question by the cladistic analyses of Schram (1989) and Ballard et al. (1992). These analyses imply that the onychophorans could be included within the Arthropoda.

The anterior end of extant adult onychophorans bears a single pair of appendages; a pair of fleshy antennae, mobile but not retractile, each with an eye at the base; two embryonic pairs of appendages become a pair of papillae alongside the mouth that contain a gland capable of producing a sticky substance utilized to capture prey (Ghiselin, 1985), and a single pair of jaws that cut through the body armor of captured prey. Their prowess in the "art" of expectoration is renowned in the animal kingdom: "They are probably the champion spitters of the animal kingdom, for they can project their secretions distances of up to 0.5 m" (Meglitsch and Schram, 1991, pp. 353–357).

The wormlike body is studded with tiny wartlike papillae covered with minute scales and sensory bristles. The papillae or tubercles are usually arranged in close-set annular rings around the body and appendages; arrangement of the rings bears no relationship to true body segmentation. The only external evidence of segmentation is the stumpy, ringed, paired limbs or legs (lobopods), which appear as saclike protrusions of the body. There are from 14 to 43 of these simple unjointed walking legs (Meglitsch and Schram, 1991), each ending in a double hooked claw, or onychium. Number of legs varies in the different species of onychophorans, both fossil and Recent (Moore, 1959). Since the legs are unjointed, locomotion is achieved by elongation of the body segments, which lifts the legs, moving them forward; contraction then depresses the leg, which exerts a force against the substratum (hydrostatic locomotion). This wavelike extension and contraction has been compared to the locomotion attained by some polychaetes (Brusca and Brusca, 1990). Small nephridia (segmental excretory organs) are located at the base of each walking leg; these, combined with the posteriorly located anus, serve to remove metabolic wastes. An increase in size is gained by a periodic molting (ecdysis) of the cuticle, a feature shared with the arthropods. An onychophoran's unhardened, deformable body allows it to squeeze through narrow openings and spaces, enabling it to escape predators. Respiration is achieved

Figure 15C.1. *Aysheaia,* modified from various sources. Scale bar = 10 mm.

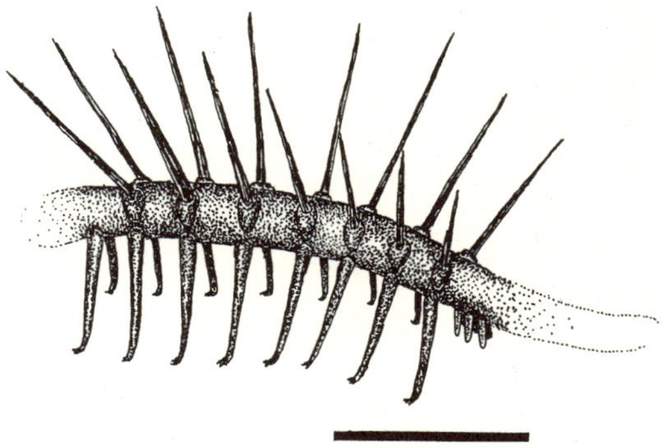

Figure 15C.2. *Hallucigenia,* modified from Ramskold and Hou (1991). Scale bar = 10 mm.

by numerous small body openings or spiracles that lead into tufts of fine tracheae. The spiracles of onychophorans cannot be closed, and this allows water to freely evaporate from the body. Water loss through the spiracles and the cuticle of the body surface restricts onychophorans to nocturnal, warm, humid environments (Meglitsch and Schram, 1991).

Extant onychophorans have a hemocoelic circulating system that maintains a hydrostatic or fluid skeleton. The body plan consists of partitioned, interconnected cavities surrounding a simple gut, internal organs, and muscles. Heart pulsations coupled with body movements circulate the body fluids between the cavities (sinuses) and subcutaneous channels.

Recent Onychophora are divided into two families, the Peripatidae and the Peripatopsidae. The peripatids inhabit tropical America, tropical western Africa, and southeast Asia. The peripatopsids are found in Chile, South Africa, Australia, Tasmania, and New Zealand (Kraus, 1974).

Fossil Record

The fossil record of onychophorans until recently was thought to be quite sparse, consisting of the Middle Cambrian Burgess Shale *Aysheaia pedunculata* Walcott (1911; also see Whittington, 1978; Figure 15C.1) and the problematic Early Cambrian *Xenusion auerswaldae* Pompeckj (1927) from the Baltic region. New discoveries of onychophorans and onychophoran-like animals have been documented by Thompson and Jones (1980) from Illinois, Rolfe et al. (1982) from France; Robison (1985) from Utah, Dzik and Krumbiegel (1989) from the Baltic area, and Ramskold and Hou (1991) from China. Reinterpretations of *Ilyodes* Scudder (1890) by Rolfe et al. (1982) and *Hallucigenia* Conway Morris (1977) by Ramskold and Hou (1991) have added specimens from Illinois and British Columbia to the list of fossil onychophorids. These recent discoveries have revived interest in the classification of the Onychophora and their relationship to the annelids and the arthropods (e.g., Dzik and Krumbiegel, 1989; Schram, 1989; Ramskold and Hou, 1991; Ballard et al., 1992).

Interesting articles by Lewin (1992) and Gould (1992) discuss the importance of solving the dilemma of the bizarre onychophoran-like Burgess Shale fossil *Hallucigenia* (Figure 15C.2), and see Collins (1989). The authors question whether evolution has expanded or decreased the diversity of life: Was there a greater diversity of body plans during the "Cambrian explosion," as exemplified by *Hallucigenia,* than there are today?

Ecdysis, an arthropod-like feature of Onychophora, is a great boon to the paleontologist; without molts the meager fossil record of onychophorans might well be nonexistent. Just as the most abundant remains of trilobites consist of their fossilized exuviae (Levi-Setti, 1975, p. 8), so too, on a greatly reduced scale, for the fossil Onychophora. Dzik and Krumbiegel (1989) infer that the specimens of the fossil onychophorans *Aysheaia pedunculata* Walcott (1911), *Xenusion auerswaldae* Pompeckj (1927), and the latest specimen of *Xenusion* (Krumbiegel et al., 1980) are fossilized exuviae or body integuments. Similarly, Robison noted that the Utah onychophoran *Aysheaia prolata* "is represented only by cuticle, cast off during molting" (Robison, 1985).

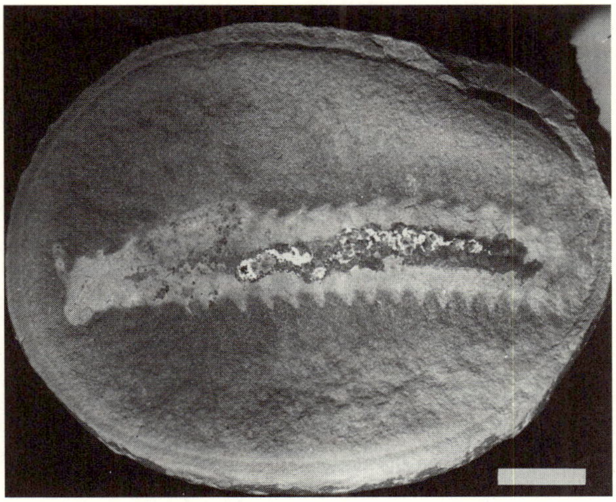

Figure 15C.3. *Ilyodes (Helenodora) inopinata,* NEIU MCP 184. Scale bar = 10 mm.

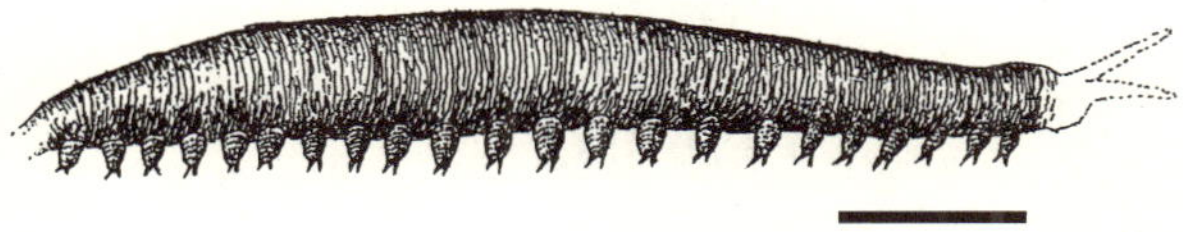

Figure 15C.4. *Ilyodes,* modified from Thompson and Jones (1980). Scale bar = 10 mm.

Figure 15C.5. *Peripatoides,* an extant onychophoran, modified from Snodgrass (1958). Individuals range up to 15 cm in length.

Mazon Creek

There has been some confusion in the Mazon Creek literature regarding multilegged (serially repeated) wormlike fossils. The affinities of two genera, *Ilyodes* Scudder (1890) and *Helenodora* Thompson and Jones (1980), have been reexamined in recent papers. Specimens discussed in the original descriptions were conceded to have been poorly preserved. Specimens of the same or a similar animal discovered at Montceau-les-Mines in France prompted a review of those earlier articles (Rolfe et al., 1982).

Scudder described *Ilyodes* (Figure 15C.3) as being chilopodiform, composed of a large number of similar segments each bearing a single pair of legs. Jointing of the short legs is uncertain. Each segment is composed of a series of about eight transverse ridges having a beaded appearance; front margins of the segments bear tubercular bosses. Scudder assigned *Ilyodes* to the Myriapoda, a classification then frequently utilized for multilegged, wormlike, invertebrates.

In the original description of Thompson and Jones, *Helenodora inopinata* (Figure 15C.4) was described as being a wormlike lobopod fossil with 21 or more pairs of short, tapered legs located ventrolaterally. The legs bear hooklike claws. The body is annulated with about nine annulations per ring. Small papillae are seen on the annulations. Specimens occasionally display a dark brown spot that may be a jaw. Assignment to the phylum Onychophora was considered tentative until details of the head (not preserved in the two originally described specimens) could be examined on fossils yet to be discovered.

Rolfe et al. (1982) in their description of the fossils found at a Carboniferous (Stephanian B) locality in Central France (Mazon Creek is Westphalian D) stated, "The Mazon Creek onychophoran *Ilyodes* Scudder, 1890 (=*Helenodora* Thompson and Jones, 1980) is matched by a virtually identical onychophoran at Montceau." Scudder, unwittingly, had described the first fossil onychophoran; it took the discovery of other fossil onychophorans and 92 years to establish the fact. The name *Ilyodes* will have to replace *Helenodora,* according to the rules of zoological nomenclature, if the more recent studies prove correct.

Extant onychophorans are terrestrial. Cambrian species are presumed to be marine, but "the habitat of *Ilyodes* is uncertain" (Robison, 1985). The type specimens of *Ilyodes* were in all probability found in the natural exposures along Mazon Creek, in whose bed and banks the majority of concretions were found during the last century and the early part of this century (Richardson, 1980). Baird et al. (1986) noted that both Braidwood (nonmarine) and Essex (estuarine) animals are found in the concretions from Mazon Creek. The type and paratype specimens of *Ilyodes* (*Helenodora* Thompson and Jones, 1980) were found at Pit

11, considered to have been an estuarine environment, although fossil terrestrial plants and animals are also found here.

The ongoing discussions of the affinities and systematics of the diverse and sometimes oddly constructed Cambrian onychophoran-like animals is in sharp contrast to the assessment of *Ilyodes.* A marked similarity between the structural plan of *Ilyodes* and that of the extant genus *Peripatoides* (Figure 15C.5) suggests that the Onychophora had become a conservative group by the Pennsylvanian; their living descendants being little changed structurally from Paleozoic ancestors.

References

Baird, G. C., C. W. Shabica, S. D. Sroka, and G. J. Kuecher. 1986. Taphonomy of Middle Pennsylvanian Mazon Creek area fossil localities, northeast Illinois. Palaios, 1:271–285.

Ballard, J. O., et al. 1992. Evidence from 12S ribosomal RNA sequences that onychophorans are modified arthropods. Science, 258:1345–1348.

Brusca, R. C., and G. J. Brusca. 1990. Invertebrates. Sinauer, Sunderland, MA.

Collins, M. 1989. *In* S. J. Gould, Wonderful Life. Norton, New York, pp. 154.

Conway Morris, S. 1977. A new metazoan from the Cambrian Burgess Shale, British Columbia. Palaeontology, 20:623–640.

Dzik, J., and G. Krumbiegel. 1989. The oldest "onychophoran" *Xenusion:* a link connecting phyla? Lethaia, 22:169–181.

Ghiselin, M. T. 1985. A movable feaster. Natural History, 94(9):54–60.

Gould, J. G. 1992. The reversal of *Hallucigenia.* Natural History, 101(1):12–20.

Kraus, O. 1974. Phylum: Onychophora, pp. 387–390. *In* Grzimek (ed.), Animal Life Encyclopedia, vol. 1. Van Nostrand Reinhold, New York.

Krumbiegel, G., H. Deichfuss, and H. Deichfuss. 1980. Ein neuer fund von *Xenusion.* Hallesches Jahrb. Geowiss., 5:97–99.

Levi-Setti, R. 1975. Trilobites. University of Chicago Press, Chicago.

Lewin, R. 1992. Whose view of life? Discover, 13(5):18–19.

Meglitsch, P. A., and F. R. Schram. 1991. Invertebrate Zoology. Oxford University Press, New York.

Moore, R. C. 1959. Protarthropoda, pp. O16–O18. *In* R. C. Moore (ed.), Treatise on Invertebrate Paleontology. Part O, Arthropoda 1. Geological Society of America and University of Kansas, Lawrence.

Pompeckj, J. F. 1927. Ein neues Zeugnis uralten Lebens. Palao. Zeit., 9:287–313.

Ramskold, L., and Hou, X. 1991. New early Cambrian animal and onychophoran affinities of enigmatic metazoans. Nature, 351:225–228.

Richardson, E. S., Jr. 1980. Life at Mazon Creek pp. 217–224. *In* R. L. Langenzheim and C. J. Mann (eds.), Middle and Late Pennsylvanian Strata on Margin of Illinois Basin (Guidebook). 10th Field Conf., Great Lakes Section, Soc. Econ. Paleo. Mineral., Danville, Illinois.

Robison, R. A. 1985. Affinities of *Aysheaia* (Onychophora), with description of a new Cambrian species. Jour. Paleo., 59:226–235.

Rolfe, W. D. I., F. R. Schram, G. Pacaud, D. Sotty, and S. Secretan. 1982. A remarkable Stephanian biota from Montceau-les-Mines, France. Jour. Paleo., 56:426–428.

Schram, F. R. 1991. Cladistic analyses of metazoan phyla and problematica. *In* Alberto M. Simonetti and Simon Conway Morris (eds.), The Early Evolution of Metazoa and Significance of Problematic Taxa. Proc. International Symposium, 27–31 March 1989, Univ. Camerino. Cambridge University Press, Cambridge.

Scudder, S. H. 1890. New Carboniferous Myriapoda from Illinois. Boston Soc. Nat. Hist. Mem., 4:417–442.

Snodgrass, R. E., 1958. Evolution of arthropod mechanisms. Smithsonian Misc. Coll., 138:1–77.

Thompson, I. D., and D. S. Jones. 1980. A possible onychophoran from the Middle Pennsylvanian Mazon Creek of northern Illinois. Jour. Paleo., 54:588–596.

Whittington, H. B. 1978. The lobopod animal *Aysheaia pedunculata* Walcott, Middle Cambrian, Burgess Shale, British Columbia. Phil. Trans. R. Soc. London B, 284:165–197.

16A

Brachiopods

Andrew A. Hay

Brachiopods are solitary, bivalved, bottom-dwelling, marine animals; their fossil record extends back to the beginning of the Cambrian period. The soft body parts are enclosed between two valves, generally dissimilar in size; these two valves give them a superficial resemblance to the bivalve molluscs (Bivalvia), whose typically oval valves are usually similar in size, oysters (Ostreidae) being one of the exceptions that prove the general rule. In brachiopods, the valves enclose the body dorsally and ventrally rather than laterally, as is the case in the bivalves. Moreover, the ventral valve of articulate brachiopods is typically the largest, and it is the ventral (pedicle) valve that is usually attached to the substratum. The majority of extant species attach themselves to the substratum by means of a fleshy stalk or pedicle.

Brachiopods are unselective suspension feeders, utilizing their lophophore, a combination respiratory-feeding structure that is basically stationary. The lophophore of brachiopods symmetrically surrounds the mouth and bears a food (brachial) groove and numerous filamentary cilia plus ciliated tentacles, which are used to direct water through the lophophore and food particles to the ciliated surface of the brachial food groove, thence along to the mouth. Although the exact diet of any species of brachiopod is unknown, food items may include planktonic animals or plants and dissolved or colloidal organic material (Foster, 1982). Depending on the species, the brachiopod lophophore has a variety of shapes (circular, looped, horseshoe shaped, or spiral) and is almost always located within the protection of the valves.

Phoronids and bryozoa are two other phyla that feed by means of a lophophore; however, they protract their lophophores into the water currents surrounding their bodies, while the brachiopods draw water into the mantle cavity that surrounds their lophophores.

The phylum Brachiopoda is divided into two classes, Inarticulata and Articulata. This division in the phylum is based on the methods used to articulate the valves. Due to the lack of a hinging device, valves of inarticulate brachiopods are operated by muscles only, a musculature more complex than those of the hinged valve articulates. Articulates bear interlocking projections along the hinge line, teeth and sockets that lock the valves together securely, allowing a gape that is far less than that obtainable by the inarticulate brachiopods. Members of this phylum were a very significant portion of marine animal life prior to a major extinction event at the end of the Permian and another, though less severe, around the end of the Triassic period. The Articulata constituted the major fraction of the numerous brachiopod species during the Paleozoic era; they compose the majority of extant brachiopods. Although the phylum is insignificant numerically in today's oceans (Rudwick, 1965), they are not insignificant in some localities. In fact, they may be the dominant element of the megainvertebrate fauna in some areas (Foster, 1982). Extant brachiopod species are most commonly found on the continental shelves, but they range from the intertidal (littoral) zone to abyssal depths. Out of the more than 12,000 species that once lived, only about 335 species survive today (Foster, 1982; Brusca and Brusca, 1990, pp. 771–800).

Rudwick (1965) and Foster (1982) mention fish, birds, starfishes, gastropods, and crustaceans as some of the possible predators upon brachiopods. Foster names *Lingula* as being eaten by humans in Japan and Australia. Richardson (1986), in a study of the ecology of existing species of brachiopods, noted that the protective shells and unpleasant taste of the small amount of edible tissue explain why brachiopods appear to be rarely disturbed by marine animals. In laboratory tests, fish rejected fresh brachiopod flesh offered to them.

Figure 16A.1 Chonetoid brachiopod, FMNH GEO 1146, Pit 11. Scale bar = 1 mm. Photograph courtesy Field Museum of Natural History.

Collections of fossils from marine Pennsylvanian sites will commonly contain a representative suite of brachiopods; this is not the case in the Essex (euryhaline) component of the Mazon Creek biota. Only two genera of brachiopods have been reported from there (Schram, 1979), their percentage of the composition of the Essex biota being given as 0.1 percent (Foster, 1979); however, M. W. Foster (personal communication) is currently studying specimens that can be assigned to the articulate suborder Chonetidina (Figure 16A.1). The two genera of Mazon Creek brachiopods described in the existing literature (*Lingula* and *Orbiculoidea*) are inarticulates, which are generally considered to have been shallow-water dwellers and capable of tolerating brief periods of varying salinity and turbidity, as do their descendants today. Nearly all other genera of extant brachiopods are incapable of surviving these conditions. Modern lingulids retreat into their burrows,

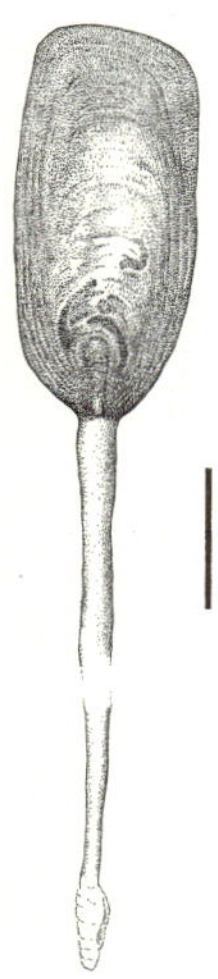

Figure 16A.2 *Lingula,* NEIU MCP 104, Pit 11. Scale bar = 5 mm.

Figure 16A.3 *Lingula* with pedicle, modified from several sources. Scale bar = 5 mm.

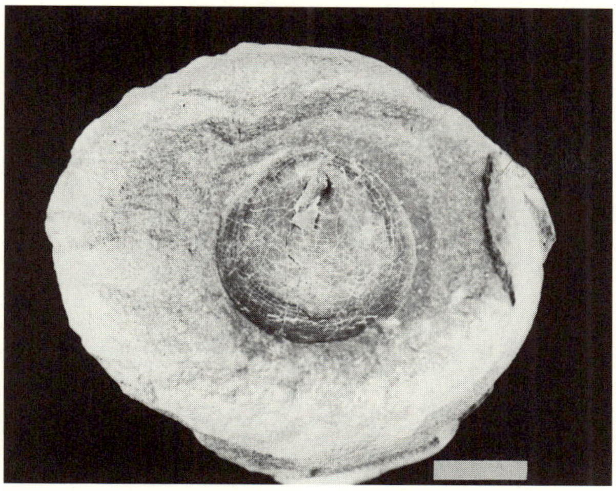

Figure 16A.4 *Orbiculoidea,* FMNH PE 22302, Pit 11. Scale bar = 5 mm.

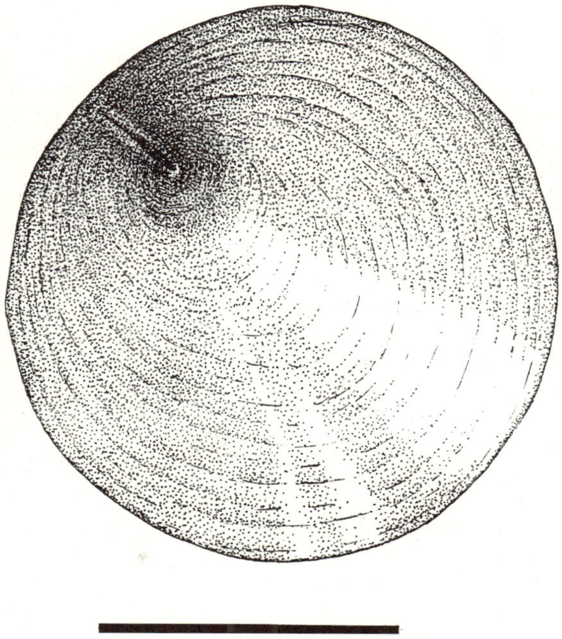

Figure 16A.5 *Orbiculoidae,* modified from several sources. Scale bar = 5 mm.

while the discinids (=orbiculids) close their shells tightly during periods of environmental stress (Rudwick, 1965).

Genus *Lingula* Bruguiere, 1797
Figures 16A.2, 16A.3

Shells are thin, chitinophosphatic, linguiform or spatulate in outline, ornamented only by concentric growth lines, and generally less than 2 cm long. Extant lingulids are known from brackish water tidal flats, live in deep vertical burrows, and are attached to the substrate at the bottom of the burrow with a long muscular stalk called a pedicle. *Lingula* is locally common in the Mazon Creek biota, notably in the southern end of Pit 11, where they were found fossilized in living position with burrow and pedicle preserved (Baird et al., 1986). The Pit 11 *Lingula* specimens, available for study by the author, have a maximum shell length of 14 mm. *Lingula* has a noteworthy geological record. It is represented in modern seas and in the fossil record of the Cambrian.

Genus *Orbiculoidea* d'Orbigny, 1847
Figures 16A.4, 16A.5

Shells are thin and have a circular to subcircular outline. Brachial valve conical to subconical, pedicle (ventral) valve slightly convex; valves ornamented with many fine concentric growth lines. Among extant discinids, and most inarticulates, a short pedicle anchors the shell to gravel, coral, or the undersides of boulders, the pedicle being securely attached to these substrates by a sticky secretion. The muscles of the pedicle support the weight of the shell and allow it to be rotated, depressed, or elevated (Rudwick, 1965). *Orbiculoidea* is found in the fossil record of the Ordovician to the Permian and is assumed to have had a lifestyle similar to that of the modern discinid brachiopods. *Orbiculoidea* is uncommon in the Mazon Creek biota.

References

Baird, G. C., S. D. Sroka, C. W. Shabica, and G. J. Kuecher. 1986. Taphonomy of Middle Pennsylvanian Mazon Creek area fossil localities. Palaios, 1:271–272.

Brusca, R. C., and G. J. Brusca. 1990. Invertebrates. Sinauer, Sunderland, Massachusetts.

Foster, M. W. 1979. Soft-bodied Coelentrates in the Pennsylvanian of Illinois, pp. 196–199. *In* M. H. Nitecki (ed.), Mazon Creek Fossils. Academic Press, New York.

———. 1982. Brachiopoda, pp. 773–780. *In* S. P. Par-

ker (ed.), Synopsis and Classification of Living Organisms. McGraw-Hill, New York.
Richardson, J. R. 1986. Brachiopods. Sci. Amer., 255:100–106.
Rudwick, M. J. S. 1965. Ecology and Paleoecology, pp. H199–H213. *In* R. C. Moore (ed.), Treatise on Invertebrate Paleontology. Part H, Brachiopoda. Geological Society of America and University of Kansas, Lawrence.
Schram, F. R. 1979. The Mazon Creek biotas in the context of a Carboniferous faunal continuum, pp. 159–190. *In* M. H. Nitecki (ed.), Mazon Creek Fossils. Academic Press, New York.

16B

Echinodermata

Steven D. Sroka and Daniel B. Blake

Echinoderms are complex and elegant marine invertebrates common today in the world's oceans. The oldest known representatives are Cambrian in age, and they have been successful ever since, including today approximately 6,000 living species assigned to five classes: the Asteroidea (sea stars), Crinoidea (sea lilies and feather stars), Holothuroidea (sea cucumbers), Echinoidea (sea biscuits, sea urchins, and sand dollars), and Ophiuroidea (basket stars, serpent stars, and brittle stars). Most echinoderms are exclusively marine, although a few are found in brackish water. They occur from intertidal zones to deep oceanic trenches, on or in all kinds of sediments, and on shells and rocks. They feed in many ways; there are predators, herbivores, filter feeders, and detritus feeders.

Most living echinoderms have pentameral symmetry, in which five similar sections are arranged around a central axis, although some, such as holothurians and several extinct classes, have predominantly bilateral or no symmetry (Sprinkle, 1987). Echinoderms have several body cavities, or coelomic pouches. One coelom is developed as a system of fluid-filled vessels called the water vascular system. Vessels extend through the body wall as tube feet and tentacles, functioning as feeding, locomotive, and respiratory organs. All echinoderms lack a head and brain, and sensory organs are limited. Another distinguishing feature of echinoderms is their internal mesodermal skeleton composed of numerous calcium carbonate plates, elements, or sclerites, imbedded in a thick or thin dermal layer. The skeleton commonly supports projecting spines or tubercles that give a warty or spiny appearance to the body, hence the name "echinoderm" or "spiny skin."

The only echinoderms found in the Mazon Creek area are a crinoid, known from a single specimen, and a whole-bodied holothurian, which is quite common. Both belong to the "Essex concretion fauna."

In other settings, crinoids have an abundant and diverse fossil record, whereas fossil holothurians, especially those preserved with the body intact, are rare. This does not necessarily mean they were rare in life, but only that their loosely linked skeletal arrangements are not readily preserved or, being microscopic, are rarely seen by the casual observer. Thus, the majority of fossil holothurians are known only from isolated skeletal elements, either so-called ring elements that are associated with the mouth region or sclerites. The sclerites are found in the body wall and range in size from 1 cm to 10 μm (Pawson, 1980). They occur in a great number of shapes, which are given descriptive names, such as wheels, sigmoid hooks, plates, and anchors. Sclerites aid in distinguishing holothurians, but Smith and Gallemi (1991) suggested that they are only poor guides to natural biological classification.

A skeletal ring supports the esophagus and serves as an insertion point for the five radial

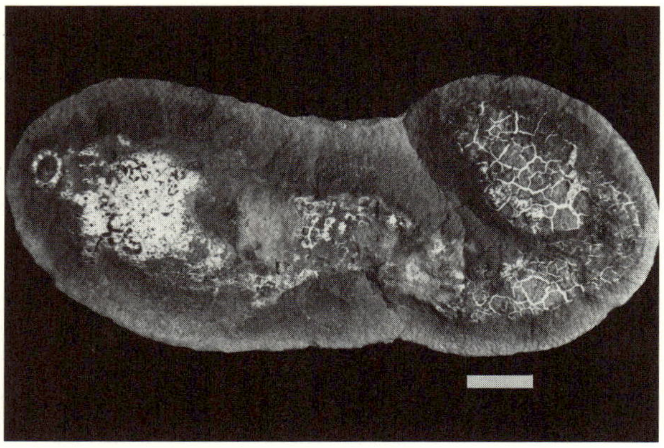

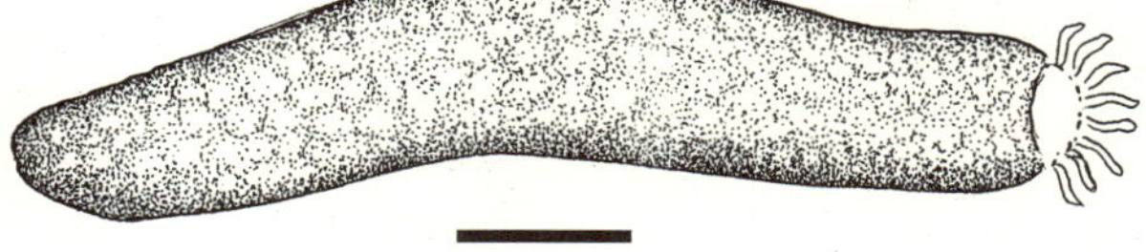

Figure 16B.1. Complete individual showing a well-preserved calcareous ring, NEIU MCP 417: Scale bar = 10 mm.

Figure 16B.2. Mazon Creek holothurian, modified from Pawson (1980). Scale bar = 10 mm.

longitudinal muscles than run along the inside of the body wall (Pawson, 1980). The ring is composed of simple plates, usually 10 in number, or alternatively may be a mosaic of tiny plates (Pawson, 1980).

Class Holothuroidea
Order Apodida
Family Achistridae
Genus *Achistrum* Etheridge, 1881,
amend. Frizzell and Exline, 1955
Figures 16B.1, 16B.2

Mazon Creek holothurian specimens contain well-preserved calcareous rings and sclerites. The sclerites take the shape of sigmoid hooks (Figure 16B.2) that occur in situ within the body wall in well-preserved specimens. Based on these sclerites, the specimens were tentatively assigned to the genus *Achistrum* (by Sroka, 1988). Hook size is bimodal, with larger sclerites somewhat more common in the posterior and smaller sclerites more numerous, scattered evenly throughout the body wall. The calcareous ring shows the 10-plate arrangement characterizing modern forms. Many Mazon Creek holothurian specimens range from complete and well preserved, with gut trace and possibly other organs present, to barely recognizable. Tube feet and tentacles have not yet been observed.

Care must be used in specimen preparation; cleaning holothurian specimens with acetic acid (vinegar) or any other acid will quickly damage and ultimately destroy both the ring and the sclerites.

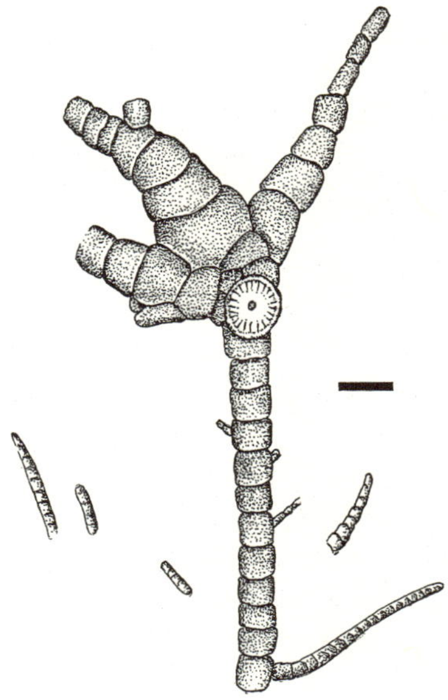

Figure 16B.3. Drawing of crinoid holotype, modified from Lane (1969). Scale bar = 1 mm.

Class Crinoidea
Subclass Inadunta
Order Cladida, Suborder Poteriocrinitia
Family, genus, and species unknown Lane, 1969
Figure 16B.3

Lane (1969) described the only known crinoid discovered from the Essex fauna. He was unable to identify the fossil below the subordinal level because it is immature, incompletely exposed, and peculiar in arm structure.

The specimen has a small and lightly constructed crown with nerve grooves preserved in some cup and arm plates. Arms branch on the third primibrach, an unusual feature for Pennsylvanian inadunate crinoids, and long slender cirri are directed toward the crown. Lane (1969, p. 155) suggested that the crinoid attached itself to floating seaweed or logs by distal cirri and may have hung upside down in the water. If accurately interpreted, crinoids would not have lived within the Essex environment, and they would only rarely have been washed in, thus perhaps explaining this sole known occurrence. Because of its apparent unusual structure, it would be desirable to isolate additional specimens. Figure 16B.3 shows a drawing of the holotype.

References

Etheridge, R., Jr. 1881. On the presence of the scattered skeletal remains of Holothuroidea in the Carboniferous Limestone Series of Scotland. Proc. R. Phil. Soc. Edinburgh, 6:183–198.

Frizzell, D. L., and H. Exline. 1955. Monograph of fossil holothurian sclerites. Bull. Univ. Missouri School Mines Metal. Tech. Ser., 89.

Lane, N. G. 1969. A crinoid from the Pennsylvanian Essex fauna of Illinois. Fieldiana Geol., 12(9):151–156.

Pawson, D. L. 1980. Holothuroidea, pp. 175–189. *In* T. W. Broadhead and J. A. Waters (eds.), Echinoderms, Notes for a Short Course. Univ. Tennessee Dept. Geol. Sci. Stud. Geol., 3.

Smith, A. B., and J. Gallemi. 1991. Middle Triassic holothurians from northern Spain. Palaeontology, 34, pt. 1:49–76.

Sprinkle, J., with P. M. Kier. 1987. Echinodermata. Part 1, Phylum overview, pp. 550–572. *In* R. S. Boardman, A. H. Cheetham, and A. J. Rowell (eds.), Fossil Invertebrates. Blackwell, Palo Alto, California.

Sroka, S. D. 1988. Preliminary studies on a complete fossil holothurian from the Middle Pennsylvanian Francis Creek Shale of Illinois, pp. 159–160. *In* R. D. Burke, P. V. Mladenov, P. Lambert, and R. L. Parsley (eds.), Echinoderm Biology. Proc. 6th International Echinoderm Conference, 23–28 August 1987, Victoria, B.C. Balkemia, Rotterdam.

17

Fishes:

"Agnatha," Acanthodii, and Osteichthyes

David Bardack

The word "fish," like the word "mollusc," includes diverse groups. The interrelationships of the fishes are complex, and their phylogeny and classification have been under intensive study in recent years (see Maisey, 1986, and earlier references in that paper). Indeed, the relationships of some components of the Mazon Creek fish assemblage remain uncertain, and in the professional literature these fishes often are left in a taxonomic limbo (Patterson, 1982). However, in order to provide a framework for the arrangement of the Mazon Creek fish assemblage and a relatively convenient reference, we have chosen to use the classification in Carroll (1988) but note, when appropriate, alternative views on the systematics of particular groups. Readers may also find it helpful to consult this vertebrate paleontology textbook for background reading on the anatomy of fishes. Four of the five major fish groups known from the Late Paleozoic are represented in the Mazon Creek fauna. These are 1) "agnathans," which are jawless forms including lampreys and hagfishes, 2) chondrichthyans, which are sharks and several genera of less certain affinity (for information on this group, see Chapter 18), 3) acanthodians, fishes of disputed affinities but probably related to the last group, 4) the osteichthyans. The last is the largest group of fishes and the predominant group today. The jawless fishes evolved possibly as early as the Cambrian, the others by the Silurian or Devonian.

During the Paleozoic, fishes evolved along many different pathways; it was a time of great morphological experimentation. In most groups, different body shapes and fin structures emerged, reflecting changes in locomotory mechanisms along with changes in jaws and related parts of the head skeleton concerned with feeding. Each of the four Paleozoic fish groups had representatives in a range of fresh and marine environments, and such is the case for the Mazon Creek fishes that are found in the Essex or Braidwood component of the Mazon Creek fauna. There are other localities of about the same age as Mazon Creek (Westphalian D) containing some but not all of the fish groups of the Illinois deposits. These include Linton, Ohio; Cannelton, Pennsylvania; and Nýřany, Czechoslovakia. The same or similar taxa also occur in other late Pennsylvanian localities in Nova Scotia, the British Isles, and the United States. For a table showing the distribution of aquatic vertebrates in major late Pennsylvanian sites, see Schultze and Maples (1992).

The Mazon Creek fauna is especially rich with in excess of 30 species in four major fish groups, in part because of the diverse types of environments represented, including areas of coal swamp, delta, and brackish marine habitat, and in part because collectors have been active for many years in a large area with many newly exposed concretions. Identification of Mazon Creek fish specimens is usually difficult unless a whole individual is preserved. In such cases one or two features may suffice for assignment of the specimen. But, in most cases, the fossils are incomplete or small, and a significantly larger number of structures must be examined particularly to differentiate genera let alone identify a species. Recognition of some taxa is difficult without extended taxonomic diagnoses. In the following sections, selected characteristics of each taxon are presented rather than complete diagnoses.

"Agnatha"

This term serves mainly to distinguish fishes without jaws from all other fishes. Several quite different and very distantly related forms are included under this term. Today there are two very different kinds of jawless fishes, hagfishes and lampreys, which although linked in many textbooks as cyclostomes (round mouths) are considered now to be only distantly related.

Paleozoic agnathans were much more diverse than today and many were heavily skeletonized. The Mazon Creek representatives are unusual in that they provide a record of soft-bodied jawless fishes not seen elsewhere.

Order Myxiniformes
Myxinikela siroka Bardack
Figure 17.1

The Myxiniformes, the most primitive of the jawless fishes, are classified separately from other living and fossil craniates based on many morphological as well as biochemical and physiological features. They are considered distinct from the vertebrates, but still craniates. Living myxiniforms, commonly known as hagfishes, are entirely marine. They usually live at moderate depths and feed as scavengers on dead or dying invertebrates and fishes. A single specimen from Pit 11 belongs to this group (Bardack, 1991). It is about 7 cm in total length, characterized by a tubular body with confluent dorsal, caudal, and anal fins. Eyes are prominent, a feature that is not well developed in most living hagfishes. The gill pouches lie closer to the head than in living forms. All hagfishes are distinguished by a series of tentacles that extend forward from the single anterior nasal opening or from above and below the mouth. Internally, skeletal elements as well as parts of the digestive and branchial (gill) circulatory system are seen. The features of this fossil hagfish are basically similar to those of living ones. These fishes thus demonstrate a long and evolutionarily conservative history. No traces of lamprey or other jawless fish morphological features are seen in the fossil. Differences between *Myxinikela* and living hagfishes, such as the position of the branchial skeleton and the somewhat minimal development of the head skeleton, suggest that this specimen may be a juvenile. There is no evidence to suggest that the individual is sexually mature while somatically immature. In general, many fish specimens from the Mazon Creek fauna are small and probably juveniles.

Holotype. NEIU MCP 126, Pit 11.

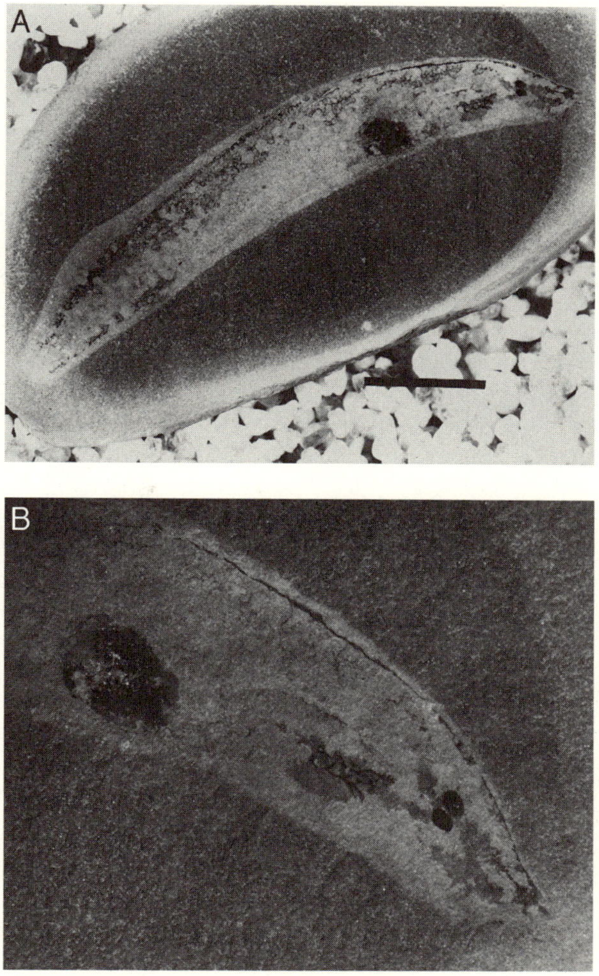

Figure 17.1. *Myxinikela siroka,* NEIU MCP 126, Pit 11. **A** Lateral view. Scale bar = 10 mm. **B** Enlarged anterior end of specimen showing eyes (two oval structures toward the anterior end), tentacles extending from mouth and nasal openings, and liver (large dark mass toward posterior end).

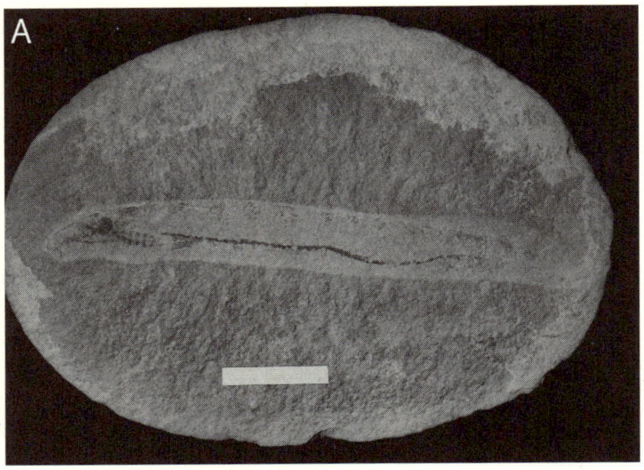

Figure 17.2. *Mayomyzon pieckoensis.* **A** FMNH PF 5687, Pit 11; lateral view. **B** FMNH PF 8167, Pit 11; dorsal view. Both scale bars = 10 mm.

Order Petromyzontiformes
Mayomyzon pieckoensis Bardack and Zangerl
Figure 17.2

This order includes the living lampreys (sometimes called "lamprey eels" but not related to true eels), some of which are confined to freshwater, while others spend part of their lives at sea before returning to freshwater for spawning. These are elongate fishes, the adults of which show large round mouths containing horny teeth. There are a number of separate gill openings. The skeleton is cartilaginous. The fossils are generally similar to living lampreys but differ 1) in the apparent absence of a large buccal (mouth) funnel, 2) in the lack of dentition, 3) in the position of the otic (ear) capsule, which lies above the first and second gill pouches, 4) in the position of the gill pouches, which, when viewed from above, extend at nearly a right angle to the body axis, and 5) in the otic capsule's not being in contact with the braincase (Bardack and Zangerl, 1971; Bardack, 1979). More than two dozen specimens have been assigned to *M. pieckoensis.* All are small, measuring less than 8 cm in total length. Some of the differences between the fossils and living adult lampreys (e.g., items 1, 2, and 3 above) suggest that these fossils may be in a juvenile rather than adult stage of development. Typical adult characters, especially the buccal dentition so important in distinguishing living forms, may not be preserved, may be too small to recognize, or may form later in ontogenetic (individual) or phylogenetic (evolutionary) development. However, the large eyes, well-developed gill chambers, and large piston cartilage (part of the unique feeding feature of lampreys) implies an adult stage. All specimens are from the Essex fauna. However, it remains a possibility that these lampreys spent part of their life history in freshwater coal swamps or that these small, possibly late larval forms were swept into the brackish marine water area of the Mazonian delta from freshwater where eggs may have been laid and underwent early development as in the case of living sea lampreys. There is one other fossil lamprey (Janvier and Lund, 1983), which differs somewhat from *Mayomyzon* especially in the caudal fin. This fossil from the Mississippian marine sediments of the Bear Gulch Limestone of Montana is not as well preserved in the head region as *Mayomyzon.*

Holotype. FMNH PF 5687, Pit 11.

Other "Agnathans"

The two fishes described below are certainly "agnathan"; that is, they are jawless. But they are quite different from both living and fossil hagfish and lamprey and any of the "agnathans" with external skeletons. Together with the fossil hagfish and fossil lamprey, they reinforce the unique character of the Mazon Creek chordate assemblage characterized by unusual diversity of its organisms, especially those with few hard skeletal elements.

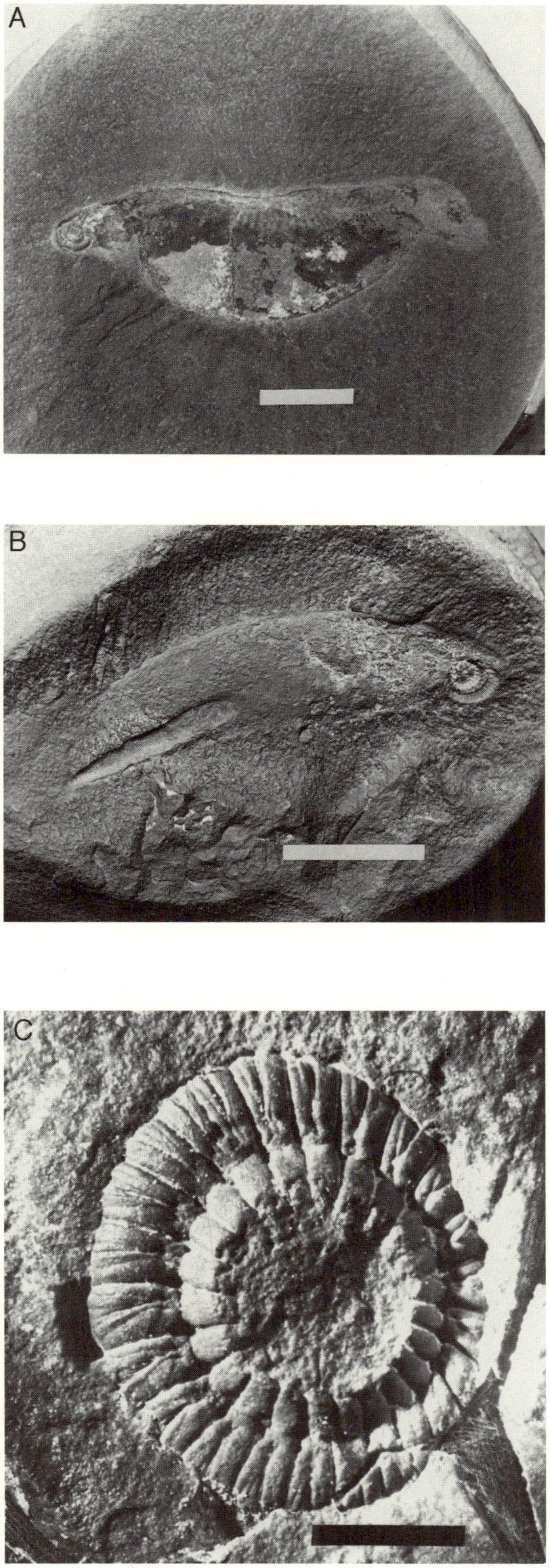

Pipiscius zangerli Bardack and Richardson
Figure 17.3

The striking feature of this fish is a round mouth formed by a ring of 23 lamellar-like plates surrounding the mouth opening and articulated to a second set of plates lying just posterior to them. These unique plates appear to have been operated by a complex musculature indicated by a series of grooves and ridges in the matrix of the specimen. These plates were probably part of a feeding system designed to utilize benthic animals and detritus. There are at least four gill clefts and a postanal dorsal fin. *Pipiscius* is an unusual fossil. It is considered a chordate based on the presence of gill slits, a well-developed eye, angular metamerism (muscle segmentation), and a postanal caudal region. No rasping tongue or dental mechanism is preserved as in lampreys or hagfishes. No similar fossils are recognized elsewhere. About a dozen specimens between 4 and 6.5 cm in length are known, all are from the Essex fauna (Bardack and Richardson, 1977).

Holotype. FMNH PF 8344, Pit 11.

Gilpichthys greenei Bardack and Richardson
Figure 17.4

An elongate agnathan with a tubular, muscular buccopharynx (mouth and pharynx) bearing teeth. At least six gill clefts. There are no fins. This is a very abundant form; in excess of 200 are known, ranging from about 4 to 15 cm in length. Originally thought to be molluscan, they were often called "nudibranchs" by collectors in the belief that the buccopharyngeal structure was a radula. *Gilpichthys* is assigned to the vertebrates because it possesses a notochord, gill slits, a dorsal hollow nerve tube, angular metamerism, and a postanal caudal region. On excellently preserved specimens, the buccopharynx shows a series of about 20 segments of decreasing thickness, some of which

Figure 17.3. *Pipiscius zangerli.* **A** FMNH MCP 266 (Douglass collection, 604), Pit 11; lateral view, anterior to left. Scale bar = 10 mm. **B** FMNH PF 8345, Pit 11; body with mouth region bent lateral, anterior to right. Scale bar = 10 mm. **C** FMNH PF 8344, Pit 11; impression of inner surface of mouth as viewed from within the animal, anterior to upper left. Scale bar = 2 mm.

Figure 17.4. *Gilpichthys greenei.* **A** FMNH PF 8475, Pit 11; lateral view, anterior to right. Scale bar = 10 mm. **B** FMNH PF 18703, Pit 11; scanning electron microscope photograph of buccopharynx showing muscle blocks, anterior to right. Scale bar = 1 mm.

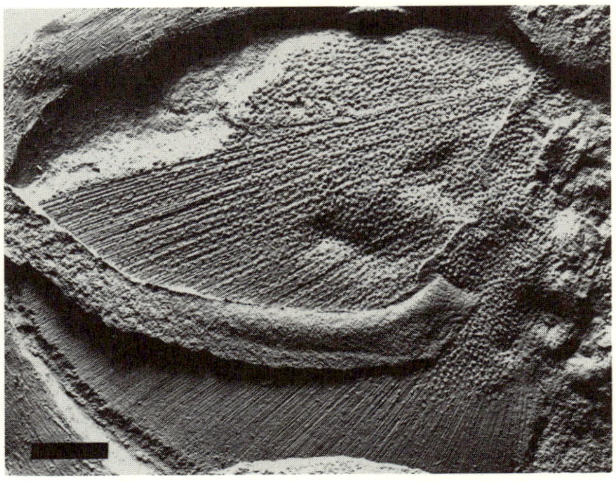

Figure 17.5. *Trichorhipis praecursor,* FMNH MCP 1581, Mazon Creek; photograph of latex peel of prepectoral spine. Scale bar = 10 mm.

support sharp, posteriorly directed fanglike teeth. Assignment to the agnatha is based on the lack of jaws and the simple metameric (body segmentation) pattern (Bardack and Richardson, 1977).

Gilpichthys occurs only in the brackish marine phase of the Mazonian delta. No fossils or living forms with the specializations of *Gilpichthys* are known elsewhere.

Holotype. FMNH PE 18703, Pit 11.

Class Chondrichthyes
See Chapter 18

Acanthodii

This is a geographically widespread group that shows relatively little taxonomic diversity. They are found in rocks from Silurian to Permian times. Unlike agnathans, jaws are present. These fishes typically have large eyes. The upper jaw (palatoquadrate) and second pharyngeal arch (hyomandibula) that braces the jaw against the braincase are closely associated. Thus the opening between them (the spiracle) is almost closed. The gills are covered with a series of plates, rather than one bone as in osteichthyans. A significant feature of acanthodians is the presence of spines that appear before the dorsal, anal, and paired fins and sometimes several more between the paired fins. Acanthodians have been placed systematically one time or another with the chondrichthyans and osteichthyans among major fish groups or treated as an independent lineage (Maisey, 1986). The consensus at present is to place them close to the osteichthyans despite the many specializations of the head skeleton, scales, and fin rays.

Gyracanthidae
Trichorhipis praecursor Zangerl
Figure 17.5

A single specimen was described (Zangerl, 1973) as a ?pectoral fin of a shark. Baird (1978) considers it a prepectoral spine and assigns it to the family Gyracanthidae. There is little diagnostic about the specimen. Gyracanthid spines are also known from Pit 15, which is equivalent in age to the Mecca Shale of Indiana.

Holotype. FMNH PF 5729, Mazon Creek, Grundy County.

Acanthodidae
Genus *Acanthodes*
Figure 17.6

The acanthodians of the Mazon Creek nodules are small and exhibit fewer distinguishing morphological features than some members of this

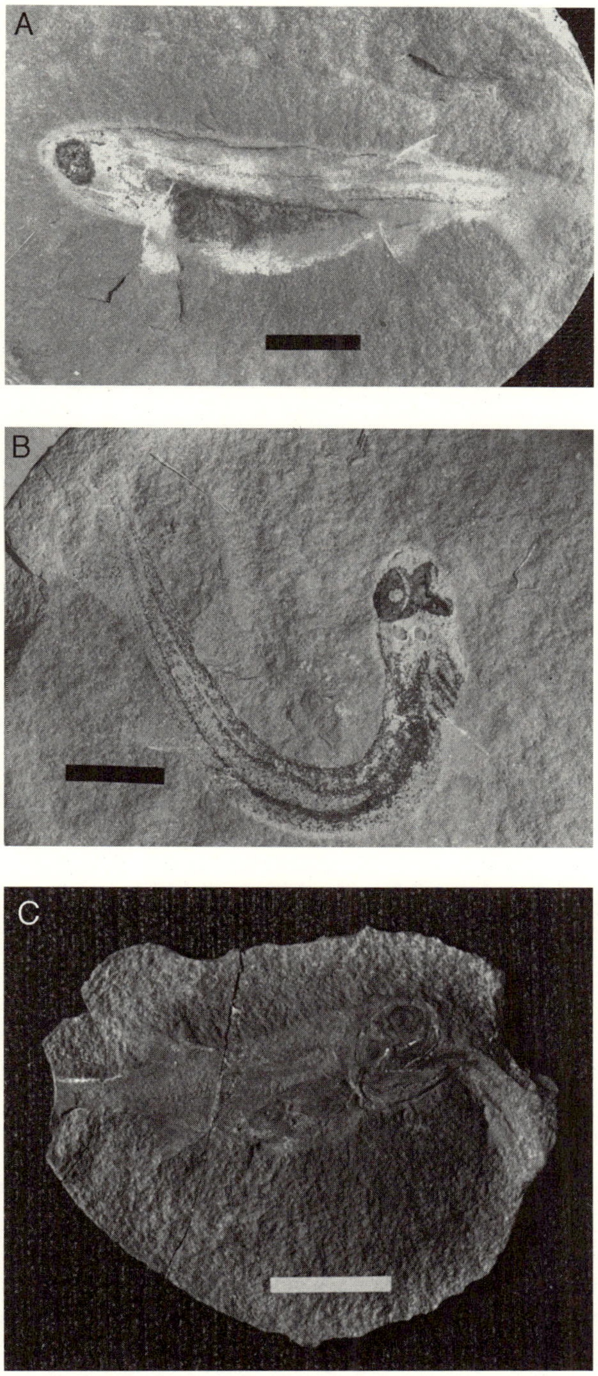

Figure 17.6. **A** *Acanthodes* sp., Pit 11; lateral view. **B** *Acanthodes* sp., FMNH MCP 2144 (Agazzi collection), Pit 11; dorsal view. **C** Paleoniscoid eating an acanthodian, NEIU MCP 358, Pit 11. All scale bars = 10 mm.

group found at other Paleozoic sites. Mazon Creek specimens usually show large, heavily pigmented eyes. Scales may be absent or appear as small, square-crowned structures. Several nodules show 3–4 cm patches of scales only. These fragmentary pieces of body covering suggest the presence of larger specimens. Ostracodes are not uncommonly found in the digestive system of Mazon Creek *Acanthodes.*

Two species, *A. beecheri* Eastman and *A. marshi* Eastman, have been described from the Mazon Creek fauna. Zidek (1976), in a review of the North American members of the genus based only on some of the older collections, argues that the types of these species and some other small specimens from this fauna are poorly preserved juveniles not warranting specific distinction. The large number of specimens now in museums and private collections must be studied in order to assess the specific diversity of this genus in the Mazonian delta deposits. *Acanthodes* occurs in both Braidwood and Essex faunas. Species assigned to this genus are found in many parts of the world as marine and freshwater fishes.

Holotype. A. marshi, YPM 2803, Mazon Creek; *A. beecheri,* YPM 630, Mazon Creek.

Class Osteichthyes

The osteichthyans are divided into two major subgroups: 1) the Actinopterygii (ray-finned fishes), constituting the largest group of bony fishes and including such living types as herring, salmon, carp, bass, and flounder, as well as the more primitive types such as sturgeons and gars, and 2) the Sarcopterygii (lobe-finned fishes), with the coelacanths, lungfishes, rhipidistians, and related forms. In the osteichthyans, the skeleton includes both dermal bone forming the outer skeleton of the head and bone that embryologically forms in a cartilaginous matrix (endochondral bone) and characterizes the vertebral and appendicular skeleton. The mouth is rimmed by dermal bones, and teeth are fused to these bones. The skull shows a characteristic pattern of canals that are part of a sensory canal system termed the lateral line. A hydrostatic bladder (swim bladder) is usually present. This structure is modified into lungs, especially in some of the Paleozoic sarcopterygians. Paleozoic actinopterygians usually show a tail form known as heterocercal and distinguished by the upturn of the body axis into the caudal fin. This contrasts with the pattern in most living actinopterygians, in which the tail is homocercal. At least 15 osteichthyan genera are known in the Mazon Creek fauna.

Subclass Actinopterygii

Specimens of actinopterygians from the Mazon Creek assemblage are relatively numerous in comparison with their occurrence at other Pennsylvanian localities (Maples and Schultze, 1989). However, the quality of preservation of actinopterygians is often better at these other sites. Many of the specimens from the Mazon Creek (both Braidwood and Essex faunas) are small and show few diagnostic characters.

Chondrostei

For much of this century, the Actinopterygii were divided into three groups, Chondrostei, Holostei, and Teleostei. As living and fossil actinopterygii were studied in greater detail, it became clear that this division, rather than reflecting phylogenetic relationships, simply grouped fishes that share similar stages of morphologic development. The Chondrostei have been reduced (Patterson, 1982) to a much smaller unit comprising sturgeons and paddlefishes, the term holostean eliminated, and the teleosteans expanded to include some of the holosteans. The character analyses that resulted in these changes depend on a detailed understanding of anatomy of fossil as well as living fishes. Many of the Paleozoic genera (including those found in the the Mazon Creek assemblage) that were included in the chondrosteans have been omitted from classifications and cladograms that focus only on a few unusually well preserved fossil forms and living fishes, which may be studied in minute detail. To use this system here would mean placing many Mazon Creek actinopterygians as "incertae sedis," at least above the family level. We have chosen instead to use a classification (Carroll, 1988) readily accessible to most readers while recognizing that the relationships among these fishes are being reexamined. Chondrosteans (in a broad sense) show an upper jaw (premaxilla and maxilla) firmly joined to the palate (ectopterygoid and dermopalatine), and the interoperculum (part of the gill cover) is absent. Members of the Chondrostei are differentiated by

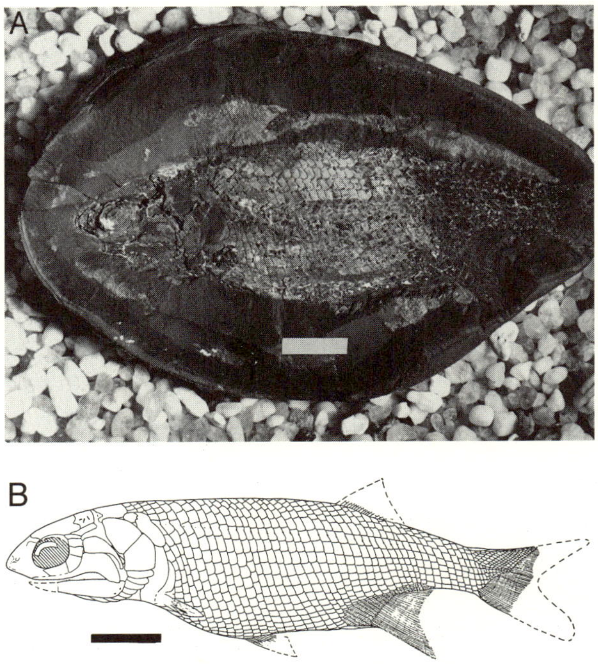

Figure 17.7. *Illinichthys cozarti,* FMNH UC 21716, Braidwood, Will County. **A** Lateral view. **B** Restoration in lateral view. Both scale bars = 10 mm.

body form, fin position, and fin ray number, as well as dermal skull pattern. Several groups of chondrosteans are represented: Paleonisciformes, platysomoids, Bobasatriiformes, and Haplolepiformes.

Order Paleonisciformes

Within the Mazon Creek paleonisciforms, there are two recently described genera, and an older genus, nominally termed *"Elonichthys."* Specimens assigned to the latter may or may not actually belong to this European genus. Until the original material is carefully reexamined, certain Mazon Creek paleonisciform fishes are retained in this genus, but the assignment is qualified by the use of quotation marks. Hook and Baird (1993) based on Five Points, Ohio, specimens believe that the American specimens should be referred to *Elonichthys* without quotation marks. They also suggest some different specific assignments for members of the genus. Most actinopterygians from the Mazon Creek assemblage belong to *"Elonichthys."* Some of the characteristics that may help to distinguish the paleonisciforms are presented below (Schultze and Bardack, 1987).

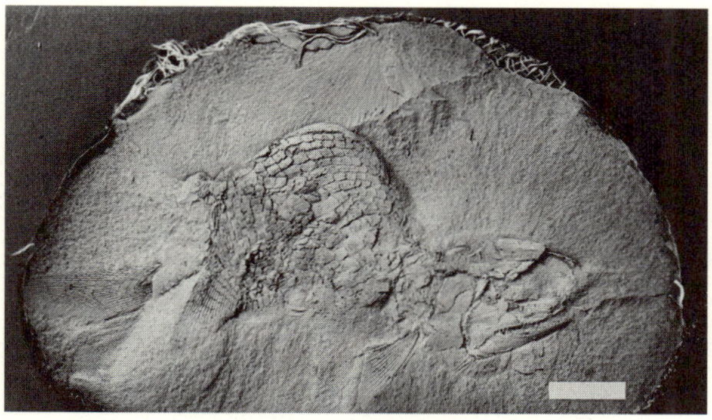

Figure 17.8. *Nozamichthys contorta,* USNM 359112, Mazon Creek, Grundy County; head and body of twisted fish. Scale bar = 10 mm.

Figure 17.9. *"Elonichthys" peltigerus,* FMNH PF 7502, Pit 11. A Lateral view. B Restoration in lateral view. Both scale bars = 10 mm.

Illinichthyes cozarti Schultze and Bardack
Figure 17.7

This is an elongate fish with a deepened abdomen. The dorsal fin is short with 19 fin rays and arises in front of the anal fin but behind the pelvic fins. The anal fin is long with 28 or 29 rays and starts opposite the posterior end of the dorsal fin. The scales have smooth surfaces and a straight posterior border. Half a dozen specimens of this species occur in the coal swamp and euryhaline deposits of the Mazonian delta.

Holotype. FMNH UC 21716, Braidwood, Will County.

Nozamichthys contorta Schultze and Bardack
Figure 17.8

This elongate, spindle-shaped fish has a tail with equally forked lobes and a triangular anal fin with 23 to 24 rays. Scales on the midbody are smooth, while those along the dorsal and ventral edges of the body show a few striae (ridges). The head is rounded anteriorly, and the posterior part of maxilla (upper jaw) is 6 times as deep as the anterior part. There is a single specimen from the Braidwood fauna.

Holotype. USNM 359112, Mazon Creek, Grundy County.

"Elonichthys" peltigerus Newberry
Figure 17.9

A spindle-shaped paleonisciform with a large triangular, long based dorsal fin with 29 to 39 rays, and an anal fin with 31 to 49 rays. In the caudal fin, the upper lobe is longer than the lower lobe. The dorsal fin starts at midbody opposite the posterior end of the pelvic fin base and extends to a point opposite the middle of the anal fin. That fin begins opposite the anterior or middle of the dorsal fin. Body scales show a few longitudinal ridges, and the posterior border is serrated. The head shows a protruding snout and the posterior part of the maxilla is 4 times as deep as the anterior part. Upward of 75 specimens are known from Pits 1, 6, 7, 11, and 13; hence this species is found in the Essex and Braidwood faunas. But whether in life these fishes really lived in both habitats is uncertain. This species also occurs at Linton, Ohio.

Holotype. See Schultze and Bardack (1987).

"Elonichthys" hypsilepus Hay
Figure 17.10

This fish shows a deep abdomen and very elongated upper lobe of the caudal fin. The long based anal fin with 32 to 50 rays is barely separated from the caudal fin. The anal fin starts opposite the middle or posterior part of the triangular dorsal fin which has 20 to 28 rays and arises from an elevated base opposite the posterior part of pelvic fin. The pectoral fin base is lobed. The first ray of this fin is elongated,

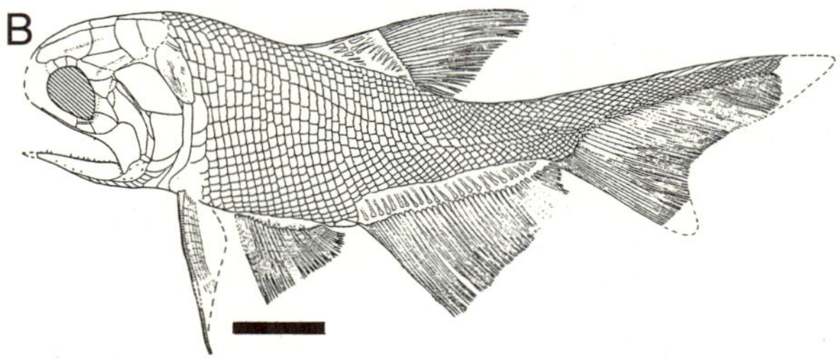

Figure 17.10. *"Elonichthys" hypsilepus.* **A** FMNH PF 7521, Pit 11; lateral view. **B** AMNH 518, Mazon Creek; restoration in lateral view. Both scale bars = 10 mm.

Figure 17.11. *"Elonichthys" disjunctus,* YPM 3260, Mazon Creek; head and body of twisted fish. Scale bar = 10 mm.

closely jointed, and covered anteriorly by small, thickly packed fringing fulcra. Small ridges are seen on the ventral parts of the scales. The front end of the head is rounded, and the posterior part of maxilla is 8 times deeper than the anterior part. There are about 25 specimens from Braidwood and Essex faunas. *"E." perpennatus* Eastman is a juvenile form and is included in *"E." hypsilepus. Amblypterus macropterus*? (Newberry and Worthen) also belongs in *"E." hypsilepus* (Schultze and Bardack, 1987).

Holotype. USNM 4848, Mazon Creek, Grundy County.

"Elonichthys" disjunctus Eastman
Figure 17.11

This is a difficult species to identify. The end of the anal fin is separated from the beginning of the caudal fin by at least nine scale rows. The caudal fin is forked, and the lobes of equal length. The anal fin is triangular with 22 rays. Scales show low ridges and small denticulations (teeth) posteriorly. A single specimen is known from the Braidwood fauna.

Holotype. YPM 3260, Mazon Creek, Grundy County.

"Elonichthys" wolffi Schultze and Bardack
Figure 17.12

This stout paleonisciform has a long dorsal fin with 51 rays, which starts opposite the beginning of the pelvic fin base and extends posteriorly to the middle of the anal fin. The anal fin with 49 rays extends along the ventral margin of the body to the lower lobe of the caudal fin. The pectoral fin base is lobed. The snout end is rounded, and the posterior part of maxilla is 7 to 8 times as deep as the anterior part. About half a dozen specimens from Pits 1, 6, and 11 are known. This species occurs in the Braidwood and Essex faunas.

Holotype. FMNH PF 7340, Pit 11.

"Elonichthys" remotus Schultze and Bardack
Figure 17.13

This last member of the *"Elonichthys"* group shows a dorsal fin that begins behind the pelvic fin and ends opposite the end of the anal fin base. The anal fin has 32 rays. The snout end of

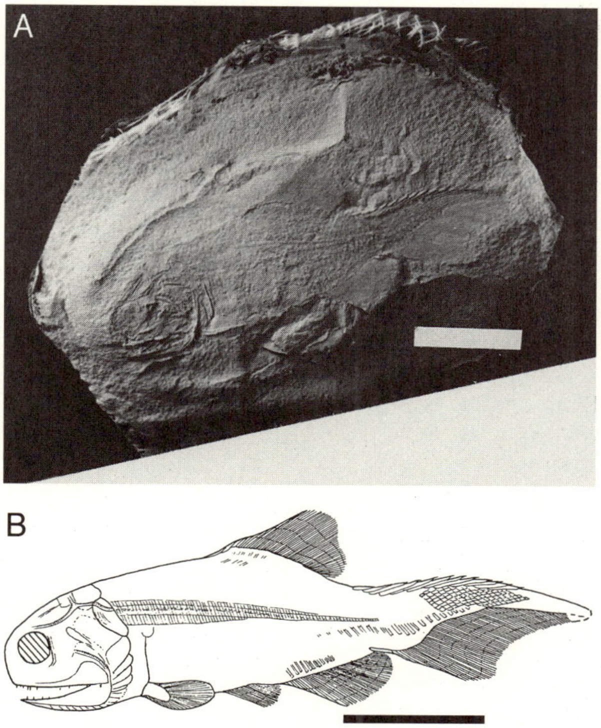

Figure 17.12. *"Elonichthys" wolfii,* FMNH PF 7340, Pit 11. **A** Lateral view of latex peel. **B** Restoration in lateral view. Both scale bars = 10 mm.

the head is somewhat pointed, and the posterior part of maxilla is about 4 times as deep as anterior part. One, and perhaps a second, specimen comes from Pit 11.

Holotype. FMNH PF 7945, Pit 11.

Suborder Platysomoidei

Deep-bodied paleonisciforms were reviewed by Zidek (1992). Two families are included, Amphicentridae and Platysomidae. The former Zidek refers to the Paleonisciformes, the latter to the Bobasatraniiformes.

Family Amphicentridae
Amphicentrum orbiculare
(Newberry and Worthen)
Figure 17.14

Few specimens (YPM 3240, YPM 3241, FMNH 7339, USNM 4404) are referred to this species (Zidek, 1992). Individuals probably reach 6 cm in length; the body is about twice as deep as long, with a pointed, middorsal peak marked by two scales. FMNH PF 7339 shows a dorsal fin

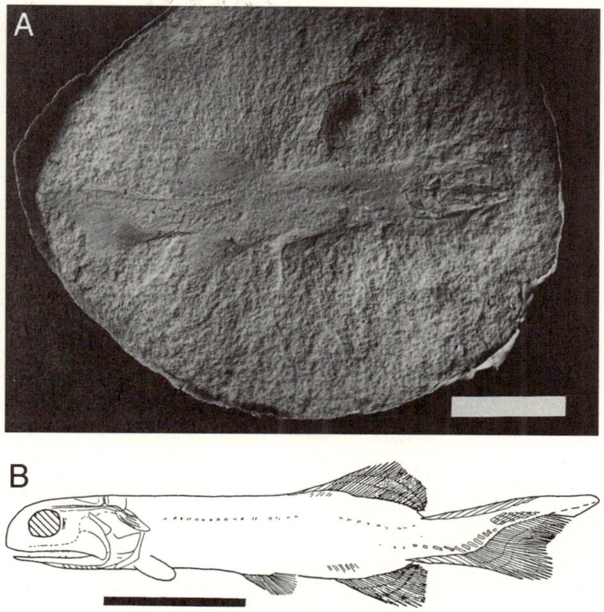

Figure 17.13. *"Elonichthys" remotus,* FMNH PF 7495, Pit 11. **A** Lateral view. **B** Restoration in lateral view. Both scale bars = 10 mm.

Figure 17.14. *Amphicentrum orbiculare,* FMNH PF 7339 (loc?); lateral view. Scale bar = 10 mm.

with 56 rays and an anal fin with 42. There are 18 to 23 vertical scale rows between the back of the shoulder girdle and start of the caudal peduncle. Specimens collected in the nineteenth century come from the classic Mazon Creek localities (Braidwood fauna), while more recent material (e.g., FMNH PF 7339) comes from Pit 11; so the species occurs in the Braidwood and Essex faunas. Additional, unstudied specimens may be in the hands of several collectors.

Holotype. Lost. Zidek (1992) made USNM 4404 the neoholotype.

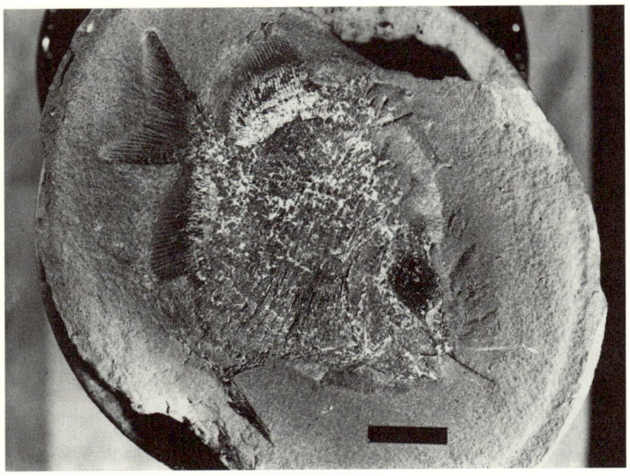

Figure 17.15. *Platysomus circularis,* FMNH PF 7333, Pit 11; lateral view. Scale bar = 10 mm.

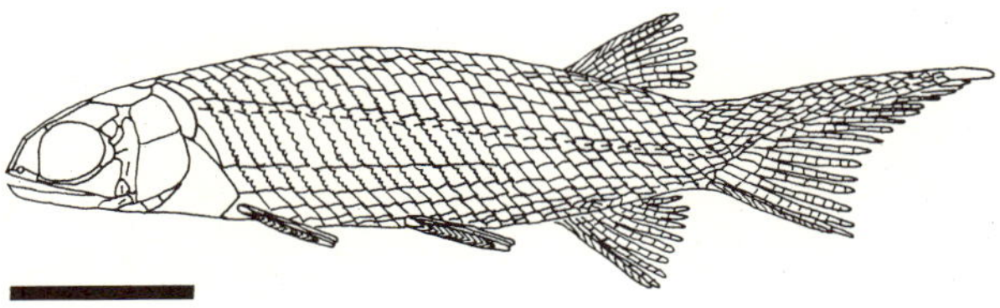

Figure 17.16. *Parahaplolepis tuberculata,* reconstruction of Linton, Ohio, specimen (from Lowney, 1983); lateral view. Scale bar = 10 mm.

Order Bobasatriiformes
Platysomus circularis Newberry and Worthen
Figure 17.15

Two species of *Platysomus, P. circularis* and *P. lacovianus,* have been described from the Mazon Creek fauna. Zidek (1992) places the latter in synonymy. He describes *P. circularis* as a small species, perhaps not over 6 cm in standard length. Specimens show approximately 40 dorsal fin rays, 30 to 35 anal fin rays, and 28 to 30 caudal fin rays. There are 20 to 22 vertical scale rows between the back of the shoulder girdle and start of the caudal peduncle. Zidek (1992) recognizes nine specimens from the Braidwood and Essex faunas.

Holotype. UI X489 (part), FMNH UF 593 (counterpart).

Order Haplolepiformes

Members of this order from the Mazon Creek biota have not been studied as a group since Westoll (1944). Lowney (1980) lists three genera, *Pyritocephalus, Parahaplolepis,* and *Microhaplolepis,* from Mazon Creek deposits. D. Baird (personal communication) suggests that there are still undescribed haplolepid genera in the fauna. The haplolepids are small (under 10 cm in length) with a flattened ventral surface. The dorsal fin is situated posteriorly, originating in front of or at the origin of the anal fin. All fins have few rays, and these are all unbranched. Scales are large and in most genera denticulated posteriorly.

Genus *Pyritocephalus* Fritsch

The characteristic of this genus, the large "fenestrae" in the parietal, is considered by Hook and Baird (1993) to have been covered by a thin tympanum-like structure. The dorsal fin arises behind the anal fin. The head is short with a large orbit. Two species of this genus have been described in the Mazon Creek assemblage: *P. gracilis* (Newberry and Worthen) and *P. comptus* Westoll, but it would be unwise to assign new fossils to these species without examination of more Mazon Creek specimens, which may be in the hands of collectors. Indeed, D. Baird (personal communication) suggests that there is a new genus among the material referred to *P. gracilis* (Newberry and Worthen).

Holotypes. P. gracilis (Newberry and Worthen), YPM 440, Mazon Creek, Grundy County; *P. comptus* Westoll, USNM 4310, from the Braidwood fauna.

Genus *Parahaplolepis* Westoll
Figure 17.16

Westoll placed one Mazon Creek specimen in this genus and suggested that it was taxonomically close to a better known species from Linton, Ohio, *P.* cf. *tuberculata* (Newberry). Here too, examination of additional specimens is essential for proper systematic evaluation of haplolepids from the Mazon Creek biota.

Genus *Microhaplolepis* Lowney
Figure 17.17

Westoll (1944) recognized a single Mazon Creek specimen, which Lowney (1980) included in her new genus as *Microhaplolepis* sp. She compared it with better preserved specimens from Linton, Ohio, and suggested that it shows features of two species from that site; hence she left the specific identification open. D. Baird

Figure 17.17. ?*Microhaplolepis* sp., FMNH PF 5540, Pit 11. Scale bar = 10 mm.

(personal communication) suggests that specimens FMNH PF 5540 and FMNH PF 7498 should be referred to *M. serrata.*

Subclass Sarcopterygii

This is the second group within the Osteichthyes. The Mazon Creek biota includes three sarcopterygians: rhipidistians (which are a component of a larger group, the crossopterygians), actinistians (or coelacanths), and dipnoi (or lungfishes). Crossopterygian fishes, unlike actinopterygians, have two dorsal fins. The posterior dorsal fin and both paired fins include muscle masses extending into the fin, hence the name "lobe-finned" for these fishes. The scales also differ from those of actinopterygians in showing a covering material called cosmine. The rhipidistian component of the crossopterygians includes those fishes from which the amphibians were derived at the end of the Devonian. The coelacanths are represented today by one species, *Latimeria chalumnae,* found in marine waters off the Comoro Islands northwest of Madagascar. Careful examination of specimens collected after World War II helped researchers to understand the anatomy of the many fossil forms that have been known since the nineteenth century (Forey 1988). The third component of the sarcopterygians, lungfishes, today are restricted to freshwaters of the Southern Hemisphere in tropical South America, Africa, and Australia, but they were greatly diversified in the past and found throughout the Northern and Southern Hemispheres.

Order Crossopterygii
Suborder Rhipidistia

Most of the fishes of this group occur in rocks of Devonian age. Mississippian and Pennsylvanian forms are fewer and less well known. They are represented in the Mazon Creek biota only by fragments including scales and some isolated large bones.

Family Osteolepidae
Genus *Megalichthys* Agassiz

A Carboniferous genus with some specimens from North America. It is closely related to *Ectosteorachis* from the Permian of North America.

Megalichthys cf. *M. hibberti* Agassiz
Figure 17.18

This species is recorded from Pits 11 and 15 based on parts of individuals including the snout region, opercule, tooth-bearing elements, and scales. The best preserved part of this species in the Mazon Creek fauna is a snout region (Tully collection, 248) described by Schultze (1974). Scales assignable to *Megalichthys* are occasionally found. The exposed area of the scale is clearly marked by grooves anteriorly and dorsally where it is overlapped by other scales. Medially there is a slight dorsal keel. Pore canal openings (part of the lateral line sensory canal system) are visible in the exposed part of the scale. Some of the large, isolated bones that appear in Mazon Creek fauna concretions probably belong to this genus or to another rhipidistian.

Genus *Rhizodopsis* Young
Rhizodopsis cf. *R. robustus* Woodward
Figure 17.19

A single specimen of a fin covered with scales at its base. Scales are rather elongate and distally pointed. On the exposed posterior surface there are concentric ridges. Thin rays extend across these ridges. Five striae also cover the anterior field. The specimen comes from Pit 11 and is in the collection of the Illinois Geological Survey. This species is also known from Cannelton, Pennsylvania; Great Britain; and Silesia.

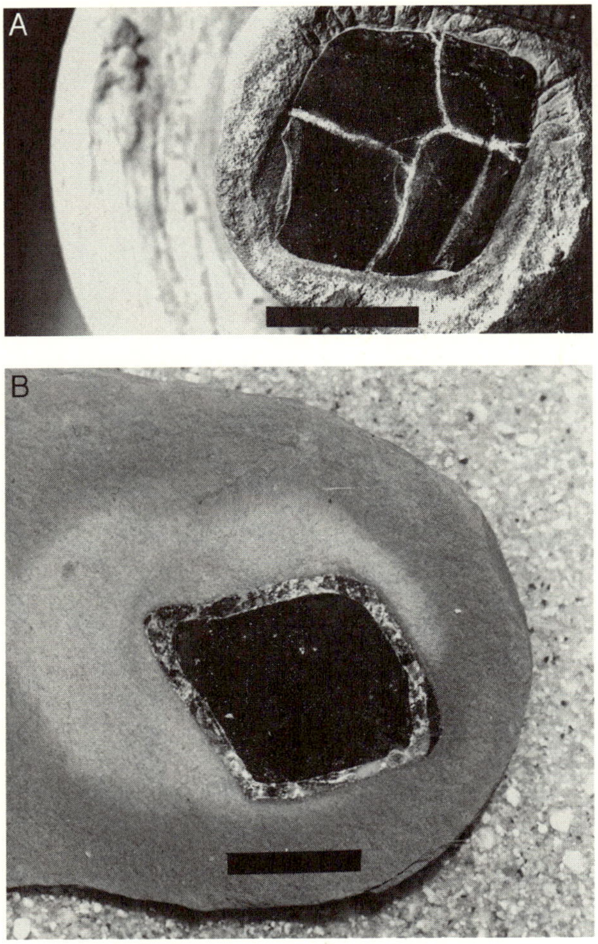

Figure 17.18. *Megalichthys* cf. *M. hibberti.* **A** FMNH PF 12375 (Herdina collection, H 502), Pit 11; external view of scale. **B** Wolff collection, 162, Pit 11; external view of scale. Both scale bars = 10 mm.

Rhizodopsis? *mazonius* Hay
Figure 17.20

A few isolated scales are referred to this species. The exposed area of the elongate scales is characterized by densely packed, fine, radially arranged striae. There is a triangular posterior field. Scales from different parts of the body vary significantly in shape and surface markings. Scales nominally from the same species may be quite distinct. Several specimens are known from Pit 11 and Mazon Creek.

Holotype. USNM 4337, Mazon Creek.

Suborder Actinistia

Coelacanths differ from rhipidistians in lacking several bones that make up the jaw and side of the head (e.g., maxilla and quadratojugal). They

Figure 17.19. *Rhizodopsis* cf. *R. robustus,* Illinois Geological Survey, Pit 11; ?dorsal fin and scales. Scale bar = 2 mm.

differ also in lacking internal choanae, that is, narial openings into the mouth cavity common to rhipidistians and tetrapods. The tail is diphycercal, which means that the body axis extends posteriorly in a straight line. The caudal fin is divided into three lobes. Specimens have been assigned to a single family, and three species have been named. Numerous specimens in museums and private collections require study.

Family Rhabdodermatidae
Genus *Rhabdoderma*

Forey (1981) and Lund and Lund (1985) consider *Rhabdoderma* to be a primitive coelcanth showing broad, partly overlapping cheek bones forming a complete cover behind the eye. The premaxilla carries few large teeth, while other tooth-bearing bones show fine granular teeth. The pelvic fin originates behind the first dorsal fin. The air bladder, where it is known, shows calcified walls, a rather easily seen character in the Mazon Creek biota specimens. The scales show a pattern of elongate ridges and tubercles, which converge posteriorly toward the midline of the scale. Specimens of juvenile individuals often show a circular expansion below the abdomen, which represents the yolk sac. Such small coelacanth fossils also possess a much elongated, central caudal lobe.

Rhabdoderma exiguum (Eastman)

An inadequately described species based on small specimens. None of the meristic data is unique to *Rhabdoderma,* and species identifi-

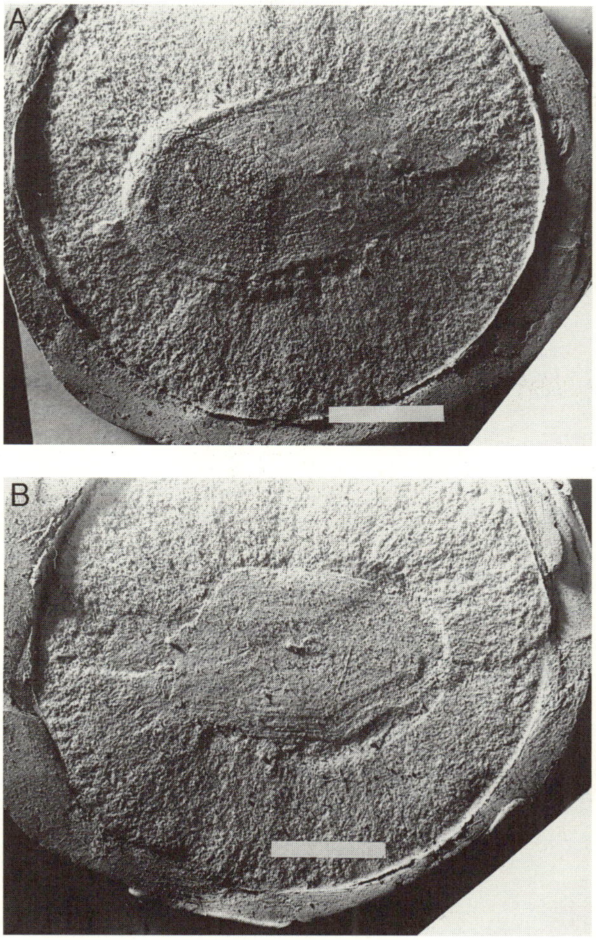

Figure 17.20. *Rhizodopsis? mazonius,* FMNH PF 3788 (Herdina collection, H 643), Pit 11. **A** External view of scale. **B** Internal view of scale. Both scale bars = 10 mm.

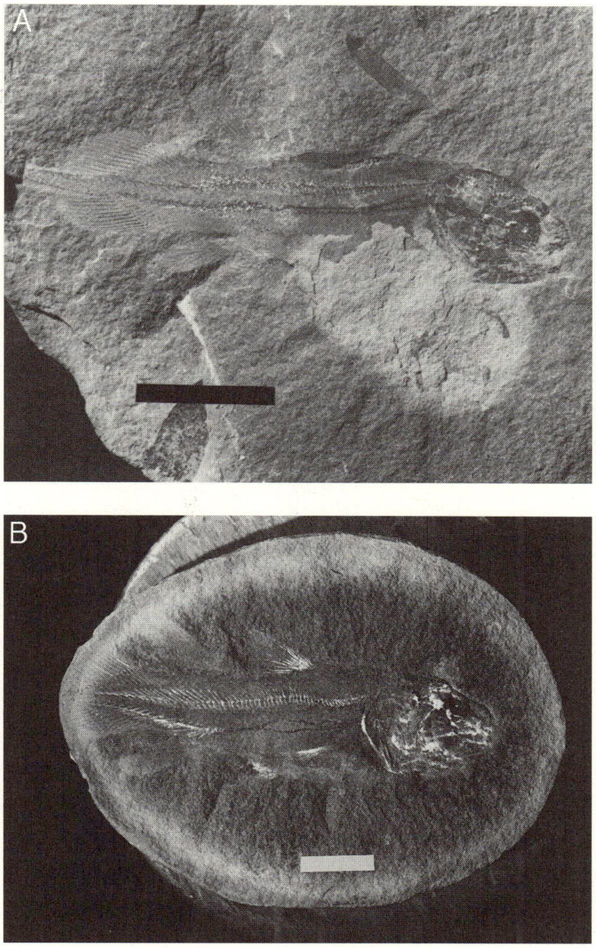

Figure 17.21. *Rhabdoderma elegans.* **A** FMNH PF 5760, Pit 11; lateral view showing yolk sac. **B** George collection, C 11–70, Pit 11; lateral view. Both scale bars = 10 mm.

cation is uncertain (Forey, 1981). Specimens are recognizable by the two dorsal fins, elongate supplementary (mid) caudal lobe, and the lobed paired fins.

Rhabdoderma robustus (Newberry)

A scale of uncertain specific affinity. Similar scales are found in the Linton, Ohio, fauna. This species probably should be included in *R. elegans* according to Lund and Lund (1985).
Holotype. USNM 4336, Mazon Creek.

Rhabdoderma elegans (Newberry)
Figure 17.21

This species also occurs at Linton, Ohio, and Cannelton, Pennsylvania, and is also the most widespread species in Europe. Mazon Creek fossils perhaps should be included here, but according to Lund and Lund (1985), the North American species *R. elegans* is specifically if not generically distinct from the English material of the same name.

Order Dipnoi

Lungfishes are numerous in both Braidwood and Essex parts of the Mazon Creek fauna. The overwhelming number of specimens are scales that often are of large size, 10+ cm in length. However, it is difficult to distinguish types of lungfishes based on the scales. Bones of the head skeleton of lungfishes are distinctly different from those of other fishes, and instead of the usual names applied to the layout of skull-roofing bones, letters of the alphabet are used to designate presumably homologous bones within the lungfishes.

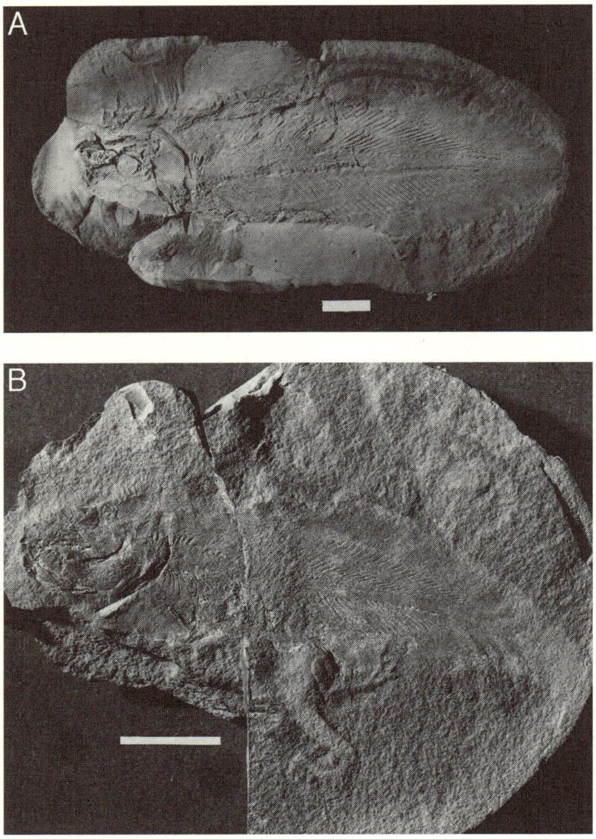

Figure 17.22. *Conchopoma edesi.* **A** Pit 15; latex peel of lateral view. **B** FMNH PF 5611, Pit 11; lateral view showing enlarged pelvic fin. Both scale bars = 10 mm.

Few Mazon Creek specimens are preserved in such a manner that the skull roof can be examined. Also, whole, intact, specimens are uncommon. Four genera, *Conchopoma, Ctenodus, Megapleuron,* and *Palaeophichthys,* are recognized from Mazon Creek deposits. One additional genus that possibly belongs among the Dipnoi, *Esconichthys,* also occurs here.

Conchopoma edesi Denison
Figure 17.22

This species is characterized by extremely long and relatively slender pectoral and pelvic fins (Denison, 1969). The caudal fin is diphycercal; that is, the body axis extends anteroposteriorly in a straight line. The dorsal, caudal, and anal fins are conjoined. The scales are thin and cycloidal (i.e., lacking marginal projections). Cone-shaped individual teeth rather than fused tooth plates lie on the roof and floor of the mouth. Marshall (1988) described the largest and most complete specimen of this species from Mazon Creek.

Conchopoma includes at least five species. In addition to *C. edesi,* there is the indeterminate (D. Baird, personal communication) *C. arctatum* from Vermilion County, Illinois, and *C. exanthematicum* from the Linton mine dump in Ohio. Hook and Baird (1993) suggest that the latter species may be the adult form of *C. edesi.* Two additional species are known from West Germany (Schultze, 1975). All except *C. edesi* are from freshwater deposits.

There are perhaps 15 specimens all from the Essex fauna (Pit 11) with one additional specimen from Pit 15.

Holotype. FMNH PF 5611, Pit 11.

Ctenodus cristatus Agassiz
Figure 17.23

According to Baird (1978), this is the common and possibly only lungfish found in the old collecting area along Mazon Creek in Grundy County as well as in coal pits within the Braidwood fauna. Most of the fossils are scales, and many of these reach 5–10 cm in length. Mazon Creek lungfish scales have been referred to *Sagenodus,* which is the common Carboniferous–Lower Permian lungfish, but these scales are indistinguishable from those of *Ctenodus* (Baird, 1978). Because other skeletal parts of lungfishes (except for the distinctive *Conchopoma* and *Megapleuron*) from the Mazon Creek fauna are considered *Ctenodus,* assignment of the scales to this genus is warranted. *Ctenodus* occurs in Nova Scotia and widely in Europe. There are probably upward of 200 to 300 scales of this species in museum and private collections. Other skeletal elements, especially the toothplates characteristic of many lungfishes, are rare. Based on scale material, this is the dominant lungfish of the Mazon Creek biota.

Megapleuron zangerli Schultze
Figure 17.24

This and *Palaeophichthys* are the only species of lungfish unique to the Mazon Creek fauna. Furthermore, this is the only record of this genus in North America, which is known else-

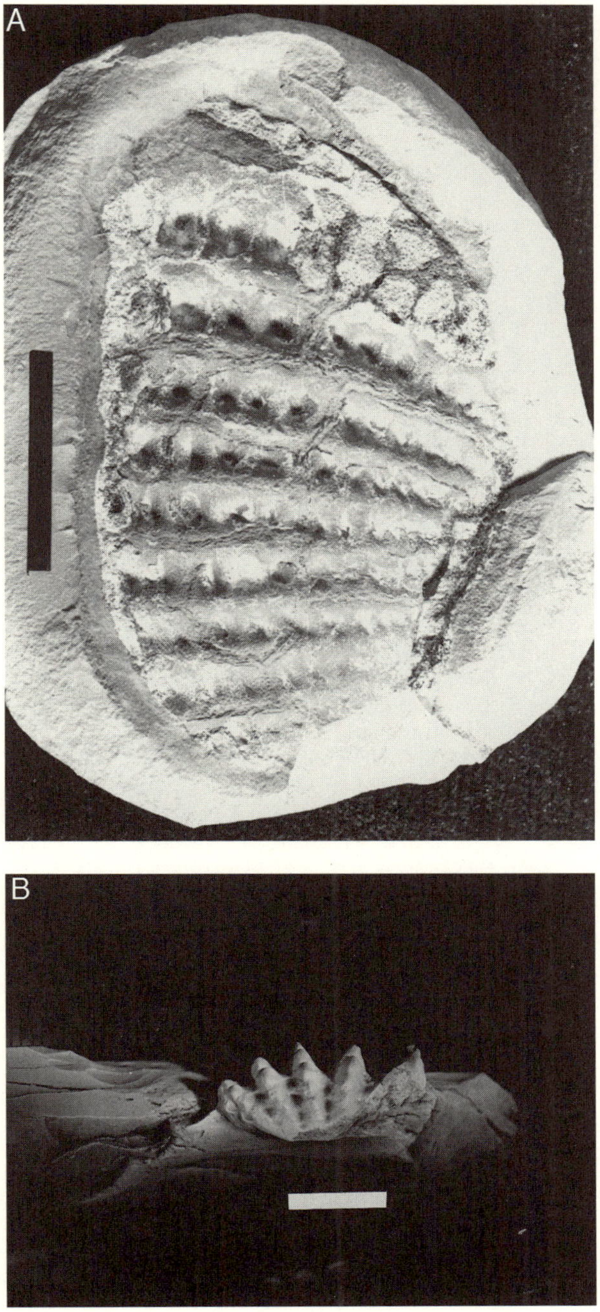

Figure 17.23. *Ctenodus cristatus.* **A** Mold of upper tooth plate. **B** Pit 15; dental plate. Both scale bars = 10 mm.

Figure 17.24. *Megapleuron zangerli,* FMNH PF 8369, Pit 11; head and anterior part of body. Scale bar = 10 mm.

where only from the Lower Permian of France as *M. rochei.* There are about 10 specimens from Pit 11. All are juvenile. Distinguishing this species from other lungfishes depends on relative sizes of the "alphabetic" bones of the skull roof, length of the dorsal fin, and shape of dental tooth plates (Schultze, 1977).

Holotype. Rockwell collection, 5323, Pit 11.

Palaeophichthys parvulus Eastman

An inadequately figured and described specimen presumably from the Mazon Creek Braidwood fauna was reported by Eastman (1908). Schultze and Baird (in preparation) have relocated the type (MCZ 5090a and b) are redescribing it. According to D. Baird (personal communication), it shows similarities to *Monongahela* Lund (1970).

Dipnoi?
Esconichthys apopyris Bardack
Figure 17.25

Specimens are preserved dorsoventrally or laterally compressed as well as partially twisted. Laterally compressed specimens typically show 25 to 30 myomeres (body muscle segments). There are two pairs of elongate, external gills; the posterior pair is usually half the length of the body. Each gill bears numerous fine filaments along the distal half of the gill bar. Two pairs of hard structures that cannot be identified (they are not otoliths) as specific parts of the skull lie posteromedially to the eyes. These specimens are probably larval based on the lack of ossification, presence of gills, and simple myomeric development. Taxonomic assignment of *Esconichthys* to a group of vertebrates is difficult.

Bardack (1974) suggested reasons why it might be a larval lungfish, and Schultze (1977) argued that it is probably not a lungfish or an

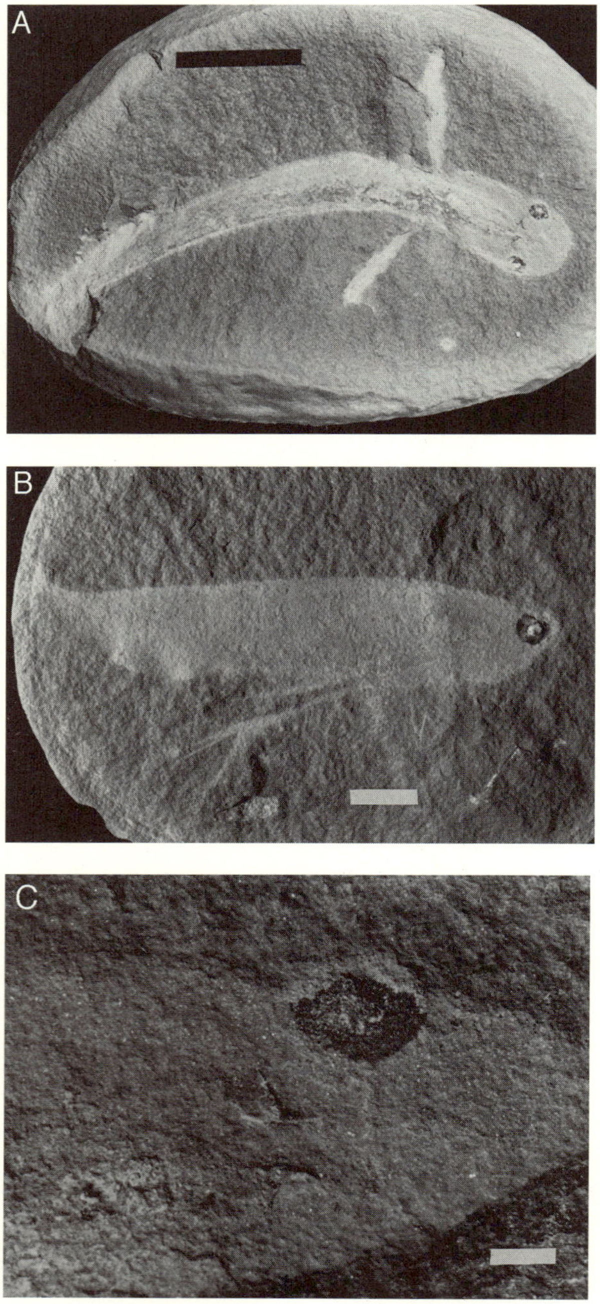

Figure 17.25. *Esconichthys apopyris.* A FMNH PF 3028, Pit 1 or 6; dorsal view. Scale bar = 10 mm. B Lateral view. Scale bar = 10 mm. C Wolff collection, 637, Pit 11; dorsal view of head with eyes and "hard structures." Scale bar = 2 mm.

amphibian without providing an alternative placement. This form, commonly called by collectors "blades," "grasshoppers," or "tadpoles" based on their preserved shapes, is the most common vertebrate in the Mazon Creek fauna, even if one counts all of the lungfish scales. All specimens are from the Essex fauna.

References

Baird, D. 1978. Studies on Carboniferous freshwater fishes. Amer. Mus. Novitates, 2641:1–22.

Bardack, D. 1974. A larval fish from the Pennsylvanian of Illinois. Jour. Paleo., 38:988–993.

———. 1991. First fossil hagfish (Myxinoidea): a record from the Pennsylvanian of Illinois. Science, 254:701–703.

Bardack, D., and E. Richardson, Jr. 1977. New agnathous fishes from the Pennsylvanian of Illinois. Fieldiana Geol., 33(26):489–510.

Bardack, D., and R. Zangerl. 1971. Lampreys in the fossil record, pp. 67–84. *In* M. W. Hardisty and I. C. Potter (eds.), The Biology of Lampreys, vol. 1. Academic Press, New York.

———. 1979. Fishes of the Mazon Creek fauna. *In* M. H. Nitecki (ed.), Mazon Creek Fossils. Academic Press, New York.

Carroll, Robert. 1988. Vertebrate Paleontology and Evolution. Freeman, San Francisco.

Denison, R. H. 1969. New Pennsylvanian lungfishes from Illinois. Fieldiana Geol., 12(12):193–211.

Eastman, C. R. 1908. Devonian fishes of Iowa. Iowa Geol. Surv. Ann. Rep., 18[1907]:252–254.

Forey, P. L. 1981. The coelacanth *Rhabdoderma* in the Carboniferous of the British Isles. Paleontology, 24, pt. 1:203–229.

———. 1988. Golden Jubilee for the coelacanth *Latimeria chalumnae.* Nature, 336:727–732.

Hook, R. W., and D. Baird. 1993. A new fish and tetrapod assemblage from the Allegheny Group "Late Westphalian, Upper Carboniferous" of eastern Ohio, U.S.A. *In* U. Heidtke, New research on Permo-Carboniferous Faunas. Pollichia, 29:143–154.

Janvier, P., and R. Lund. 1983. *Hardiestella montanensis* n. gen. et sp. (Petromyzontida) from the Lower Carboniferous of Montana with remarks on the affinities of the lampreys. Jour. Vert. Paleont., 2:407–413.

Lowney, K. A. 1980. A revision of the family Haplolepidae (Actinopterygii, Paleonisciformes) from Linton, Ohio (Westphalian D, Pennsylvanian). Jour. Paleo., 54:942–953.

———. 1983. The earliest (Namurian, A, E) haplolepids (Osteichthyes: Actinopterygii). Trans. R. Soc. Edinburgh Earth Sci., 74:69–78.

Lund, R. 1970. Fossil fishes from southwestern Pennsylvania. Part 1, Fishes from Duquesne Limestone, Conemaugh, Pennsylvanian. Ann. Carnegie Mus., 41(8):231–261.

Lund, R., and W. L. Lund. 1985. Coelacanths from the Bear Gulch Limestone (Namurian) of Montana and the evolution of the coelacanthiformes. Bull. Carnegie Mus., 25:1–74.

Maisey, J. G. 1986. Heads and tails: a chordate phylogeny. Cladistics, 2(3):201–256.

Maples, C. G., and H.-P. Schultze. 1989. Preliminary comparison of the Pennsylvanian assemblage of Hamilton, Kansas, with marine and non-marine contemporaneous assemblages. Kansas Geol. Surv. Handb. Ser., 6:253–273.

Marshall, C. R. 1988. A large, well preserved specimen of the middle Pennsylvanian lungfish *Conchopoma edesi* (Osteichthyes: dipnoi) from Mazon Creek, Illinois, U.S.A. Jour. Vert. Paleont., 8:383–394.

Patterson, C. 1982. Morphology and interrelationships of primitive actinopterygian fishes. Amer. Zool., 22:241–259.

Schultze, H.-P. S. 1974. Osteolepididae Rhipidistia (Pisces) aus dem Pennsylvanian von Illinois U.S.A. Neues Jahrb. Geol. Paleo. Abh., 146:29–50.

———. 1975. Die Lungenfisch Gattung *Conchopoma* (Pisces, Dipnoi). Senckenbergiana Lethaea, 56:191–231.

———. 1977. *Megapleuron zangerli,* a new dipnoan from the Pennsylvanian, Illinois. Fieldiana Geol., 33(21):375–396.

Schultz, H.-P., and D. Bardack. 1987. Diversity and size changes in palaeonisciform fishes (Actinopterygii, Pisces) from the Pennsylvanian Mazon Creek fauna of Illinois, U.S.A. Jour. Vert. Paleont., 7:1–23.

Schultz, H.-P., and C. G. Maples. 1992. Comparison of the Late Pennsylvanian faunal assemblage of Kinney Brick Company Quarry, New Mexico, with other late Pennsylvanian Lagerstatten. New Mexico Bur. Mines Min. Res. Bull., 138:231–242.

Westoll, T. S. 1944. The Haplolepidae, a new family of Late Carboniferous bony fishes. Bull. Amer. Mus. Nat. Hist., 83:1–122.

Zangerl, R. 1973. Interrelationships of Early Chondrichthyans. *In* P. H. Greenwood, R. S. Miles, and C. Patterson (eds.), Interrelationships of Fishes. Jour. Linn. Soc. London Zool., 53, Suppl. 1:1–14.

Zidek, J. 1976. Kansas Hamilton Quarry (Upper Pennsylvanian) *Acanthodes,* with remarks on the previously reported North American occurrences of the genus. Univ. Kansas Paleo. Contr., 83:1–41.

———. 1992. Late Pennsylvanian Chondrichthyii, Acanthodii, and deep-bodied actinopterygii from the Kinney Quarry, Manzanita Mountains, New Mexico. New Mexico Bur. Mines Min. Res. Bull., 138:145–182.

18

Cartilaginous Fishes (Chondrichthyes)

Rainer Zangerl
Including a Description of a New Fossil Shark by David Bardack

Cartilaginous fishes, including sharks and their allies, are rare in the Francis Creek Shale. Yet both groups of chondrichthyans, Elasmobranchii (sharks) and the Subterbranchialia (which includes the still extant chimaeroids), are represented in the Mazon Creek fauna.

Because the majority of Mazon Creek specimens are very young individuals lacking well-calcified skeletons, their relationship to other Paleozoic chondrichthyans like those of the Mecca fauna can be determined only in broad terms. It can be stated with confidence, however, that most of the chondrichthyans of the Mazon Creek fauna are not simply the young of those preserved in the Mecca Quarry Shale member, a thin, black, sheety organic-rich shore unit deposited elsewhere in the Illinois Basin at about the same time, but apparently represent the young of animals occupying another habitat. The following account is based mainly on Zangerl (1979, 1981).

Subclass Subterbranchialia

Most of the Subterbranchialia are "exotic"-looking fishes; they do not resemble any living or fossil sharks. Instead, they tend to have the shape of tadpoles possessing large heads with relatively slim bodies. The cartilaginous shoulder girdle is close behind the head, with crowded gill arches in front of the girdle and mostly beneath the braincase, just as they are in the modern chimaeroids. Unfortunately, specimens with perfectly intact gill baskets have not yet been found among any of the fossil species; hence, the above definition of the group must be considered preliminary.

Polysentor gorbairdi Zangerl, 1979
Figure 18.1

The holotype specimen of this species is a nearly entire male skeleton, preserved in lateral body position, lacking only the tail fin (FMNH PF 8445). The specimen was found near Braceville, Grundy County, where the Francis Creek Shale contains fossils of the Essex fauna. The braincase, flattened into a near two-dimensional state, yields little useful information. It is relatively large and blunt in front. The palatoquadrates (upper jaws) are free (not fused to the braincase), and unfortunately, no teeth are visible (Figure 18.2). The lower jaws (Meckel's cartilages) are rather nondescript pieces of cartilage, although both are still in articulation with the posterior ends of the palatoquadrates. The gill arches are located beneath the back half of the braincase and are closely followed by a very prominent shoulder girdle (scapulocoracoid) whose dorsal end is attenuated and points backward. This cartilage provides an approximate measure of the dorsoven-

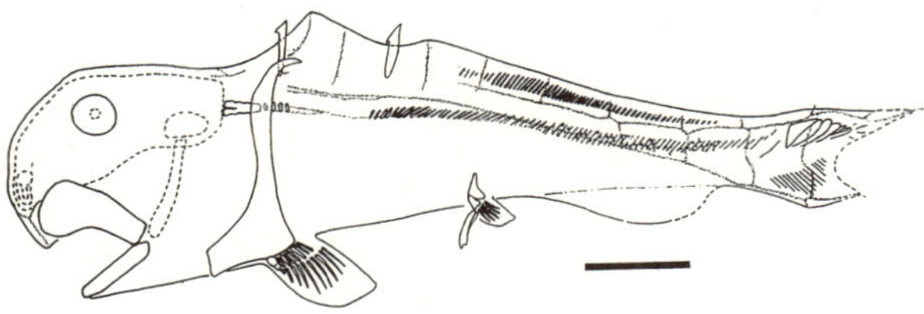

Figure 18.1. *Polysentor gorbairdi,* holotype, FMNH PF 8445 (coll. Gordon Baird). Scale bar = 10 mm.

Figure 18.2. *Polysentor gorbairdi,* reconstruction based on FMNH PF 8445 and PF 8548 (coll. Jerry Herdina). Scale bar = 10 mm.

tral dimension of the body in life. The pectoral fins are simple: a row of about nine basal rods (attached directly to the shoulder girdle) is distally followed by 10 radials. The pelvic fin complex is subject to different interpretations depending on whether the structures visible represent the right and left elements superimposed or only one side is preserved. The pelvic cartilages are roughly triangular, the fins consist of very few radials, and the clasper resembles somewhat the terminal element of the clasper complex of Paleozoic sharks such as *Cobelodus* or *Denaea* (Zangerl and Case, 1976; Williams, 1985).

The vertebral column consists of a series of short cartilage rods that represent the neural arch pieces; between braincase and shoulder girdle these pieces are fused to form a synarcual (Figure 18.2). At the other end of the column, in the tail region, the neural arch pieces are much expanded to form plates in the plane of the fin.

Confluent with the dorsal lobe of the tail fin there is a long dorsal fin extending forward to the back of the head. For part of the way this is supported by calcified radials, but anterior to the level of the pelvic fins, none are visible and presumably were uncalcified. That a fin was present in this region is indicated by the presence of two spines and three linear strands of dermal denticles, which mark the position of sensory canals. There is some evidence that the animal had an extended anal fin that was apparently unsupported by calcified radials.

The spines are unusual: they possess pronounced barbs at the upper ends and the barbs face forward. A larger spine of this construction has been collected in the Excello Black Shale of southern Indiana.

The lateral line system of the body is most interesting, and the generic name *Polysentor* alludes to its condition. The lateral (sensory) lines are marked by double rows of tiny denticles. Two such strands (on the right and left sides of the body wall; Figure 18.2) run parallel to the vertebral column then descend into the lower lobe of the tail fin and follow its ventral border; two very short strands run to the very edge of the fin membrane. From the lateral lines at least six vertical commissures connect with an unpaired strand that probably runs along the base of the dorsal fin. From this middorsal strand no fewer than nine vertical branches extend up into the web of the dorsal fin and the dorsal lobe of the tail fin.

The denticles that follow the lateral lines and other sensory tracts vary slightly in shape but mainly are simple, hollow cones (lepidomoria), many virtually identical with those illustrated by Zangerl (1966, 1981) of *Eugeneodus richardsoni* (in 1966 thought to belong to *Agassizodus*). A thin spread of dermal denticles—hollow cones with widely flaring, hollow bases—occurs in FMNH PF 8548 in the area of the somewhat displaced vertebral column, but other than that the body is free of denticle cover.

The recovery of a much larger spine of the type described above from the Excello Shale at the Bethel Quarry, Pike County, Indiana, suggests that this fish is also a member of the Mecca fauna and that fully grown specimens may be found in the black shale deposits.

Behind the pectoral fins of the holotype specimen there is a small quantity of stomach or in-

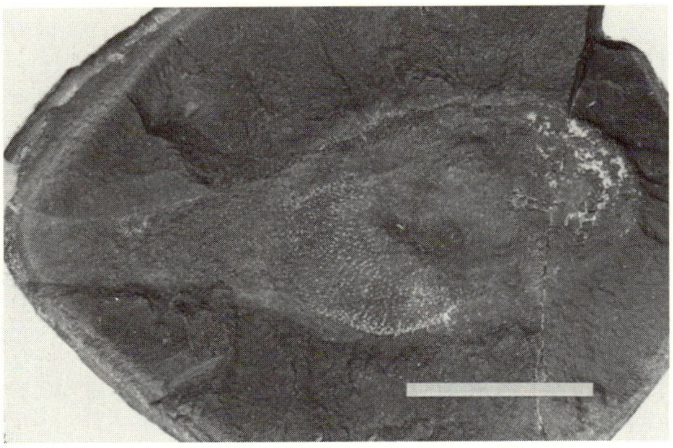

Figure 18.3. *Jimpohlia erinacea,* holotype, FMNH PF 8539 (donated by James Pohl). Scale bar = 10 mm.

testinal remains of food, including two ostracode shells. There is little doubt that these have been eaten; indeed ostracodes may have been the chief food of this species during infancy.

Jimpohlia erinacea Zangerl, 1979
Figure 18.3

One very immature specimen of this species has been reported. The holotype, FMNH PF 8539, is a nearly whole animal, lacking only the tail. It is preserved approximately in lateral burial position on both halves of a split concretion.

Excepting the outline of the body, dermal denticles and the dentition are the only structures that are visible; but the attitudes and the spacing of the denticles in the abdominal region permit the inference that the animal's body cavity became bloated after death.

The whole body, except the head region with its dentition, is covered with needle-sharp denticles that were anchored in the skin by sturdy little tripods, giving the animal an appearance similar to that of the modern porcupine fish *Diodon.* Along the middorsal line of the neck region there are three enlarged denticles with large internal cavities and flaring bases (Figure 18.4). The middle one of these is nearly twice the size of the others and may have grown to become a spine in the adult condition.

What makes this specimen most interesting is its dentition. The individual teeth are preserved as negatives in the siderite matrix, and they stand in a variety of attitudes; but in general the crowns are seen on one half of the concretion, the basal views of the teeth on the other half. The teeth are small, the largest being about 1 mm in anteroposterior dimension (as oriented in the dentition), and many of them no longer occupy their respective places in the original dentition. For these reasons there remain uncertainties in my interpretation of this dentition, and the reader is referred to the original description for the details (Zangerl, 1979).

There are two sets of teeth, which probably belong to the right and left upper jaws (Figure 18.5). These teeth appear to have retained, fairly well, their original positions in the dentition, which apparently consisted of four tooth families (or tooth files) on each side, each consisting of three teeth of differing degrees of ontogenetic development, including size. The largest tooth in each family is the outermost one, followed medially by a smaller, less mature one, which in turn is followed by a still smaller, clearly immature tooth. All of the teeth, including the largest ones, are strongly concave on the undersides, meaning that they all are immature teeth.

In addition, there are four conical teeth of differing sizes located between the right and left sets of tooth families. These teeth have widely flaring bases and are best interpreted as forming a symphyseal tooth row with the largest tooth again being the oldest.

In modern sharks, teeth similarly occur in tooth families with the younger teeth standing medial to the older ones; but once a tooth is fully developed it becomes functional on the biting crest of the jaw and stops growing. After a time the functioning tooth is replaced by the next one in line and falls off. In a young, growing individual the replacement teeth are larger than the ones they replace because the dentition has to keep pace with the growth of the mouth.

In *Jimpohlia* the replacement teeth located in the middle of each family have rather well differentiated crowns and may have been functional, but they are decidedly smaller than their lateral neighbors. This differs drastically

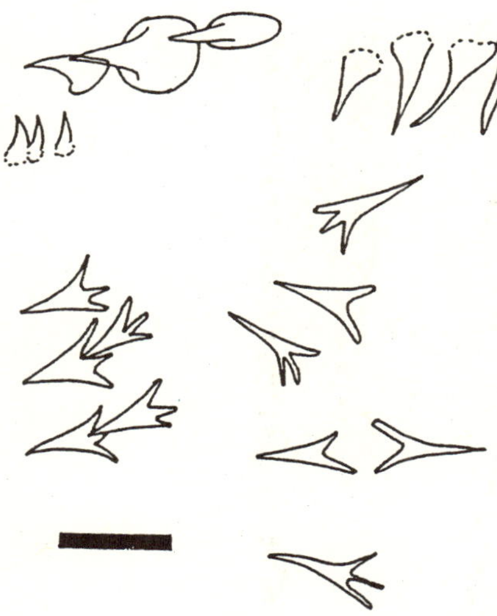

Figure 18.4. *Jimpohlia erinacea: upper left,* enhanced dermal denticles along mid-dorsal line of neck region; *upper right,* denticles along ventral edge of abdomen; *lower left,* denticles from less bloated side of abdomen; *lower right,* denticles from more distended side of abdomen. Scale bar = 1 mm.

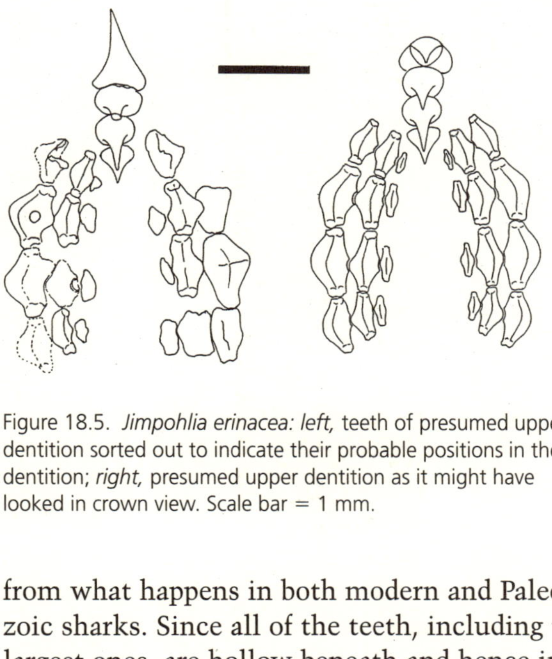

Figure 18.5. *Jimpohlia erinacea: left,* teeth of presumed upper dentition sorted out to indicate their probable positions in the dentition; *right,* presumed upper dentition as it might have looked in crown view. Scale bar = 1 mm.

from what happens in both modern and Paleozoic sharks. Since all of the teeth, including the largest ones, are hollow beneath and hence immature, we may conclude that in this species tooth growth did not stop with the achievement of apical position on the jaw and hence function, but that the teeth kept on growing in concert with the growing jaws. Tooth replacement was thus probably very slow. The consequences of such a replacement mode would be rather profound changes in the morphology of the crown relief with age, not to mention wear and perhaps even dissolution of crown surface tissues by stomach acid accompanying the periodically regurgitated gastric residues. Indeed, Woodward (1921) has suggested very slow tooth replacement in a group of Paleozoic chondrichthyans that he called bradyodonts, and Patterson (1968) has pointed out that no bradyodont teeth have ever been seen to possess a coat of orthodentine (pallial dentine) over the characteristic tubular trabecular dentine, let alone a superficial layer of vitrodentine, both typical tissues in the architecture of shark teeth.

Based on the above considerations *Jimpohlia* has been tentatively placed among the bradyodonts.

Similihariotta dabasinskasi Zangerl, 1979
Figure 18.6

The holotype of this species is the complete outline of a juvenile specimen in lateral burial position, FMNH PF 8547. Only one side of a split concretion containing the fish has been recovered, but it was returned to its original owner and its present whereabouts are unknown.

The body angle of this specimen (Figure 18.7) resembles rather strikingly that of the modern chimaerid *Harriotta,* with its elongated snout, stout thorax, and slender, pointed tail. However, it lacks a spine in the neck region, and there are no tooth plates in the area of the mouth. Two small calcifications near the eye spot might be parts of slightly dislocated tooth plates, but the substance gives no indication of that. The absence of any other calcified remains of the skeleton is probably attributable to the immaturity of the individual.

A chimaerid has been described from the Late Mississippian Bear Gulch fauna of Montana by Lund (1977), hence the plausible presence of a chimaerid in the Essex fauna. Whether it is a Paleozoic relative of the extant genus *Harriotta* is uncertain and cannot be decided on the basis of this single, immature specimen. Chondrichthyans with elongated snouts were evidently common in the Pennsylvanian (e.g., *Caseodus basalis* and *C. eatoni, Ornithoprion hertwigi, Bandringa rayi,* and *B. herdinae*), as well as another fish in the Mazon

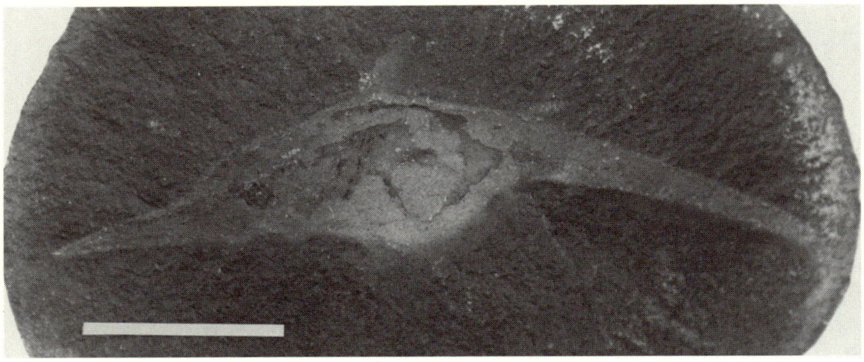

Figure 18.6. *Similihariotta dabasinskasi,* holotype, FMNH PF 8547. Scale bar = 20 mm.

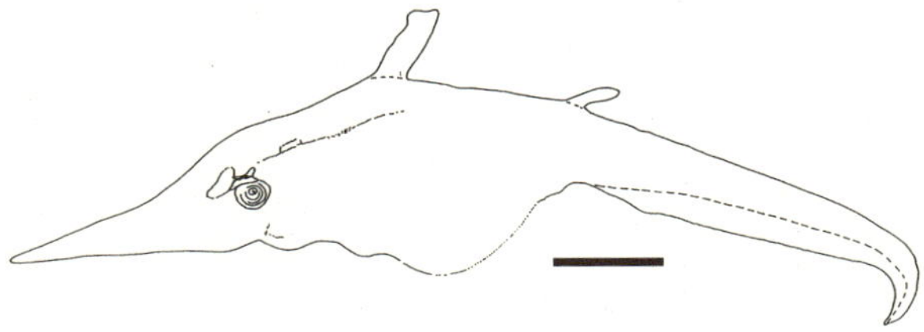

Figure 18.7. *Similihariotta dabasinskasi,* drawing made with a Wild stereo microscope drawing tube. Scale bar = 10 mm.

Creek fauna of which only a relatively small fragment of a very large snout is preserved that might be allied to the Mesozoic *Raineria.* Under these circumstances it is possible that the similarity between this animal and the modern *Harriotta* is due to convergence.

Subclass Elasmobranchii
Bandringa rayi Zangerl, 1969
Figure 18.8

The holotype of this species, FMNH PF 5686, is a sharp imprint of a young shark, displayed on both halves of a split concretion. Only the very tip of the snout and the tip of the tail fin are missing; these extended beyond the "reach" of the concretion.

The most striking aspect of this animal is its enormously elongated snout. Probably because of immaturity, the skeleton was uncalcified at death and hence not preserved. But behind the base of the long rostrum there are both eye spots consisting of a dark brown film surrounding tiny vugs of white calcite crystals that probably formed in place of the vitreous bodies of the eyeballs.

Behind the circular eye spot (and partly overlapping it) minute pin pricks representing cladodont type teeth are arranged in a V pattern and preserved as negatives only (Figure 18.9). The largest of these teeth measures 30.4 μm (less than 0.5 mm) from its base to the tip of the crown.

The specimen has a dorsal spine in front of each of the two dorsal fins. The first spine is smaller than the second and stands at a steeper angle (55°) to the horizontal axis of the body than the second spine (52°; Figure 18.9). This arrangement is the opposite of what is usually seen in ctenacanthoid and hybodontoid sharks. Ventral to the second dorsal fin is a substantial anal fin; immediately behind it lies the dorsal lobe of the tail fin, the ventral lobe being essentially indistinguishable from the dorsal lobe.

The paired fins of the right and left sides are either perfectly superposed, or one set was lost before the concretion could preserve it. The pectoral fins are much larger than the pelvics (Figure 18.9).

Since the description of the type specimen in 1969 several additional specimens have been collected, only one of which is as nicely preserved as the type, and none is significantly larger. Baird (1978), however, described a much larger individual, from the Upper Kittanning Coal at Cannelton, Pennsylvania, with a calcified head and rostrum, and cladodont teeth with stout, recurved, labially fluted cusps. While the type specimen only shows minute, monocuspid teeth, this more mature individual has teeth with side cusps. Baird could show that the rostrum consists of a central cultriform process and paired, upward-curved, ventral selinoid processes (having the cross section of celery stalks), which in the type specimen are represented by the "fringes" that border the main rostral process. The orbits are relatively

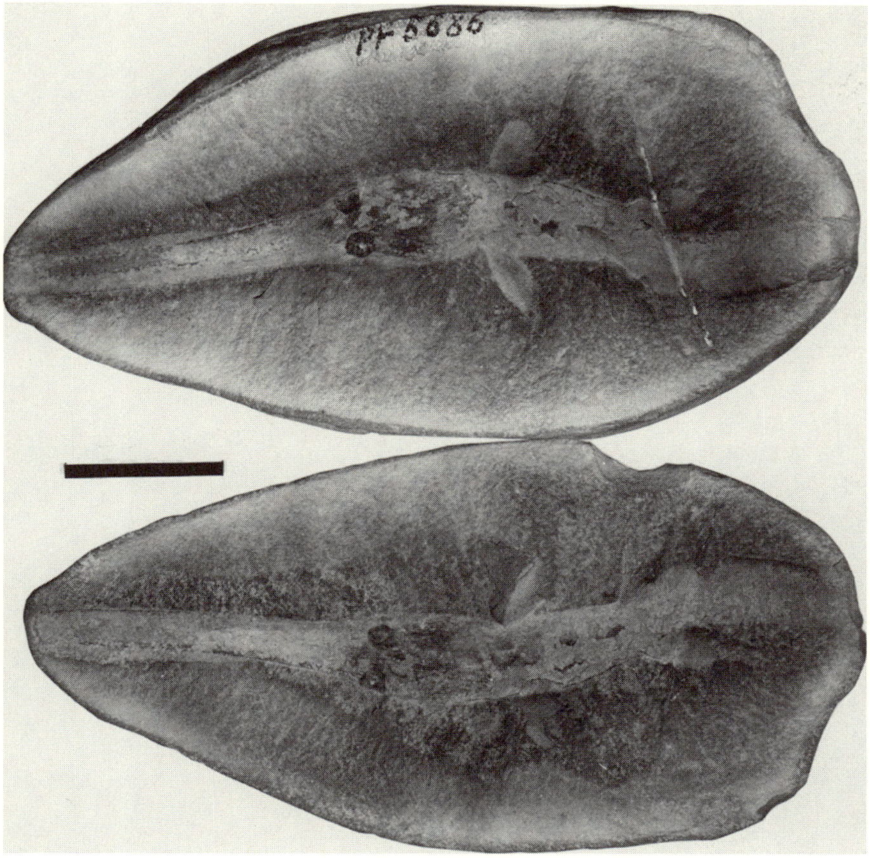

Figure 18.8. *Bandringa rayi,* holotype, FMNH PF 5686 (donated by Ray Bandringa). Scale bar = 20 mm.

small and rather far forward in the braincase. The palatoquadrate does not have a pronounced dorsal (otic) process, hence does not have the appearance (in side view) of a meat cleaver (Figure 18.10).

The Cannelton specimen of *B. rayi* was preserved in an unquestionable freshwater deposit, whereas the type (and to my knowledge all subsequent finds of this species) came from Pit 11, that is, from the predominantly marine Essex concretion fauna (Johnson and Richardson, 1966). By contrast, all but one of the specimens of *B. herdinae* originated from Pit 1 or 6 of the old strip mine area containing the predominantly freshwater Braidwood concretion fauna.

The most likely explanation of this apparent difference in environmental preference of these animals is that they were euryhaline, that is, capable of tolerating both salt and freshwater (as do many modern sharks). It is possible that in the Illinois Basin the young of *B. rayi* and *B. herdinae* may have preferred the salt and freshwater environments, respectively.

Among the Paleozoic sharks *Bandringa* belongs to the ctenacanthoid sharks as defined in Zangerl (1981).

Bandringa herdinae Zangerl, 1979
Figure 18.11

When the late Jerry Herdina donated his fabulous Mazon Creek fossil collection to the Field Museum of Natural History, it was found to contain five specimens of a very immature shark, apparently a close relative of *B. rayi.* Mr. Herdina had collected all five specimens in either Pit 1 or Pit 6 of the old strip mine area east of Mazon Creek. In view of the large number of collectors searching the spoil heaps of those old pits over many years it seems odd that this fish has not turned up in any other collections.

The five specimens are not of equal size and range, as reconstructed; they range from about 50 mm (FMNH PF 8542) to about 70 mm (PF 8541) to about 130 mm (PF 8439, PF 8436, PF 8442; Figure 18.12).

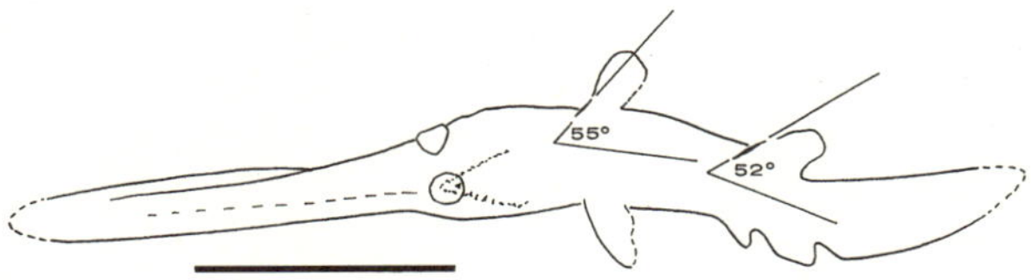

Figure 18.9. *Bandringa rayi,* drawing of holotype as preserved, showing eye spots, part of the dentition and two teeth, and attitudes of fin spines. Drawing made with a Wild stereo microscope drawing tube. Scale bar = 30 mm.

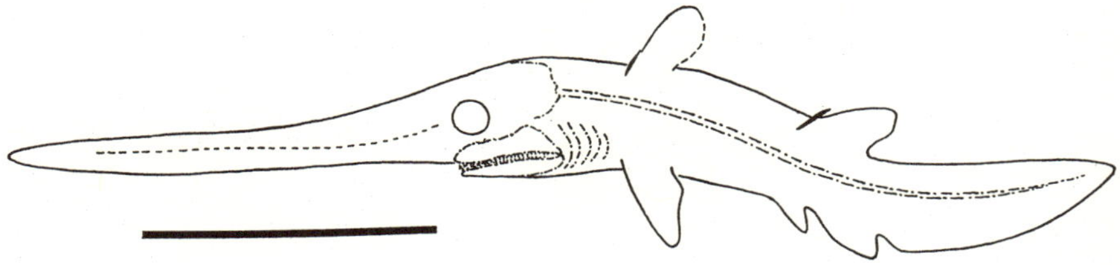

Figure 18.10. *Bandringa rayi,* reconstruction with addition of presumed jaws and gill arches. Scale bar = 30 mm.

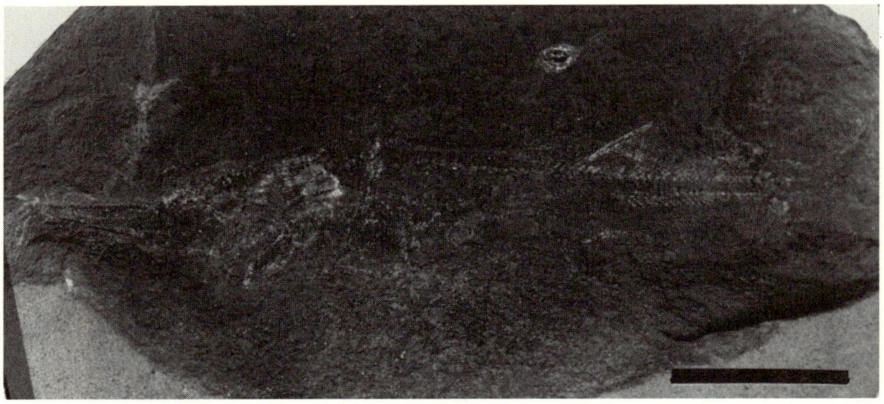

Figure 18.11. *Bandringa herdinae,* holotype, FMNH PF 8442. Scale bar = 20 mm.

Another very small specimen (about 50 mm as reconstructed), PF 8540, was collected in Pit 11 and is from the marine Essex fauna.

The holotype of this species is PF 8442, one of the largest specimens contained on both halves of a split nodule. It consists of the weakly calcified skeleton except for the tail, and the very tip of the rostrum. It is preserved in side position.

Bandringa herdinae has an elongated rostrum, but this is much more slender and only about half as long as in the other species, *B. rayi,* at about the same overall size of the body (Figure 18.12). The braincase is represented in some of the specimens by bulges in the surface relief between the eye spots and the shoulder girdle, but it is not clear what this surface relief represents in anatomical terms. What can be said about the braincase with some confidence is that the portion behind the postorbital processes was much elongated, as in *"Cladodus"* (Gross, 1937) and the xenacanths.

Of the visceral arches only the mandibular arch (palatoquadrate and Meckel's cartilage) is calcified, and this is seen only in the type specimen. Because right and left elements are often superposed in lateral burial position but are not perfectly congruous, the determination of the exact shape of these cartilages is difficult. As preserved in the holotype, the palatoquadrate resembles very much that of *B. rayi* as determined in the much larger Cannelton specimen (see above) by Baird (1978). It is a dorsoventrally expanded cartilage that narrows slightly toward the front and lacks a pronounced otic process. The lower jaw is dorsoventrally expanded and strongly curved, which is almost certainly an artifact of preservation. No teeth are visible in any specimen. They may have been very small and delicate and subject to dissolution in the acidic Braidwood burial environment.

The vertebral column is relatively well calcified; anterior to the second spine only the neural arch rodlets are calcified, but in the tail peduncle there are additionally haemal arch pieces. The structural supports of the tail fin are not calcified (Figure 18.12d–18.12f).

The shoulder girdle consists of a long scapu-

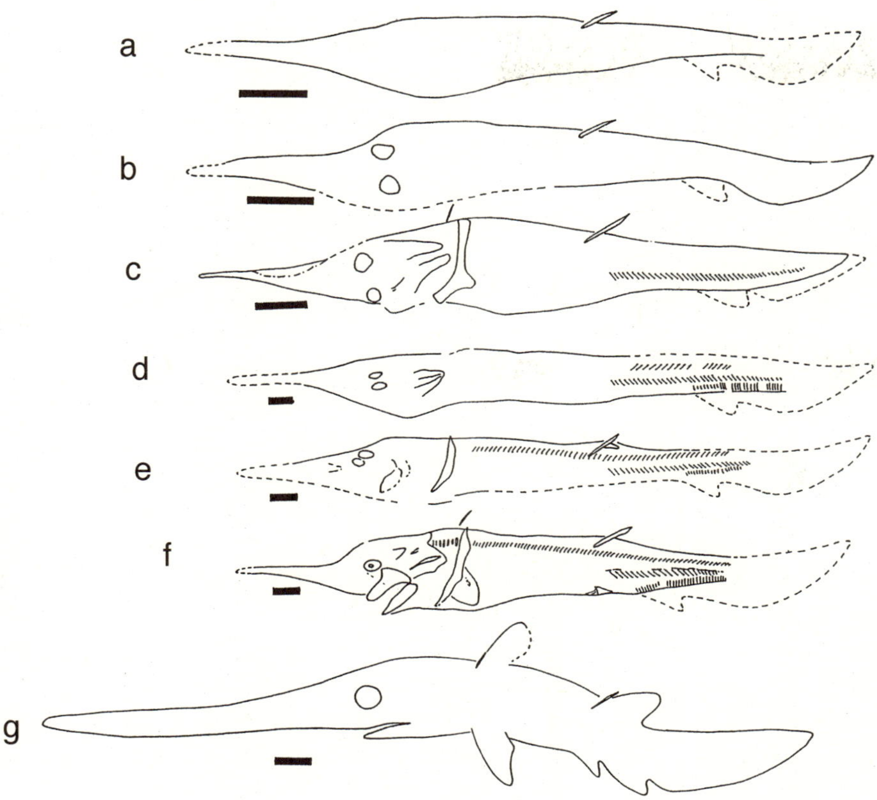

Figure 18.12. Specimens of *Bandringa herdinae:* **a** FMNH PF 8542; **b** FMNH PF 8540 (donated by Mr. and Mrs. F. A. Wolff); **c** FMNH PF 8541; **d** FMNH PF 843; **e** FMNH PF 8536; **f** holotype, FMNH PF 8442; **g** outline of *B. rayi* for comparison. All scale bars = 5 mm.

lar shaft, a short and blunt-ending coracoidal process, and a pronounced joint projection for articulation with the pectoral fin. In the type specimen there is a smooth area adjacent to the shoulder girdle that might represent the fin membrane, but no radials are visible.

Nothing is seen of the dorsal fins, though their positions are marked by the location of the spines. The anterior spine, just above the shoulder girdle, is much smaller than the second one and stands at an angle of about 60° to the horizontal body axis. The second spine, however, is relatively robust and larger than in *B. rayi.* About three-fourths of this spine is implanted in the body, and it is associated with a triangular cartilage that is attached to it posteriorly. It stands at an angle of about 25° to the longitudinal axis of the body.

A short distance behind the pelvic cartilage fragment of the type specimen there are two series of short rodlets separated by a small gap. The anterior set probably supported the anal fin, while the larger set behind the gap belongs to the tail fin.

The dermal skeleton consists of a fairly dense spread of conical, hollow denticles with flaring bases; their crowns are strongly bent backward. In one specimen, FMNH PF 8439, there are several longitudinal rows of enlarged denticles apparently located on the dorsal surface of the head region. These denticles are of the same structure as the ordinary dermal denticles elsewhere on the body, but their bases are more widely flaring, forming wide-open cones. Seen from beneath these denticles look like little funnels. In another specimen, PF 8541, a few larger denticles are irregularly distributed over the entire body.

While all specimens of *Bandringa* in the Mazon Creek fauna are unquestionably young individuals, the onset of calcification of the skeleton clearly occurred at an earlier age in *B. herdinae* than in *B. rayi.*

Dabasacanthus inskasi Zangerl, 1979
Figure 18.13

Another ctenacanthoid shark, but one showing some hybodont characters, is *D. inskasi.* The holotype is FMNH PF 8546, a nearly entire specimen, lacking the tail fin. It is a very young

Figure 18.13. *Dabasacanthus inskasi,* holotype, FMNH PF 8546. Scale bar = 10 mm.

individual seen on both halves of a split nodule. A second specimen has since been found.

The holotype is mainly preserved in lateral burial position with its head region sharply bent downward. The internal skeleton is uncalcified except for the upper and lower jaws, which made sharp impressions. Both large dorsal fins are preceded by large, nearly straight spines. Nothing is visible of the pectoral and pelvic girdles and fins, but an anal fin seems to have been preserved (Figure 18.14a).

The upper outline of the palatal portion of the palatoquadrate is uncertain, but it is clear that the otic process is much smaller than in the symmoriid *Cobelodus aculeatus* (Figure 18.14c). Meckel's cartilage is dorsoventrally expanded. The two elements of the mandibular arch apparently were articulated with double joint facets, which is a characteristic feature in many different sharks (Figure 18.14d).

A heap of teeth, piled up in front of and below the lower jaw probably represents the entire dentition; the few teeth that lie separate from the pile are preserved as negative impressions. Hence nothing can be said about the dentition as a whole, and the form of the teeth has to be based essentially on a single tooth. The teeth are tiny, the one furnishing most information measures 0.5 mm in greatest dimension. It has a low crown with a sharp crest extending the length of the tooth. There is a crown point, but this is barely higher than the crest at either end (Figure 18.14e). On the lingual tooth side there is a bulge beneath the crown point, and a blunt ridge connects the two. The labiolingual diameter across the crown point is about one-third of the greatest dimension of the tooth. The crown surface appears smooth at ordinary magnification.

Behind the teeth there are a few linearly arranged denticles that vaguely resemble the shape of *Petrodus* denticles (Figure 18.14f). These are probably mucous membrane denticles that lined the branchial arches.

The dermal skeleton consists of loosely spaced, hollow, simple cones with posteriorly directed, needle-sharp points, and widely flaring, hollow bases (Figure 18.14g).

The two fin spines are almost identical in length, shape, and ornamentation (Figure 18.14b). As preserved they stand nearly parallel to each other, but this is due to the strong post mortem flexure of the head and neck region. If the specimen is "straightened out" as has been done in Figure 18.14h, the anterior spine becomes less steeply inclined than the posterior spine, conforming with other ctenacanths where this feature is known. Both spines are preserved as negative impressions except for a small piece near the tip of the anterior spine. They are nearly straight, and about one-fifth of their length is implanted in the body of the fish. The protruding portion is laterally ornamented with three prominent ribs that lack any finer differentiations (Figure 18.14b). The posterolateral edges lack denticles or serrations of any kind, and the cross sections of the spines are very probably oval. A worn spine fragment essentially indistinguishable from the spines of *Dabasacanthus* except for size and slightly more pronounced posterior curvature was described by St. John and Worthen (1875) as *Asteroptychius keokuk* on the assumption that an unworn specimen would show *Asteroptychius*-like ornamentation and posterior denticles, neither of which is certain. There remains the possibility that this unsatisfactory spine fragment belongs to the ctenacanth genus *Dabasacanthus.*

The fins behind the spines are very large and may even have extended dorsally beyond the break line of the concretion. Both fins have a smooth surface and bear a scattering of dermal denticles.

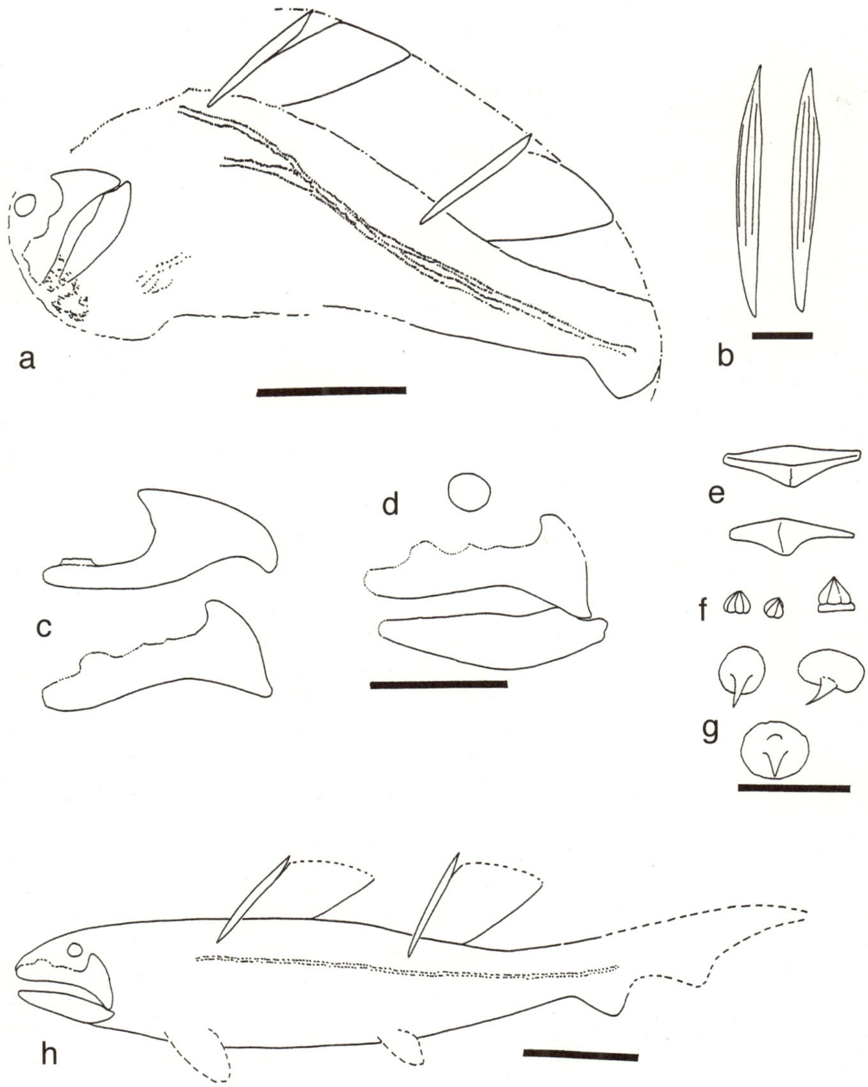

Figure 18.14. *Dabasacanthus inskasi:* **a** Drawing of holotype as preserved. Scale bar = 10 mm. **b** Anterior and posterior fin spines. Scale bar = 2 mm. **c** Upper jaw cartilage compared to that of *Cobelodus aculeatus* (Cope). **d** Upper and lower jaws in articulation and in relation to eye spot. Scale bar = 5 mm. **e** Best preserved dentition tooth in coronal view (*top*) and obliquely from lingual side. **f** Presumed mucous membrane denticles along gill arches, one shown reconstructed in side view. **g** Dermal denticles from body below first spine. Scale bar (e, f, g) = 0.5 mm. **h** Reconstruction of *D. inskasi.* Scale bar = 10 mm.

The decision as to whether *Dabasacanthus* should be placed among the ctenacanths or the hybodonts depends on interpretation of the most fundamental fossil character. The jaw cartilages show a basically cladodont configuration (Schaeffer's 1967 terminology) and hence point toward the ctenacanths. The teeth, on the other hand, have a distinctly hybodont aspect and are even very similar to those of the Mesozoic hybodont *Lonchidion breve* (Patterson, 1966). The other potentially important structures, the spines and the dermal denticles, do not yield useful information, partly because of the immaturity of the specimen, and partly because the internal structure of the spines is not preserved.

Holmacanthus keithi new genus and species
Figures 18.15, 18.16

Diagnosis. Probable hybodontoid (Zangerl, 1981) with two dorsal fins, the first triangular, the second about twice as long and half as high. Each is preceded by short, barely emergent fin spines in this juvenile specimen. Spines vertically ridged. Short based, distally elongate anal fin. Palatoquadrate deep with elongate otic process.

Holotype. FMNH PE 14309, one half of part and counterpart, complete specimen of young individual, Pit 11.

Etymology. Generic name derived from the last name of the collector, Keith Holm, and

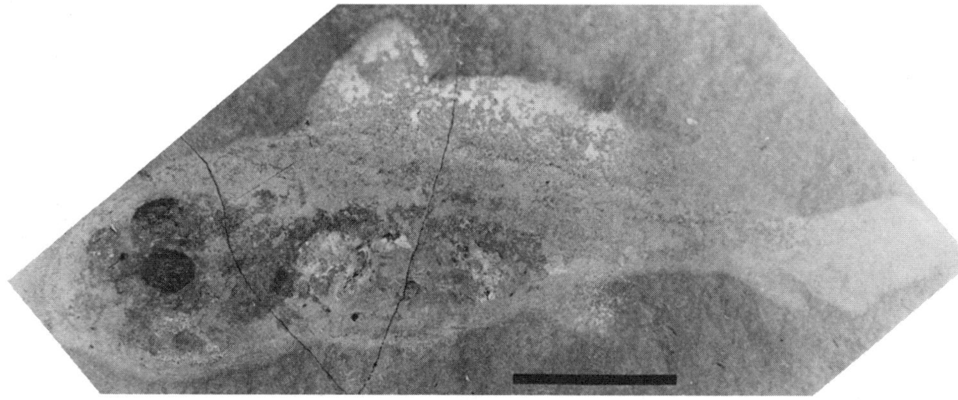

Figure 18.15. *Holmacanthus keithi* n. gen. n. sp., holotype, FMNH PF 14309.
Scale bar = 10 mm.

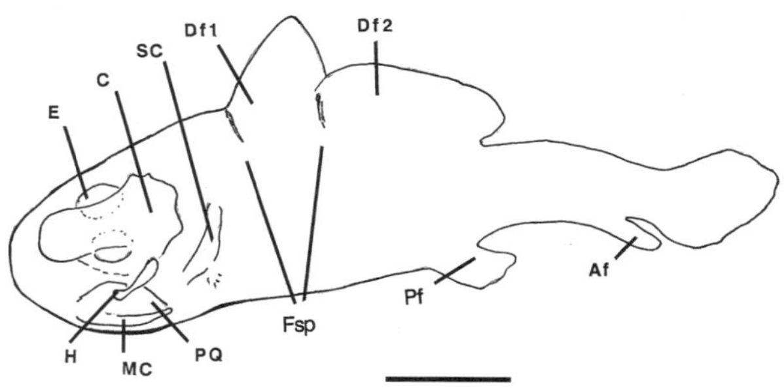

Figure 18.16. *Holmacanthus keithi* n. gen. n. sp., holotype FMNH PF 14309; modified camera lucida drawing. Scale bar = 10 mm. Abbreviations: Af, anal fin; C, chondrocranium; Df1, first dorsal fin; Df2, second dorsal fin; E, eye; Fsp, fin spine; H, hyomandibular; MC, Meckel's cartilage; Pf, pelvic fin; PQ, palatoquadrate; SC, scapulocoracoid.

"acanthus," common name component of sharks. Specific name derived from given name of the collector.

Description. This species known at present from a single specimen, measures 5.7 cm in length from the tip of the snout to the end of the tail. The individual is viewed from a three-quarter dorsal perspective anteriorly and from a lateral aspect more posteriorly. Depth at the first dorsal fin is 1.3 cm. The head is bluntly rounded, and from its anterior tip to the rear of the scapulocoracoid measures one-third of the length of the fish. The two eyes are seen as prominent, slightly ovate, very darkly pigmented structures in about their natural position relative to the chondrocranium. Anterior to the eyes and around as well as below them and more posteriorly are less distinctive and less dark shadows representing the chondrocranium, which still has not calcified. Obtusely rounded at the anterior end, this structure narrows between the eyes and then expands broadly behind them. Paired dark bodies behind the eyes but within the chondrocranium probably represent the otic capsules. The chondrocranium appears to be fairly deep as shown by continuous pigment patterns below and around the eyes. Posterior to the braincase several dark, somewhat ill-defined structures may be parts of the branchial arches. The mandibular arch (palatoquadrate and Meckel's cartilage) is distinguished by irregularly shaped polygonal remains indicating an early stage of cartilage calcification. The palatoquadrate has a deep, anteroposteriorly extended otic region and probably tapers anterior to this process more gradually than in ctenacanthid sharks. But the anterior end of this arch is not well preserved. The palatoquadrate is closely applied to the Meckel's cartilage, which seems most distinct where these cartilages articulate. The location of the mandibular arch, even without well-preserved anterior ends, suggests that the mouth was subterminal in position. Some faint impressions along the anterior end of the Meckel's cartilage suggest the early development of dentition. An elongate, rather flat, calcified bar extends from near the posterior end of the chondrocranium across the upper part of the palatoquadrate. This is probably the slightly displaced hyomandibular. Behind the chondrocranium and palatoquadrate, a large, dark, gently curved bar represents the scapulocoracoid. At the bend of this structure, a few indistinct, dark stains presumably are basipterygia of the pectoral fin.

There are two dorsal fins. The one in front is gently triangular in outline; the one behind is half its height but twice its length. Each is preceded by a fin spine; the anterior spine is shorter than the posterior one. Apparently, only the small, slightly falcate tips of these spines emerged from the muscle in which they were embedded. No supporting basil cartilages are evident. The caudal fin is asymmetric. The anal fin, narrow and about twice the length of its base, is nearly in contact with the caudal fin. Of the paired fins, the pectoral is not preserved. The pelvic fin arises below and near the end of the second dorsal fin. Dermal denticles are possibly indicated by impressions along the body surface, but these have, at this stage of development, no distinctive feature and are about the same size as the graininess of the matrix surface.

The abdominal cavity contains several fragments some of which may be bony, but none is identifiable. These remnants do demonstrate that *Holmacanthus* even at this small size was actively feeding.

Discussion. The new elasmobranch, *Holmacanthus,* adds another species to the growing list of Mazon Creek chondrichthyans. Although small in size and providing limited diagnostic features, it clearly differs from other chondrichthyans such as *Polysentor, Similihariotta, Jimpohlia,* and the elasmobranch *Bandringa* from the Mazon Creek biota in overall shape, fin characteristics, and head features as far as these may be examined among these genera. A fifth Mazon Creek elasmobranch, *Dabasacanthus,* has, broadly speaking, a similar body shape, but its dorsal fins and spines are much larger with coarser longitudinal costae. Also, in contrast to *Holmacanthus,* the otic portion of the palatoquadrate in *Dabasacanthus* is much lower and shorter. Few young individuals of Pennsylvanian chondrichthyans have been described. Preserved elasmobranchs from localities of Pennsylvanian age (Maples and Schultze, 1988) such as the Kinney Quarry in New Mexico or Hamilton Quarry in Kansas, and the Indiana black shale location (Zangerl, 1981) are represented by larger, postlarval or juvenile stages with distinctive dentitions, fin spines, and often chondrocranial features commonly used for comparing elasmobranchs.

References

Baird, Donald. 1978. Studies on Carboniferous freshwater fishes, Amer. Mus. Novitates, 1641:1–22.

Gross, Walter. 1937. Das Kopfskelett von *Cladodus wildungensis* Jaekel. Part 1, Endocranium und Palatoquadratum. Senckenbergiana Frankfurt, 19:80–107.

Johnson, R. G., and E. S. Richardson, Jr. 1966. A remarkable Pennsylvanian fauna from the Mazon Creek area, Illinois. Jour. Geol., 74:626–631.

Lund, Richard. 1977. *Echinochimaera meltoni,* a new genus and species (Chimaeriformes) from the Mississippian of Montana. Ann. Carnegie Mus. Nat. Hist., 46:195–221.

Maples, C. G., and H.-P. Schultze. 1988. Preliminary comparison of the Pennsylvanian assemblage of Hamilton, Kansas, with marine and nonmarine contemporaneous assemblages, pp. 253–272. *In* G. Mapes and R. Mapes (eds.), Regional geology and paleontology of upper Paleozoic Hamilton Quarry area in southeastern Kansas. Kansas Geol. Surv. Guideb., 6.

Patterson, Colin. 1966. British Wealden sharks. Bull. British Mus. (Nat. Hist.) Geol., 11(7):283–350.

———. 1968. *Menaspis* and the bradyodonts, pp. 171–205. *In* T. Orvig (ed.), Current Problems of Lower Vertebrate Phyogeny. Almqvist and Wiksell, Stockholm.

Schaeffer, Bobb. 1967. Comments on elasmobranch evolution, pp. 1–35. *In* P. W. Gilbert, R. F. Mathewson, and D. P. Rall (eds.), Sharks, Skates and Rays. John Hopkins University Press, Baltimore.

St. John, O., and A. H. Worthen. 1875. Descriptions of fossil fishes. Illinois State Geol. Surv., Paleo., 6:245–488.

Williams, M. E. 1985. The "Cladodont level" sharks of the Pennsylvanian black shales of central North America. Palaeontographica A, 190:83–158.

Woodward, A. S. 1921. Observations on some extinct elasmobranch fishes. Proc. Linnean Soc. London, 133:29–39.

Zangerl, Rainer. 1966. A new shark of the family Edestidae, *Ornithoprion hertwigi,* from the Pennsylvanian Mecca and Logan Quarry shales of Indiana. Fieldiana Geol., 16(1):1–43.

———. 1969. *Bandringa rayi,* a new ctenacanthoid shark from the Pennsylvanian Essex fauna of Illinois. Fieldiana Geol., 12(10):157–169.

———. 1979. New Chondrichthyes from the Mazon Creek fauna (Pennsylvanian) of Illinois, pp. 449–500. *In* M. H. Nitecki (ed.), Mazon Creek Fossils. Academic Press, New York.

———. 1981. Chondrichthyes. 1, Paleozoic Elasmobranchii. Handbook of Paleoichthyology, vol. 3A. Gustav Fischer Verlag, Stuttgart, New York, 115 pp.

Zangerl, Rainer, and G. R. Case. 1976. *Cobelodus aculeatus* (Cope), an anacanthous shark from Pennsylvanian black shales of North America. Palaeontographica A, 154(4–6):107–157.

19

Tetrapods (Amphibians and Reptiles)

Stephen J. Godfrey

Although the first tetrapod from Mazon Creek was discovered over 100 years ago, their remains are still exceedingly rare when compared to the vast numbers of other organisms found in this assemblage. Of the 24 published tetrapod specimens from the Francis Creek Shale, 23 are of amphibians, and only one is of a reptile (other specimens exist in both private and museum collections but remain inadequately studied). The amphibians include Temnospondyli—*Saurerpeton obtusum, Amphibamus grandiceps, ?Branchiosaurus* species indeterminate; Anthracosauria—*Spondylerpeton spinatum;* Aïstopoda—*Aornerpeton (Phlegethontia) mazonense, Ophiderpeton* sp.; Nectridea—*Ptyonius marshii;* Lysorophia—*Brachydectes newberryi;* and Microsauria—species incertae sedis. The Reptilia is represented by the Captorhina—*Cephalerpeton ventriarmatum.*

Like soft-bodied organisms, the fossil record of early tetrapods is very incomplete. Many forms are so specialized when they first appear in the fossil record that it is often difficult or impossible to identify the lineage from which they evolved.

About 50 million years before Mazon Creek amphibians trundled about, *Acanthostega* and *Icthyostega,* the earliest well-known Upper Devonian tetrapods, were living in the permanent waters of East Greenland (Clack and Coats 1995; Coats and Clack 1995, and the many references within). These and the much more fragmentary remains of other Devonian tetrapods from Australia, North America, and Latvia possess many very primitive features which indicate that they and all other tetrapods are derived from panderichthyid osteolepiform fish (Schultze and Arsenault, 1985; Schultze, 1986; Panchen and Smithson, 1987; Vorobyeve and Schultze, 1991).

It is not until the Upper Mississippian that we gain another glimpse of the tetrapod radiation. A sprinkling of about 20 localities, restricted almost entirely to the British Isles and eastern North America, preserve a handful of primarily aquatic, large-bodied (1–2 m long) species (Milner et al., 1986; Carroll, 1992).

One of a number of conspicuous differences between Mississippian and Pennsylvanian tetrapod fossil localities is the great change in the diversity of their constituent faunas. Approximately 40 Pennsylvanian coal measure deposits have produced remains of both large- and small-bodied tetrapods from no less than 30 families (Carroll, 1977).

The earliest reptiles are found within fossilized stumps of the mid-Pennsylvanian lycopods *Sigillaria* and *Calamites* at Joggins, Nova Scotia, Canada (Carroll, 1964). The single reptile specimen from Mazon Creek is a member of the family Protorothyrididae. All modern birds and reptiles are derived from lineages whose ancestry can be traced back to the protorothyridids.

Even though the Mazon Creek biota ranks very high among the world's invertebrate fossil assemblages, its tetrapod fauna is eclipsed by the younger coal swamp fauna at Linton, Ohio, which boasts over 600 tetrapod fossils representing about 27 species (Hook and Baird, 1986). All currently known tetrapods from Mazon Creek are represented by either the same or a closely related species at Linton. However, each locality provides a unique picture of North America's Middle Pennsylvanian equatorial biota. With so few specimens (and species) known from the Francis Creek Shale, chances are very good that additional discoveries will yield new species or additional anatomical information on known forms.

The skeletons of Mazon Creek tetrapods are sometimes preserved as natural molds, from which high-resolution latex casts can be made. The resulting latex peels are often of such high quality that they may be studied under a scanning electron microscope (Bolt, 1979).

All but one of the known specimens preserves the remains of juvenile or small adult tetrapods. This bias may reflect the conditions affecting preservation within the depositional environment. Alternatively, the bias might be one of habitat selection. Milner (1982) proposed that these small forms were adapted to life on the levees of the delta fan lobes or freshwater pools in the delta.

In summary, the study of Mazon Creek tetrapods is very important because 1) it includes several unique species, *Amphibamus grandiceps* and *Aornerpeton (Phlegethontia) mazonense*; 2) it provides additional detail on the skeletal and soft body morphology of these forms; and 3) it provides information on the faunal associations between tetrapods and the terrestrial and freshwater vertebrates and invertebrates of the Braidwood fauna.

Class Amphibia
Order Temnospondyli
Saurerpeton obtusum (Cope)
Figures 19.1–19.3

Locality. Peabody Coal Company Pit 11 spoil heap, Will-Kankakee Counties.

Horizon. Francis Creek Shale, Carbondale Formation, Desmoinesian.

The small, presumably juvenile specimens of *S. obtusum* from the Francis Creek Shale exhibit a dorsoventrally compressed skull that is broadly parabolic in dorsal view. They possess proportionately large orbits, nares, pineal foramen, anterior palatal fenestrae, and interpterygoid vacuities (Milner, 1982). The antorbital portion of the skull table is equal to approximately one-half the overall skull length. The postparietals, which form the posteromedial margin of the skull table, are roughly L-shaped. The dermal bones of the skull roof are only lightly sculptured and retain traces of lateral line canals on the frontals, postfrontals, and supratemporals (Milner, 1982). Lateral line canals in aquatic vertebrates house sensory cells that are sensitive to movements of water. The squamosal embayment (a concave embayment along the posterior margin of the squamosal bone) is only weakly developed in *S. obtusum.* No denticles are known to occur on any of the palatal bones, and the cultriform process is broad, reaching far forward between the vomers.

In common with most other Francis Creek amphibians, each orbit may contain a black disc that appears to represent the stable degraded pigment of the retina (Richardson and Johnson, 1971). The delicate series of tiny ossicles forming the sclerotic ring may also be preserved in each orbit.

In FMNH PR 1036, four rows of tooth-bearing ossicles are preserved in the branchial region on each side of the pectoral girdle. These denticulate ossicles may have been attached to four cartilaginous ceratobranchials bordering three gill slits and could have functioned as gill rakers filtering small foodstuffs from the water before being expelled through the gill slits (Milner, 1982). This hypothesis is supported by the observation that in one unidentified amphibian from this locality, ostracods have been found in its abdominal region. It would have been difficult to consume these tiny crustaceans without gill rakers (Richardson and Johnson, 1971). This specimen also preserves carbonized remains of three pairs of filamentous external gills (Milner, 1982).

Because larger individuals of *S. obtusum* are known from Linton, the specimens from Mazon Creek are believed to be of juveniles. The poorly ossified postcranial skeleton of *S. obtusum* substantiates this belief. Nevertheless, the lack of ossification may be used to distinguish between *S. obtusum* and some of the other amphibians that, with comparably sized skulls, possess more highly ossified postcranial skeletons.

In *S. obtusum* the vertebral column consists of approximately 24 presacral vertebrae (Milner, 1982). The number of caudal (tail) vertebrae is unknown. All that is preserved of the multipartite temnospondylous vertebrae are the paired halves of low neural arches. The ribs are short and essentially straight with slightly expanded proximal and distal ends.

The large, lightly sculptured dermal pectoral girdle consists of a roughly rhomboidal interclavicle that is more than half the length of the skull and a pair of triangular clavicles which rabbet onto the ventrolateral margins of the former element. Bones of the pectoral and pelvic limbs are known but are usually poorly ossified and will not be described here.

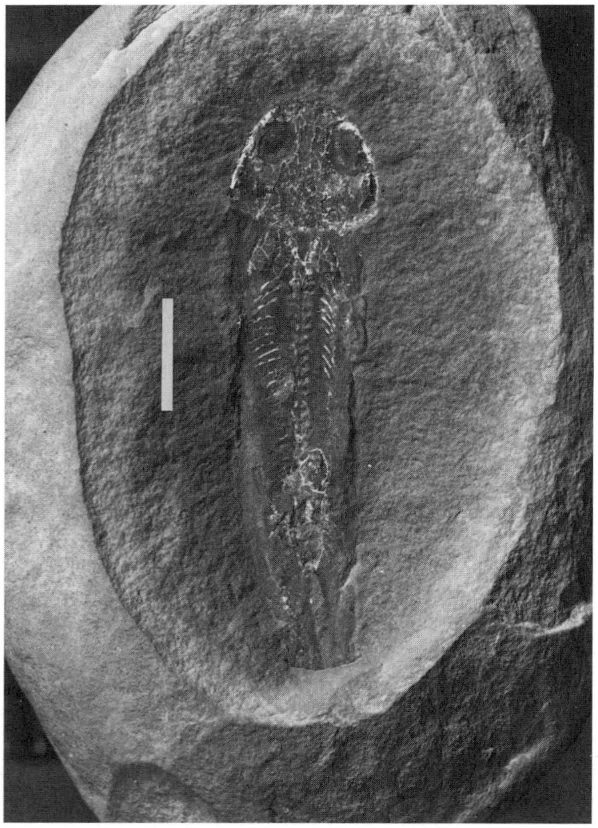

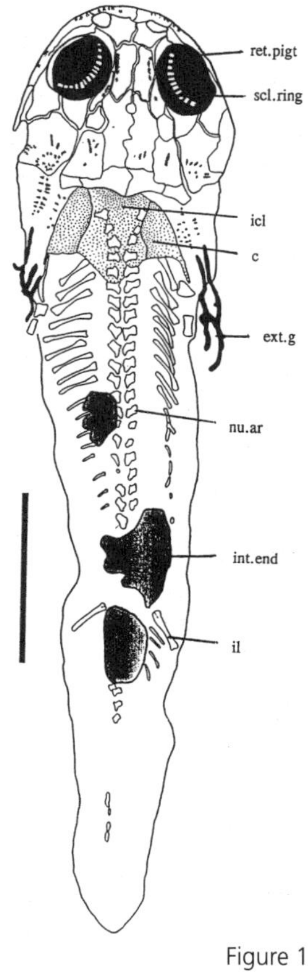

Figure 19.1. *Saurerpeton obtusum,* FMNH PR 1036. Scale bar = 10 mm.

Figure 19.2. *Saurerpeton obtusum,* FMNH PR 1036; specimen drawing in dorsal view, redrawn from Milner (1982). Scale bar = 10 mm. Abbreviations: c, clavicle; ext. g, external gills; icl, interclavicle; il, ilium; int. end, intestinal endocasts; nu.ar, neural arch; ret. pigt, retinal pigment; scl. ring, sclerotic ring.

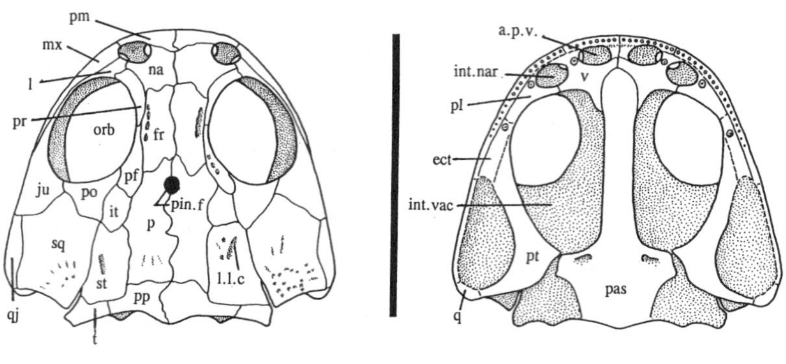

Figure 19.3. *Saurerpeton obtusum,* reconstruction of the skull in dorsal and ventral views, respectively, redrawn from Milner (1982). Lower jaw not included. Scale bar = 10 mm. Abbreviations: a.p.v, anterior palatal vacuity; ect, ectopterygoid; fr, frontal; int.nar, internal nares; int. vac, interpterygoid vacuities; it, intertemporal; ju, jugal; l, lacrimal; l.l.c, lateral line canals; mx, maxilla; na, nasal; orb, orbit; p, parietal; pas, parasphenoid; pf, postfrontal; pin.f, pineal foramen; pl, palatine; pm, premaxilla; po, postorbital; pp, postparietal; pr, prefrontal; pt, pterygoid; q, quadrate; qj, quadratojugal; sq, squamosal; st, supratemporal; v, vomer.

In life, the aquatic larvae of *S. obtusum* may have resembled in their general proportions Hellbender larvae (*Cryptobranchus*).

Amphibamus grandiceps Cope
Figures 19.4–19.6

Locality. Mazon Creek, near Morris, Grundy County, and Pit 11 spoil heap, Will-Kankakee Counties.

Horizon. Francis Creek Shale, Carbondale Formation, Desmoinesian.

Amphibamus grandiceps, represented by nine specimens (six "adult" and three larval specimens) is the most common amphibian encountered within the Francis Creek Shale.

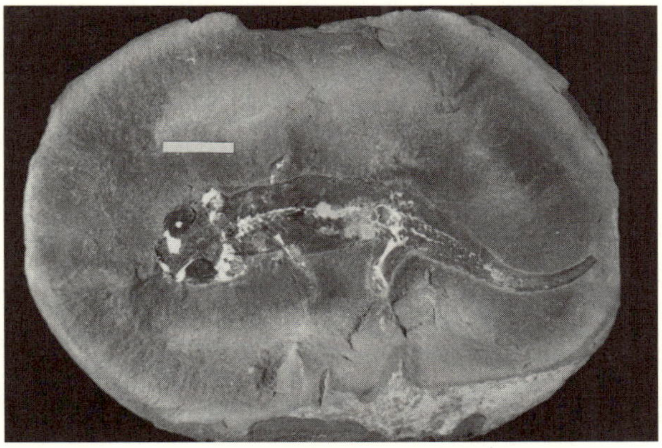

Figure 19.4. *Amphibamus grandiceps,* NEIU MCP 342. Scale bar = 10 mm.

As in *S. obtusum,* the skull of *A. grandiceps* is dorsoventrally compressed and broadly parabolic in dorsal view. *Amphibamus grandiceps* also exhibits large orbits, external nares, and interpterygoid vacuities, but unlike *S. obtusum,* it lacks anterior palatal fenestrae (Milner, 1982). There are a number of other conspicuous differences that help distinguish between these not too distantly related forms. In *A. grandiceps,* the snout is not nearly as abbreviated, whereas the skull table is shorter than in *S. obtusum.* With a shorter skull table, the occiput lies well anterior to the level of the quadrates. *Amphibamus grandiceps* has lost the intertemporal bones from the skull table and possesses enormous squamosal embayments, each of which probably held a tympanum (eardrum). The dermal bone of the skull roof appears to be more heavily sculptured in *A. grandiceps,* but no lateral line grooves are present.

On the palate of *A. grandiceps,* the vomers are much larger than those of *S. obtusum,* and the palatines are proportionately much longer. A posterolateral flange of the pterygoid alters the configuration of the subtemporal fossa from that seen in *S. obtusum.* The vomers, palatines, and pterygoids bear a shagreen of denticles. The cultriform process is slender, displaying a single median row of denticles. The occipital margin of the parasphenoid extends laterally be-

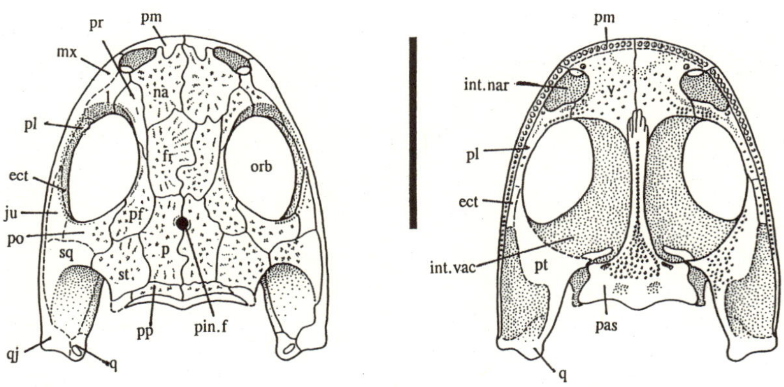

Figure 19.5. *Amphibamus grandiceps,* reconstruction of the skull in dorsal and ventral views, respectively, redrawn from Milner (1982). Lower jaw not included. Scale bar = 10 mm. Abbreviations: ect, ectopterygoid; fr, frontal; int.nar, internal nares; int.vac, interpterygoid vacuities; ju, jugal; l, lacrimal; mx, maxilla; na, nasal; orb, orbit; p, parietal; pas, parasphenoid; pf, postfrontal; pin.f, pineal foramen; pl, palatine; pm, premaxilla; po, postorbital; pp, postparietal; pr, prefrontal; pt, pterygoid; q, quadrate; qj, quadratojugal; sq, squamosal; st, supratemporal; v, vomer.

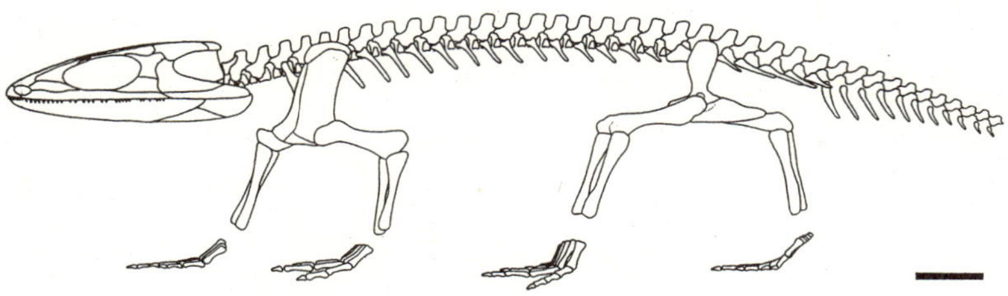

Figure 19.6. *Amphibamus grandiceps,* reconstruction of the skeleton in left lateral view, based in part on *A. lyelli* by Hook and Baird (1986). Scale bar = 10 mm.

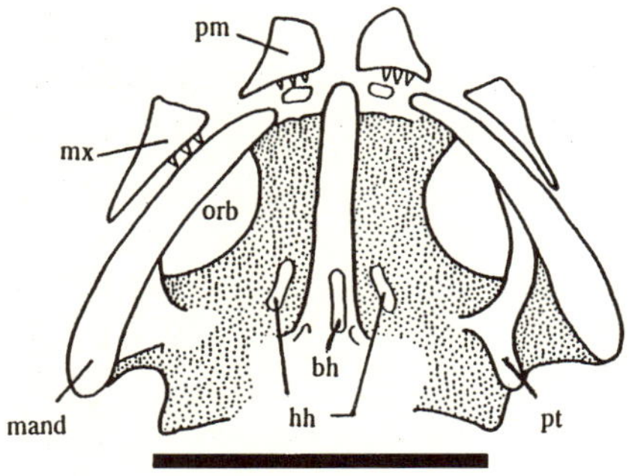

Figure 19.7. *Branchiosaurus,* YPM 803; specimen drawing in ventral view, redrawn from Milner (1982). Scale bar = 5 mm. Abbreviations: bh, basihyal; hh, hypohyal; mand, mandible (lower jaw); mx, maxilla; orb, orbit; pm, premaxilla; pt, pterygoid.

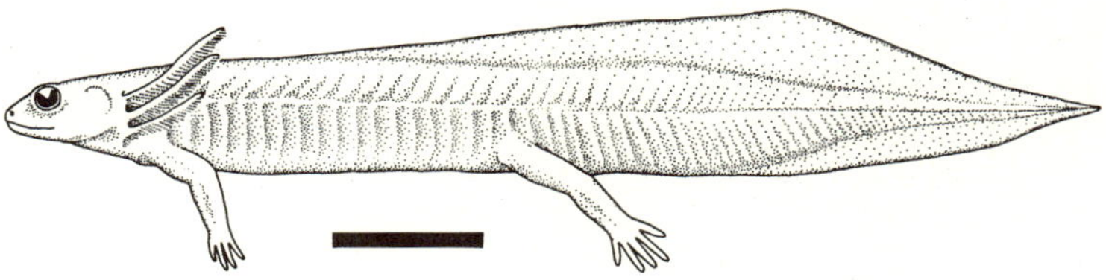

Figure 19.8. *Branchiosaurus,* reconstruction based on specimen KC 106, Odernheim, Saarpfalz, Germany; redrawn from Milner (1982). Scale bar = 10 mm.

yond the basipterygoid process, whereas in *S. obtusum* it is constricted mediolaterally.

There are at least 20 presacral vertebrae in *A. grandiceps* (Bolt, 1979) and possibly as many as 25 caudal vertebrae (Milner, 1982). The bicipital (two-headed) ribs exhibit little curvature, are very short, and extend almost directly laterally.

Compared to the length of the body, the limbs of *A. grandiceps* are relatively long. The closely related species *A. lyelli* from Linton has a phalangeal formula of 2,2,3,4,4 on the pes (foot; Hook and Baird, 1984). Judging from the proportions of its long limbs, *A. grandiceps* may have been, as an adult, the most adept terrestrial amphibian known from the Francis Creek Shale.

Mazon Creek larvae attributed to *A. grandiceps* bear at least two pairs of external gills. One remarkable specimen preserves casts of the stomach and intestine (Milner, 1982). Donald Baird believes that at least one of the specimens attributed to *Amphibamus* may in fact belong to *Stegops,* a genus hitherto unknown from this area but represented by several good specimens from Linton. He notes that specimens attributed to *Amphibamus* should have no gastralia, whereas those assigned to *Stegops* would as well as rugose ornament on the bones of the skull.

Although *Amphibamus* would not be mistaken for a frog, it exhibits a number of skeletal features, including an abbreviated skull table, pedicellate bicuspid teeth, reduced ribs, a pectoral girdle incorporating a small interclavicle, and narrow barlike clavicles, which strongly suggest that frogs evolved from an *Amphibamus*-like amphibian (Bolt, 1979; Milner, 1982; Milner et al., 1986). Furthermore, the structure of the stapes and palate in the Peliontidae (the family to which *Amphibamus* belongs) and frogs supports their close relationship (Bolt and Lombard, 1985; Milner, 1990).

?*Branchiosaurus* sp. indeterminate
Figures 19.7, 19.8

Locality. Mazon Creek, near Morris, Grundy County.

Horizon. Francis Creek Shale, Carbondale Formation, Desmoinesian.

Two specimens from the Francis Creek Shale were tentatively assigned by Milner (1982) to the genus *Branchiosaurus.* On the basis of these specimens, it is very difficult to distinguish between *Branchiosaurus* and *Amphibamus* larvae. Nevertheless, these specimens appear to exhibit several features that ally them more closely with *Branchiosaurus* than *Amphibamus.* They include a denticle-devoid parasphenoid with conspicuous foramina for the internal carotid arteries and an ossified basihyal and hypohyals (Boy, 1978; Milner, 1982). As in *A. grandiceps,* one specimen of *Branchiosaurus* retains casts of the stomach and intestine.

Branchiosaurids are believed to have been paedomorphic (i.e., retaining larval characteristics) throughout life.

Order Anthracosauria
Spondylerpeton spinatum Moodie
Figures 19.9, 19.10

Locality. Shale above Morris Coal seam, Mazon Creek, near Morris, Grundy County.

Figure 19.9. *Spondylerpeton spinatum,* YPM 793, near Morris; photograph of the holotype. Across the bottom of the nodule are the poorly preserved remains of three caudal vertebrae. The impression of two other caudal vertebrae are preserved along the top left-hand margin of the nodule. Scale bar = 10 mm.

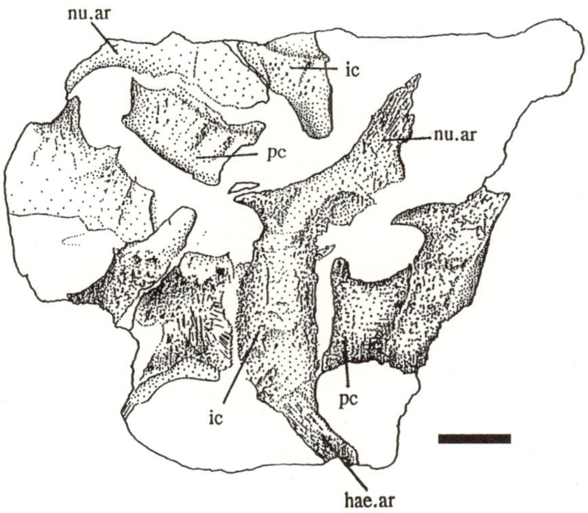

Figure 19.10. *Spondylerpeton spinatum,* YPM 793; specimen drawing of the holotype. Scale bar = 10 mm. Abbreviations: hae.ar, haemal arch; ic, intercentrum; nu.ar, neural arch; pc, pleurocentrum.

Horizon. Francis Creek Shale, Carbondale Formation, Desmoinesian.

A single specimen consisting of at least four caudal vertebrae of an embolomerous anthracosaur represents the sole remains of the largest amphibian known from the Francis Creek Shale. Embolomeres were large carnivorous crocodile-like amphibians with long and flexible bilaterally compressed tails (Panchen, 1970).

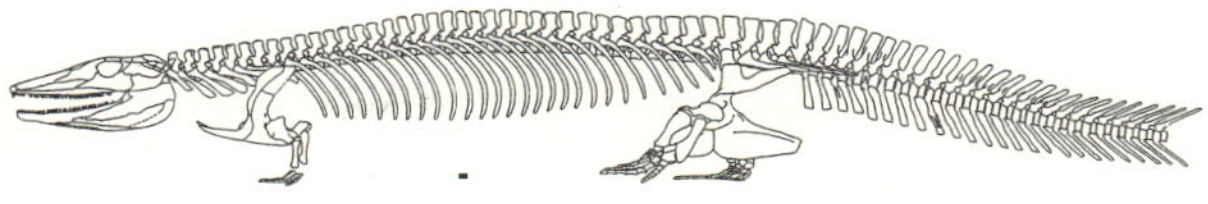

Figure 19.11. *Archeria crassidisca,* reconstruction of skeleton in left lateral view. Scale bar = 10 mm. Courtesy of Robert Holmes.

This scanty material is assignable to the infraorder Embolomeri on the basis of the structure of the vertebral centra. Each centrum is composed of two donut-shaped elements, an intercentrum anteriorly followed by a pleurocentrum. These vertebrae are known to come from the posterior caudal region because each intercentrum 1) bears a haemal arch and 2) provides most of the support for the neural arch. Although it is impossible to assign these diplospondylous caudal vertebrae to any particular family, *S. spinatum* may have resembled the Permian embolomere *Archeria crassidisca* (Holmes, 1989; Figure 19.11).

Order Aïstopoda
Aornerpeton (Phlegethontia) mazonense
(Gregory)
Figures 19.12–19.14

Locality. Mazon Creek, near Morris, Grundy County, and Braidwood, Will County.

Horizon. Francis Creek Shale, Carbondale Formation, Desmoinesian.

Aornerpeton mazonense, represented by five specimens, is the second most common amphibian encountered at Mazon Creek. These limbless, snakelike amphibians have highly specialized skulls that are lanceolate in dorsal view. The fused frontals and parietals form the dermal skull roof above the large orbital and temporal fenestrae. A sclerotic ring surrounds each eye. There are no interpterygoid vacuities. The braincase bears large ventrolaterally directed fenestrae ovales.

The lower jaw articulates with the quadrate anterior to the posterior margin of the occiput. Each long and slender mandible apparently consists of only two elements, the dentary and a posterior element (McGinnis, 1967). A pair of delicate, roughly sickle-shaped hyoids can also be preserved.

Approximately 90 precaudal and 50 caudal

vertebrae are known in one specimen of *A. mazonense* (McGinnis, 1967). The holospondylous vertebrae are characterized by low neural spines, moderately well developed transverse processes, and large intravertebral foramina for the spinal nerves.

The ribs are long and very thin and may possess up to two stiletto-like costal processes that emanate from the proximal end of the shaft (Lund, 1978). The processes form an overlapping pattern between adjacent ribs. The unicipital ribs appear to attach only to the transverse process. *Aornerpeton mazonense* lacks any trace of either pectoral or pelvic girdles and limbs.

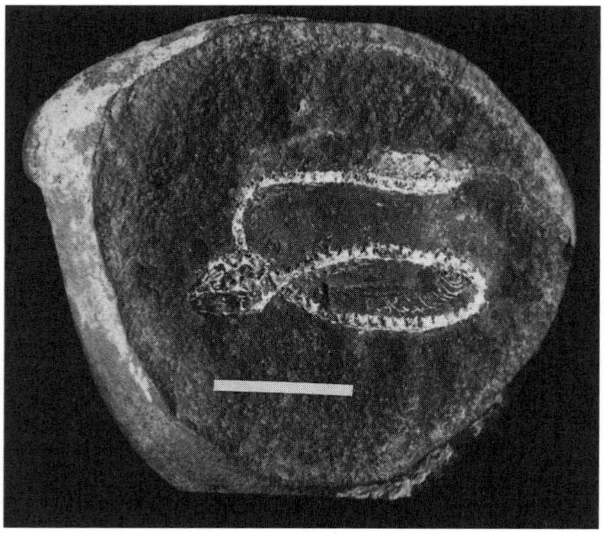

Figure 19.12. *Aornerpeton (Phlegethontia) mazonense,* NEIU MCP 367, Will and Grundy Counties. Scale bar = 10 mm.

The ventral surface of the body is clad in a series of delicate needlelike gastralia (ventral scales). Individual scales are only about 1.2 mm long and 0.03 mm in diameter (Gregory, 1948). The overall pattern is *en chevron.* One specimen of *A. mazonense* appears to retain what Lund (1978) interprets as a pigment pattern consisting of dark transverse bars and spots distributed over the head, neck, and throat region. These markings may correspond to cryptic or disruptive coloration comparable to that seen on many diurnal predacious snakes (Lund, 1978).

Ophiderpeton sp.: unnamed species
Figures 19.15–19.17

Locality. Pit 11 spoil heap, Kankakee County.

Horizon. Francis Creek Shale, Carbondale Formation, Desmoinesian.

A second aïstopod is known from a single specimen that preserves the skull and cervical region. *Ophiderpeton* resembles *Aornerpeton mazonense* in that both are long-bodied snakelike forms that lack limbs or girdles and pos-

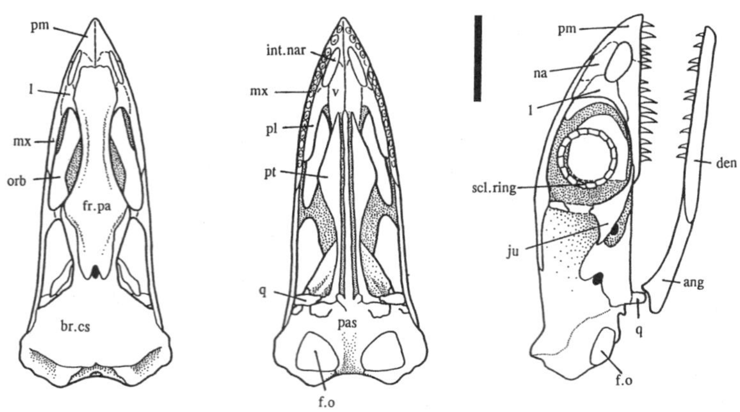

Figure 19.13. *Aornerpeton mazonense,* reconstruction of the skull after Gregory (1948) and Lund (1978) in dorsal, ventral, and right lateral views, respectively. Scale bar = 2 mm. Abbreviations: ang, angular; br.cs, braincase; den, dentary; f.o, fenestra ovalis; fr.pa, frontaparietal; int.nar, internal nares; ju, jugal; l, lacrimal; mx, maxilla; na, nasal; orb, orbit; pas, parasphenoid; pl, palatine; pm, premaxilla; pt, pterygoid; q, quadrate; scl.ring, sclerotic ring; v, vomer.

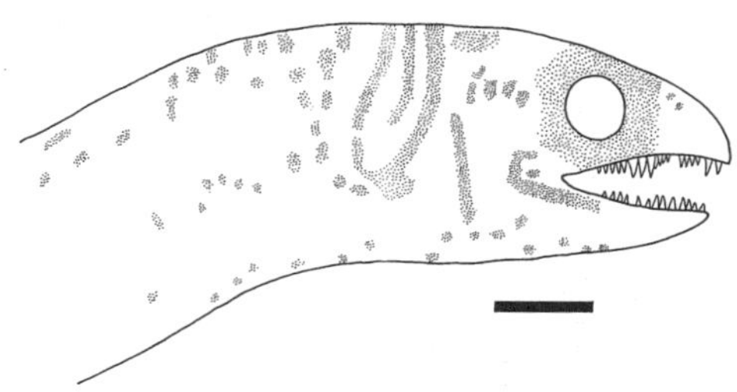

Figure 19.14. *Aornerpeton mazonense,* restoration based on FMNH MCP 501 of the pigment pattern on the head and neck; redrawn from Lund (1978). Scale bar = 2 mm (estimate).

Figure 19.15. *Ophiderpeton* sp., NEIU MCP 323 (HTP 5006), Pit 11, Kankakee County; left dorsolateral view of the skull and cervical region. Scale bar = 10 mm.

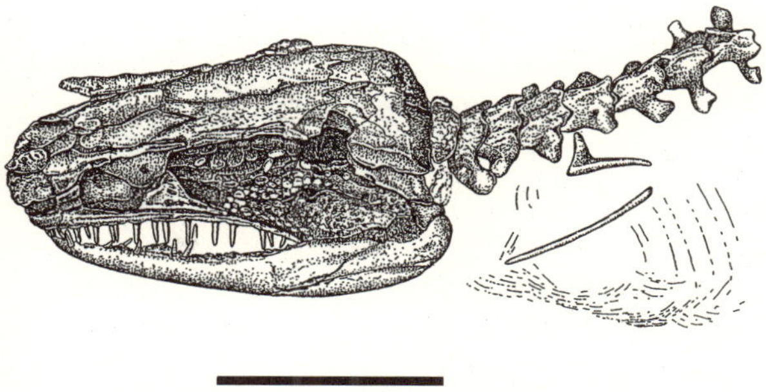

Figure 19.16. *Ophiderpeton;* specimen drawing by Pamela Gaskill of the skull and cervical region in a left dorsolateral view. Scale bar = 10 mm. Courtesy of Robert L. Carroll.

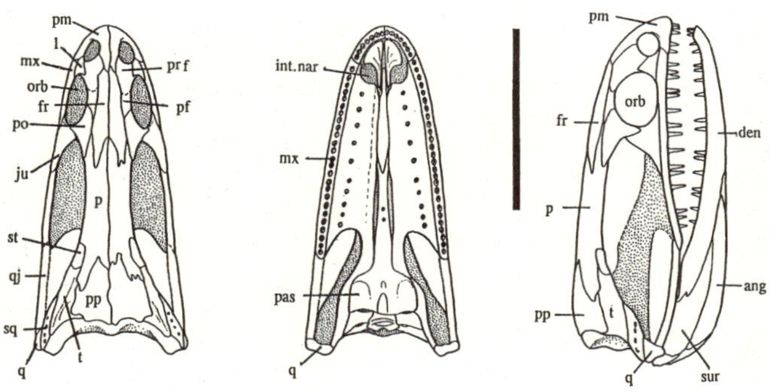

Figure 19.17. *Ophiderpeton,* reconstruction of the skull based in part on Bossy (1976), in dorsal, ventral, and right lateral views, respectively. Scale bar = 10 mm. Abbreviations: ang, angular; den, dentary; fr, frontal; int.nar, internal nares; ju, jugal; l, lacrimal; mx, maxilla; orb, orbit; p, parietal; pas, parasphenoid; pf, postfrontal; pm, premaxilla; po, postorbital; pp, postparietal; pr, prefrontal; q, quadrate; qj, quadratojugal; sq, squamosal; st, supratemporal; sur, surangular; t, tabular; v, vomer.

sess holospondylous centra with intravertebral foramina and unicipital tetraradiate ribs. The most conspicuous differences between these genera are found in the proportions of the skull and pattern of body scales.

Ophiderpeton retains a more primitive pattern of dermal bones on the skull roof. The frontals and parietals are separate elements and paired along the midline. In *Ophiderpeton,* the antorbital portion of the skull is proportionately shorter and the postorbital region is much longer than in *A. mazonense.* A large temporal fenestra is present in *Ophiderpeton.*

Ophiderpeton exhibits a well-developed ventral squamation composed of numerous, closely packed gastralia forming a continuous plastron. In *A. mazonense,* the scales are more widely spaced. Unlike *A. mazonense, Ophiderpeton* possesses dorsal and lateral scalation of pebble-shaped osteoderms. These osteoderms appear to have covered the large temporal fenestrae superficially.

Order Nectridea
Ptyonius marshii Cope
Figures 19.18–19.20

Locality. Mazon Creek, Grundy County.

Horizon. Francis Creek Shale, Carbondale Formation, Desmoinesian.

Ptyonius marshii is known only from a single incomplete specimen. Both the anterior part of the skull and distal end of the tail are missing, and little structure can be determined beyond the highly characteristic vertebrae and gastralia. *Ptyonius marshii* is known from more completely preserved material from Linton.

These small amphibians have a trowel-shaped skull in dorsal view, which reaches a known maximum length of only 19 mm (Bossy, 1976). The orbits and the interpterygoid vacuities are both fairly large. A median tongue-shaped flange from the postparietals projects beyond the posterior margin of the skull table and occiput. The cultriform process of the parasphenoid is narrow anteriorly but expands abruptly just ahead of the basal articulation to form a broad triangle-shaped plate posteriorly.

The trunk in *P. marshii* is fairly short, with only 22 presacral vertebrae (Bossy, 1976). These holospondylous vertebrae bear long transverse processes and broad subrectangular neural spines in the trunk region but fan-shaped spines with narrow bases in the tail. Unlike any other Paleozoic amphibian, nectridean vertebrae bear two sets of accessory articulations above their zygapophyses. The neural and haemal arches are co-ossified with the centra. One of the diagnostic features of nectrideans is the peculiar nature of their caudal vertebrae. The distal ends of the caudal neural spines do not abut as in the trunk region. The haemal spines are roughly fan-shaped with very broad bases such that adjacent spines overlap to articulate with each other by two interdigitations (Bossy, 1976). All the neural and haemal spines are grooved, or crenulated distally, or both. The long deep tail, with over 60 vertebrae, is about twice the length of the rest of the animal. The ribs are conspicuously bicipital, slender, and gently curved. The ventral dermal scales are large overlapping plates about equal in length to a trunk centrum. The overall chevron pattern is formed by three scales per row.

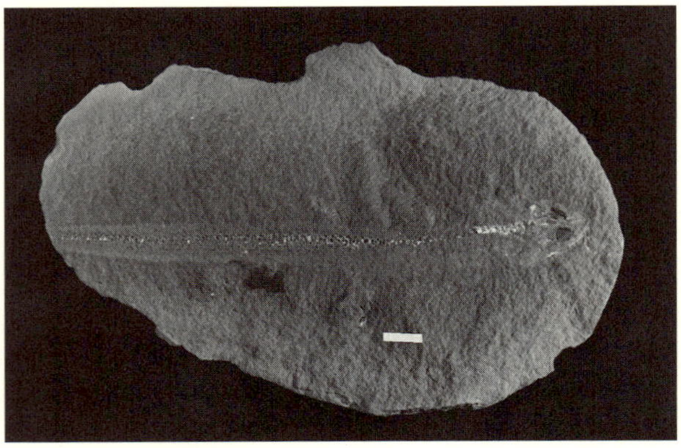

Figure 19.18. *Ptyonius marshii,* FMNH PR 1353. Scale bar = 10 mm (estimate).

The dermal pectoral girdle is relatively massive. The delicate limbs each bear four digits with a phalangeal formula of 2,3,4,4 (Bossy, 1976).

The closest living analogues to nectrideans are the newts (*Notophthalmus*), although the latter do not possess tails twice as long as the trunk. Bossy (1976) concluded that these nectrideans were powerful swimmers.

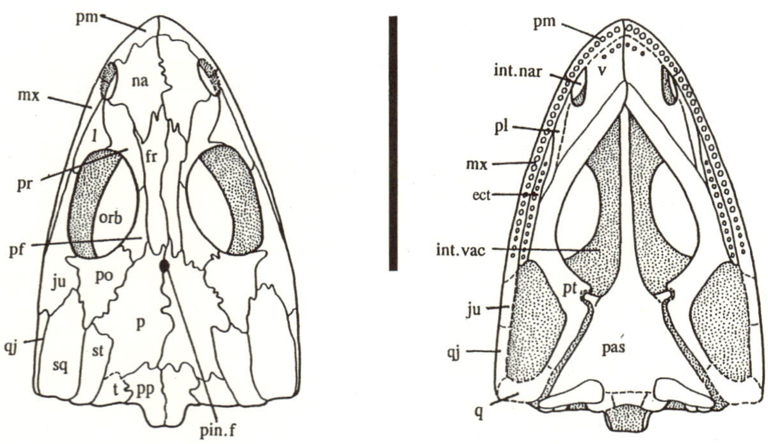

Figure 19.19. *Ptyonius marshii,* reconstruction of the skull in dorsal and ventral views, respectively, redrawn from Bossy (1976). Lower jaw not included. Scale bar = 10 mm. Abbreviations: ect, ectopterygoid; fr, frontal; int.nar, internal nares; int.vac, interpterygoid vacuities; ju, jugal; l, lacrimal; mx, maxilla; na, nasal; orb, orbit; p, parietal; pas, parasphenoid; pf, postfrontal; pin.f, pineal foramen; pl, palatine; pm, premaxilla; po, postorbital; pp, postparietal; pr, prefrontal; pt, pterygoid; q, quadrate; qj, quadratojugal; sq, squamosal; st, supratemporal; t, tabular; v, vomer.

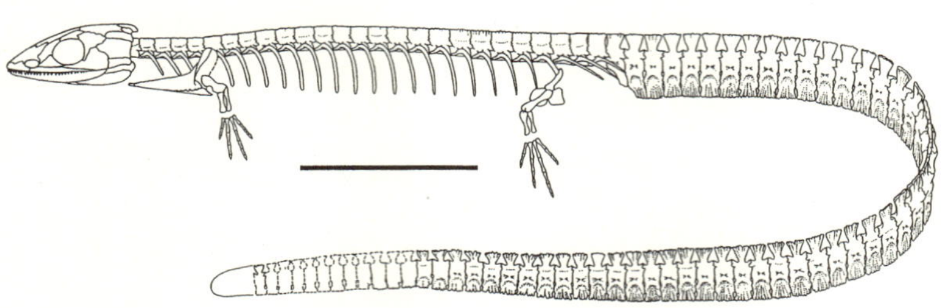

Figure 19.20. *Ptyonius marshii,* reconstruction of the skeleton in left lateral view, redrawn from Bossy (1976). Scale bar = 10 mm.

Order Lysorophia
Brachydectes newberryi Cope
Figures 19.21–19.24

Locality. Mazon Creek, 10 km southeast of Morris, Grundy County.
Horizon. Francis Creek Shale, Carbondale Formation, Desmoinesian.

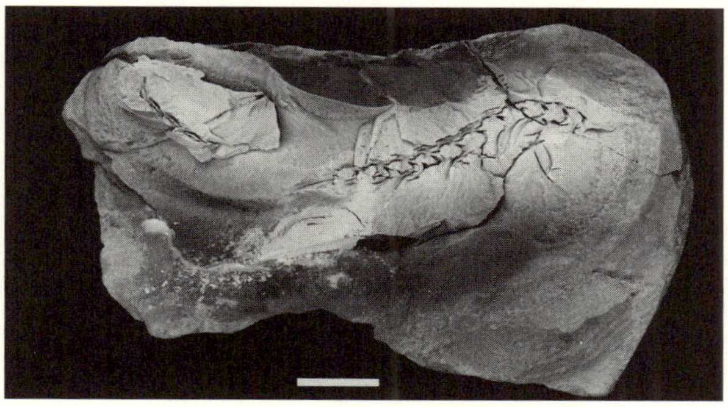

Figure 19.21. *Brachydectes newberryi* (*Coctyinus gyrinoides*), USNM 4313; photograph. Scale bar = 10 mm.

A single incomplete specimen consisting of 19 presacral vertebrae and some associated ribs is assigned to *B. newberryi* (Wellstead, 1991, fig. 12A-B; formerly *Cocytinus gyrinoides* Cope). The anatomical description provided here is based on more completely preserved specimens from Linton.

The skull of *B. newberryi* exhibits large external nares and very large orbitotemporal fenestrae. The tabular, squamosal, and quadrate form an anteroventrally slanting strut that places the craniomandibular articulation conspicuously ahead of the occiput. As a result, the lower jaw, which possesses an elongate lateral mandibular fenestra, is only about two-thirds the length of the skull. The hyoid apparatus, unlike other Mazon Creek amphibians, is well developed in *B. newberryi.* It consists of paired hypo- and ceratohyals plus four paired branchial arches. As in some living amphibians, the well-developed hyobranchial skeleton

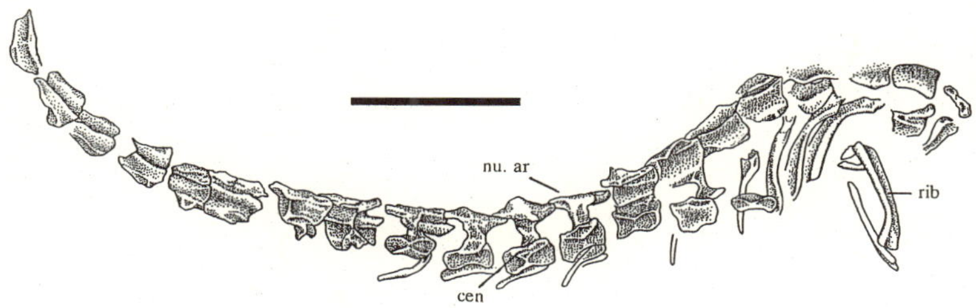

Figure 19.22. *Brachydectes newberryi,* USNM 4313; specimen drawing. Scale bar = 10 mm. Abbreviations: cen, centrum; nu.ar, neural arch.

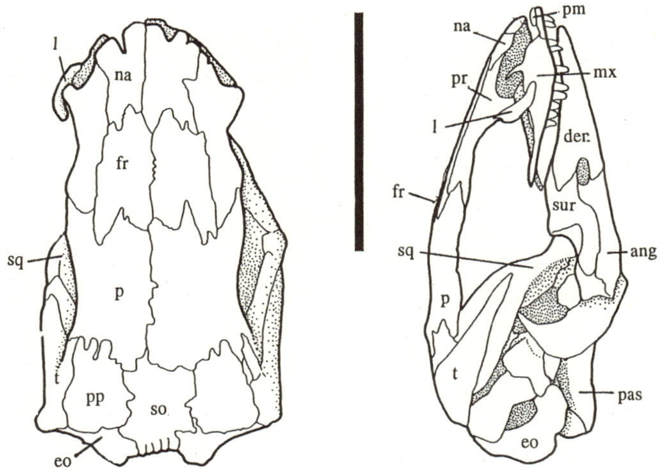

Figure 19.23. *Brachydectes newberryi,* AMNH 6172; drawing of the skull in dorsal and right lateral views, respectively, redrawn from Carroll and Gaskill (1978). Scale bar = 10 mm. Abbreviations: ang, angular; den, dentary; eo, exoccipital; fr, frontal; l, lacrimal; mx, maxilla; na, nasal; p, parietal; pas, parasphenoid; pm, premaxilla; pp, postparietal; pr, prefrontal; so, supraoccipital; sq, squamosal; sur, surangular; t, tabular.

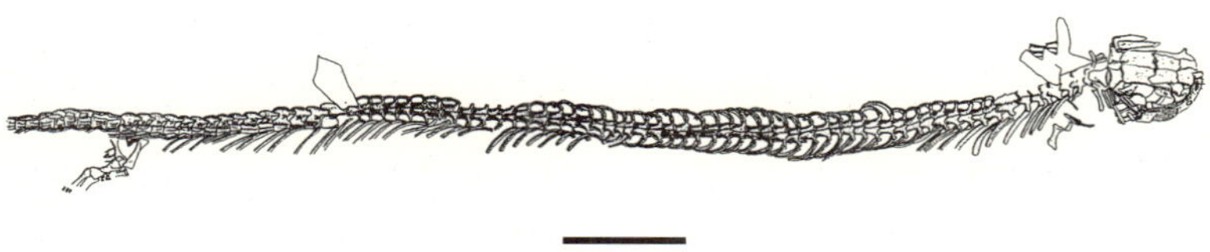

Figure 19.24. *Brachydectes newberryi,* BM(NH) R.2554; specimen drawing by Pamela Gaskill of a virtually complete skeleton of this amphibian from Linton, from Carroll and Gaskill (1978). Scale bar = 10 mm.

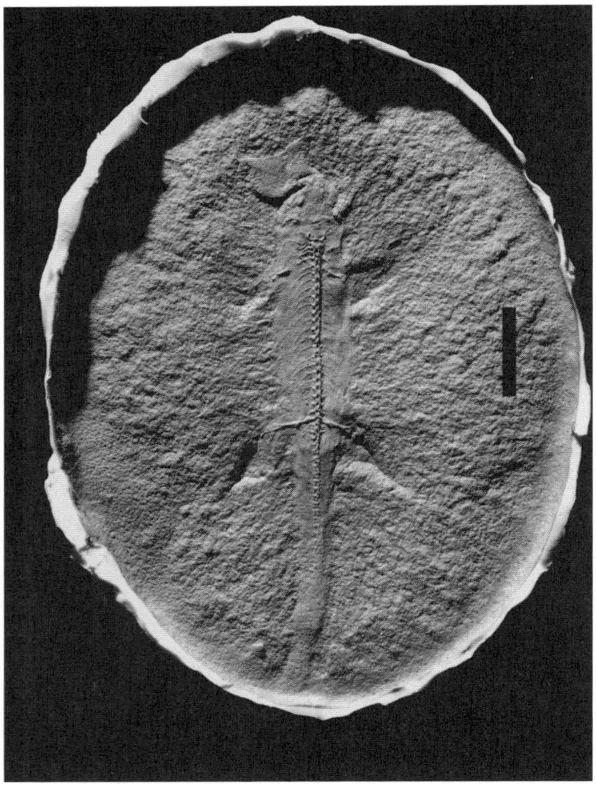

Figure 19.25. Unnamed microsaur, FMNH PR 981, Pit 1 or 6, near the county line between Will and Grundy Counties; photograph of a latex cast. Dorsal view of the virtually complete skeleton. Scale bar = 10 mm.

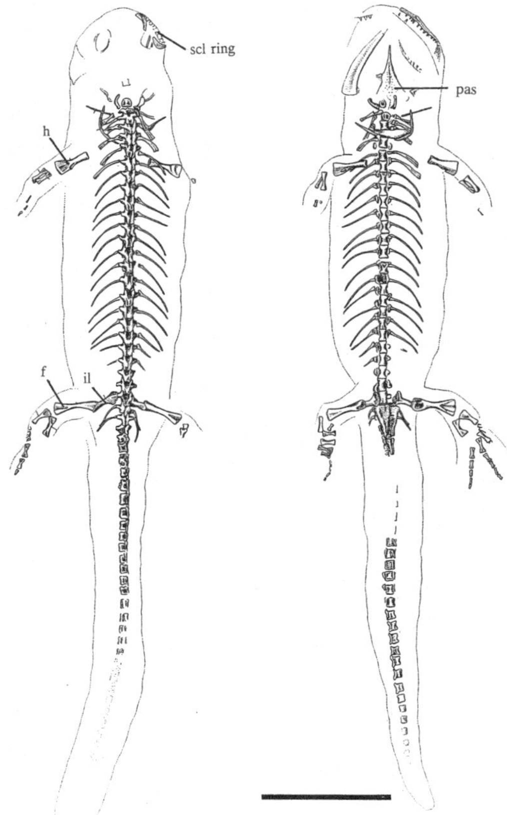

Figure 19.26. Unnamed microsaur, FMNH PR 981; specimen drawings by Pamela Gaskill of the skeleton in dorsal and ventral views, respectively, from Carroll and Gaskill (1978). Scale bar = 1 mm. Abbreviations: f, femur; h, humerus; il, ilium; pas, parasphenoid; scl.ring, sclerotic ring.

in *B. newberryi* would have been essential for its mode of feeding. A relatively large volume of food-laden water could have been sucked into the mouth by the rapid expansion of the throat region through the contraction of large muscles attaching to the hyobranchial apparatus (Wellstead, 1991).

Brachydectes newberryi possesses an elongate presacral vertebral column consisting of about 69 vertebrae. The neural arches are sutured to deeply amphicoelous centra. Only 10 caudal vertebrae are known in *B. newberryi,* and although there were probably more, the tail was much shorter than the presacral column. The ribs are curved, bicipital, and rather robust. No scalation is known in this group.

Based on the proportions of their tiny limbs and elongate vertebral column, Wellstead (1991) has suggested that lysorophoids were probably permanently aquatic, living much like aquatic salamander do today.

Order Microsauria
Unnamed specimen: family incertae sedis
Figures 19.25, 19.26

Locality. Peabody Coal Companies Pit 1 or 6, approximately at the county line between Will and Grundy Counties.

Horizon. Francis Creek Shale, Carbondale Formation, Desmoinesian.

A virtually complete and articulated skeleton, only 55 mm long, is the only described specimen attributable to the order Microsauria (Carroll and Gaskill, 1978). Unfortunately, the skull and the number of digits in the hand of this specimen are incompletely preserved. Without the diagnostic features of the skull and manus it is impossible to assign this specimen to any family. Therefore, a definitive statement on the familial assignment of this microsaur must await the discovery of more completely preserved specimens.

A few comments on the cranial elements that are preserved are warranted however. Well-preserved sclerotic plates are found in the large widely separated orbits. The parasphenoid is shaped like an acute triangle with its adjacent sides interrupted only by the bluntly ending, anterolaterally directed basipterygoid processes (Carroll and Gaskill, 1978). The parasphenoid is covered with scattered patches of denticles.

There are 23 presacral vertebrae, one sacral, and approximately 25 caudals, all of which are holospondylous and well ossified (Carroll and Gaskill, 1978). The first four pairs of ribs are short and flattened, forming wide blades. All but the first pair of these ribs are conspicuously bicipital. The remaining ribs are longer, very narrow, cylindrical, and gently curved. Only the first two caudal vertebrae possess ribs, although others may have been present in life. Oval scales, approximately 0.6 mm in diameter, are also preserved in this specimen. The overall pattern of overlap is *en chevron* with the exposed portion of each scale displaying delicate ridges radiating from its posterior border (Carroll and Gaskill, 1978).

The minuscule elements forming the pectoral and pelvic limbs are remarkably well ossified. The phalangeal formula of the pes appears to have been 2,3,4,5,4 (Carroll and Gaskill, 1978).

In life this microsaur may have resembled salamanders of the genus *Ambystoma.*

?Amphibian Spawn

Godfrey (1995) described a string of fossilized eggs that was collected from a Pit 11 spoil heap. Although some living fishes are known to extrude egg strings, they are more common among living amphibians. Unfortunately, this egg strand exhibits no currently known features that would permit it to be assigned to one of the taxa from Mazon Creek.

Class Reptilia
Order Captorhinida
Cephalerpeton ventriarmatum Moodie
Figures 19.27–19.30

Locality. Mazon Creek, Grundy County.

Horizon. Francis Creek Shale, Carbondale Formation, Desmoinesian.

Cephalerpeton ventriarmatum is known only from a single immature specimen that preserves most of the skull and anterior presacral skeleton. Unlike most Mazon Creek tetrapods, the center of the very large circular orbits in *C. ventriarmatum* lies just behind the midpoint of the length of the skull. Furthermore, the frontal bones contribute to the dorsomedial margin of each orbit (Carroll and Baird, 1972). A series of sclerotic plates is preserved in the right orbit. The parietals are well developed, occupying the entire width of the skull table. All the teeth in *C. ventriarmatum* are conical and bluntly pointed, with room for three in each premaxilla and 16 in each maxilla (Carroll and Baird, 1972; Reisz and Baird, 1983). Other protorothyridids have many small teeth, suggesting that the fewer, relatively large teeth in *Cephalerpeton* were adapted to some as yet unknown dietary specialization.

The palatal bones are covered by a shagreen of denticles, which, on the vomers and palatines, lie in uniform rows. The pterygoids bear a denticulate transverse flange. There appears to have been room for approximately 24 teeth along each ramus of the lower jaw.

Remains of the axial skeleton include 23 presacral vertebrae. Each vertebra consists of a high-spined neural arch, suturally attached to a cylindrical pleurocentrum. Intercentra are preserved as thin crescentic wedges sandwiched between adjacent pleurocentra. Anterior ribs have flattened shafts, whereas more posterior ones are cylindrical along their entire length. All the ribs are clearly bicipital. A series of elongate, overlapping gastralia are also preserved with this individual.

Elements forming the pectoral girdle are small and incompletely ossified. The humerus, radius, and ulna are slender bones with poorly defined articulating surfaces (Carroll and Baird, 1972).

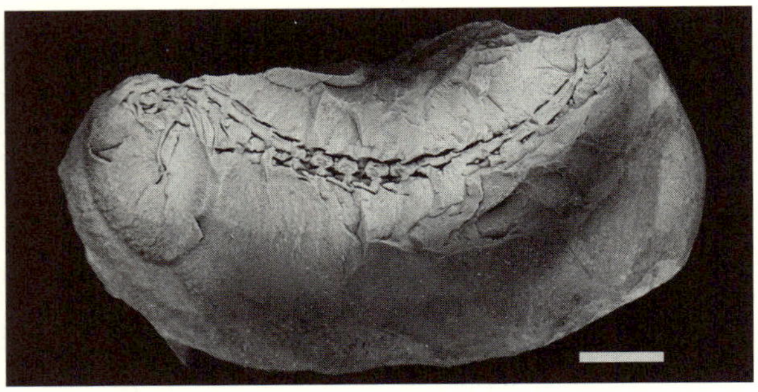

Figure 19.27. *Cephalerpeton ventriarmatum,* YPM 796, above the Morris (No. 2) Coal, Grundy County; photograph of a latex cast. Partial skeleton in a left ventrolateral view. Scale bar = 10 mm.

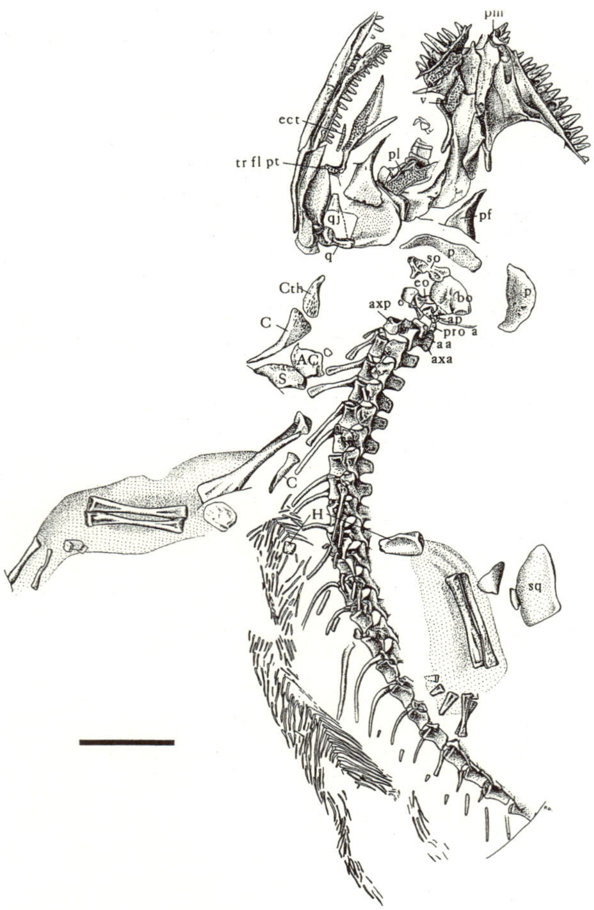

Figure 19.28. *Cephalerpeton ventriarmatum,* YPM 796; specimen drawing by Pamela Gaskill for Carroll and Baird (1972). Most of the vertebral column is exposed in a left lateral view. Scale bar = 10 mm. Abbreviations: aa, atlas arch; ac, anterior coracoid; ap, atlas pleurocentrum; axa, axis arch; axp, axis pleurocentrum; bo, basioccipital; c, clavicle; cth, cleithrum; ect, extopterygoid; eo, exoccipital; h, humerus; p, parietal; pf, postfrontal; pl, palatine; pm, premaxilla; pro.a, proatlas q, quadrate; qj, quadratojugal; s, scapulocoracoid; so, supraoccipital; sq, squamosal; tr.fl.pt, transverse flange of the pterygoid; v, vomer.

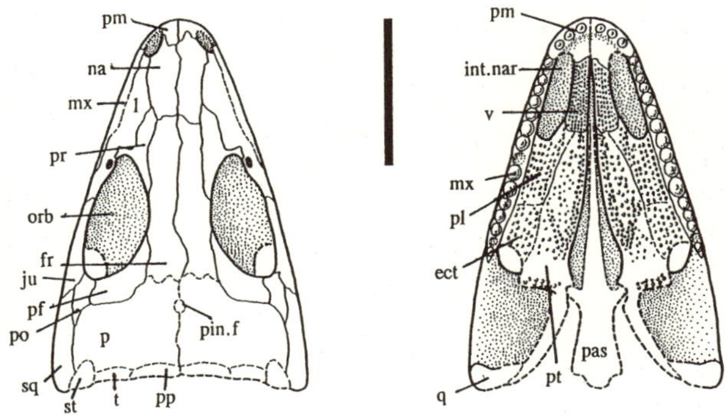

Figure 19.29. *Cephalerpeton ventriarmatum,* reconstruction of the skull in dorsal and ventral views, respectively, redrawn from Carroll and Baird (1972). Lower jaw not included. Scale bar = 10 mm. Abbreviations: ect, ectopterygoid; fr, frontal; int.nar, internal nares; ju, jugal; l, lacrimal; mx, maxilla; na, nasal; orb, orbit; p, parietal; pas, parasphenoid; pf, postfrontal; pin.f, pineal foramen; pl, palatine; pm, premaxilla; po, postorbital; pp, postparietal; pr, prefrontal; pt, pterygoid; q, quadrate; sq, squamosal; st, supratemporal; t, tabular; v, vomer.

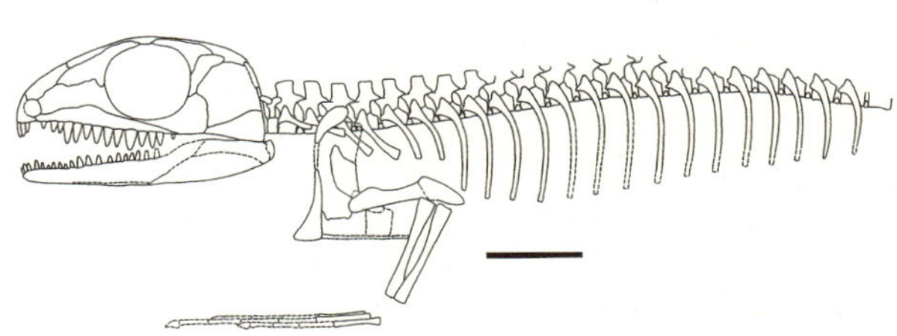

Figure 19.30. *Cephalerpeton ventriarmatum,* reconstruction of the skeleton in a left lateral view, redrawn from Carroll and Baird (1972). Scale bar = 10 mm.

Judging from the skeletal proportions of *Cephalerpeton,* it could have resembled, at least superficially, a small iguanid lizard.

Acknowledgments

I wish to thank Drs. R. L. Carroll, D. Baird, and A. R. Milner for their help in the preparation of this chapter.

References

Bolt, J. R. 1979. *Amphibamus grandiceps* as a juvenile dissorophid: evidence and implications, pp. 529–563. *In* M. H. Nitecki (ed.), Mazon Creek Fossils. Academic Press, New York.

Bolt, J. R., and R. E. Lombard. 1985. Evolution of the amphibian tympanic ear and the origin of frogs. Biol. Jour. Linnean Soc., 24:83–99.

Bossy, K. V. 1976. Morphology, paleoecology and evolutionary relationships of the Pennsylvanian urocordylid nectrideans (Subclass Lepospondyli, Class Amphibia). Ph.D. thesis, Yale University, New Haven, Connecticut.

Boy, J. A. 1978. Die tetrapodenfauna (Amphibia, Reptilia) des saarpfalzischen Rotliegenden (Unter-Perm: SW-Deutschland). 1, *Branchiosaurus.* Mainzer Geowiss. Mitt., 7:27–76.

Carroll, R. L. 1964. The earliest reptiles. Jour. Linnean Soc. London Zool., 45:61–83.

———. 1977. Patterns of amphibian evolution: an extended example of the incompleteness of the fossil record. *In* A. Hallam (ed.), Patterns of Evolution. Elsevier, Amsterdam.

———. 1992. The primary radiation of terrestrial vertebrates. Ann. Rev. Earth Planet. Sci., 20:45–84.

Carroll, R. L., and D. Baird. 1972. Carboniferous stem-reptiles of the family Romeriidae. Bull. Mus. Comp. Zool., 143(5):321–364.

Carroll, R. L., and P. Gaskill. 1978. The Order Microsauria. Mem. Amer. Phil. Soc., 126:1–211.

Clack, J. A., and M. I. Coats. 1995. *Acanthostega gunnari,* a primitive, aquatic tetrapod? Bull. Mus. Natl. Hist. Nat., Paris, 4th ser., sec. C, no. 1–4, 17:359–372.

Coats, M. I., and J. A. Clack. 1995. Romer's gap: tetrapod origins and terrestriality. Bull. Mus. Natl. Hist. Nat., Paris, 4th ser., sec. C, no. 1–4, 17:373–388.

Godfrey, S. J. 1995. Fossilized eggs from the Pennsylvanian of Illinois. Ichnos, 4:71–75.

Gregory, J. T. 1948. A new limbless vertebrate from the Pennsylvanian of Mazon Creek, Illinois. Amer. Jour. Sci., 246:636–663.

Holmes, R. 1989. The skull and axial skeleton of the Lower Permian anthracosauroid amphibian *Archeria crassidisca* Cope. Palaeontographica A, 207:161–206.

Hook, R. W., and D. Baird. 1984. *Ichthycanthus platypus* Cope, 1877, reidentified as the dissorophoid amphibian *Amphibamus lyelli.* Jour. Paleo., 58:697–702.

———. 1986. The Diamond Coal Mine of Linton, Ohio, and its Pennsylvanian-age vertebrates. Jour. Vert. Paleo., 6(2):174–190.

Lund, R. 1978. Anatomy and relationships of the family Phlegethontiidae (Amphibia, Aïstopoda). Ann. Carnegie Mus., 47:53–79.

McGinnis, H. J. 1967. The osteology of *Phlegethontia,* a Carboniferous and Permian aïstopod amphibian. Univ. California Pub. Geol. Sci., 71:1–46.

Milner, A. R. 1982. Small temnospondyl amphibians from the Middle Pennsylvanian of Illinois. Palaeontology, 25:635–664.

———. 1990. The radiations of temnospondyl amphibians, pp. 321–349. *In* P. D. Taylor and G. P. Larwood (eds.), Major Evolutionary Radiations. Syst. Assoc. Spec. Vol., 42. Clarendon, Oxford.

Milner, A. R., T. R. Smithson, A. C. Milner, M. I. Coates, and W. D. I. Rolfe. 1986. The search for early tetrapods. Mod. Geol., 10:1–28.

Panchen, A. L. 1970. Anthracosauria. *In* O. Kuhn (ed.), Encyclopedia of Paleoherpetology. Gustav Fischer Verlag, Stuttgart.

Panchen, A. L., and T. R. Smithson. 1987. Character diagnosis, fossils and the origin of tetrapods. Biol. Rev., 62:341–438.

Reisz, R., and D. Baird. 1983. Captorhinomorph "stem" reptiles from the Pennsylvanian coal-swamp deposit of Linton, Ohio. Ann. Carnegie Mus., 52:393–411.

Richardson, E. S., Jr., and R. G. Johnson. 1971. The Mazon Creek faunas, pp. 1222–1235. Proc. North American Paleo. Conv. 5–7 September 1969, Field Mus. Nat. Hist., Vol. 1. Allen, Lawrence, Kansas.

Schultze, H.-P. 1986. Dipnoans as sarcopterygians. Jour. Morphol., suppl. 1:39–74.

Schultze, H.-P., and M. Arsenault. 1985. The panderichthyid fish *Elpistostege:* a close relative of tetrapods? Palaeontology, 28:293–309.

Vorobyeve, E., and H.-P. Schultze. 1991. Description and systematics of panderichthyid fishes with comments on their relationship to tetrapods, pp. 68–109. *In* H.-P. Schultze and L. Trueb (eds.), Origins of the Higher Groups of Tetrapods, Comstock. Cornell University Press, Ithaca, New York.

Wellstead, C. F. 1991. Taxonomic revision of the Lysorophia, Permo-Carboniferous lepospondyl amphibians. Bull. Amer. Mus. Nat. Hist., 209:1–90.

20

Problematica

Steven D. Sroka and Eugene S. Richardson, Jr.

A problematic fossil is one whose phylum-level affinity is unknown (Bengtson, 1986). Compared to other soft-bodied fossil localities, there are few problematic fossils within the Mazon Creek fauna. These problematic fossils can be divided into two basic groups: organisms whose anatomy is well known and organisms with little anatomical detail (e.g., *Etacystis communis*). Since the first specimens collected do not always clarify aspects of anatomy and taxonomic relationships, extensive collecting may be required before enough evidence is in to warrant a description and a name. The best example of this is the Tully Monster, an animal with a reasonably complex anatomy. Perhaps 40,000 Tully Monster specimens have been collected. It is very well preserved and reliably described, yet its anatomy is unlike that of any known organism. If cladistic analyses are accepted, *Tullimonstrum* could graduate from its present problematic status and be placed into modern high-level groups. Eventually, other problematic fossils may also be referable to modern phyla or classes. *Escumasia roryi* (commonly called "the Y") may be a cnidarian, and *Etacystis communis* ("the H") a hemichordate or hydrozoan.

Perhaps specimens from other soft-bodied fossil localities will have critical pieces of anatomy preserved that will end the problematic status of Mazon Creek animals with unknown affinities. We suppose the day will come when all of the Mazon Creek problematica will be identified. Problematica are interesting, and the same rate of collecting that may bring us answers to the ones we have now will also provide future problematica.

Chondrichthyan Egg Cases

Palaeoxyris and *Vetacapsula* have variously been assigned to both plants and animals. The current opinion is that they are ancient chondrichthyan egg cases; see Crookall (1932) or Zidek (1976) for an excellent discussion on the affinities of these fossils. Geologically, these fossils are most common in the Carboniferous, although *Palaeoxyris* ranges into the Mesozoic.

Palaeoxyris Brongniart, 1828
Figures 20.1, 20.2

Palaeoxyris has a slender fusiform body tapering at both ends; at one end a pointed beak occurs, at the other a pedicle. Fine spiral lines mark the surface of the body and pedicle; the lines become straighter on the beak and converge toward its tip. On compressed specimens, the intersection of spiral markings on opposite sides of the body produces rhomboidal fields. Many specimens are known.

Vetacapsula cooperi Mackie *rev.* Crookall, 1928
Figures 20.3, 20.4

McGhee and Richardson (1982) described *V. cooperi* as a taxon with a broadly fusiform body

Figure 20.1. *Palaeoxyris,* NEIU MCP 143; photograph. Scale bar = 10 mm.

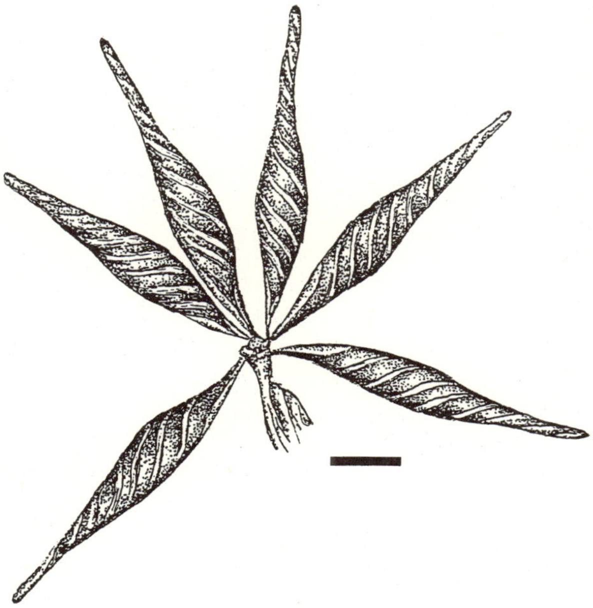

Figure 20.2. *Palaeoxyris,* modified from Langford (1958). Scale bar = 10 mm.

Figure 20.3. *Vetacapsula cooperi,* FMNH PF 8961; photograph. Scale bar = 10 mm.

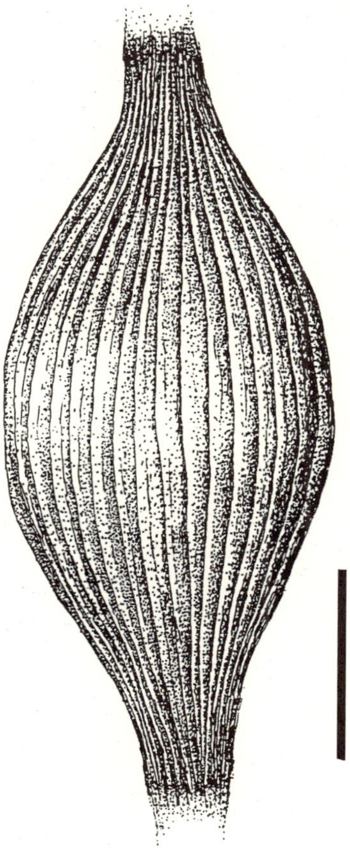

Figure 20.4. *Vetacapsula cooperi,* modified from McGhee and Richardson (1982). Scale bar = 10 mm.

marked by longitudinal pleatlike ribs, approximately 1 mm apart; a strong median ridge transverses the body from beak to pedicle. This description was based on a single specimen from Pit 11. In contrast to *Palaeoxyris,* spiral twisting on *V. cooperi* was not observed (McGhee and Richardson, 1982).

"H" Animal

Etacystis communis Nitecki and Schram, 1976
See Figures 6.14, 6.15

Known to collectors as "the H," *Etacystis* is a relatively large asymmetrical H-shaped animal

Figure 20.5. *Escusmasia roryi,* NEIU MCP 56; photograph. Scale bar = 10 mm.

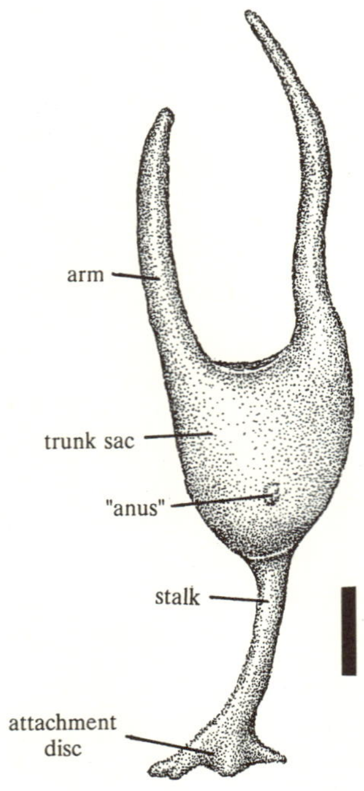

Figure 20.6. *Escusmasia roryi,* modified from Nitecki and Solem (1973). Scale bar = 10 mm.

or part of an animal with two unequal arms and a sac connected to a crossbar (peduncle). For a more detailed description of *E. communis,* see Chapter 6. Unfortunately, detailed internal morphology is largely lacking.

The affinities of *E. communis* have been debated since its discovery. Nitecki and Schram (1976), with some uncertainty, considered it to be similar to pterobranch hemichordates such as *Rhabdopleura,* which has paired tentaculate lophophore arms. Foster (1979b) assigned *Etacystis* to the Hydrozoa (order Siphonophora) because of its possible planktic nature, morphology, and the great range in size in specimens lacking appreciable variation in form of the peduncle, sac, and arms.

Alternatively, Schram (1991), using cladistic analysis, argued that *Etacystis* is closely allied with the Ectoprocta, Phoronida, and the higher deuterostome/lophophorates.

"Y" Animal

Escumasia roryi Nitecki and Solem, 1973
Figures 20.5, 20.6

Escumasia roryi is a bilaterally symmetrical animal with a Y-shaped body supported on a circular disc. The body bears two long, round and slender arms or tentacles, with a possible mouth located between the branches of the arms; an anus may have opened on the side of the body. Nitecki and Solem (1973), using measurements from 19 specimens, determined *Escumasia* had a median body length of 120 mm, with a range of 75–205 mm. The absence of crucial anatomical details makes speculation on the affinities of *Escumasia* extremely difficult. However, Nitecki and Solem (1973) hinted toward coelenterate relationships, but there is little evidence at this time to support this theory.

Tully Monster

Tullimonstrum gregarium Richardson, 1966
Figures 20.7–20.10

Probably the most fascinating of all Mazon Creek fossils is *T. gregarium,* popularly known as the "Tully Monster." Tully Monster fossils are restricted to the Essex fauna of Will and Kankakee Counties, at Chowder Flats near Mor-

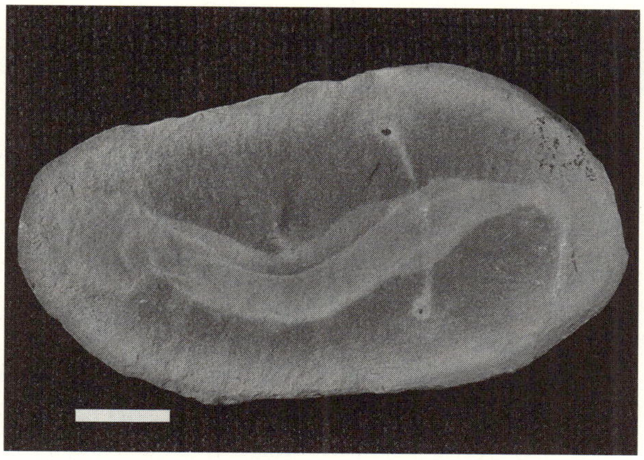

Figure 20.7. *Tullimonstrum gregarium,* complete individual, NEIU MCP 421; photograph. Scale bar = 10 mm.

Figure 20.8. *Tullimonstrum,* head region; photograph. Scale bar = 10 mm. Photograph courtesy Field Museum of Natural History.

ris in Grundy County, and from the Sunspot Mine near Astoria, Fulton County. Appreciation of the uniqueness of the Tully Monster goes beyond scientists and collectors because it was recently chosen as the state fossil of Illinois. Johnson and Richardson (1969), Foster (1979a), and Beall (1991) all provided excellent descriptions of *T. gregarium,* but their interpretation of anatomical details and functional morphologies differ.

Tullimonstrum gregarium is an elongated, bilaterally symmetric fossil ranging in length, in known specimens, from approximately 7.5 cm (3 in.) to just over 30 cm (1 ft.). The body is divided into three distinct regions: 1) an anterior or head end, which includes the eye stalks (= transverse bar), eyes (= bar organs), and a long flexible proboscis with a toothed "claw"; 2) a trunk; and 3) a posterior end with its bluntly pointed tail with lateral fins.

The anterior proboscis bears a terminal claw or jaw apparatus with up to 14 stylets (teeth). The stylets are usually about 0.05 mm in length, and the jaw impressions are sharp, clear, and commonly covered with the clay mineral kaolinite. Foster believed the tip area represents a buccal mass like that in gastropods, whereas Johnson and Richardson suggested that the jaws served as a grasping organ. There is no evidence that the proboscis could be withdrawn into the head. A crescent-shaped impression near the base of the proboscis may represent the hinge area between the trunk and the proboscis (Foster, 1979a). The eye bars and eyes, termed the transverse bar and bar organs by Richardson and Johnson, are diagnostic features of *Tullimonstrum.* The bar consists of a medial plate (central swelling) and somewhat rigid stalks. At the end of the stalks are small oval optic organs or eyes.

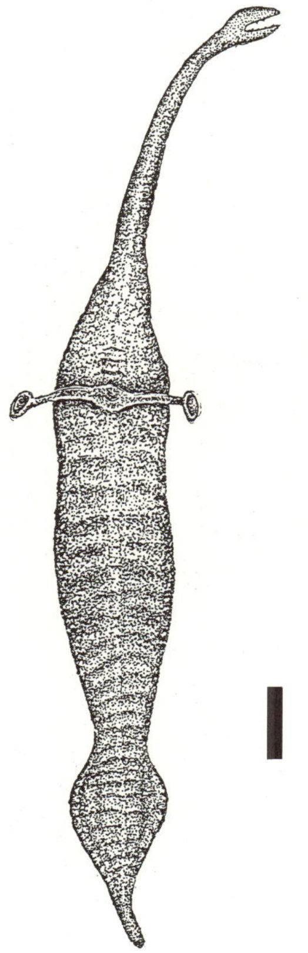

Figure 20.9. *Tullimonstrum,* modified from several sources. Scale bar = 10 mm.

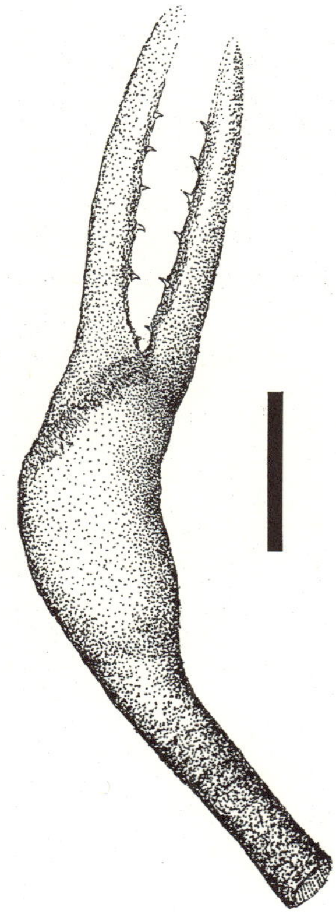

Figure 20.10. *Tullimonstrum,* reconstruction of "claw." Scale bar = 5 mm.

Knowledge of the trunk region is rather limited. The linear gut track is distinguishable from the rest of the body by its lighter color. Regularly spaced dark-colored bands occur in the head area and tail and may represent true body segments (Johnson and Richardson, 1969) or muscles (Foster, 1979a).

The posterior region of the body is distinguished by having a bluntly pointed tail and two triangular lateral fins. Almost all reconstructions and photographs show stubby incomplete tails because the end of the tail usually is not preserved. Complete specimens or tail sections show that the tail is more elongated than previously thought, resulting in a wormlike appearance.

Beall (1991) believed that several morphological aspects of *Tullimonstrum* can be reinterpreted. Most significant is that *Tullimonstrum* had posterior, slightly asymmetrical fins oriented dorsoventrally, not laterally (horizontally). According to Beall, this asymmetry would interfere with swimming if the fins were oriented laterally. Another reinterpretation involves the transverse bar organ and organs, which Beall suggested function as paired copulatory organs, or highly modified setae, or as benthic support structures. Last, in Beall's description, the proboscis bore simple cone-shaped stylets, each with a basal cavity. This interpretation becomes important because it allows for comparison with conodont elements.

Johnson and Richardson (1969) found no reason to assign *Tullimonstrum* to any known phylum because critical information needed for classification was lacking. Foster (1979a) believed *T. gregarium* is related to heteropod gastropods because of the similarities in overall shape, proboscis morphology, complex eye structure, and tooth form. Cladistic analyses by Beall (1991) and Schram (1991) did not resolve the problem of the Tully Monster's affinities but only complicated the picture by suggesting contrasting relationships. Schram's analysis resulted in two possibilities: 1) that *Tullimonstrum* affinities lie in relationships with Annelida and higher annulates or 2) that *Tullimonstrum* is in close proximity to Nermertinea. Beall suggested close affinities with either the Conodonta or Mollusca.

References

Beall, B. S. 1991. The Tully Monster and a new approach to analyzing problematica, pp. 271–285. *In* Alberto M. Simonetta and Simon Conway Morris (eds.), The Early Evolution of Metazoa and Significance of Problematic Taxa. Proc. International Symposium, 27–31 March 1989, Univ. Camerino. Cambridge University Press, Cambridge.

Bengtson, S. 1986. Introduction: the problem of problematica, pp. 3–11. *In* A. Hoffman and M. H. Nitecki (eds.), Problematic Fossil Taxa. Oxford University Press, New York.

Brongniart, A. 1828. Essai d'une flore du bigarré. Sci., nat., le ser., Ann., 15:456.

Crookall, R. 1928. Palaeozoic species of *Vetacapsula* and *Palaeoxyris*. Geol. Surv. Great Britain Mus. Pract. Geol. Summ. Progr., [1927], pt. 2:87–107.

———. 1932. Further morphological studies in *Palaeoxyris*, etc. Geol. Surv. Great Britain Mus. Pract. Geol. Summ. Progr., [1931], pt. 2:122–140.

Foster, M. W. 1979a. A reappraisal of *Tullimonstrum gregarium*, pp. 269–302. *In* M. H. Nitecki (ed.), Mazon Creek Fossils. Academic Press, New York.

———. 1979b. Soft-bodied coelenterates in the Pennsylvanian of Illinois, pp. 191–268. *In* M. H. Nitecki (ed.), Mazon Creek Fossils. Academic Press, New York.

Johnson, R. G., and E. S. Richardson, Jr. 1969. Pennsylvanian invertebrates of the Mazon Creek area, Illinois: the morphology and affinities of *Tullimonstrum*. Fieldiana Geol., 12(8):119–149.

Langford, G. 1958. The Wilmington Coal Flora from a Pennsylvanian Deposit in Will County, Illinois. ESCONI, Downers Grove, Illinois, 360 pp.

McGhee, G. R., Jr., and E. S. Richardson, Jr. 1982. First occurrence of the problematical fossil *Vetacapsula* in North America. Jour. Paleo., 56:1295–1296.

Nitecki, M. H., and F. R. Schram. 1976. *Etacystis communis*, a fossil of uncertain affinities from the Mazon Creek fauna (Pennsylvanian of Illinois). Jour. Paleo., 50:1157–1161.

Nitecki, M. H., and A. Solem. 1973. A problematic organism from the Mazon Creek (Pennsylvanian) of Illinois. Jour. Paleo., 47:903–907.

Richardson, E. S., Jr. 1966. Wormlike fossil from the Pennsylvanian of Illinois. Science, 151:75–76.

———. 1991. Cladistic analysis of metazoan phyla and the placement of fossil problematica, pp. 35–46. *In* Alberto M. Simonetta and Simon Conway Morris (eds.), The Early Evolution of Metazoa and Significance of Problematic Taxa. Proc. International Symposium, 27–31 March 1989, University of Camerino. Cambridge University Press, Cambridge.

Zidek, J. 1976. A new shark egg case from the Pennsylvanian of Oklahoma, and remarks on the chondrichthyan egg capsules in general. Jour. Paleo., 50:907–915.

21

Coprolites and Trace Fossils (Biogenic Structures)

Charles W. Shabica
Including a Section on Fossilized Eggs
by Stephen J. Godfrey

Coprolites and trace fossils, although generally not as aesthetically pleasing as the Mazon Creek body fossils, are particularly useful to paleoecologists for inferring environmental conditions and modes of life of the once living animals in the Mazonian delta region. These traces of animal activity can be tracks, trails, feeding or dwelling burrows, or sediment mottling, in addition to the ubiquitous fossil feces termed either "coprolites" or "fecal pellets."

Coprolites and Fecal Pellets

The name "coprolite," from the Greek *copros,* was coined by William Buckland in 1829. Amateur collectors generally shy away from collecting coprolites, since human values denigrate this fossil form; most are thrown away without close inspection. This is to be regretted because the coprolites may contain undigested fragments of an animal's last meal. Further, because coprolites rapidly decompose, their presence is generally considered evidence that the animal producing them was living at or near the location where the coprolites were buried.

Coprolites sometimes include faunal elements from other environments suggesting that feeding may have occurred elsewhere. For example, a crinoid fragment representing a fully marine environment was reported in a Mazon Creek coprolite by Richardson and Johnson (1969). As only one other crinoid is reported from the Mazon Creek biota (Figure 16B.3), a likely explanation is the fish or shark that produced this unique specimen may have fed farther out to sea before returning to the Mazonian estuary to deposit the evidence.

Regarding the eating habits of coprolite producers, Bradley (1946) noted that the most abundant mineral in modern carnivore feces and most fossil coprolites is apatite (calcium phosphate). Modern omnivore and herbivore feces by comparison have a substantially lower percentage of apatite. Bradley concluded that the apatite may have been necessary for preservation, as only carnivore coprolites are preserved or unmistakably identified as such. While it is probable that Mazon Creek coprolites were also produced by carnivores, well-preserved fecal pellets produced by filter feeders and deposit feeders (omnivores) also occur in Mazon Creek concretions.

Spiral coprolite. Buckland (1829) first described spiral coprolites and noted their similarity to the intestines of sharks and rays. Most authors now agree that spiral coprolites are produced by the spiral intestinal valves of vertebrates. The term "spiral coprolite" is loosely applied to any spiral-shaped mass that might be fossil feces, fossil fecal matter preserved in the intestine ("cololites"; Agassiz *in* Buckland, 1837), or fossil intestines ("enterospirae"; Fritsch, 1907). A discussion of features differ-

Figure 21.1. Spiral coprolite probably made by a shark or lungfish, NEIU MCP 224. Scale bar = 10 mm.

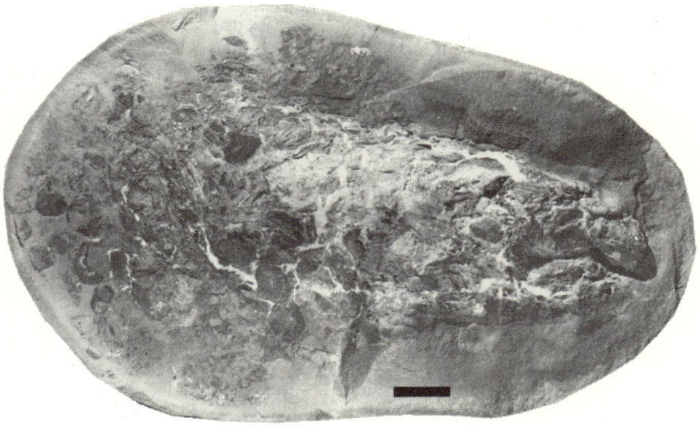

Figure 21.2. Common coprolite with bone fragments, possibly made by a shark or fish, Sroka collection. Scale bar = 10 mm.

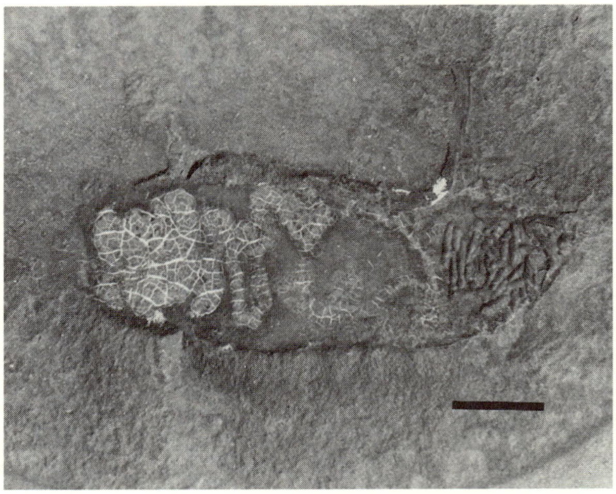

Figure 21.3. Fecal pellets in *Coprinoscolex,* private collection. Scale bar = 10 mm.

entiating the three forms is found in McAllister (1985).

Genera preserved in the Mazon Creek known to possess spiral valves include the shark *Xenacanthus* and the lungfish *Sagenodus.* However, these specimens are all juveniles and are too small to have produced the coprolites described below. Larger shark fossils are, however, found in associated strata. Zangerl and Richardson (1963) reported a "spiral fecal mass" in the pelvic area of a mutilated adult shark specimen from the Mecca Quarry Shale. The Mecca Quarry Shale overlies the thin facies of the Francis Creek Shale and represents a contemporaneous environment to those of the Mazon Creek.

Figure 21.1 shows a Mazon Creek spiral coprolite probably from a shark or lungfish.

Common coprolite. Common coprolites are abundant elements of the Mazon Creek and are loosely defined as undifferentiated masses containing apatite with no identifiable morphology (see Figure 21.2). Their general shape within the siderite concretion is that of an oblate spheroid (slightly squashed sphere, less than 0.5 cm to more than 8 cm in diameter). Some contain septaria-like cracks that are commonly found filled with sphalerite or calcite. This is probably due to dewatering of the coprolite after formation of the enclosing concretion. A small percentage of Mazon Creek coprolites contain identifiable organic traces, mainly bone fragments, fish scales, ostracodes, fragments of the clams *Myalina* and *Solenomya radiata,* and fragments of *Dithyrocaris* and other fossil crustaceans. The presence of common and spiral coprolites is evidence that large species inhabited the Mazonian estuary in spite of the fact that no large body fossils are found.

Fecal pellets. Fecal pellets are small 1–5 mm rod- or football-shaped grains often found inside body fossils or burrows. These are remarkably similar to fecal pellets produced by modern filter- and deposit-feeding organisms. *Coprinoscolex* (Figure 21.3) was named because of the common presence of fecal pellets within its gut. As with the larger coprolites, fecal pellets are evidence for on-site metabolic activities by elements of the fauna.

Trace Fossils

Trace fossils (also called lebenspurren or ichnofossils) are structures or imprints in sediment caused by animal activity. These "biogenic" sedimentary structures have unique properties that differentiate them from other fossils and make them well suited to paleoecological study. Unlike body fossils, trace fossils are usually composed of sediment grains and are thus nearly always found where they were made (autochthonous). Sediment reworking by currents or storm waves will generally destroy any evidence of trace fossils. Reworked or transported structures are rare; none have been observed by the author among the Mazon Creek fossils.

In most cases, the morphology of a particular trace fossil depends more on the animal's behavior than on its shape or species (Seilacher, 1964). Study of modern marine ecosystems indicate that very similar structures may be made by different animals reacting in the same way to similar environmental conditions. For example, high sedimentary rates may cause infaunal clams, crustaceans, and polychete worms to "flee" upward, leaving "escape burrows" that are indistinguishable.

Trace fossils are rarely found associated with Mazon Creek fossils, whether they be in siderite concretions or simply preserved as color variations in the associated sedimentary rocks. Rocks enclosing Mazon Creek nodules are typically laminated, showing little evidence of biologic activity. Thus the rarity of trace fossils suggests that the organisms were not well established at these localities. However, in areas where burrowing organisms like *Edmondia* are found, burrows and burrow mottles are sometimes seen at bedding planes in the concretions. Regions where Francis Creek Shale sediments contain burrows and burrow mottles include parts of Pit 11, the southern part of Pit 13, and Pit 15. Poor preservation of fossils was observed at these sites by Baird et al. (1986) and is thought to be the result of sediment reworking by scavengers (or possibly deposit feeders). See Chapter 4C for further discussion of Mazon Creek sediment reworking by burrowing organisms.

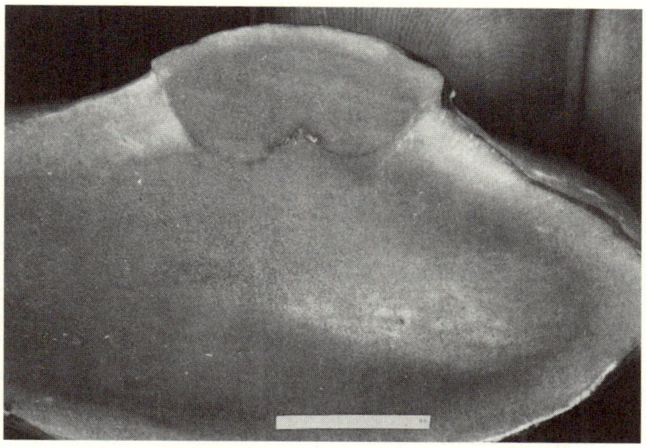

Figure 21.4. *Rusophycus,* arthropod resting track; vertical cross section. Scale bar = 10 mm. Specimen courtesy Field Museum of Natural History.

Rusophycus Hall

Rusophycus is a trace fossil that is considered to be the shallow resting tracks of an arthropod. Found on horizontal surfaces or bedding planes, it is similar in appearance to *Cruziana* D'Orbigny, which is described as shallow pocketlike pits, passages, or pocked burrows shoveled or scratched by trilobites. Figure 21.4 shows a vertical cross section cut through *Rusophycus* preserved in a Mazon Creek siderite concretion. The size and W shape of the structure strongly suggests an arthropod origin. The most likely candidates from the Mazon Creek are shrimp that may have made this characteristic structure while feeding on detritus on the sediment surface.

Arenicolites Salter

Described by Hantzschel (1962), *Arenicolites* are U-shaped, rounded or compressed burrows that are preserved perpendicular to sediment bedding planes. Figure 21.5.A shows a fecal-pellet-filled structure believed to be the base of a U-shaped *Arenicolites* burrow. A vertical cross section through the concretion shows the shape of the burrow (Figure 21.5.B). *Arenicolites* is one of the traces included by Seilacher (1964) in his *Cruziana* facies. Traces in this facies are thought to have been made by mainly suspension-feeding animals living in well-circulated environments characterized by oscillation ripples and well-sorted sands and silts.

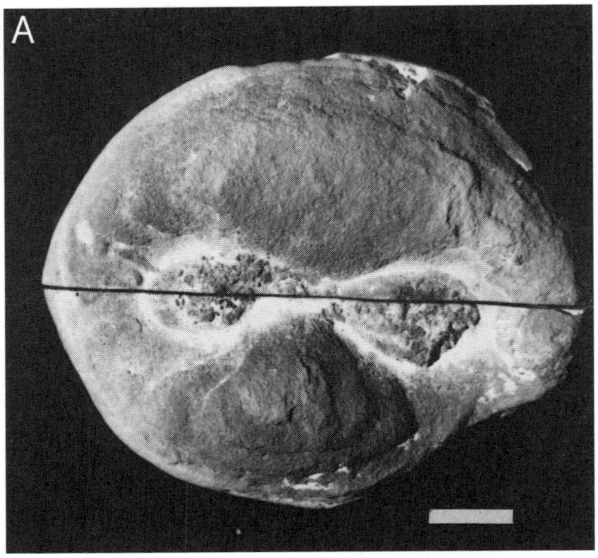

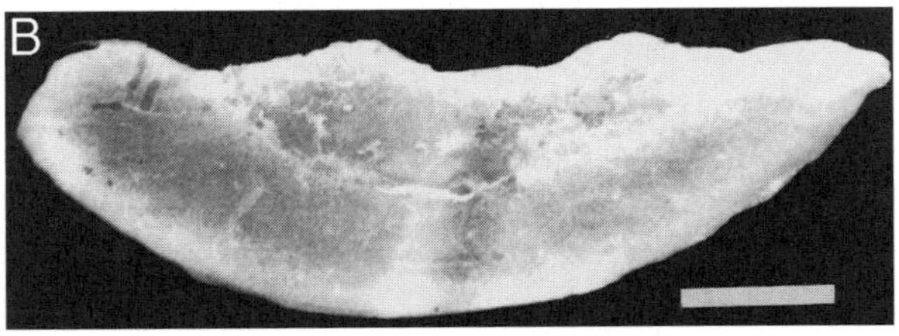

Figure 21.5. *Arenicolites,* U-shaped burrow probably made by a filter-feeding polychete or crustacean, NEIU MCP 7. Structure contains fecal pellets. Base only is preserved. **A** Top view. **B** Vertical cross section. Scale bars = 10 mm.

Named after the marine polychete *Arenicola, Arenicolites* may have been made by a polychete worm or a small crustacean.

Escape structures. As the name implies, escape structures record the activity of an organism in flight, reacting to environmental stress. In a few cases, the organism is preserved within the structure. Figure 21.6 shows the bivalve *Edmondia* in "life position" at the end of a burrow. More often, the structure is the only remnant, and its identification as a biogenic structure can be problematic.

A large number of vertical rod-shaped concretions (Figure 21.7) have been observed in the Francis Creek Shale beds that also contain Mazon Creek fossil concretions. The "rods" may have been formed by iron-rich water flowing up from the compacted underlying peat. Unless fecal pellets or indications of sediment manipulation are found associated with these structures,

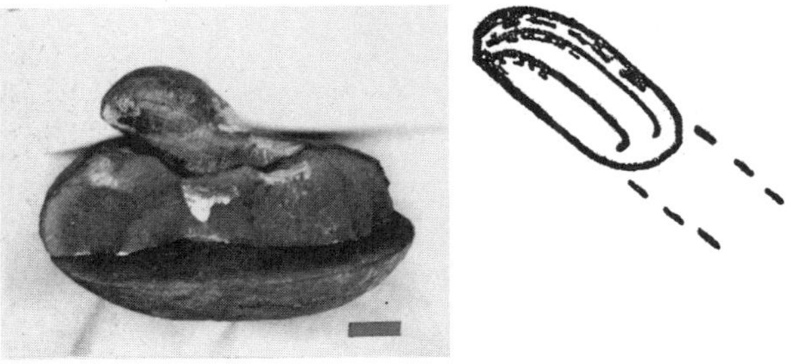

Figure 21.6. Bivalve *Edmondia* at end of escape burrow, NEIU MCP 8. It is unusual in the fossil record for the body fossil to be preserved along with the trace fossil. Scale bar = 10 mm.

a physical process like water or gas escape or slumping may be the probable cause.

Fossilized Eggs from Pit 11

Mazonova helmichnus

Apart from the egg capsules of chondrichthyans (sharks and their kin; Zidek, 1976), fossilized eggs from the Paleozoic are exceedingly rare. Nine nodules from the spoil heaps within the Pit 11 area are an exception, as they preserve strings of eggs that closely resemble those spawned by some living fish and amphibians. To my knowledge, fossilized eggs of this type have never been found elsewhere. The known specimens are currently held by the Royal Ontario Museum (Toronto, Canada), Mary Pat Smith (Chicago, Illinois), and the Mazon Creek Project (Northeastern Illinois University, Chicago, Illinois). These unique trace fossils have been named *Mazonova helmichnus* (Godfrey, 1995).

Differential staining of the iron carbonate (Figure 21.8) immediately around the egg impressions indicates that they were originally spawned within a gelatinous sheath containing one or two rows of eggs. The preserved remnants of the sheath are almost always a lighter color than the eggs, ranging from tan to white. What is now interpreted as the sheath remnant does not appear to be a preservational artifact. In ROM 47490, for example (Figure 21.8), some of the egg impressions are seen to zig-zag within the ribbon. Had the decomposing eggs altered the appearance of the surrounding matrix, one would expect the outer margin of the sheath to follow more closely the alignment of

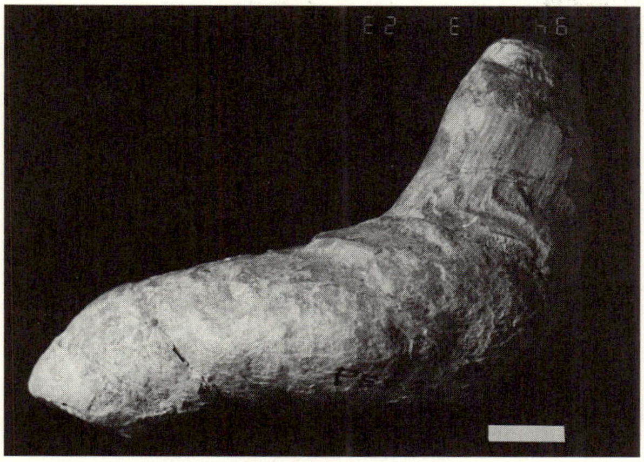

Figure 21.7. "Water escape structure," NEIU MCP 4. Unless there is some evidence of bioturbation, this will be interpreted as a physical sedimentary structure rather than a trace fossil. Scale bar = 10 mm.

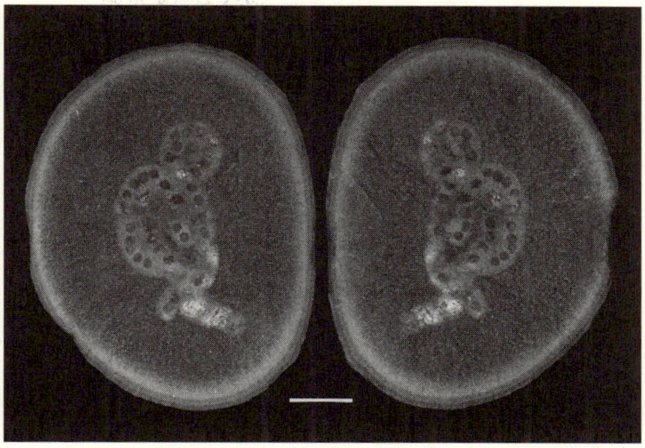

Figure 21.8. ?Vertebrate eggs, ROM 47490, Pit 11. Scale bar = 10 mm. Photograph S. Godfrey.

the eggs within. Furthermore, in some specimens, short segments of the sheath are devoid of eggs. In the known specimens, the diameter of the sheath ranges from 3.0 to 6.0 mm.

The number of eggs per nodule varies from 6 to 50. The diameter of the eggs ranges from 1.4 to 1.7 mm. Adjacent ova frequently abut; others are separated by 1–2 mm. In most specimens, the ova are about the same color as the iron carbonate matrix, although they can be either darker or lighter. Embryological differentiation within the egg impressions has not been observed.

Discussion. The Pit 11 fauna consist primarily of the marine Essex type, but freshwater Braidwood elements are also found. In the Pit 11 area, a 2–5 km wide zone of mixing occurs between the Braidwood and Essex, particularly at the northern end (Schram, 1979). Baird (1979) has suggested that the Braidwood faunal constituents within this area were washed into the sea during distributary flooding of the neighboring swamps and delta levees. Unfortunately, as a result of this localized faunal mixing, it is uncertain whether *M. helmichnus* eggs were spawned by a marine or freshwater organism. Nevertheless, on the basis of the spawn morphology in living organisms, the following observations are given: Among the invertebrates, some opisthobranch molluscs extrude eggs in gelatinous ribbons or sheets (Thompson, 1976). However, opisthobranch eggs range in diameter from about 40 to 400 μm, with most about 100 μm in diameter, much smaller than those described here. The diameters of both the eggs and surrounding sheaths lie within the vertebrate range. Among extant vertebrates, at least five species within the Teleostei spawn egg ropes or ribbons (Breder and Rosen, 1966). A greater number of living amphibian species spawn egg strings that resemble those figured here (Duellman and Trueb, 1986). One should note however that amphibians assignable to the extant orders and teleosts do not appear in the fossil record until the Mesozoic. Nevertheless, the Mazon Creek fauna is not without fish or amphibians that could have spawned these eggs.

Acknowledgments

S. J. G. thanks Dr. Desmond Collins and Janet Waddington, Royal Ontario Museum, and Mary Pat Smith for permission to borrow and study specimens from their respective collections.

References

Baird, G. C. 1979. Lithology and fossil distribution, Francis Creek Shale in northeastern Illinois, pp. 41–67. *In* M. H. Nitecki (ed.), Mazon Creek Fossils. Academic Press, New York.

Baird, G. C., S. D. Sroka, C. W. Shabica, and G. J. Kuecher. 1986. Taphonomy of Middle Pennsylva-

nian Mazon Creek area fossil localities, northeast Illinois: significance of exceptional fossil preservation in syngenetic concretions. Palaios, 1:271–285.
Breder, C. M., Jr., and D. E. Rosen. 1966. Modes of Reproduction in Fish. Natural History Press, New York, 941 pp.
Buckland, W. 1829. On the discovery of coprolites, or fossil feces, in the Lias at Lyme Regis, and in other formations. Trans. Geol. Soc. London, 2nd ser., 2, 3, pt. 1:223–236.
———. 1837. Geology and Mineralogy Considered with Reference to Natural Theology, vol. 1. Carey, Lea, and Blanchard, Philadelphia, 443 pp.
Duellman, W. E., and L. Trueb. 1986. Biology of Amphibians. McGraw-Hill, New York, 670 pp.
Fritsch, Anton. 1907. Miscellanea Paleontologica. 1, Palaeozoica. Rivnac, Prague, 23 pp.
Godfrey, S. J. 1995. Fossilized eggs from the Pennsylvanian of Illinois. Ichnos, 4:71–75.
Hantzschel, W. 1962. Trace fossils and problematica, pp. W177–W245, 246–249. *In* E. C. Moore (ed.), Treatise on Invertebrate Paleontology. Part W, Miscellanea, Conodonts, Conoidal Shells of Uncertain Affinities, Worms, Trace Fossils and Problematica. Geological Society of America and University of Kansas Press, Lawrence.
McAllister, James A. 1985. Reevaluation of the formation of spiral coprolites. Univ. Kansas Paleo. Contrib., 114, 12 pp.
Richardson, E. S., and Johnson, R. G. 1969. The Mazon Creek Fauna, pp. 1222–1235. Proc. North American Paleo. Conv., 5–7 September 1969, Field Mus. Nat. Hist., vol. 1. Allen, Lawrence, Kansas.
Schram, F. R. 1979. The Mazon Creek biotas in the context of a Carboniferous faunal continuum, pp. 159–190. *In* M. H. Nitecki (ed.), Mazon Creek Fossils. Academic Press, New York.
Seilacher, A. 1964. Biogenic sedimentary structures, pp. 296–316. *In* J. Imbrie and N. Newell (eds.), Approaches to Paleoecology. Wiley, New York.
Thompson, T. E. 1976. Biology of Opisthobranch Molluscs. Ray Society, London, 207 pp.
Zangerl, R., and E. S. Richardson, Jr. 1963. The paleoecological history of two Pennsylvanian black shales, Fieldiana Geol. Mem., 4, 352 pp.
Zidek, J. 1976. A new shark egg case from the Pennsylvanian of Oklahoma, and remarks on the chondrichthyan egg capsules in general. Jour. Paleo., 50:907–915.

APPENDIX A

Freeze-Thawing Technique for Opening Nodules

Steven D. Sroka and Gordon C. Baird

Concretions are collected and processed in a variety of ways depending on the experience, knowledge, and individual preferences of the collector. Field collecting in strip mined areas involves picking up exposed nodules that have weathered out of backpiled Francis Creek Shale overburden. Typically, most concretionary siderite on dump surfaces consists of irregular plates, chips, and tubular masses; this debris is usually unfossiliferous except for trace fossils. Fossil-bearing concretions in contrast display a relatively smooth surface and more regular shape. Nodules on or immediately below dump surfaces are subjected to seasonal frost-wedging and when struck a sharp blow may split naturally along the fossil plane, though others containing mineral-filled fractures may shatter, or resist fracturing almost entirely. In most collecting areas only a low percentage of concretions are found open; the remainder must be split by the collector. In rare sites a majority of concretions are found open; typically such places yield numerous and sometimes spectacular fossils. Because fossils are abundant, autosplitting is natural and common. A reverse corollary applies to places where no concretions are found open; these sites yield barren "nuckers," and such areas are remembered and avoided by collectors.

Whole concretions must be opened; it is the bane of Mazon Creek paleontologists to hold a tantalizing nodule in hand and still not "have the fossil." The collector must either risk it all by pounding nodules with hammers on flat rock "anvils" in the mine or choose to wait it out by using a slower method. Too often, adrenalin gets the better of the novice and he pounds away only to get "half the Tully" or a decapitated amphibian. This experience instills wisdom and increased appreciation for patience. The best results in splitting concretions mimic the natural freeze-thawing process, and many good collectors long ago opted for the waiting game, using a freeze-and-thaw approach at home. Historically, freeze-thawing of nodules has been reported by Moysey (1908) and Foster (1979, pp. 195, 201). Moysey (1908), who used nodules collected from a clay pit near Ilkeston (England), first heated them until red hot and then plunged them into water; this process resulted in the nodules' exploding. Then he tried soaking nodules in water, placing them in a tin canister, and placing the tin in a ice and salt mixture for 48 hours. Upon thawing some had cracked, and a light hammer tap split them. Finally, he gently heated nodules that had not split and, while they were still hot, plunged them into water.

Foster (1979) used the freeze-thawing technique on concretions from Astoria, Illinois; he repeatedly froze and thawed concretions in water-filled containers and recorded the number of cycles required to split them. Foster's experiments showed that it took an average of 20 freeze-thawings to open his concretions with a range from 2 to 51. Foster also tried this technique with Pit 11 concretions. Although he did not record the number of freezing cycles for the Essex-area material, he suggested (based on preliminary observation) that the average opening time might be as much as 10 times that required for Astoria concretions. He also left concretions from both localities outside in the natural weather cycles to experience climate extremes, concluding that this method is effective but slow.

The common freeze-thawing technique used by large-volume collectors is to put the concretions into water-filled plastic containers and leave them outside over the winter. This method is prohibitively slow, taking several seasons to succeed. The method outlined below (successfully used by us since the mid-1970s) can be accomplished all year using a refrigerator or freezer.

1) Whole nodules are placed in water-filled plastic containers for two days, so that the concretions become water saturated.
2) The containers are placed in a freezer for two days to ensure complete freezing.
3) The containers are removed from the freezer and allowed to thaw; then the freezing is repeated.
4) Remove open concretions. During thawing cycles gently tap remaining unopened concretions before refreezing.
5) After 12 freeze-thaws, nodules are removed from container and allowed to dry thoroughly.
6) Dry nodules are again tapped lightly with a hammer or tapped together side to side.
7) Nodules that do not split are returned to a clean container with new water and the freeze-thaw cycle is repeated.
8) Experience shows that after approximately 25 freeze-thaw cycles, few added nodules will open; these can be hammered, but remember rock or steel fragments are dangerous! Eyes, in particular, must be protected with shatterproof goggles.

Results using our freeze-thawing method vary from locality to locality. It is found that the majority of localities from the Will County part of Pit 11 had approximately 30 percent of the nodules open after eight freeze-thaws, but certain areas of Pits 11, 13, and 15 in Kankakee County produce results as low as 10 percent yield after 20 cycles. Furthermore, concretions from old spoil piles have already undergone variable exposure to the weather, and average "openability" of nodules is related to other individual properties, such as silt and pyrite content. Concretions cored by pyrite resist freeze splitting and, in many instances, even the hammer.

References

Foster, M. W. 1979. Soft-bodied coelenterates in the Pennsylvanian of Illinois, pp. 191–267. *In* M. H. Nitecki (ed.), Mazon Creek Fossils. Academic Press, New York.

Moysey, L. 1908. On a method of splitting ironstone nodules by means of an artificial freezing mixture. Geological Magazine, 5(5):220–222.

APPENDIX B

Taxonomy of Mazon Creek Fauna

Andrew A. Hay

Fossil Classification

Classification entails the division of organisms into a hierarchical sequence of groups, an identification key. Systematics is the science of the diversity of organisms. Taxonomy comprises the study of classification and description and the relationship between organisms.

Ideally, the grouping of organisms into a formal listing is based on their evolutionary relationships (phylogenetic relationships). Practically and historically, the classification of fossil and living organisms has been based mainly on morphological similarities and differences (i.e., the more similar the organism, the more closely related; related organisms must share at least one derived [= specialized] feature).

As our knowledge of biological relationships has changed, so have our classification schemes. Cladists (phylogenetic taxonomists), in an effort to more objectively arrive at a biological classification, have developed computer programs to construct graphic genealogical trees (cladograms). Molecular biology at the gene level has been used to establish phylogenetic relations in some groups; however, this new discipline is in its infancy and has limited applicability to paleontology because biochemicals are seldom preserved in Paleozoic fossils. Traditional (evolutionary) taxonomists group species in hierarchical categories that tend to reflect their evolutionary relationships. In this book, we use the systems that we feel are best suited for each group.

An example is shown below for the species *Ptyonius marshii.*

Phylum (Phyla, pl.)	Chordata
Class	Amphibia
Order	Nectridea
Family	Urocordylidae
Genus (Genera, pl.)	*Ptyonius*
Species	*Ptyonius marshii*

Subdivision of the above taxa into smaller units is done by adding the prefix "sub" or "super."

Rules for classifying organisms are set out in the International Code of Zoological Nomenclature (W. D. L. Ride et al., 1985). Some useful rules are listed below.

1) A scientific name for a species is binomial (consists of two names): for example, genus *Essexella* and species modifier *asherae.*
2) The name of the author and date of the first description of the species are often added at the end of the scientific name: *Essexella asherae* Foster, 1979.
3) In order to be valid, a newly named species must be described and illustrated in a formal publication (book or scientific journal).

   The description of the "type" specimen should be that of an actual specimen designated as the "holotype" and perhaps variants ("paratypes"). Other "types" include figured specimens, referred specimens, and so forth.

Type material should be available for examination in museum or university collections.

4) When two or more authors give different names to the same species, the "law of priority" applies, whereby the first name published is considered valid and later names are called synonyms.

Faunal List

The following Mazon Creek faunal list is presented as an aid to the reader—a means of keeping track of those animals that are generally accepted by the senior reviewer of each phylum as being present in the Mazon Creek biota. We have opted to utilize the hierarchical structure of traditional (formal) classification to serve the needs of our readers, amateur and professional. Above the family level, this list is of necessity a compromise between that of the original describer(s), that suggested by the senior reviewer, and the arrangement of the groups in such works as Meglitsch and Schram (1991) and Carroll (1988), among others. To avoid an over-long list, synonyms have been omitted. In the few instances they are given, the synonym appears in parentheses. In the case of insects and millipedes, several groups are represented by species typical of that group and are so noted in the list.

Kingdom Animalia

Subkingdom Eumetazoa
Subsection Coelenterata

Phylum Cnidaria
CLASS CUBOZOA (sea wasps and box jellyfish)
Anthracomedusa turnbulli
unknown cubomedusan[1]
CLASS SCYPHOZOA (jellyfish)
Essexella asherae
Lascoa mesostaurata
Octomedusa pieckorum
Reticulomedusa greenei
CLASS HYDROZOA (hydras, siphonophores)
Drevotella proteana
Mazohydra megabertha
CLASS ANTHOZOA (sea anemones, corals)
possible cerianthid (tube anemone)[2]
11 unknown medusans[3]

Subsection Bilateralia

Phylum Nemertea = Rhynchocoela (ribbon worms)
Archisymplectes rhothon

Phylum Nematoda (roundworms or thread worms)
Nemavermes mackeei

Phylum Priapulida
Priapulites konecniorum

Phylum Chaetognatha (arrow worms)
Paucijaculum samamithion

1. M. W. Foster, *in* M. H. Nitecki (ed.), Mazon Creek Fossils (Academic Press, New York, 1979); also M. W. Foster, written communication, 1992.
2. F. R. Schram, K. Fitzhugh, and K. Fauchald, written communications to A. A. Hay concerning drawings and photograph of specimen in private collection, 1991.
3. M. W. Foster, written communication, 1992.

Phylum Annelida (segmented worms)
CLASS POLYCHAETA (bristleworms)
Order Amphinomida
Rhaphidiophorus hystrix ("Oliver Hardy worm")
Order Eunicida
Esconites zelus
Order Flabelligerida
Flabelligeridae sp. ("fanworm")
Order Phyllodocida
Fastuoscolex gemmatus ("papillae worm")
Hystriciola delicatula ("baby tooth")
Dryptoscolex matthiesae ("rat worm")
Fossundecima konecniorum ("simple jaw worm")
Pieckonia helenae ("fish worm")
Rutellifrons wolfforum ("shovel nose worm")
Astreptoscolex anasillosus ("plaine worm")
Didontogaster cordylina ("tummy tooth worm")
Levisettius campylonectus ("Riccardo worm")
undescribed palmyrid[4] ("felt worm")
Order Sabellida
Spirorbis sp. (a tubeworm)
Order incertae sedis
Palaeocampa anthrax
Order Opheliida
undescribed opheliid[5] ("mud worm")

Phylum Echiura (spoon worms)
Coprinoscolex ellogimus ("leech")

Phylum Onychophora (velvet worms)
Ilyodes (=Helenodora) inopinata

Phylum Arthropoda
Subphylum Cheliceriformes
CLASS CHELICERATA
Subclass Merostomata
Order Eurypterida (water scorpions)
Adelophthalmus mazonensis
Order Xiphosura (horseshoe crabs)
Euproops danae
Euproops laevicula
Euproops sp.
Pringlia bispinosa(?)
Streptocyclus langfordi(?)
Liomesaspis laevis
Palaeolimulus sp.
Subclass Arachnida (spiders and spiderlike)
Order Phalangiotarbida
Phalangiotarbus lacoei
Bicarinitarbus pieckorum
Paratarbus carbonarius
Mesotarbus extraneus
Architarbus rotundatus
Mazonitarbus minor
Hadrachne horribilis
Order Anthracomartida
undescribed anthracomartids[6]
Order Trigonotarbida
Aphantomartus (=Trigonomartus) pustulatus
Pleophrynus ensifer

4. I. Thompson, Errant polychaetes (Annelida) from the Pennsylvanian Essex fauna of northern Illinois, Paleontographica A 163 [1979]:194.
5. Ibid.
6. B. S. Beall, Chapter 11.

Lissomartus schucherti
Order Uropygida (whip scorpions)
Geralinura carbonaria
Prothelyphonus sp.
Order Amblypygida (tailless whip scorpions and whip spiders)
Graeophonus carbonarius
Protophrynus carbonarius
Thelyphrynus elongatus
Order Araneida
Arthrolycosa danielsi
Arthrolycosa sp.
Order Ricinuleida
Curculioides mcluckiei (=Curculioides ansticii)
Curculioides scaber
Curculioides gigas
Amarixys sulcata (=Curculioides sulcatus)
Amarixys gracilis (=Curculioides gracilis)
Amarixys stellaris
Poliochera punctulata
Poliochera glabra
Poliochera gibbsi
Poliochera incertae sedis
Palaeoricinulei incertae sedis
Order Opilionida (daddy long-legs)
Nemastomoides longipes
Order Solpugida (wind scorpions)
Protosolpuga carbonaria
Order Scorpionida
Eoctenus miniatus
Palaeobuthus distinctus
Kronoscorpio danielsi
Palaeopisthocanthus schucherti
Subphylum Atelocerata
Superclass Myriapoda
CLASS EUTHYCARCINOIDEA
Order Euthycarcinoida
Pieckoxerxes (=Sottyxerxes) pieckoae
Smithixerxes juliarum
Kottixerxes gloriosus
CLASS DIPLODA (millipedes)
Order Amynilyspedida
Amynilyspes wortheni
unpublished species[7]
Order Euphoberiida
Euphoberia armigera (plus about nine additional species)
Acantherpestes horridus
Acantherpestes major
Acantherpestes sp.
Myriacantherpestes hystricosus
Myriacantherpestes inequalis
Xyloiulus sp.
Order uncertain
Archiulus glomeratus
CLASS CHILOPODA (centipedes)
Order Scolopendromorpha
Mazoscolopendra richardsoni
Palenarthrus impressus
Order Scutigeromorpha
Latzelia primordialis
CLASS ARTHROPLEURIDEA
Arthropleura cristata
Arthropleura sp.

7. J. Hannibal, written communication, 1991.

?Myriapoda, class and order uncertain
Eileticus anthracinus
Superclass Hexapoda
CLASS INSECTA (Solely in the interest of conserving space, we have chosen not to list all of the more than 150 Mazon Creek insect species, many known by wings alone.)
Order Diplura
Testajapyx thomasi
Order Monura
Dasyleptus sp.
Order Thysanura (bristletails, silverfish)
Ramsdelepidion schusteri
Subclass Pterygota (winged insects)
Infraclass Palaeoptera (wings primitively flexed or extended)
Order Palaeodictyoptera (about 30 species described; frequently found representatives are listed)
Eubleptus danielsi
Homaloneura dabasinskasi
Thesoneura americana
Order Megasecoptera
Eubrodia dabasinskasi
Mischoptera douglassi
Order Diaphanopterodea
Prochoroptera calopteryx
Order Protodonata
Oligotypus makowskii
Paralogopsis longipes
Order Ephemeroptera (mayflies)
Lithoneura clayesi
Lithoneura lameeri
Lithoneura mirifica
Lithoneura piecko
Infraclass Neoptera (wings flexed)
Order "Protorthoptera" (artificial "wastebasket" group; about 65 species have been described; frequently found representatives are listed)
Eucaenus ovalis
Gerarus danielsi
Protecticus ambulans
Order "Blattodea" (ancestors of cockroaches; more than 30 species have been described; representatives of the order are listed)
Archimylacris paucinervis
Archoblattina beecheri
Order Blattodea (cockroaches)
Mylacris elongata
Subphylum Crustacea
CLASS REMIPEDIA
Order Enantiopoda
Cryptocaris hootchi
CLASS MALACOSTRACA ("higher" crustaceans)
Subclass Hoplocarida
Order Aeschronectida
Kallidecthes richardsoni
Order Stomatopoda
Tyrannophontes theridion
Subclass Eumalacostraca
Order Syncarida
Acanthotelson stimpsoni
Palaeocaris typus
Palaeosyncaris micra
Order Belotelsonidea
Belotelson magister
Order Pygocephalomorpha
Anthracaris gracilis
Mamayocaris jaskoskii
Order Lophogastrida
Peachocaris strongi
Incertae sedis (possible mysidaceans)

Anthracophausia ingelsorum
Essoidia epiceron
Order Hemicaridea
Suborder Tanaidacea
Eucryptocaris asherorum
Order Edriophthalma
Suborder Isopoda
Hesslerella shermani
CLASS PHYLLOPODA
Subclass Phyllocarida
Order Archaeostraca
Dithyrocaris n. sp.
Order Hoplostraca
Kellibrooksia macrogaster
Subclass Branchiopoda
Order Conchostraca
Leaia tricarinata
Pemphilimnadiopsis ortoni
CLASS MAXILLOPODA
Subclass Ostracoda (seed shrimp)
Order Podocopida
Geisina (= Hastifaba) robusta[8]
Paraparchites mazonensis
Subclass Cirripedia (barnacles and allies)
Illilepas (= Praelepas) damrowi
Subclass Halicyna
Cyclus americanus
Cyclus obesus
Halicyne max n. sp.
Apionicon apioides n. gen. and sp.
Subphylum uncertain
CLASS THYLACOCEPHALA
Order Concavicarida
Concavicaris georgeorum (common flea shrimp)
Concavicaris remipes
Order Conchyliocarida
Convexicaris mazonensis

Phylum Mollusca
CLASS POLYPLACOPHORA (chitons and allies)
Glaphurochiton concinnus
CLASS GASTROPODA (snails and slugs)
Euphemites richardsoni
Hypselentoma aff. *H. perhumerosa*
Straparollus (Euomphalus) sp. indet.
Naticopsis (Naticopsis) sp. indet.
Strobeus cf. *S. primigenius*
CLASS BIVALVIA (pelecypods)
Subclass Palaeotaxodonta
Order Solemyoida
Undescribed genus and species[9] ("clam-clam")
"Solemya" radiata
Subclass Pteriormorphia
Order Pterioida
Myalinella meeki
Myalinella sp.
Anthraconaia ohioensis
Anthraconauta sp.
Leptodesma ohioense
Posidonia fracta

8. G. Langford, The Wilmington Coal Fauna (ESCONI, Downers Grove, Illinois, 1963), 69: Langford states that this species is abundant in the Pennsylvanian of Illinois and should be looked for in the Wilmington (Mazon Creek) area.
9. J. B. Bailey, Chapter 8A.

Order Pectinoida
Aviculopecten mazonensis
Aviculopecten janewiedlini
Acanthopecten sp.
Euchondria pellucida
Dunbarella striata
Order Limoida
Palaeolima retifera
Subclass Isofilibranchia
Order Modiomorphoida
Permophorus spinulosa
Permophorus sp.
Subclass Heteroconchia
Order Veneroida
Astartella concentrica
Order Trigonioida
Schizodus cf. *S. wheeleri*
Schizodus cf. *S. affinis*
Subclass Anomalodesmata
Order Pholadomyoida
Edmondia ovata
Edmondia aspinwallensis
Edmondia oblonga
Sedgwickia sp.
Grammysioidea hayi
CLASS CEPHALOPODA
Subclass Bactritoidea
Bactrites sp.
Subclass Ammonoidea
Wiedeyoceras sp.
Schistoceratid indet.
Subclass Nautiloidea
Stearoceras sp.
Titanoceras sp.
Paleocadmus herdinae
Paleocadmus pohli
Paleocadmus sp.
unnamed nautiloid[10]
Subclass Coleoidea (squids, octopi, and kin)
Order Sepioidea
Pohlsepia mazonensis
Order Teuthoidea
Jeletzkya douglassae
undescribed cephalopod, genus and species unknown[11]

Phylum Brachiopoda
CLASS INARTICULATA
Order Lingulida
Lingula sp.
Order Acrotretida
Orbiculoidea sp.
CLASS ARTICULATA
Order Strophomenida
chonetoids (undescribed genera and species)[12]

Phylum Echinodermata (crinoids, sea urchins, starfish, etc.)
CLASS HOLOTHUROIDEA (sea cucumbers)
Achistrum sp.

10. J. Kluessendorf, Chapter 8D.
11. Ibid.
12. M. W. Foster, written communication, 1992.
13. N. G. Lane, A crinoid from the Pennsylvanian Essex fauna of Illinois. Fieldiana Geol., 12, no. 9 [1969]:151–156.

CLASS CRINOIDEA
Order Cladida
genus and species unknown[13]

Phylum Hemichordata
CLASS ENTEROPNEUSTA
Mazoglossus ramsdelli (acorn worm)

Phylum Chordata
Infraphylum Myxinoidea
Order Myxiniformes (hagfishes)
Myxinikela siroka
Infraphylum Vertebrata
Order Petromyzontiformes (lampreys)
Mayomyzon pieckoensis
"agnatha" incertae sedis
Pipiscius zangerli
Gilpichthys greenei
CLASS CHONDRICHTHYES (cartilaginous jawed fishes)
Subclass Subterbranchialia
Order unnamed
Polysentor gorbairdi
?Order Bradyodontida
Jimpohlia erinacea
Order Holocephali
Similihariotta dabasinskasi
Subclass Elasmobranchii
Order Euselachii
Bandringa rayi
Bandringa herdinae
Dabasacanthus inskasi
Holmacanthus keithi n. gen. and sp.
Order Xenacanthida
Xenacanthus sp.
Class inceretae sedis Acanthodii (spiny sharks)
Trichorhipis praecursor
Acanthodes beecheri
Acanthodes marshi
Acanthodes sp.
CLASS OSTEICHTHYES (bony fish)
Subclass Actinopterygii (spiny-rayed fishes)
Order Paleonisciformes
Illinichthys cozarti
Nozamichthys contorta
"Elonichthys" peltigerus
"Elonichthys" hypsilepus
"Elonichthys" disjunctus
"Elonichthys" wolffi
"Elonichthys" remotus
Amphicentrum orbiculare
Order Bobasatriiformes
Platysomus circularis
Order Haplolepiformes
Pyritocephalus comptus
Pyritocephalus gracilis
Parahaplolepis sp.
Microhaplolepis sp.
Actinopterygii incertae sedis
Paleoniscoid sps.
Subclass Sarcopterygii (fleshy-finned fishes)
Order Crossopterygii (coelacanths, lobe-finned fishes)
Megalichthys cf. *hibberti*
Rhizodopsis cf. *robustus*
Rhizodopsis? mazonius
Rhabdoderma exiguum

Rhabdoderma robustus
Rhabdoderma elegans
Order Dipnoi (lungfish)
Conchopoma edesi
Conchopoma arctatum
Ctenodus cristatus
Megapleuron zangerli
Palaeophichthys parvulus
CLASS AMPHIBIA
Order Temnospondyli
Saurerpeton obtusum
Amphibamus grandiceps
?*Branchiosaurus* sp. indet.
Order Anthracosauria
Spondylerpeton spinatum
Order Aïstopoda
Aornerpeton mazonense (=Phlegethontia mazonensis)
Ophiderpeton sp.
Order Nectridea
Ptyonius marshii
Order Lysorophia
Brachydectes newberryi (=Cocytinus gyrinoides)
Order Microsauria
unnamed specimen
CLASS REPTILIA
Order Captorhinida
Cephalerpeton ventriarmatum

Taxonomy uncertain
Tullimonstrum gregarium (phylum?; mollusc? nemertine? annelid?)
Esconichthys apopyris (larval lungfish or amphibian)
Escumasia roryi (coelenterate or - ?)
Palaeoxyris (shark egg case or plant)
Vetacapsula cooperi (shark egg case or plant)
Etacystis communis ("the H") (hemichordate?)
Erierpeton branchialis[14] (tetrapod?)
Erpetobrachium mazonensis (tetrapod?)
Mazonerpeton costatum (tetrapod?)

Trace fossils
Arenicolites
Diplocraterion
Mazonova helmichnus
Rusophycus
escape burrows
insect bites on flora
ova

References

Carroll, R. L. 1988. Vertebrate Paleontology and Evolution. Freeman, New York.
Meglitsch, P. A., and F. R. Schram. 1991. Invertebrate Zoology. Oxford University Press, New York.

14. This and the following two fossils are dubious tetrapod taxa; their names have appeared in the literature occasionally. These taxa are indeterminate, and Milner has rated them as "garbage" (written communication, 1994).

APPENDIX C

Names of Mazon Creek Fossils Honoring Collectors

Adolarryia bairdi, Gordon C. Baird and Larry Osterberger
Bandringa herdinae, Jerry Herdina and Ray Bandringa
Bandringa rayi, Ray Bandringa
Bicarinitarbus pieckorum, Helen and Ted Piecko
Caulopteris mcluckie, Lucy and John McLuckie
Caulopteris langfordi, George Langford, Sr. and Jr
Concavicaris georgeorum, Harriet and Calvin George
Conchopoma edesi, Thomas Edes
Curculioides mcluckei, Lucy and John McLuckie
Dabascanthus inskasi, Walter Dabasinskas
Dasyonthus pohli, Joe Pohl
Dryptoscolex matthiesae, Catherine and Harry Matthies
"Elonichthys" wolffi, Terry and Francis Wolff
Esconichthys apopyris, Earth Science Club of Northern Illinois (ESCONI)
Esconites zelus, Earth Science Club of Northern Illinois (ESCONI)
Escumasia roryi, "Escumas Rory," a pen name of Eugene S. Richardson, Jr.
Essexella asherae, Helen Asher
Eubrodia dabasinskasi, Walter Dabasinskas
Eucryptocaris asherorum, Helen Asher
Fossundecima konecniorum, Sylvia and James Konecny
Gilpichthys greenei, Frank Greene
Grammysioidea hayi, Josephine and Andrew Hay
Herdina mirificus, Jerry Herdina
Hesslerella shermani, Robert Hessler and Levi Sherman
Heterologus langfordum, George Langford, Sr. and Jr.
Holmacanthus keithi, Keith Holm
Homaloneura dabasinskasi, Walter Dabasinskas
Illinichthy cozarti, Chris Cozart
Illilepas damrowi, Dan Damrow
Jeletzkya douglassae, June Douglass
Jimpohlia erinacea, Jim Pohl
Kallidecthes richardsoni, Eugene S. Richardson, Jr.
Kottixerxes gloriosus, A. W. Kott
Larryia osterbergi, Larry Osterberger
Levisettius campylonectus, Riccardo Levi-Setti
Lithoneura piecko, Helen and Ted Piecko
Lithoneura clayesi, Walter Clay
Mayomyzon pieckoensis, Helen and Ted Piecko and Stephen LeMay
Mazoglossus ramsdelli, Karlene Ramsdell
Mazonopterum wolfforum, Terry and Francis Wolff
Mcluckiepteron luciae, Lucy McLuckie
Megapleuron zangerli, Rainer Zangerl
Mischoptera douglassi, Lincoln Douglass
Myxinikela siroka, Daniel Sroka
Nemavermes mackeei, Melbourne A. McKee
Neofouquea suzanneae, Suzanne Dabasinskas
Notorachis wolfforrum, Terry and Francis Wolff
Octomedusa pieckorum, Helen and Ted Piecko
Paleocadmus herdinae, Jerry Herdina
Paleocadmus pohli, Joe Pohl
Pherusa prinosi (in prep.), John Prinos
Pieckonia helenae, Helen Piecko
Pieckoxerxes pieckoae, Helen Piecko
Pohlsepia mazonensis, Jim Pohl
Polysentor gorbairdi, Gordon C. Baird
Priapulites konecniorum, Sylvia and James Konecny
Ramsdelepidion schusteri, Karlene and Stephen Ramsdell and Tony Schuster
Reticulomedusa greenei, Frank Greene
Rossites inopinus, Herbert M. Ross
Rutellifrons wolfforum, Terry and Francis Wolff
Similihariotta dabasinskasi, Walter Dabasinskas
Smithixerxes juliarum, Julia Smith
Srokalarva berthei, JoLynn Berthe-Traub and Steven D. Sroka
Stigmarioides holmesi, Mrs. George (née Sidney Holmes) Langford
Streptocyclus langfordi, George Langford, Sr.
Testajapyx thomasi, Thomas Testa
Titanoscorpio douglassi, David Douglass
Tullimonstrum gregarium, Francis J. Tully
a tubeworm (in preparation) John Prinos

Abbreviations for Fossil Collections

AMNH	American Museum of Natural History, New York
BM(NH)	British Museum (Natural History), London
CMNH	Cleveland Museum of Natural History, Cleveland, Ohio
FMNH	Field Museum of Natural History, Chicago
FMNH MCP	Mazon Creek Photograph
FMNH UC	University of Chicago Collection
ISM	Illinois State Museum, Springfield
KC	Department of Geology, University of Newcastle upon Tyne
MCZ	Museum of Comparative Zoology, Harvard University, Cambridge, Massachusetts
NEIU MCP	Mazon Creek Project, Northeastern Illinois University, Chicago
ROM	Royal Ontario Museum, Toronto
UI	University of Illinois, Urbana
UMMP	University of Michigan Museum of Paleontology, Ann Arbor
USNM	U.S. National Museum, Smithsonian Institution, Washington, D.C.
WIU	Museum of Geology, Department of Geology, Western Illinois University, Macomb
YPM	Yale Peabody Museum, Yale University, New Haven, Connecticut

Authors

John L. Anderson is a computer scientist at Argonne National Laboratory, operated by the University of Chicago for the U.S. Department of Energy. His responsibilities have included mathematical analysis, systems analysis, and computer programming of a wide variety of scientific applications. He obtained his B.S. in engineering and M.S. in mathematics from the University of Illinois. An avid "amateur" fossil collector at Pit 11 since 1971, he has been deeply involved in several projects to ensure the future of collecting at the Pit and has provided data processing assistance to paleontologists associated with the Field Museum and various universities.

Jack Bowman Bailey is professor of geology at Western Illinois University and fellow of the Geological Society of America. He received his Ph.D. in paleontology in 1975 from the University of Illinois at Urbana-Champaign under Prof. Daniel B. Blake. He is founder and director of WIU's Museum of Geology. He has authored numerous papers on the systematics, ecology, and biogeography of Paleozoic bivalves. He was honored as Western Illinois University's Outstanding Teacher of 1981 and has since been the recipient of other honors including Western Illinois Faculty Excellence Awards for 1978, 1987, and 1991 and WIU Student Government Association's Select Faculty of 1989.

Gordon C. Baird has held a college teaching post at the State University of New York at Fredonia since 1982. After receiving his Ph.D. from the University of Rochester in 1975, he secured a position as assistant curator at the Field Museum of Natural History, working closely with the late Gene Richardson on various Mazon Creek related projects. His current research interests include stratigraphy and sedimentology of Ordovician and Devonian deposits in the eastern United States, paleocommensalism of ancient organisms, and the study of potential parameters for measuring heat-of-burial changes in fossils and associated sediments.

David Bardack has been professor of biological sciences at the University of Illinois at Chicago since 1965. He has taught vertebrate paleontology, comparative anatomy, evolution, biogeography, and at various times, ichthyology and herpetology. His research interests focus on fishes from various times, including Pennsylvanian, Cretaceous, and Tertiary. He has done field work in the United States, South America, Yugoslavia, Mexico, Arctic Canada, and Pakistan.

Bret S. Beall is former curatorial coordinator of Mazon Creek Paleontology at the Field Museum of Natural History, where he curated and reorganized the world's largest collection of Mazon Creek fossils. He earned his B.S. at the University of Missouri–Columbia and his M.S. and

Ph.D. at the University of Michigan. His research covers the functional morphology and general evolution of arachnids, the early history of terrestrial arthropods (arachnids, myriapods, and insects), and the structure of Paleozoic terrestrial communities. Current projects include an analysis of the affinities of *Tullimonstrum,* an investigation of the ecological relationships of epibiotic algae on some Mazon Creek pectens, and (with Conrad C. Labandiera) clarifying the macroevolutionary significance of early plant-animal interactions. In addition, he does freelance consultation, writing, and scientific illustration.

Daniel B. Blake is a director of the Museum of Natural History and professor of geology at the University of Illinois at Urbana-Champaign. He specializes primarily in the taxonomy, functional morphology, and phylogeny of stelleroid echinoderms of all ages and of Paleozoic bryozoans.

Frank M. Carpenter was born in 1902, in Boston. He became interested in insects, particularly fossil insects, when in high school as a result of reading some of Samuel Scudder's publications. He entered Harvard College in 1922 to study under Professor W. M. Wheeler and took his D.Sc., his thesis being on the fossil ants of North America. After a few years of assisting, he was appointed curator of Fossil Insects at the Museum of Comparative Zoology, and later Alexander Agassiz Professor of Zoology, and still later Fisher Professor of Natural History. He retired in 1973 but was permitted to keep his research space as honorary curator of Fossil Insects.

Kirk Fitzhugh has been the curator of polychaetes at the Los Angeles County Museum of Natural History since 1990. He received his Ph.D. in 1988 from George Washington University, in conjunction with the Smithsonian Institution's National Museum of Natural History, and was Thorne Research Fellow at the American Museum of Natural History in New York City (1988–90). His research has concentrated on the systematics and phylogenetic relationships of living polychaetes from a number of families.

Stephen J. Godfrey obtained his Ph.D. in vertebrate paleontology from McGill University, specializing in the skeletal anatomy of early tetrapods. Although most of his work has dealt with Paleozoic tetrapods, more recently he has broadened his paleontological horizons by studying Carboniferous rhizodontid fish and Cretaceous dinosaurs and pterosaurs.

Joseph T. Hannibal is curator of Invertebrate Paleontology at the Cleveland Museum of Natural History. He has studied fossil millipedes, especially pill millipedes, found at Mazon Creek. He received his doctoral degree from Kent State University.

Andrew A. Hay is a retired tool and die maker and designer. He was president of the world's second oldest astronomical society, the Chicago Astronomical Society. He is also a member of professional and nonprofessional geological and paleontological societies. His interest in Mazon Creek fossils prompted him to write articles about the Mazon Creek fauna and to assist in the publication of two popular books: *Keys to Identify Pennsylvanian Fossil Plants of the Mazon Creek Area* and *Keys to Identify Pennsylvanian Fossil Animals of the Mazon Creek Area.* He is at present a Northeastern Illinois University research fellow.

Michael D. Henderson obtained his B.S. and M.S. in geology from Southern Illinois University at Carbondale. He is currently serving as curator of collections for the Burpee Museum of Natural History in Rockford, Illinois. His interests include paleobiology and the study of Konservat Lagerstätten.

Joanne Kluessendorf is a native of Milwaukee, Wisconsin. After attending the University of Wisconsin–Milwaukee and Oregon State University, she received her B.S. degree and her Ph.D. in geology from the University of Illinois at Urbana-Champaign. Her research focuses on Paleozoic fossil Konservat Lagerstätten, ancient reefs, trace fossils, and the history of geology. She is currently a curator at the University of Illinois Museum of Natural History and a research associate in the University of Illinois Department of Geology.

Samuel Kruty was educated at Northwestern University in Evanston, Illinois (B.S. chemical engineering) and Washington University, St. Louis. World War II interrupted his career, and he flew 35 missions as a first lieutenant (bombardier) in B17s. After the war, he worked as a chemical engineer for Monsanto and W. R. Grace Corporation, and he holds several patents in the area of pipeline coatings. His strong interest in paleontology began while supervising the wrapping of an oil transmission pipeline in the Cretaceous strata of west Texas. So many fossil molluscs awakened a desire to collect, study, and write about fossils, an avocation that has continued to today.

Jarmila Kukalova-Peck is research adjunct professor in earth science at Carleton University. She studied geology at Charles University, where she became interested in Paleozoic insects and in the evolutionary clues they offered. After receiving her Ph.D., she taught invertebrate paleontology at Charles University. Her publications on Carboniferous and Permian insects brought her two invitations from Harvard University as an Alexander Agassiz lecturer. She has traveled widely and has published more than 60 original papers on insect evolution and the interpretation of major arthropod structures. In 1992 she became a Fellow of the Royal Society of Canada.

Christopher T. Ledvina has had nearly 20 years experience in the coal industry of Illinois. After graduating from the University of Illinois in geology in 1974, he became a geologist in the Coal Section of the Illinois State Geological Survey. From the Survey, he went on to become assistant to the superintendent of Freeman United Coal Mining Company's Crown II Mine near Virden, Illinois. Then he became an examiner at Old Ben Coal Company's Mine 26 near Sesser, Illinois. On leave from Old Ben as the result of a mining accident, in 1988 he received an M.S. in geology from Northeastern Illinois University and in 1991 received a Ph.D. from Northwestern University in mineral resource engineering. He is currently an assistant professor in the Earth Science Department at Northeastern Illinois University and is curator of the Mazon Creek Project. He is finishing a book entitled *The Pictorial History of Coal Mining in Illinois* and continues his research in coal mining geology and mining history. He is also chairman of the National Museum of Coal Mining in West Frankfort, Illinois, and serves as chairman of the Coal Committee of the National Mining Hall of Fame and Museum in Leadville, Colorado.

Christopher G. Maples is a geologist-paleontologist and chief of the Geologic Investigations Section at the Kansas Geological Survey, University of Kansas, Lawrence. He received his B.S. in geology at West Georgia College and his M.S. and Ph.D. degrees in paleontology at Indiana University. His main research interests are Paleozoic fossil Lagerstätten, invertebrate paleoecology, echinoderms, trace fossils, and brachiopods.

Donald G. Mikulic is a geologist at the Illinois State Geological Survey. A Wisconsin native, he received his B.A. in geology at the University of Wisconsin–Milwaukee. His Ph.D. in paleontology was conferred by Oregon State University. His main research interests are Paleozoic fossil Konservat Lagerstätten, trilobites, reefs, and the history of paleontology.

Roy E. Plotnick is an associate professor in the Department of Geological Sciences at the University of Illinois at Chicago, where he teaches geology and invertebrate paleontology. His doctoral thesis (1983) at the University of Chicago examined evolutionary patterns in the eurypterids. Since then, he has published on a wide range of topics, including eurypterid and crinoid functional morphology, arthropod fossilization, and applications of fractals to paleontology and geology.

W. D. Ian Rolfe has been keeper of geology, National Museums of Scotland, since 1986. He received his Ph.D. from Birmingham University, England, in 1960. After a year (1961–62) at the Museum of Comparative Zoology, Harvard University, he spent the next 24 years at the Hunterian Museum of the University of Glasgow, ultimately serving as deputy director. He was a visiting scientist at the Field Museum of Natural History in 1981.

Frederick R. Schram received his Ph.D. from the University of Chicago in 1968. His doctoral dissertation was on the crustaceans of the Mazon Creek biotas and involved close work with Dr. E. S. Richardson, Jr., and Prof. Ralph Gordon Johnson. Since that time he has gone on to establish an acknowledged expertise in the field of crustacean phylogenetics and systematics. He is the author of more than 130 scientific and popular articles and four books and is the general editor of the series *Crustacean Issues.* After a productive career affiliated with universities and museums in North America, including the Los Angeles County Museum of Natural History, he is now professor of systematic zoology and zoogeography in the Institute for Taxonomic Zoology at the University of Amsterdam.

Paul A. Selden is senior lecturer in Palaeontology at the University of Manchester, England. His fascination with arthropods began with a childhood interest in Lepidoptera. He read geology and zoology at Manchester and then studied for a Ph.D. on eurypterids under Prof. H. B. Whittington at Cambridge. After a post at the University of London, he returned to Manchester, where he works on fossil chelicerates. He has published on eurypterid functional morphology, evolution in the Xiphosura, fossil arachnids, and terrestrialization.

Charles W. Shabica received his Ph.D. from the University of Chicago in 1971. His doctoral dissertation was on the depositional environment of the Francis Creek Shale. Supported by the Field Museum and the Committee on Evolutionary Biology at the University of Chicago, he was assigned the task of researching the life and death environments of the Mazon Creek fossils by Dr. E. S. Richardson, Jr., and Prof. Ralph Gordon Johnson. He is currently professor of coastal geology and engineering at Northeastern Illinois University.

Steven D. Sroka, a native Chicagoan, became interested in studying Mazon Creek fossils when collecting with his parents during the early 1970s. He received his M.S. degree in 1984 from Northeastern Illinois University under the supervision of Dr. C. W. Shabica. His M.S. dissertation involved the taxonomy and paleoecology of Mazon Creek bivalves. He completed his Ph.D. in January 1996. In the past, he has curated Mazon Creek collections at each institution he has attended, as well as working at the Field Museum with Dr. G. C. Baird on censusing the Mazon Creek area. Publication and research interests include the Mazon Creek fauna and its relationship to other Carboniferous Lagerstätte. Currently, he is employed at the Museum of Natural History at the University of Illinois at Urbana-Champaign.

Rainer Zangerl was the chairman of the Department of Geology, Field Museum of Natural History. His special interests include reptiles, Paleozoic cartilaginous fishes, and the depositional environment of Pennsylvanian black shales and his 69 papers and 11 monographs deal primarily with fossil Paleozoic sharks and their relatives and the paleoecology of Mecca Quarry Shale. He has done field work in Switzerland, Austria, Utah, Wyoming, Texas, Indiana, Arkansas, and Arizona.

Reviewers

Chapter(s) reviewed by each person is listed in parentheses.

Jack Bailey, Western Illinois University, Macomb (App. B)
Donald Baird, Emeritus, Princeton University, Princeton, New Jersey (17)
Gordon C. Baird, SUNY-Fredonia, Fredonia, New York (10)
David Bardack, University of Illinois at Chicago, Chicago (App. B)
Bret Beall, Field Museum of Natural History, Chicago (8B, 8C, 8D, App. A)
William B. N. Berry, University of California, Berkeley (7C)
Gale A. Bishop, Georgia Southern University, Statesboro (9, 12)
Daniel B. Blake, University of Illinois, Urbana-Champaign (App. A)
Carlton E. Brett, University of Rochester, Rochester, New York (1)
Robert L. Carroll, McGill University, Montreal (19)
Desmond H. Collins, Royal Ontario Museum, Toronto (20)
Heinz Damberger, Illinois Geological Survey, Champaign (2)
Salim Sh. Dautov, Academy of Sciences of Russia, Vladivostok (7C)
Howard R. Feldman, Kansas Geological Survey, Lawrence (1)
Rodney M. Feldmann, Kent State University, Kent, Ohio (9, 12)
Daniel C. Fisher, University of Michigan, Ann Arbor (11)
Merrill W. Foster, Bradley University, Peoria, Illinois (16A)
Robert A. Gastaldo, Auburn University, Auburn, Alabama (3, 4A, 4B, 4C, 5)
Henri Goulet, Centre for Land and Biological Resources, Ottawa (14)
Thomas E. Guensburg, Rock Valley College, Rockford, Illinois (20)
Raymond C. Gutschick, Emeritus, University of Notre Dame, South Bend, Indiana (6)
Joseph T. Hannibal, Cleveland Museum of Natural History, Cleveland, Ohio (App. B)
Michael C. Hansen, Ohio Department of Natural Resources, Columbus (18)
Phillip H. Heckel, University of Iowa, Iowa City (3, 4A, 4B, 4C, 5)
Douglas S. Jones, University of Florida, Gainesville (16A)
Joanne Kluessendorf, University of Illinois, Urbana-Champaign (10)
Jerry D. Kudenov, University of Alaska, Anchorage (7A, 7B)
Jarmila Kukalova-Peck, Carleton University, Ottawa (App. B)
Conrad C. Labandeira, U.S. National Museum, Washington, D.C. (11, 15A)
N. Gary Lane, Indiana University, Bloomington (16B)
Chris Maples, Kansas Geological Survey, Lawrence (6, App. B)
Andrew R. Milner, Birkbeck College, London (19, App. B)
Ken J. McNamara, Western Australian Museum, Perth (15B)
John Nelson, Illinois State Geological Survey, Champaign (2)
Stewart B. Peck, Carleton University, Ottawa (14, 15C)
Thomas H. Perkins, Florida Department of Natural Resources, St. Petersburg (7A, 7B)
Daniel D. Petzold, Indiana University, Bloomington (12)
John Pojeta, Jr., U.S. Geological Survey, Washington, D.C. (8A)
Richard A. Robison, University of Kansas, Lawrence (15C)
W. D. Ian Rolfe, National Museums of Scotland, Edinburgh (9)
Bruce Runnegar, University of California, Los Angeles (8A)
W. Bruce Saunders, Bryn Mawr College, Bryn Mawr, Pennsylvania (8B, 8C, 8D)
Frederick R. Schram, Universiteit van Amsterdam, Amsterdam (15A, App. B)

Paul A. Selden, University of Manchester, Manchester, England (15B)
William Shear, Hampden-Sydney College, Hampden-Sydney, Virginia (13)
Barbara J. Stahl, St. Anselm College, Manchester, New Hampshire (18)
Johnny A. Waters, West Georgia College, Carrollton (16B)
Ronald R. West, Kansas State University, Manhattan (7C, 21)
Michael E. Williams, Cleveland Museum of Natural History, Cleveland, Ohio (18)
Keith H. Woodwick, California State University, Fresno (7C)
Rainer Zangerl, Field Museum of Natural History, Chicago (App. B)

Taxonomic Index

Page numbers in italics indicate illustrations.

Index

Page numbers in italics indicate illustrations.

The text for *Richardson's Guide to the Fossil Fauna of Mazon Creek* is composed in Trump Mediäval; the captions in Frutiger. Trump was designed by George Trump and first issued in 1954 by the Weber Foundry, Stuttgart. It is a strong roman and italic font with a humanist axis; Frutiger was designed by Adrian Frutiger in 1975 and issued by Mergenthaler. It was first designed for the signage at Paris-Roissy Airport. The design of the book is by Cameron Poulter; composition by Graphic Composition; printing and binding by Thomson-Shore.